SCIENCE AND TECHNOLOGY

DESK REFERENCE

SCIENCE AND
TECHNOLOGY

DESK REFERENCE

1,500 Answers to
Frequently-Asked or
Difficult-To-Answer
Questions

THE CARNEGIE

LIBRARY OF

PITTSBURGH

SCIENCE AND

TECHNOLOGY

DEPARTMENT

 Gale Research Inc. DETROIT • WASHINGTON D.C. • LONDON

Edited by The Carnegie Library of Pittsburgh, Science and Technology Department, under the direction of James E. Bobick and with the assistance of Margery Peffer.

Gale Research Staff

RITA RUNCHOCK,
Coordinating Editor
with the assistance of
JENNIFER J. DYE, and with
special thanks to
KATHLEEN DROSTE,
MONICA M. HUBBARD,
and PHYLLIS SPINELLI

LINDA METZGER,
Senior Editor

MARY BETH TRIMPER,
Production Director

EVI SEOUD,
Assistant Production Manager

SHANNA HEILVEIL,
Production Assistant

CYNTHIA BALDWIN,
Art Director

MARK C. HOWELL,
Cover and Page Design

BARBARA J. YARROW,
Graphic Services Supervisor

WILLIE MATHIS,
Camera Operator

NICHOLAS JAKUBIAK,
Keyliner

BENITA L. SPIGHT,
Data Entry Supervisor

GWENDOLYN S. TUCKER,
Data Entry Group Leader

KENNETH D. BENSON, Jr.,
Data Entry Associate

JEANNE A. GOUGH,
*Permissions and
Production Manager*

MARGARET A. CHAMBERLAIN,
*Permissions Supervisor
(Pictures)*

PAMELA A. HAYES,
Permissions Associate

AMY LYNN EMRICH,
KARLA KULKIS,
NANCY M. RATTENBURY,
and KEITH REED,
Permissions Assistants

While every effort has been made to ensure the reliability of the information presented in this publication, Gale Research Inc. does not guarantee the accuracy of the data contained herein. Gale accepts no payment for listing; and inclusion in the publication of any organization, agency, institution, publication, service, or individual does not imply endorsement of the editors or publisher. Errors brought to the attention of the publisher and verified to the satisfaction of the publisher will be corrected in future editions.

This book is printed on acid-free paper that meets the minimum requirements of American National Standard for Information Sciences—Permanence Paper for Printed Library Materials, ANSI Z39.48-1984. ∞™

ISBN 0-8103-8884-7   Library of Congress Catalog Card Number 92-075423

Printed in the United States of America

Published in the United States by Gale Research Inc.
Published simultaneously in the United Kingdom
by Gale Research International Limited
(an affiliated company of Gale Research Inc.)

CONTENTS

PREFACE

In 1991, the staff of the Science and Technology Department at The Carnegie Library of Pittsburgh received more than 95,000 reference questions and inquiries by the telephone, through the mail, and from personal in-house contact. Even with 390,000 books, 425,000 bound periodicals, and thousands of government documents, technical reports, and microfiche available, the Department has maintained an extensive quick reference file to facilitate reference service. For the most part, the file contains questions and answers to unusual, interesting, or frequently-asked questions as well as answers to inquiries where the information was difficult to locate. The purpose of this book is to share with others some of this wealth of cumulative information that has been in existence for at least 42 years—as far back as any of the current Carnegie staff can remember. It may even date back to 1905, the year when The Carnegie Library of Pittsburgh became the first major public library in the United States to establish a separate Science and Technology Department.

Science and Technology Desk Reference is designed as a convenient resource to give reference librarians and other users information quickly for typical or time-consuming questions they most likely will encounter in the sci-tech area. Especially helpful to small libraries having neither the range of sci-tech resources nor the subject specialists with the expertise to field some of these questions, *Science and Technology Desk Reference* becomes an invaluable one-volume tool that can act as a ready reference file in the area of science and technology.

SCOPE AND COVERAGE

The 1,500 questions in *Science and Technology Desk Reference* were selected either from the Science and Technology Department's ready-reference file or from appropriate questions that were recently answered by the Department's reference staff or forwarded to the Department from the 18 branch libraries. Questions having wide appeal to patrons of public, school, and academic libraries form the base of *Science and Technology Desk Reference*. Although the Department answers a sizeable number of highly specialized and technical questions, these have been weeded out, since they would be of interest to only a small number of professionals.

The Science and Technology Department not only provides reference service in areas of science and technology from astronomy to zoology, but also in such areas as food technology, health, and pet care. Since the Department houses material cataloged in the Library of Congress classification system for Classes Q (Science), R (Medicine), S (Agriculture), T (Technology), U (Military Science), and V (Naval Science), high interest or difficult questions on these subjects are represented here. In some ways, science touches so much of our lives—whether it be our environment, our homes, our workplaces, right down to our physical bodies themselves—that it can become difficult to categorize what actually constitutes science.

Because *Science and Technology Desk Reference* focuses on only those questions that have achieved noteworthiness either through their popularity, their time-consuming nature, or their uniqueness as determined by the Science and Technology Department, the data has not been manipulated to ensure that all topics within the sci-tech boundaries have been equally covered. Such skewing of questions would defeat its original timesaving intent of providing reference librarians with high interest questions or time-consuming questions.

Science and Technology Desk Reference is meant to be a "desk reference" in the sense that its size, format, and coverage make it a "handy" resource to keep available at or near the reference desk for quick access in answering both phone and in-house questions. Users are invited to submit questions that they believe fit the selection criteria for use in future editions. A convenient form is provided for this purpose at the end of the index.

ARRANGEMENT

Following the chronology highlighting the technological developments throughout recorded history, the questions are divided into 20 broad sci-tech subject areas starting with THE ANIMAL WORLD and ending with WEIGHTS, MEASURES, AND TIME. Within each chapter, the entries are arranged under specific topics (such as "Birds," "Mammals," etc. in the chapter THE ANIMAL WORLD). Cross-references within the chapters lead users to other relevant aspects of the topics. For example, in the section "Birds" a cross reference directs the user to the section "Pets and Pests," where a question on the best caged birds as pets appears. Each question is referenced with an entry number to facilitate quick retrieval of the information through the subject index.

SAMPLE ENTRY

The following sample entry is from the chapter THE ANIMAL WORLD under the topic "Physical characteristics":

76 ❶

❷ *Do all animals have red blood?*

❸   The color of blood is related to the compounds which transport oxygen. Hemoglobin, containing iron, is red and is found in all vertebrates (animals having a backbone) and a few invertebrates (animals lacking a backbone). Annelids (segmented worms) have either a green pigment chlorocruorin or a red pigment, hemerythrin. Some crustaceans (arthropods having divided bodies and generally having gills) have a blue pigment, hemocyanin, in their blood.

❹   **Sources:** *New Encyclopedia Britannica*, vol. 2, p. 290.

❶ Sequential entry number arranged under a topic within a broad subject chapter. Eye-catching typography to locate the information on the page quickly.

❷ Question based on patron inquiry. Sometimes the scope of the question has been broadened to include more physical examples, especially in the areas of animal characteristics, earth features, and planet and star characteristics. For example, "What is the largest mammal?" has been expanded into "What are the largest and smallest animals?"

❸ Answer with additional details. Some answers contain definitions, U.S. customary and metric measurements, scientific names, or tabular data or illustrations.

❹ Source citation(s) lead the user to additional data, if desired. Each entry provides a maximum of three used resources (books, magazines, and journals). Source citation includes author of book, title of book or journal, volume number if applicable, date of journal, and page number. In a few cases, no page number appears because the information was culled from the entire source. Full bibliographic particulars can be obtained from the bibliography following chapter 21.

FEATURES

- The answers are written in non-technical language and provide either a succinct response or a more elaborate explanation, depending on the nature of the question. A general reader will be able to understand the concept, process, or term described, and the librarian will not have to "interpret" the answer for the user.

- Definitions to scientific terminology are given within the answer itself, use of scientific names for plants and animals and dual (U.S. customary and metric) measurements.

- Figures and dates have been verified. Even in science, figures can seem to be in conflict; many times such discrepancies may be attributable to the authority's perspective, or more commonly, to the results of simple mathematical rounding of figures. Sometimes in Science and Technology Desk Reference the figure or date listed is a consensus of the consulted sources; other times the discrepancy is noted and alternatives given. Science and Technology Desk Reference rounds off figures whenever it seems that such precision is unnecessary.

- 108 illustrations and many tables augment the text whenever graphic depiction would benefit the entry.

- Partial bibliographical information for each source consulted appears at the end of each entry. Full citations for each of the 892 books and 87 journals and magazines consulted can be located in the bibliography following chapter 21. The bibliography directs users, who wish to find additional information on a topic, to the appropriate resource.

- A chronology of significant inventions and discoveries unfolds the history of the development of science and technology.

- A comprehensive subject index includes names of inventors and discoverers, terms and concepts that are defined, and topics covered within the text of the answers. Since specific inventions, discoveries, and patents have been categorized generally under relevant topics, the index contains the general headings "Discoveries," "Inventions," "Patents" under which these entries have been grouped. However, no attempt has been made to index the "Chronology" in chapter 1, since the information there is very cursory.

COMMENTS ARE WELCOME

Users are encouraged to send their comments and suggestions to

Editor,
Science and Technology Desk Reference
Gale Research Inc.
835 Penobscot Bldg.
Detroit, Michigan 48226-4094

Fax: 313-961-6815

▼ACKNOWLEDGMENTS

There have been many people who have made significant contributions to *Science and Technology Desk Reference* from its initial stage of conception to its realization as a reference work. Margery Peffer kept the work going day after day for almost seven months, as well as answering hundreds of questions, checking leads, verifying sources, assisting other staff members, and generally being one of the finest science reference librarians that I have ever worked with and have known.

The staff librarians in the Science and Technology Department of The Carnegie Library of Pittsburgh know how much I have appreciated both their individual and collective efforts in gathering, reviewing, answering, revising, and verifying many more questions than the number included in this volume. I want to thank Joan Anderson, Naomi Balaban, Jan Comfort, Diane Eldridge, Susan Horvath, Dorothy Melamed, Kristine Mielcarek, David Murdock, Gregory Pomrenke, Gertrude Ross, and Donna Strawbridge for their sustained endeavors. These librarians were remarkable at balancing the never-ending needs of our patrons with the frequent deadlines required for submitting batches of questions. Special thanks should be extended to Gregory Pomrenke for his work in compiling the chronology, which appears as chapter 1.

Among those who need to be thanked are Marilyn Macevic, who typed most of the questions in the first draft; Helen Yee, who formatted every question and typed the final copy; Mary Shields, who read and proofed most of the questions, verified citations, and solved numerous problems relating to the citations. Staff members throughout The Carnegie Library of Pittsburgh system submitted numerous questions for possible inclusion in this book. I want to thank the following for their response to my request: Ernestine Audin, Judy Beczak, Elaine Bisiada, Dallas Clautice, Lisa Dennis, Clare DiDominicis, Cheryl Engel, Phoebe Ferguson, Constance J. Galbraith, Mary Ellen Gildroy, Mildred Glenn, Esther Hamiel, Kathy Herrin, Thomas Holmes, Meeghan Humphrey, Sheila Jackson, Patricia James, Nancy Jessup, Andrea Jones, Amy Korman, Rebecca Kosanovich, Pamela Kuchta, Kathryn Logan, Mary Long, James Lutton, Stephanie Messina, Charmaine Spaniel Mozlack, David Murray, Anne New, Julianna Posch, Barbara Preusser, Barbara Rogers, Barbara Rush, Marsha Schafer, Deborah Silverman, Christine Stein, Viktoria Strod, Helene Tremaine, Pamela Van Eman, and Gloria Zuckerman.

I thank Robert B. Croneberger, Director of The Carnegie Library of Pittsburgh, for initially approving this project and signing the contract and Joseph Falgione, Gladys Shapera, and Loretta O'Brien, senior library administrators at The Carnegie Library of Pittsburgh, who offered encouragement as well as the submission of various questions.

In addition I want to thank the staff at Gale Research, in particular David Salamie, Senior Editor, New Product Development, who shaped my book idea into a feasible project, and Rita Runchock, Editor, Dictionaries and General Reference Division, and her staff, who worked with me to transform my idea into a reality. Also I wish to thank both the independent researchers, Len Bruno and Alana Wilson, for their expertise and subject knowledge and Monica Leach and Marco DiVita for their assistance in making *Science and Technology Desk Reference* a reality.

Finally, thanks to my wife, Sandi, and sons, Andrew and Michael, for their patience and understanding as well as for their interesting discussions of the various questions and answers that were "hot off the press."

James E. Bobick
Head, Science and Technology Department
The Carnegie Library of Pittsburgh

CREDITS

Cover illustrations: Illustration of earthquake waves from *The Golden Guide to Geology*, by Frank H.T. Rhodes. Illustration by Raymond Perlman. Copyright © 1972 by Western Publishing Company, Inc. Used by permission. Illustration of moon crater from *A Field Guide to Stars and Planets*, by Donald H. Menzel and Jay M. Pasachoff. Houghton Mifflin Company, 1992. Illustration courtesy of Questar Corporation. Copyright © 1983 by the Estate of Donald H. Menzel and by Jay M. Pasachoff. Reprinted by permission of Jay M. Pasachoff. Illustration of alligator from *Everglades Wildguide*, by Jean Craighead George. Illustration by Betty Fraser. U.S. Department of the Interior, Washington, DC, 1988. Reproduced by the Division of Publications, National Park Service.

page 20 and **front cover:** Illustration from *The Science Teacher*, April 1992. Copyright © 1992 by the National Science Teachers Association, 1742 Connecticut Avenue, NW, Washington, DC, 20009. Reproduced by permission of National Science Teachers Association; **p. 23:** Illustration by Anne Senechal Faust; **pp. 26, 233, 608:** Illustrations from *Environmental Trends*, prepared by Executive Office of the President, Council on Environmental Quality, Washington, DC, 1989; **p. 36:** K.Fink/ARDEA LONDON LTD.; **p. 38:** Photograph by Ernest P. Walker; **pp. 41, 270:** Illustrations by Betty Fraser from *Everglades Wildguide*, by Jean Craighead George. U.S. Department of the Interior, Washington, DC, 1988. Reproduced by the Division of Publications, U.S. National Park Service; **p. 46:** Illustration from *An Introduction to the World's Oceans*, 3rd edition, by Alyn C. Duxbury and Alison B. Duxbury. Copyright © 1991 by Wm. C. Brown Communications, Inc., Dubuque, Iowa. All Rights Reserved. Reprinted by permission; **p. 65:** Illustration from *United States Patents*. Patent Number 587, 607, August 3, 1897; **p. 80:** Illustration from *Popular Science*, Volume 49, November 1946. Reprinted by permission of L.A. Times Syndicate; **pp. 84, 214, 369:** Illustrations from *What's What: A Visual Glossary of the Physical World*, by Reginald Bragonier, Jr. and David Fisher. Copyright © 1981 by Reginald Bragonier, Jr. and David Fisher. Reprinted by permission of Hammond Incorporated, Maplewood, NJ; **pp. 123, 377, 378, 437:** Illustrations by Peter Bull from *The Random House Book of How Things Work*, by Steve Parker. Kingfisher Books, 1990. Copyright © 1990 by Grisewood & Dempsey Ltd. Reprinted by per-

Copyright © 1991 by William Ashworth. Reprinted with the permission of Facts on File, Inc., New York; **p. 281:** Illustration from *Environmental Quality*, edited by Dale Curtis and Barry Walden Walsh. Council on Environmental Quality, March 1992; **p. 294:** U.S. Environmental Protection Agency; **p. 298:** National Solid Wastes Management Association, 1990; **p. 338:** Courtesy of the U.S. Department of Agriculture, Washington, DC; **p. 348:** © Rube Goldberg from King Features Syndicates, Inc., 1978. Reprinted with special permission of King Features Syndicates; **p. 352:** Illustration from *United States Patents*, Patent Number 6,469, May 22, 1849; **p. 361:** Illustration by Neil Bulpit from *The Random House Book of How Things Work*, by Steve Parker. Kingfisher Books, 1990. Copyright © 1990 by Grisewood & Dempsey Ltd. Reprinted by permission of Grisewood & Dempsey Ltd.; **p. 380:** Illustration from *Working Safely With Video Display Terminals*, prepared by U.S. Department of Labor, and Occupational Safety and Health Administration, revised 1991; **pp. 393, 430, 436, 457:** Illustrations by Fred Maschall from *First Aid Book*, prepared by U.S. Department of Labor, and Mine Safety and Health Administration, reprinted 1991; **p. 402:** Illustration from *Van Nostrand's Scientific Encyclopedia*, edited by Douglas M. Considine, P.E. Copyright © 1983 by Van Nostrand Reinhold Company, Inc. Illustration by A.M. Winchester; **pp. 408 right, 455:** Illustrations from *Harver Ready Reference Encyclopedia*, International Edition. Illustrations copyright © 1969 by N.V. Uitgeversmaatschappij Elsevier, Amsterdam-Brussels. Reprinted by permission of Elsevier Science Publishers BV Academic Publishing Division; **p. 414:** Illustration from *The Injury Fact Book*, by Susan P. Baker, Brian O'Neill, Marvin J. Ginsburg and Guohua Li. Copyright © 1992 by Oxford University Press, Inc. Reprinted by permission of Oxford University Press, Inc.; **p. 421:** Illustration from *Eating to Lower Your High Blood Cholesterol*, prepared by U.S. Department of Health and Human Services, Public Health Services, National Institutes of Health, reprinted June 1989; **p. 445:** Illustration from *The University Desk Encyclopedia*. Copyright © 1977 by Elsevier Publishing Projects S.A., Lausanne. Reprinted by permission of Elsevier Trading & Copyrights; **p. 453:** Illustration from *The Work of the Science Film Maker*, by Alex Strasser, F.R.P.S. Copyright © 1972 by Focal Press Ltd. Reprinted by permission of Eagle Publishing Company; **p. 467:** Illustration from *What's What: A Visual Glossary of the Physical World*, by Reginald Bragonier, Jr., and David Fisher. Copyright © 1981 by Reginald Bragonier, Jr., and David Fisher. Illustration copyright © 1981 by International Business Machines Corporation. Reprinted by permission of International Business Machines Corporation; **p. 491** and **back cover:** Illustration from *The Almanac of Science and Technology: What's New and What's Known*, edited by Richard Golob and Eric Brus. Copyright © 1990 by Harcourt Brace Jovanovich, Inc. Reprinted by permission of Florante A. Quiocho, Rice University; **pp. 531, 532:** Illustrations from *Growing Vegetables in the Home Garden*, prepared by Agricultural Research Service and Extension Service, and United States Department of Agriculture. Issued December 1972, slightly

SCIENCE AND TECHNOLOGY

DESK REFERENCE

▼ CHRONOLOGY

Lists highlights of inventions and discoveries in science and technology.

Year	Invention or Discovery	Origin
4500 BCE	**Copper** The technique of copper smelting is developed.	Yugoslavia
	Cast Copper Artifacts made of melted and cast copper are in use at Sialk.	Iran
4400 BCE	**Horizontal Loom** Depiction of a horizontal loom on pottery piece found from this era.	Egypt
	Beer and Wine Alcoholic beverages such as beer and wine are produced.	The Near East
4000 BCE	**Writing** First known written language is called cuneiform.	Sumer
3500 BCE	**Wheeled Vehicles** Carts with copper nail studs on wheel rim are in use.	Sumer
3200 BCE	**Soap** Soap made from alkalies and fats is produced.	Sumer
	Sail Boats Boats propelled by square-shaped sails are in use.	Egypt
3000 BCE	**Plow** A scratch plow drawn by oxen is being used in agriculture.	Sumer
	Bronze Production of bronze with 8-10% tin indicates a developed alloying technology.	Iran
2900 BCE	**Stonehenge** The basis of an astronomical calculator is constructed on Salisbury Plain.	England
	Great Pyramid A tomb for the Pharaoh Cheops is built.	Egypt
2700 BCE	**Horse Domesticated** Horses are tamed for transportation.	Ukraine

2200 BCE	**Stonehenge**	England
	80 bluestones are set in the form of two concentric circles, completing the observatory.	

1700 BCE	**Mathematics**	Ahmes
	A manuscript by an Egyptian priest depicts problems in algebra, geometry, and arithmetic and uses the equal sign (probably a copy of an earlier work).	

1500 BCE	**Contraceptive**	Egypt
	A plug of lint, ground acacia branches and honey is used as a contraceptive device.	

1450 BCE	**Stonehenge**	England
	Stonehenge assumes present form.	

406 BCE	**Catapult**	Dionysus of Syracuse
	A device for hurling objects in warfare is produced.	

45 CE	**Medicines**	Pedanius Dioscorides
	Greek physician describes 600 plants and nearly 1000 drugs in *De Materia Medica*.	

50	**History of Nature**	Pliny the Elder
	A summary of all knowledge on astronomy, geography and zoology is written down in *Naturalis Historia*.	

100	**Paper**	China
	Paper is being produced.	

140	**Geocentricity**	Claudius Ptolemaeus (Ptolemy)
	A summary of astronomical knowledge, *Megale Syntaxis Tes Astronomias*, (also called *Almagest*) posits an earth centered universe.	

250	**Algebra Textbook**	Diophatus
	A text of mathematical problems deals mostly with whole numbers.	

275	**Stirrups and Saddle**	China
	Rigid saddle and metal stirrups make horseback travel easier.	

400	**Waterwheel**	France
	A water-driven wheel grinds grain.	

535	**Hagia Sophia**	Constantinople, Eastern Roman Empire
	The church's large dome with numerous windows at the bottom is engineered without loss of structural strength.	

673	**Greek Fire**	Callinicus
	Greeks use a chemical mixture, which will ignite and burn on water in warfare against the Arabs.	

750	**Arabic Alchemy**	Jabir Ibn Hayyan, also
	Arab alchemist distills vinegar to make acetic acid.	known as Geber

875   **Timekeeping Candles**   England
Calibrated candles mark the passage of time.

974   **Canal Locks**   Chiao Wei-Yo
A system of canal locks with variable waterlevels is introduced in China.

1054   **Supernova**   China, Japan, Arabic lands
A supernova is observed on July 4.

1137   **Flying Buttresses**   Paris, France
The abbey church of St. Denis is completed using flying buttresses for support.

1240   **Medical Regulations**   Frederick II, Holy Roman
Regulation of the study and practice of medicine,   Emperor
licensing of physicians and the preparation of drugs
becomes law.

1269   **Compass**   Petrus Pereginus De
360° graduation and movable sights are features of this   Maricourt
early compass.

1285   **Spectacles**   De Spina, Alessandro
A convex lens for the correction of myopia is invented by
a Dominican monk.

1330   **Ockham's Razor**   William of Ockham
Publication of *Summa Totius Logicae* argues that when
several explanations of a phenomena are proposed, the
simplest should be used.

1388   **Pollution Control**   England
An act of Parliament forbids dumping of garbage in the
ditches, rivers, or water systems.

1410   **Wire**   Rodolph of Nuremberg
A process for the drawing out of wire is developed in
Germany.

1451   **Printing press**   Gutenberg, Johannes
Printing press with movable type makes mass   (Gansfleisch Zur Laden
production of books possible.   Sum)

1492   **Globe of the Earth**   Behaim, Martin
The first representational globe of the earth is made in
which the Americas and the Pacific Ocean area are absent.

| 1530 | **Natural Medicine** | Paracelsus, Philippus |
| | *Paragranum* argues that medicine should be based on natural physical laws, and suggests useful medicinal minerals and chemicals for treatment of diseases. | Aureolus |

| 1543 | **Accurate Anatomy** | Vesalius, Andreas |
| | *De Humani Corporis Fabrica* depicts human anatomy accurately. | |

| | **Heliocentric System** | Copernicus, Nicolaus |
| | *On the Revolutions of the Heavenly Spheres* contends that the planets orbit around the sun. | |

| 1568 | **Bottled Beer** | Nowell, Alexander |
| | Dean of St. Paul's in London invents bottled beer. | |

| 1610 | **Observations with Telescope** | Galilei, Galileo |
| | Using the telescope invented by Hans Lippershey of the Netherlands in 1608, Galileo discovers the moons of Jupiter and the mountains of the Earth's moon. | |

| 1614 | **Logarithms** | Napier, John |
| | *Mirifici Logarithmorum Canonis Descriptio* describes logarithms and gives rules and tables for their use. | |

| 1637 | **Scientific Method** | Descartes, Rene |
| | *Discourse on Method* advocates formulating results by using step-by-step deduction from self-evident truths rather than by inductive reasoning. | |

| 1665 | **Philosophical Transactions** | Royal Society, England |
| | The first issue of the *Philosophical Transactions of the Royal Society*, the first scientific journal in England, is published. | |

| 1676 | **Microorganisms** | Van Leeuwenhoek, Anton |
| | Tiny living creatures are discovered in water using a microscope. | |

| 1686 | **Meteorological Map** | Halley, Edmund |
| | English astronomer draws first meteorological map. | |

| 1687 | **Principia** | Newton, Isaac |
| | The *Philosophiae Naturalis Principia Mathematica* codifies the three laws of motion. | |

| 1700 | **Binary Numbers** | Leibnitz, Gottfried |
| | The binary system of notation, essential for computer programming, is devised. | |

| 1709 | **Coke Smelting** | Derby, Abraham |
| | The use of coke in iron smelting increases production of iron. | |

1712	**Newcomen's Engine**	Newcomen, Thomas

A practical steam engine for water pumping and mine drainage is built near Dudley Castle in Staffordshire, England.

1714	**Mercury Thermometer**	Fahrenheit, Gabriel

The first accurate thermometer has temperature scale.

1735	**Classification of Species**	Linnaeus, Carolus

The modern classification system of plants and animals by genus and species is developed.

1742	**Celsius Scale**	Celsius, Anders

A temperature scale used throughout the world (except the United States) in which 0° is water's freezing point and 100° is water's boiling point.

1753	**Lightning Rod**	Franklin, Benjamin

A device conducts lightning to the earth, protecting buildings from damage.

1768	**Watt's Steam Engine**	Watts, James

After creating three versions of his steam engine (1765, 1765, 1768), Watts applies for a patent.

1781	**Uranus**	Herschel, William

The seventh planet from the sun is discovered.

1783	**Balloon**	Montgolfier, Jacques
		Montgolfier, Joseph

The first manned flight travels about 5 miles.

1784	**Shrapnel**	Shrapnel, Sir Henry

The idea of a metal canister shell filled with musket balls and an explosive charge that would explode in flight is proposed.

1790	**Dentist's Chair**	Flagg, Josiah

Chair has an adjustable headrest and arm for holding instruments.

1794	**Cotton Gin**	Whitney, Eli

A machine removes seeds from cotton fibers mechanically, dramatically increasing production.

1799	**Metric System**	French Academy of Science

A uniform system of weights and measures is officially proposed and becomes worldwide standard except in the United States.

1800	**Electric Battery**	Volta, Alessandro

The first practical battery has multiple layers of copper, zinc, and cardboard soaked in salt solution.

| 1803 | **Morphine** | Serturner, Frederich |

1803   **Morphine**   Serturner, Frederich
This narcotic pain-killer is derived from opium.

1813   **Chemical Symbols**   Berzelius, Jons
A notation system uses an abbreviation of the Latin name rather than pictorial representation.

1814   **Steam Powered Printing Press**   Konig, Friedrich
The steam press allows faster mass production printing.

1823   **Macadam Road**   McAdam, John
Crushed stone road-building methods are adopted for smoother, more durable roads.

1824   **Electromagnet**   Sturgeon, William
First practical demonstration of electrically-induced magnetism lifts a weight.

1826   **Photography**   Niepce, Nicephore
The first permanent photograph is recorded on a pewter plate coated with asphalt that requires an 8 hour exposure time.

1827   **Ohm's Law**   Ohm, George
Electromagnetic force (volts) equals current (amperes) multiplied by resistance (ohms) or $E = IR$.

1831   **Electric Motor**   Henry, Joseph
The first practical electric motor turns a wheel.

1836   **Revolver**   Colt, Samuel
A handgun with a revolving cylinder holding six rounds is patented.

1837   **DC Electric Motor**   Davenport, Thomas
The first practical direct current motor is used in woodworking.

1838   **Morse Code**   Morse, Samuel
A code of "dots and dashes" for use in telegraphy is patented; 2 years later it becomes operational.

1839   **Daguerreotype**   Daguerre, L.J.M.
This commercial high-quality photographic process is announced to Academie Des Beaux-Arts on August 19.

Vulcanized Rubber   Goodyear, Charles
Accidentally discovered when rubber and sulfur were heated, this material remains flexible over a wide range of temperatures.

1842   **Fertilizer**   Lawes, John Bennet
Superphosphate, the first manufactured fertilizer, is patented in England.

1844	**Cocaine**	Merck, Heinrich

The alkaloid cocaine, derived from coca leaves, is first isolated, but commercial production does not begin until
1862.

1846	**Ether**	Morton, William T.G.

First public demonstration of use during surgery is given.

Neptune — Galle, Johann / Adams, John / Leverrier, Urbain

This distant planet is first observed by Galle, but had
been mathematically predicted earlier by Adams
and Leverrier.

1849	**Safety Pin**	Hunt, Walter

This useful device is patented on April 10.

1851	**Foucault's Pendulum**	Foucault, Jean

In a public demonstration, the device proves that the
earth rotates.

1852	**Airship**	Giffard, Henri

The flight of the first powered airship is successful.

1857	**Bunsen Burner**	Bunsen, Robert

After years of development this chemical heating apparatus is introduced.

1859	**Oil Well**	Drake, Edwin

On August 29, a 70-foot well produces oil in
Pennsylvania.

Theory of Evolution — Darwin, Charles

Publication of *On the origin of species by means of natural
selection...* describes evolution on the basis of natural
selection.

1860	**Black Body**	Kirchoff, Gustav

The concept of a body that would absorb all light and
reflect none, which when heated would emit all wave
lengths of light, is introduced.

1861	**Broca's Area**	Broca, Pierre Paul

The relationship of a specific area of the brain to a specific faculty is proven.

1862	**Machine Gun**	Gatling, Richard

This weapon, having ten barrels, rotated by hand and
gravity-fed, is patented.

1865	**Genetics**	Mendel, Gregor

First paper on laws of genetic inheritance is published.

1866	**Dynamite**	Nobel, Alfred

An accidental discovery leads to the development of this
safer form of nitro-glycerine.

1869	**Air Brake**	Westinghouse, George

The air brake dramatically improves train safety.

	Chemical Elements Table	Mendeleyev, Dmitri

The first periodic table arranges 63 known elements by atomic weight (later arranged by atomic number).

1871	**Dental Drill**	Morrison, James

The first modern dental drill (foot-powered) is patented.

1873	**Levi Jeans**	Strauss, Levi
		Davis (or Youphes), Jacob

Two varieties of riveted-seam canvas trousers are made; one is made of blue denim by an American tailor, Levi Strauss.

1874	**Barbed Wire**	Glidden, Joseph

This fencing is patented and will be used to enclose range lands.

1876	**Carpet Sweeper**	Bissell, Melville

This device to clean carpets faster and easier is patented.

1879	**Cash Register**	Ritty, James

This accounting device is invented by an American saloon owner.

	Electric Light	Edison, Thomas

The first commercially successful electric lamp uses a filament of scorched cotton thread.

1885	**Rabies Vaccine**	Pasteur, Louis

The vaccine cures rabies in nine year old Joseph Meister.

1886	**Linotype**	Mergenthaler, Ottmar

Machine-set type comes into use.

1887	**Pneumatic Tire**	Dunlop, John

An air-filled rubber tire is first used on a child's cycle.

1888	**AC Electric Motor**	Tesla, Nikola

The alternating current induction motor is patented.

1893	**Zipper**	Judson, Whitcomb

A "clasp locker" or unlocker for boots is patented.

1895	**Chiropractic**	Palmer, Daniel

A healing method using spinal adjustment is established by an osteopath.

	X-Rays	Roentgen, Wilhelm

X-rays are discovered by accident on November 5 by a German physicist.

1897	**Diesel Engine**	Diesel, Rudolf
	Compression ignition allows a more efficient internal compression engine.	
	Electron	Thomson, Joseph
	First subatomic particle is discovered.	
1899	**Aspirin**	Bayer; Kolbe, Adolph
	Acetylsalicylic acid is produced by the company Bayer AG, using a process discovered by Adolph Kolbe.	
1900	**Blood Groups**	Landsteiner, Karl
	Three groups (A, B, O) are postulated; a fourth (AB) is found two years later.	
	Quantum Theory	Planck, Max
	This theory of energy flow marks the end of classical physics and the beginning of modern physics.	
1901	**Vacuum Cleaner**	Booth, Hubert
	This cleaning aid is patented in England where it is first mounted on horse-drawn carriages.	
	Transoceanic Radio	Marconi, Guglielmo
	The first radio transmission across an ocean transmits the letter "S" in Morse Code from Cornwall, England to Newfoundland, Canada on December 12.	
1902	**Hormones**	Bayliss, William
	Secretin, controller of the pancreas, is found to be secreted by the small intestine.	Starling, Ernest
	Ionosphere	Heaviside, Oliver
	A layer of charged particles in the earth's atmosphere that can reflect radio waves is discovered.	Kennelly, Arthur
1903	**Airplane**	Wright, Wilbur
	The first successfully powered flight that is manned is made on December 17 at Kitty Hawk, North Carolina.	Wright, Orville
	Electrocardiograph	Einthoven, Willem
	A string galvanometer is devised to record the electric potential of a beating heart.	
1905	**Special Theory of Relativity**	Einstein, Albert
	Formulates that the velocity of light is absolute and not relative to the observer, leading to $E = mc^2$.	
1906	**AM Radio**	Fessenden, Reginald
	Amplitude Modulation of radio waves is used to broadcast music from Massachusetts.	
1908	**Geiger Counter**	Geiger, Hans
	The first primitive version of this detector of ionizing radiation is produced.	Rutherford, Ernest

1909	**Bakelite** The first synthetic plastic is introduced.	Baekeland, Leo
1911	**Air Conditioner** The first complete air conditioning system is invented.	Carrier, Willis
	Superconductivity Mercury is found to have no resistance to electric current at very low temperature; this phenomena, has potential energy-saving inplications.	Kamerlingh-Onnes, Heike
1912	**Automobile Self-Starter** An electric self-starter is first provided as standard equipment on Cadillac automobiles.	Kettering, Charles
1913	**Theory of Atomic Structure** Electrons orbit a nucleus in higher or lower orbits depending on the energy present.	Bohr, Neils
1916	**General Theory of Relativity** Theory on gravity and the curvature of space is published.	Einstein, Albert
1920	**Commercial Radio Broadcasts** Westinghouse station KDKA in Pittsburgh, Pennsylvania, and WWJ in Detroit, Michigan begin broadcasting.	KDKA, WWJ
1925	**Leica Camera** After 12 years of development, the first modern 35mm camera is produced.	Barnack, Oscar
1926	**Liquid Fuel Rocket** A liquid fuel rocket first flies on March 16. Liquid fuel allows more thrust and better control than solid fuel.	Goddard, Robert
1927	**Uncertainty Principle** It is impossible to accurately determine two variables (i.e. position and momentum of an electron) simultaneously.	Heisenberg, Walter
1928	**Penicillin** The first antibiotic agent is developed, but not used therapeutically until 1940.	Fleming, Alexander
1929	**Electroencephalography** The results of the first studies of the human brain's electrical activity are published.	Berger, Hans
	Hubble's Law The more distant a galaxy is, the faster it is receding from the earth.	Hubble, Edwin
1930	**Cyclotron** A circular particle accelerator is developed.	Lawrence, Ernest

Frozen Food   Birdseye, Clarence
A method of fast-freezing of food is patented.

Pluto   Tombaugh, Clyde
The ninth planet from the sun is discovered at 2300
hours GMT on February 18, but a public announcement
is delayed until March 13.

1931 **Neutrino**   Pauli, Wolfgang
A particle having a neutral electric charge and little or no
mass is proposed.

1932 **Neutron**   Chadwick, James
An elementary particle with no electric charge occurring
in the nucleus of every atom except the lightest isotope
of hydrogen.

Splitting of the Atom   Cockcroft, John
A particle accelerator is used to bombard lithium with   Walton, Ernest
protons, producing two alpha particles.

Positron   Anderson, Carl
The anti-electron or positive electron is discovered.

1933 **FM Radio**   Armstrong, Edwin
Frequency modulation (FM) radio is developed.

1934 **Bathysphere**   Beebe, William
A device capable of deep dives while connected to a sur-
face ship is developed.

1938 **Ballpoint Pen**   Biro, Georg
An ink cartridge on a rolling ball combination is   Biro, Ladislo
invented.

Fluorescent Lighting   Compton, Arthur
This energy-efficient form of lighting is introduced   Inman, George
at the World's Fair.

Nuclear Fission   Hahn, Otto
A uranium atom is split by bombardment with   Meitner, Lise
neutrons.   Strassmann, Fritz

Xerography   Carlson, Chester
A prototype model using electrostatic/thermal process
produces its first copy on October 22.

1939 **Black Holes**   Oppenheimer, J. Robert
The existence of a collapsed neutron star possessing a   Volkoff, George
gravity field so strong light cannot escape is proposed.

DDT   Muller, Paul
DDT is proposed as a pesticide harmless to plants and
animals (later it is found to be unsafe).

1942 **Nuclear Reactor**   Fermi, Enrico
On December 2, at the University of Chicago, a controlled nuclear fission chain reaction is created.

1943 **Aqualung**   Cousteau, Jacques
A device with a special valve to match the breathing of   Gagnan, Emile
the diver to his pressurized air supply is developed.

        LSD   Hoffmann, Albert
On April 16, lysergic acid diethylamide's perception alter-
ing properties are first discovered by accidental exposure.

1944 **Antihistamines**   Bouet, Daniel
Pyrilamine is introduced therapeutically. The compounds
were first discovered in 1937.

        V1 Flying Bomb   Von Braun, Wernher
Thousands of these pulse jet-powered craft containing   Gosslau, Ing Fritz
one ton bombs fall on London.

        V2 Rocket   Von Braun, Wernher
The first modern liquid fuel rocket is used against   Dornberger, Walter
England. Over 3000 are launched during an 8 month
period.

1945 **Tupperware**   Tupper, Earl
Food storage items and bowls made of polythylene are
introduced.

        Nuclear Fission Bomb   United States
On July 16, 60 miles from Alamogordo, New Mexico a 20
kiloton device is exploded.

1946 **Eniac**   Maughly, John
The Electronic Numerical Integrator And Computer,   Eckart, John, Jr.
the first all electronic computer, is completed.

1947 **Carbon Dating**   Libby, William
Dating of organic substances by measuring the amount
of the isotope carbon 14 remaining in them is developed.

        Polaroid Camera   Land, Edwin
A camera allowing "instant" personal photography is
introduced.

        Sound Barrier Broken   Yeager, Charles
On October 14, at 670 mph (Mach 1.015) the Bell X-1
becomes the first manned craft to break the sound barrier.

1948 **Bathyscaphe**   Piccard, August
A device that can move independently to explore the
deepest waters is developed.

Big Bang Theory   Gamow, George

A theory of the creation of the universe explaining the
formation of the chemical elements is presented.

Transistor   Shockley, William

The first transistor uses a semiconductor, germanium,   Brattain, Walter
as a rectifier and an amplifier.   Bardeen, John

1951   **Birth Control Pill**   Pincus, Gregory

This form of contraception for women is developed,   Chang, Min Chuch
but not approved for public use until 1961.   Rock, John
                                                          Djerassi, Carl

1952   **Artificial Heart Valve**   Hufnagel, Charles

The artificial heart valve is developed at Georgetown
University.

Jet Airliner   England

The De Havilland Comet, first flown in July 1949, enters
commercial service in May 1952 with British Overseas
Airways.

Nuclear Fusion Bomb   United States

On November 1, at Eniwetok Atoll, a 10 megaton device
is exploded.

1953   **DNA Structure**   Crick, Francis

The double helix form is proposed as a model of the DNA   Watson, James
molecule structure.

1954   **Polio Vaccine**   Salk, Jonas

This vaccine is produced successfully after testing in
1953.

1957   **Nuclear Power Plant**   Duquesne Light Company

The first full-scale commercial nuclear power plant
becomes operational December 2, at Shippingport,
Pennsylvania.

Small Transistor Radio   Morita, Akio

Pocket-size Sony model six is introduced.   Ibuka,Masaru

Sputnik I   U.S.S.R.

First manmade object to orbit the earth is launched on
October 4.

Sputnik II   U.S.S.R.

The first living creature, Laika, a dog, orbits the earth
November 3, but dies in space 7 days later.

1959   **Luna 2 Impacts on the Moon**   U.S.S.R.

Luna 2 becomes the first manmade object to (crash) land
on another celestial object on September 13.

Moon Photographed   U.S.S.R.
On October 7, at 6:30 a.m. the Luna 3 spacecraft begins
photographing the far side of the moon.

1960   **Laser**   Maiman, Theodore
An optical maser utilizing a ruby cylinder is developed.

Tiros I   NASA
Tiros I (United States) is launched April 1, followed by
Tiros II in November.

1961   **Man in Space**   Gagarin, Yuri
First man in space orbits the earth April 12, aboard
Vostok I.

First American in Space   Shepard, Alan
On May 5, a 15 minute flight reaches an altitude of 191
kilometers.

Chaos   Lorenz, Edward
The first mathematical system with chaotic behavior is cre-
ated in a computer model of the atmosphere's behavior.

1962   **Telstar**   United States
The first active telecommunications satellite is launched,
enabling transatlantic television transmissions.

***Silent Spring* Published**   Carson, Rachel
Warning of impending environmental disasters raises pub-
lic awareness of dangerous substances in the ecosystem.

First American in Orbit   Glenn, John
On February 20, three orbits of the earth are made in the
"Friendship 7" spacecraft.

1963   **Audio Cassette**   Philips Company
The tape cassette for audio recording and playback is
introduced.

Pop Top Can   Fraze, Ermal
Patent for rivet device makes pop top beverage can feasible.

Liver Transplant   Starzl, Thomas
The patient lives for 23 days after the operation.

Quasars   Spandage, Allan
Quasi-stellar objects with large red shifts are discovered.   Matthews, Thomas

1964   **Moon's Surface Photographed**   Ranger 7
4316 good-quality, close-range photographs of the
moon's surface are relayed to earth.

1965   **Spacewalk**   Leonov, Aleksei
The first man, wearing only a space-suit, makes first trip
outside space satellite, lasting 20 minutes on March 18.

| 1967 | **Heart Transplant** | Barnard, Christiaan |

The first successful human heart transplant is performed on Louis Washkansky who lives 18 days after the operation.

Pulsar — Bell, Jocelyn / Hewish, Anthony

Pulsating stars are discovered by radio astronomy.

Soyuz I — Komarov, Vladimir

The Soyuz 1 spacecraft crash lands on earth, killing its Russian occupant, after a successful orbital mission. The cosmonaut is the first human casualty of an actual space flight.

1969 **Man on the Moon** — Armstrong, Neil / Aldrin, Edwin "Buzz"

"That's one small step for man...one giant leap for mankind" on July 20 in Apollo 11.

1971 **Microprocessor** — Intel

Intel produces its 4004 chip, incorporating logic, program and data functions into one chip.

1977 **Personal Computer** — Wozniak, Stephen / Jobs, Steven

The Apple II personal computer is introduced.

1978 **Test Tube Baby** — Steptoe, Patrick / Edwards, Robert

Louise Brown is born in July in England as a result of *in vitro* fertilization.

1979 **U.S. Nuclear Accident** — General Public Utilities Corporation

A partial meltdown of the core of Three Mile Island nuclear reactor in Pennsylvania occurs the night of March 23.

1981 **Personal Computer by IBM** — International Business Machines (IBM)

Personal computer, using Microsoft DOS, is introduced.

Proton/Anti-Proton Collision — CERN (European Organization for Nuclear Research)

A circular particle accelerator four miles long is used to create a proton/anti-proton collision.

Space Shuttle — NASA

The first successful flight of a reusable space vehicle, the STS-1 *Columbia*, occurs on April 12.

1982 **Artificial Heart** — Jarvik, Robert

Barney Clark, a dentist, lives 112 days with a permanent artificial heart.

1986 **Space Station** — U.S.S.R.

The launch of the first module of the space station *Mir* ("Peace") is successful

***Challenger* Disaster** — NASA

The space shuttle *Challenger* STS-51L destructs 60 seconds after launch, and its crew of seven perish, on January 28.

Chernobyl Nuclear Accident   U.S.S.R.
The worst nuclear power accident occurs when the explosion of the reactor on April 26, 1:30 a.m. releases radioactive debris into the air.

Nonstop Global Flight   Rutan, Dick
The *Voyager,* an aircraft of composite construction, flies   Yeager, Jeana
around the world December 14-23, without landing.

1988   **Vertebrate Patented**   Harvard Medical School
A U.S. patent is issued for a mouse "developed" by   Leder, Phillip
genetic engineering.   Stewart, Timothy

1989   **Gene Therapy**   National Institutes of
First approved transfer of a gene to a melanoma patient   Health
on May 22.

1990   **Hubble Space Telescope**   NASA
Launched April 25, the telescope is designed to see further into space than earth-based telescopes.

1991   **Most Distant Object Found**   Mt. Palomar
A quasar 12 billion light years away is found by   Observatory
astronomers at Mt. Palomar.

THE ANIMAL WORLD

AQUATIC LIFE *See also: THE ANIMAL WORLD—MAMMALS; THE ANIMAL WORLD—REPTILES AND AMPHIBIANS; AROUND THE HOUSE AND GARAGE—PETS AND PESTS; BIOLOGY—FUNGI, BACTERIA, ALGAE, ETC.*

▽ 1 What are the life stages of the salmon?

Alevin—A newly hatched translucent salmon with a yolk sac. Less than 1 inch (2.5 centimeters) long, it remains in the redd (gravel nest on the stream bottom) for several weeks until it exhausts the yolk sac.

Fry—One- to two-inch (2.5- to 5-centimeter) salmon that has depleted its yolk sac and has emerged from the redd.

Parr or Fingerling—Young two-inch (5-centimeter) salmon with visible markings known as parr marks. It feeds voraciously on insects and small freshwater crustaceans.

Smolt—Young salmon about two months old that has undergone physiological changes and is ready to migrate to sea. Only 30% of the smolts make it to the ocean.

Grilse—Atlantic salmon that spends only one year at sea before returning to spawn (reproduce).

Salmon—The mature stage of the fish when it returns to the river to spawn (reproduce), usually in the fall. It stores up oils and fats to sustain it on its upstream trip.

Kelt—Atlantic salmon that has spawned, and returned to the ocean. Pacific salmon die after spawning.

Sources: Frew, Timothy. *Salmon*, p. 18.

▽ 2 How is the age of fish determined?

One way to determine the age of a fish is by its scales, which have growth rings just as trees do. Scales have concentric bony ridges, called "circuli," which reflect the growth patterns of the individual fish. The portion of the scale that is embedded in the skin contains clusters of these ridges (called "annuli"); each cluster marks one year's growth cycle.

Sources: *Academic American Encyclopedia*, vol. 8, p. 113; Grzimek, Bernhard. *Grzimek's Animal Life Encyclopedia*, vol. 4, p. 49.

3 ▽ At what speeds do fish swim?

The maximum swimming speed of a fish is somewhat determined by the shape of its body and its tail and by its internal temperature. The cosmopolitan sailfish (*Istiophorus platypterus*) is considered to be the fastest fish species, at least for short distances, swimming at greater than 60 miles (95 kilometers) per hour. However some American fishermen believe that the bluefin tuna (*Thunnus thynnus*) is the fastest, but the fastest speed recorded so far is 43.4 miles (69.8 kilometers) per hour. Data is extremely difficult to secure because of the practical difficulties in measuring the speeds. The yellowfin tuna (*Thunnus albacares*) and the wahoe (*Acanthocybium solandri*) are also fast, timed at 46.35 miles (74.5 kilometers) per hour and 47.88 miles (77 kilometers) per hour during 10 to 20 second sprints. Flying fish swim at 40+ miles (64+ kilometers) per hour, dolphins at 37 miles (60 kilometers) per hour, trout at 15 miles (24 kilometers) per hour, and blenny at 5 miles (8 kilometers) per hour. Humans can swim 5.19 miles (8.3 kilometers) per hour.

Sources: *Academic American Encyclopedia*, vol. 8, p. 118; The Diagram Group. *Comparisons*, p. 188; *Guinness Book of Records 1992*, p. 94.

4 ▽ How much electricity does an electric eel generate?

An electric eel (*Electrophorus electricus*) has current-producing organs made up of electric plates on both sides of its vertebral column running almost its entire body length. The charge—on the average of 350 volts, but as great as 550 volts—is released by the central nervous system. The shock consists of four to eight separate charges, which last only two- to three-thousandths of a second each. These shocks, used as a defense mechanism, can be repeated up to 150 times per hour without any visible fatigue to the eel. The most powerful electric eel, found in the rivers of Brazil, Colombia, Venezuela, and Peru, produces a shock of 400 to 650 volts.

Sources: *Grzimek's Animal Life Encyclopedia*, vol. 4, pp. 296-297; *Guinness Book of Records 1992*, p. 95; Taber, Robert W. *1001 Questions Answered About the Oceans and Oceanography*, p. 143.

5 ▽ What are giant tube worms?

These worms were found in 1977 when the submersible *Alvin* was exploring the ocean floor of the Galapagos Ridge (located 1.5 miles or 2.4 kilometers below the

Pacific Ocean surface and 200 miles or 321.8 kilometers from the Galapagos Islands). *Riftia pachyptila Jones,* named after worm expert Meredith Jones of the Smithsonian Museum of Natural History, were discovered near the hydrothermal (hot water) ocean vents. Growing to lengths of five feet (1.5 meters), the worms lack both mouth and gut, and are topped with feathery plumes composed of over 200 thousand tiny tentacles. The phenomenal growth of these worms is due to their internal food source—symbiotic bacteria, over 100 billion per ounce of tissue—that live within the worms' troposome tissues. To these troposome tissues, the tube worms transport absorbed oxygen from the water, together with carbon dioxide and hydrogen sulfide. Utilizing this supply, the bacteria living there in turn produce carbohydrates and proteins that the worms need to thrive.

This was only one of *Alvin's* discoveries during its historic voyage. Scientists expected to find a "desert" at these ocean depths where no light penetrated. Most of the world's organisms have photosynthesis (the use of light to make organic compounds) at the base of their food chains. But in these depths, giant tube worms, vent crabs, and mollusks thrive because these vent communities depend on chemoautotropic (chemically self-feeding) bacteria, which derive their life-sustaining energy from the oxidation of substances spewing from the vents, or in symbiotic relationships, such as that with the giant tube worms.

Sources: *The Almanac of Science and Technology,* pp. 290-292.

6 / What is krill?

Krill (Euphausiids) is the name applied to planktonic crustaceans that constitute the diet of many whales, particularly baleen whales, and invertebrates such as fish, penguins, and seabirds. Blue whales, largest of the world's animals, live on euphausiids, and a whale's stomach can contain four tons of krill. Euphausiids are all marine. Most of the 85 species feed by filtering plankton from the water. Generally, they resemble primitive decapods such as shrimp and lobsters.

Sources: *Living Invertebrates,* p. 517; *The McGraw-Hill Dictionary of Scientific and Technical Terms,* p. 1036.

7 / How are coral reefs formed?

Coral reefs grow only in warm, shallow water. The calcium carbonate skeletons of dead corals serve as a framework upon which layers of successively-younger animals attach. Such accumulations, combined with rising water levels, slowly lead to the formation of reefs that can be hundreds of meters deep and long. The coral animal, or polyp, has a columnar form; its lower end is attached to the hard floor of the reef; the upper end is free to extend into the water. A whole colony consists of thousands of individuals. There are two kinds of corals, hard and soft, depending on the type of skeleton secreted. The polyps of hard corals deposit around themselves a solid skeleton of calcium carbonate (chalk), thus, most swimmers see only the skeleton of the coral; the animal is in a cup-like formation into which it withdraws during the daytime.

Sources: Kaplan, Eugene H. *Field Guide to Coral Reefs*, pp. 78–79; *The New Book of Popular Science*, vol. 2, p. 270.

8 / What are the parts of a whelk shell?

Belonging to the family of Gastopods (meaning stomach foot), a whelk is one of the 70 thousand kinds of living mollusks (from the Latin *mollis*, meaning soft-bodied) that produce shells of calcium carbonate. Its body has a prominent foot with a covering called a mantle. This mantle will eventually line the shell or exoskeleton the whelk builds.

The glands of the mantle secrete calcium carbonate (lime), which is formed into the protective shell. The calcium carbonate comes from the food or from the water that the whelk ingests. An active predator, feeding on other gastropods, worms, fish, and decaying vegetable matter, a whelk usually lives from two to five years, but some live for decades.

Sources: George, J. David, and Jennifer J. George. *Marine Life*, pp. 84, 87, 91; *The Science Teacher*, vol. 59, no. 4 (April 1992), p. 35.

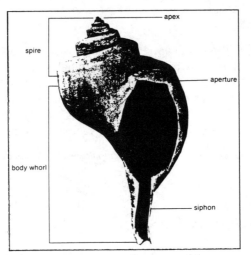

Whelk shell parts

BIRDS *See also:* THE ANIMAL WORLD—NAMES; AROUND THE HOUSE AND GARAGE—PETS AND PESTS

9 / How does a homing pigeon find its way home?

Scientists currently have two hypotheses to explain the homing flight of pigeons. Neither has been proved to the satisfaction of all the experts. The first hypothesis involves an "odor map." The theory proposes that young pigeons learn this map by smelling different odors that reach their home in the winds from varying directions. They would, for example, learn that a certain odor is carried on winds blowing from the east. If a pigeon were transported eastward, the odor would tell it to fly westward to return home. The second hypothesis proposes that a bird may be able to extract its home's latitude and longitude from the earth's magnetic field. It may prove in the future that neither theory explains the pigeon's navigational abilities or that some synthesis of the two theories would be plausible.

Sources: *The Birds Around Us,* pp. 160-161.

10 / Do birds have knees?

The bird's knee is the joint between the thigh and the "drumstick" and is usually concealed by feathers in living birds. Many species have a kneecap or patella.

Sources: *The Cambridge Encyclopedia of Ornithology,* p. 15.

11 / What are the natural predators of the penguin?

The leopard seal is the principal predator of both the adult and juvenile king penguin. The penguin may also be caught by a killer whale while swimming in open water. Eggs and chicks that are not properly guarded by adults are often devoured by skuas and sheathbills.

Sources: Pfeffer, Pierre. *Predators and Predation,* pp. 210-211.

12 / When was the bald eagle adopted as the national bird of the United States?

On June 20, 1782, the citizens of the newly independent United States of America adopted the bald or "American" eagle as their national emblem. At first the heraldic artists depicted a bird that could have been a member of any of the larger species, but by 1902, the bird portrayed on the seal of the United States of

America had assumed its proper white plumage on head and tail. The choice of the bald eagle was not unanimous; Benjamin Franklin (1706-1790) preferred the wild turkey.

Sources: *Marshall Cavendish International Wildlife Encyclopedia*, vol. 3, p. 278.

13 / When is Buzzard Day in Hinckley, Ohio?

Since 1957, Buzzard Day is celebrated on the first Sunday after March 15. The buzzards are actually turkey vultures which end their yearly migration north in Hinckley where they spend their summer. March 15 is usually the day the first "scouting" birds arrive.

Sources: *Pittsburgh Press* (March 12, 1992), p. D1.

14 / Which birds lay the largest and smallest eggs?

The elephant bird (*Aepyornis maximus*), an extinct flightless bird of Madagascar, also known as the giant bird or roc, laid the largest known bird eggs. Some of these eggs measured as much as 13.5 inches (34.29 centimeters) in length and 9.5 inches (24.13 centimeters) in diameter. The largest egg produced by any living bird is that of the North African ostrich (*Struthio camelus*). The average size is 6 to 8 inches (15.24 to 20.32 centimeters) in length and 4 to 6 inches (5 to 15 centimeters) in diameter.

The smallest mature egg, measuring less than 0.39 inches (1 centimeter) in length, is that of the vervain hummingbird (*Mellisuga minima*) of Jamaica.

Sources: Cruickshank, Allan D. *1001 Questions Answered About Birds*, p. 174; Wood, Gerald L. *The Guinness Book of Animal Facts and Feats*, pp. 90-91.

15 / What can an orphaned wild bird eat?

An orphaned songbird needs to be fed every 20 minutes during daylight hours for several weeks. The food should be placed deep in its throat. A soft-billed bird (such as a warbler or catbird) may be given grated carrots, chopped hard-boiled eggs, cottage cheese, fresh fruit, or custard. A young hard-billed bird (such as a sparrow or finch) may be given the same food, but rape, millet, and sunflower seeds should be added to this diet when the bird becomes well-developed. A mixture of dry baby cereal and the yolk of a hard-boiled egg moistened with milk can also be given.

Sources: Cruickshank, Allan D. *1001 Questions Answered About Birds*, p. 274; McElroy, Thomas P. *The New Handbook of Attracting Birds*, p. 194.

16
What is the proper size for the hole in a bird house for the purple martin?

The opening should be 2½ inches (6.35 centimeters) in diameter. Native Americans were the first to attract the martins to artificial nesting sites by hanging dried gourds from trees. In the southeastern United States today, this method remains popular. As with all martin houses, the nesting box or gourd should be mounted on a pole at least 15 feet (4.5 meters) from any buildings or overhanging branches. Purple martins avoid nests that are vulnerable to climbing predators. They like to be near water. Purple martins are highly social; they readily accept multiple housing units. In the eastern United States these units are popular structures to attract colonies of birds.

Sources: Kress, Stephen W. *The Audubon Society Guide to Attracting Birds*, p. 211; Schutz, Walter E. *How to Attract, House, and Feed Birds*, p. 175.

Purple martin birdhouse

17
Is it best to stop feeding hummingbirds after Labor Day?

There is no evidence that feeding hummingbirds after Labor Day will prevent them from migrating to avoid the fatal effects of winter weather. In fact, it may help a weakened straggler "refuel" for the long migration to warmer climates. Nectar feeders can be left out until the birds stop coming.

Sources: U.S. Fish and Wildlife Service. *Backyard Bird Feeding*, p. 19.

18
When do the swallows come back to Capistrano in California?

The swallows arrive on March 19th and depart on October 23rd when they migrate to the Southern Hemisphere.

Sources: *Encyclopedia Americana*, vol. 24, p. 210.

19
Do all birds fly?

There are flightless birds, among which the penguins and the ratites are the best known. Ratites includes emus, kiwis, ostriches, rheas, and cassowaries. They are called ratite, because they lack a keel on the breastbone. All of these birds have wings but lost their power to fly millions of years ago. Many birds that live isolated on oceanic islands (for example, the great auk) apparently became flightless in the absence of predators and the consequent gradual disuse of their wings for escape.

Sources: *Grzimek's Animal Life Encyclopedia*, vol. 7, p. 89; Palmer, Ephram L. *Fieldbook of Natural History*, p. 550; Terres, John K. *The Audobon Society Encyclopedia of North American Birds*, p. 379.

20
How fast does a hummingbird fly?

Hummingbirds fly at speeds up to 71 miles (80.5 kilometers) per hour. Small species beat their wings 50 to 80 times per second, higher in courtship displays. For comparison, the following table lists some other birds:

Bird	Speed	
	Miles per hour	Kilometers per hour
Peregrine falcon	168–217	270.3–349.1
Swift	105.6	169.9
Merganser	65	104.6
Golden plover	50–70	80.5–112.6
Mallard	40.6	65.3
Wandering albatross	33.6	54.1
Carrion crow	31.3	50.4
Herring gull	22.3–24.6	35.9–39.6
House sparrow	17.9–31.3	28.8–50.4
Woodcock	5	8

Sources: *The Cambridge Encyclopedia of Ornithology*, pp. 59-60; *Guinness Book of Records 1992*, pp. 80-81; *Marshall Cavendish International Wildlife Encyclopedia*, vol. 11, p. 1268.

21
How can bluebirds be encouraged to nest in a particular location?

Bluebirds may be attracted by providing nesting boxes and perches, having an area of low or sparse vegetation, and planting nearby trees, vines, or shrubs such as blueberries, honeysuckle, and crabapples. Bluebirds prefer open countryside with low undergrowth. Parks, golf courses, and open lawns are their preferred habitats. In the last 40 years eastern bluebird populations have declined 90% coinciding with the disappearing farmland, widespread use of pesticides, and an increase in nest competitors (house sparrows and European starlings). Artificial nesting boxes sometimes provides more secure nesting places than do natural nest sites, because artificial structures can be built to resist predators. Bluebird boxes can be made with entrance holes (1½ inches or 3.8 centimeters in diameter) small enough to exclude starlings and can have special raccoon guards on mounting poles. Mounted 3 to 6 feet (1 to 2 meters) above ground (to discourage predators) with the nesting boxes being no closer than 100 feet (30 meters). The box should have a tree within 50 feet (15 meters) so that the fledglings can perch. The box should have a 4 inch by 4 inch floor (10.6 by 10.6 centimeters), 8 to 12 inches (20.3 to 30.5 centimeters) in height, with the 1½ inch hole 6 to 10 inches (15.3 to 25.4 centimeters) above the floor.

Sources: Kress, Stephen W. *The Audubon Society Guide to Attracting Birds*, pp. 185, 204-211; Stokes, Donald. *The Bluebird Book*, pp. 30-34.

22 / What is the name of the bird that perches on the black rhinoceros' back?

The bird, a relative of the starling, is called an oxpecker (a member of the Sturnidae family). Found only in Africa, the yellow-billed oxpecker (*Buphagus africanus*) is widespread over much of western and central Africa, while the red-billed oxpecker (*Buphagus erythrorhynchus*) lives in eastern Africa from the Red Sea to Natal.

Seven to eight inches long (17 to 20 centimeters) with a coffee-brown body, the oxpecker feeds on more than 20 species of ticks that live in the hide of the black rhinoceros (*Diceros bicornis*), also called the hook-lipped rhino. The bird spends most of its time on the rhinoceros or on other animals, such as the antelope, zebra, giraffe, buffalo, etc. The bird has been known to even roost on the body of its host.

The relationship between the oxpecker and the rhinoceros is a type of symbiosis (a close association between two organisms in which at least one of them benefits) called mutualism. The rhinoceros' relief of its ticks and the bird's feeding clearly demonstrates mutualism (a condition in which both organisms benefit). In addition, the oxpecker, having much better eyesight than the nearsighted rhinoceros, alerts its host with its shrill cries and flight when danger approaches.

Sources: Burton, Maurice. *Encyclopedia of Animals in Colour*, pp. 209-210; Penny, Malcolm. *Rhinos: Endangered Species*, p. 32.

23 / Why do geese fly in formation?

Aerodynamicists have suspected that long-distance migratory birds, such as geese and swans, adapt the "V" formation in order to reduce the amount of energy needed for such long flights. According to theoretical calculations, birds flying in a "V" formation can fly some 10% farther than a lone bird can. Formation flying lessens the drag (the air pressure that pushes against the wings). The effect is similar to flying in a thermal upcurrent, where less total lift power is needed. In addition, when flying, each bird creates behind it a small area of disturbed air. Any bird flying directly behind it would be caught in this turbulence. In the "V" formation of the Canadian geese, each bird flies not directly behind the other, but aside or above the bird in front.

Sources: Callahan, Philip S. *Bird Behavior*, pp. 133-134; Terres, John K. *Audubon Society Encyclopedia of North American Birds*, p. 377.

24
Why do birds migrate annually?

Hormonal changes caused by natural breeding cycles triggers migration patterns. The major wintering areas for North American migrating birds are the southern United States and Central America. For migrating ducks there are four major flyways south: the Atlantic flyway, the Mississippi flyway, the central flyway, and the Pacific flyway. Some bird experts propose that the birds return north to breed for several reasons: (1) Birds return to nest, because there is a huge insect supply for their young. (2) The higher the earth's latitude in the summer in the Northern Hemisphere, the longer the daylight available to the parents to find food for their young. (3) There is less competition for food and nesting sites in the north. (4) In the north, there are fewer mammal predators for nesting birds (which are particularly vulnerable during the nesting stage). (5) Birds migrate south to escape the cold weather, so they return north when the weather improves. It also becomes much hotter, longer in the south during this time.

Sources: Feldman, David. *Do Penguins Have Knees?*, pp. 147-148.

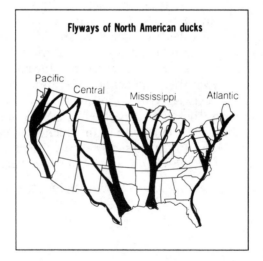

Flyways of North American ducks

25
What names are used for groups of birds?

A group of birds in general is called a congregation, flight, flock, volery, or volley.

Bird	Group name
Bitterns	Siege or sedge
Budgerigars	Chatter
Chickens	Flock, run, brood, or clutch
Coots	Fleet or pod
Cormorants	Flight

Bird	Group name
Cranes	Herd or siege
Crows	Murder, clan, or hover
Curlews	Herd
Doves	Flight, flock, or dole
Ducks	Paddling, bed, brace, flock, flight, or raft
Eagles	Convocation
Geese	Caggle or plump (on water), flock (on land), skein (in flight), or covert
Goldfinches	Charm, chattering, chirp, or drum
Grouses	Pack or brood
Gulls	Colony
Hawks	Cast
Hens	Brood or flock
Herons	Siege, sege, scattering, or sedge
Jays	Band
Larks	Exaltation, flight, or ascension
Magpies	Tiding or tittering
Mallards	Flush, sord, or sute
Nightingales	Watch
Partridges	Covey
Peacocks	Muster, ostentation, or pride
Penguins	Colony
Pheasants	Nye, brood, or nide
Pigeons	Flock or flight
Plovers	Stand, congregation, flock, or flight
Quails	Covey or bevy
Sparrows	Host
Starlings	Chattering or murmuration
Stork	Mustering
Swallows	Flight
Swans	Herd, team, bank, wedge, or bevy
Teals	Spring
Turkeys	Rafter
Turtle doves	Dule
Woodpeckers	Descent
Wrens	Herd

Sources: Gray, Peter. *The Encyclopedia of the Biological Sciences*, p. 594; *The Guinness Book of Answers*, pp. 137-139; *The World Almanac and Book of Facts 1992*, p. 573.

DINOSAURS *See: THE ENVIRONMENT—EXTINCT AND ENDAN-GERED PLANTS AND ANIMALS*

ENDANGERED OR EXTINCT ANIMALS *See: THE ENVIRONMENT—EXTINCT AND ENDANGERED PLANTS AND ANIMALS*

HUMANS *See: THE ANIMAL WORLD—MAMMALS; THE HUMAN BODY*

INSECTS, SPIDERS, ETC.

26
How many species of insects are there?

Estimates of the number of recognized insect species range from about 750 thousand to upward of one million—but it is thought by some experts that this represents less than half of the number that exists in the world. About seven thousand new insect species are described each year, but unknown numbers are lost annually from the destruction of their habitats, mainly tropical forests.

Sources: *The Encyclopedia of Insects*, pp. 2, 12; Parker, Sybil P. *Synopsis & Classification of Living Organisms*, vol. 2, p. 327.

27
What is the most destructive insect in the world?

The most destructive insect is the desert locust (*Schistocera gregaria*), the locust of the Bible, whose habitat ranges from the dry and semi-arid regions of Africa and the Middle East, through Pakistan and northern India. This short-horn grasshopper can eat its own weight in food a day, and during long migratory flights a large swarm can consume 20 thousand tons (18,144 thousand kilograms) of grain and vegetation a day.

Sources: Wood, Gerald L. *The Guinness Book of Animal Facts and Feats*, p. 174.

28
What is a lepidopterist?

A lepidopterist is a person who studies the order of insects called Lepidoptera, a large order including butterflies and moths.

Sources: *Funk & Wagnalls New Standard Dictionary of the English Language*, p. 1418.

29
Who discovered the "dance of the bees"?

In 1943, Karl von Frisch (1886-1982) published his study on the dance of the bees. It is a precise pattern of movements performed by returning forager (worker) honeybees in order to communicate the direction and distance of a food source to the other workers in the hive. The dance is performed on the vertical surface of the hive and two kinds of dances have been recognized: the round dance (performed when food is nearby) and the waggle dance (done when food is further away). In the waggle dance the bee moves in a figure 8. The angle between the figure 8 and the vertical indicates the direction. Since then, some doubt has been cast on the interpretation of the dance.

Sources: Toothill, Elizabeth. *The Facts On File Dictionary of Biology*, p. 83.

30 / What are "killer bees"?

Africanized honeybees—the term entomologists prefer rather than killer bees—are a hybrid originating in Brazil where African honeybees were imported in 1956. The breeders, hoping to produce a bee better-suited to producing more honey in the tropics, instead found that African bees soon hybridized with and mostly displaced the familiar honeybees. Although they produce more honey, the Africanized honeybees also are more dangerous than European bees because they attack intruders in greater numbers. It is expected that the Africanized bees will spread into the United States but the extent will be limited by cold weather. Some bee experts predict the bees will become less aggressive as they mingle and interbreed with North American bees.

Sources: *Audubon Magazine*, vol. 93, no. 2 (March 1991), pp. 14, 16.

31 / How does a butterfly differ from a moth?

Characteristic	Butterflies	Moths
Antennae	Knobbed	Unknobbed
Active-time of day	Day	Night
Coloration	Bright	Dull
Resting position	Vertically above body	Horizontally beside body

Note: While these guidelines generally hold true, there are exceptions. Moths have hairy bodies, and most have tiny hooks or bristles linking the fore-wing to the hind-wing; butterflies do not have either characteristic.

Sources: Feltwell, John. *The Natural History of Butterflies*, pp. 8-9.

32 / Are there any butterfly gardens available for the public to visit?

There are a number of such gardens with special plants that attract the various species of butterfly. The Cecil B. Day Butterfly Center at Callaway Gardens in Pine Mountain, Georgia, is perhaps the largest. Over 100 species visit the outdoor garden. There is also Butterfly World at Tradewinds Park South in Coconut Creek, Florida, and at Marine World Africa U.S.A. in Vallejo, California; the Cincinnati Zoo; and the Des Moines Botanical Gardens with its tropical butterfly house. The Butterfly World in Florida offers two screened aviaries having to two to three thousand butterflies, a collection of preserved specimens, and other displays; whereas the Butterfly World in California has an enclosed rain-forest environment in a 5,500 square-foot glass greenhouse.

Sources: Schneck, Marcus. *Butterflies*, pp. 20-21; Sedenko, Jerry. *The Butterfly Garden*, pp. 8, 138.

33 / How are ants distinguished from termites?

These insects—ants (order Hymenoptera) and termites (order Isoptera)—have segmented bodies with multi-jointed legs. Listed below are some differences.

Characteristic	Ant	Termite
Wings	Two pairs with the front pair being much longer than the back pair	Two pairs of equal length
Antenna	Bends at right angle	Straight
Abdomen	Wasp-waist (pinched in)	No wasp-waist

Sources: Rackensky, Stanley. *Getting Pests to Bug Off*, p. 36.

34 / Is there a national insect?

A group of people have petitioned the United States Congress to name the monarch butterfly as the national insect, but, to date, they have not been successful.

Sources: *Ranger Rick*, vol. 25 (April 1991), p. 11.

35 / Do male mosquitoes bite humans?

No, male mosquitoes live on plant juices, sugary saps, and liquids arising from decomposition. They do not have a biting mouth that can penetrate human skin. But the females do. In some species, the females need blood to lay their eggs. Blood lipids are converted into the needed iron and protein to produce great numbers (200) of eggs.

Sources: Feldman, David. *Do Penguins Have Knees?*, p. 190; Passarin d'Entreves, P. *The Secret Life of Insects*, p. 28.

36 / What is a "daddy longlegs"?

The name applies to two different kinds of invertebrates. The first is a harmless, non-biting long-legged arachnid which preys on small insects, dead insects, and fallen fruit. Also called a harvestman, it is often mistaken for a spider, but it lacks the segmented body shape that a spider has. Although it has the same number of legs (eight) as a spider, the harvestman's legs are far longer and thinner. These very long legs enable it to raise its body high enough to avoid ants or other small enemies. The term "daddy longlegs" also is used for a cranefly—a thin-bodied insect with long thin legs that has a snout-like proboscis with which it sucks water and nectar.

Sources: Burton, Maurice, and Robert Burton. *Encyclopedia of Insects and Arachnids*, p. 63; Palmer, Ephram Laurence. *Fieldbook of Natural History*, p. 433; *World Book Encyclopedia*, vol. 5, p. 3.

37 / How many eggs does a spider lay?

The number of eggs varies according to the species. Some larger spiders lay over 2,000 eggs, but many tiny spiders lay one or two and perhaps no more than a dozen during their lifetime. Spiders of average size probably lay a hundred or so. Most spiders lay all their eggs at one time and enclose them in a single egg sac; others lay eggs over a period of time and enclose them in a number of egg sacs.

Sources: Headstrom, Richard. *Spiders of the United States*, p. 34.

38 / Which is stronger—steel or silk from a spider's web?

The only man-made material that comes close to matching the strength of a spider's silk is steel, but the strongest silk has a tensile strength five times greater than that of steel having the equivalent weight. Tensile strength is the longitudinal stress that a substance can bear without tearing apart.

Sources: *How Things Work: Structures*, p. 32.

39 / How long does it take the average spider to weave a complete web?

The average orb-weaver spider takes 30 to 60 minutes to completely spin its web. The order Araneae (the spiders) constitutes the largest division in the class Arachnida, containing about 32 thousand species. These species of spiders use silk to capture their food in a variety of ways ranging from the simple trip wires used by the large bird-eating spiders to the complicated and beautiful webs spun by the orb spiders. Some species produce funnel-shaped webs and there are also communal webs built by communities of spiders.

The completed web has several spokes leading from the initial structure. The number and nature of the spokes depend on the species. The spider replaces any damaged threads by gathering up the thread in front of it and producing a new one behind it. The orb web must be replaced every few days because it loses its stickiness (and its ability to entrap food).

The largest aerial webs are spun by the tropical orb weavers of the genus *Nephila*, measuring up to 18 feet 9 inches (5.72 meters) in circumference. The smallest webs are done by the *Glyphesis cottonae* which covers about 0.75 square inches (4.84 square centimeters).

Sources: *Guinness Book of Records 1992*, p. 98; *The Illustrated Encyclopedia of Wildlife*, vol. 9, pp. 2264-2265; *Scientific American*, vol. 202 (April 1960), p. 116.

40 How long have cockroaches been on the earth?

The earliest cockroach fossils are about 280 million years old. Some cockroaches measured 3 to 4 inches (7.62 to 10.16 centimeters) long. Cockroaches (order Dictyoptera) are nocturnal scavenging insects, that eat not only human food but book-bindings, ink, and whitewash as well.

Sources: Fenton, Carroll L. *The Fossil Book*, p. 234; *Marshall Cavendish International Wildlife Encyclopedia*, vol. 6, pp. 606-607.

41 How do fleas jump so far?

The jumping power of fleas comes both from strong leg muscles and from pads of a rubber-like protein called resilin. The resilin is located above the flea's hind legs. To jump, the flea crouches, squeezing the resilin, and then it relaxes certain muscles. Stored energy from the resilin works like a spring, launching the flea. A flea can jump well both vertically and horizontally. Some species can jump 150 times their own length. To match that record, a human would have to spring over the length of 2¼ football fields—or the height of a 100-story building—in a single bound. The common flea (*Pulex irritans*) has been known to jump 13 inches (33 centimeters) in length and 7¼ inches (18.4 centimeters) in height.

Sources: *Guinness Book of Records 1992*, p. 102; *National Geographic World*, vol. 179 (July 1990), p. 3.

42 How fast is a snail's pace?

The speed of a land snail was recorded in 1970 as 0.00758 miles (40 feet or 12.2 meters) per hour. The fastest moving species of land snail is probably the common garden snail (*Helix aspersa*) which moves at 0.03 miles (158.5 feet or 48.3 meters) per hour.

Sources: *Guinness Book of Records 1992*, pp. 57, 106; Wallace, Irving. *The Book of Lists #2*, p. 142.

43 ▽ *How is the light in fireflies produced?*

The light produced by fireflies (*Photinus pyroles*) or lightning bugs, is a kind of heatless light called bioluminescence caused by a chemical reaction in which the substance, luciferin, undergoes oxidation, when the enzyme luciferase is present. The flash is a photon of visible light that radiates, when the oxidating chemicals produce a high–energy state, and revert back to their normal state. The flashing is controlled by the nervous system and takes place in special cells called photocytes. The nervous system, photocytes, and the tracheal end organs control the flashing rate. Scientists are uncertain as to why this occurs. The rhythmic flashes could be a means of attracting prey or enabling mating fireflies to signal in heliographic codes (that differ from one species to another) or perhaps even be a warning signal. The air temperature seems to be correlated with the flashing rate. The higher the temperature, the shorter the interval between flashes—8 seconds at 65°F (18.3°C) and 4 seconds at 82°F (27.7°C).

Sources: Duensing, Edward. *Talking to Fireflies, Shrinking the Moon*, p. 44; Farb, Peter. *The Insects*, p. 119; Sutton, Caroline, and Duncan M. Anderson. *How Do They Do That?*, pp. 206-207.

44 ▽ *What are the stages of insect metamorphoses?*

There are two types of metamorphoses (marked structural changes in the growth processes): complete and incomplete. In *complete metamorphoses*, the insect (such as ants, moths, butterflies, termites, wasps, and beetles) goes through all the distinct stages of growth to reach adulthood. In *incomplete metamorphoses*, the insect (such as grasshoppers, crickets, and louse) does not go through all the stages of complete metamorphoses.

Complete Metamorphoses

Egg—One egg is laid at a time or many (as much as ten thousand).

Larva—What hatches from the eggs is called "larva." A larva can look like a worm.

Pupa—After reaching its full growth, the larva hibernates, developing a shell or "pupal case" for protection. A few insects (e.g., the moth) spin a hard covering called a "cocoon." The resting insect is called a pupa (except the butterfly is called a chrysalis), and remains in the hibernation state for several weeks or months.

Adult—During hibernation, the insect develops its adult body parts. When it has matured physically, the fully-grown insect emerges from its case or cocoon.

Incomplete Metamorphoses

Egg—One egg or many eggs are laid.

Early-stage nymph—Hatched insect resembles an adult, but smaller in size. However, those insects that would normally have wings have not yet developed them.

Late-stage nymph—At this time, the skin begins to molt (shed), and the wings begin to bud.

Adult—The insect is now fully grown.

Sources: Rovin, Jeff. *The Laws of Order*, pp. 3-4.

45
What causes the Mexican jumping bean to move?

The bean moth (*Carpocapa saltitans*) lays its eggs in the flower or in the seed pod of the spurge, a bush known as *Euphorbia sebastiana*. The egg hatches inside the seed pod, producing a larva or caterpillar. The jumping of the bean is caused by the active shifting of weight inside the shell as the caterpillar moves. The jumps of the bean are stimulated by sunshine or by heat from the palm of the hand.

Sources: *Nature Magazine*, vol. 38 (1945), pp. 473-474.

MAMMALS *See also:* THE ANIMAL WORLD—NAMES; THE ANIMAL WORLD—PHYSICAL CHARACTERISTICS; AROUND THE HOUSE AND GARAGE—PETS AND PESTS

46
Why do nine-banded armadillos always have four offspring of the same sex?

The one feature that distinguishes the nine-banded armadillo (*Dasypus novemcinctus*) from all others in its offspring is that the female almost always gives birth to four young of the same sex. This consistency results from the division of the one fertilized egg into four parts to produce quadruplets.

Sources: *Treasures of the Tide*, p. 42.

47
How many quills does a porcupine have?

For its defensive weapon, the average porcupine has about 30 thousand quills or specialized hairs, comparable in hardness and flexibility to slivers of celluloid and so sharply pointed, they can penetrate any hide. The quills that do the most damage are the short ones that stud the porcupine's muscular tail. With a few lashes, the porcupine can send a rain of quills that have tiny scale-like barbs into the skin of its adversary. The quills work their way inward because of its barbs and the involuntary muscular action of the victim. Sometimes the quills can work themselves out, but other times the quills pierce vital organs, and the victim dies.

Slow-footed and stocky, porcupines spend much of their time in the trees, using their formidable incisors to strip off bark and foliage for their food, and supplement their diets with fruits, grasses, etc. Porcupines have a ravenous appetite for salt; as herbivores (plant-eating animals) their diets have insufficient salt. So natural salt licks, animal bones left by carnivores (meat-eating animals), yellow pond lilies, and other items having a high salt content (including paints, plywood adhesives and sweated-on clothing, etc. of humans) have a strong appeal to porcupines.

Sources: Costello, David F. *The World of the Porcupine*, p. 13; *Smithsonian*, vol. 23, no. 2 (May 1992), pp. 56-67.

48 / What is the difference between an African elephant and an Indian elephant?

The African species (*Loxodonta africana*) of this family Elephantidae has a larger body, ears and tusks, a concave rather than convex back, usually three toenails instead of four on its hind feet, and two finger-like lips rather than one at the tip of its trunk. The ear tops of the Asian elephant (*Elephas maximus*) turn forward whereas those of the African elephant turn backwards.

Sources: Voelker, William. *The Natural History of Living Mammals*, pp. 239-240.

49 / How many horses are there in the world?

According to Dr. D. Fielding of the Edinburgh School of Agriculture (in Edinburgh, United Kingdom) the number of horses worldwide is 65,292 thousand, the number of donkeys is 41,599 thousand, and the number of mules is 15,462 thousand.

Sources: *Cornell Animal Health Newsletter*, vol. 19, no. 1 (March 1992), p. 1.

50 / How many letters are permissible in the name of a thoroughbred horse?

Before the name of a thoroughbred can become "official" it must be submitted to the Jockey Club for approval. One of their requirements limits the name to not more than three pronounceable words with a maximum of eighteen letters.

Sources: Pettingill, O.E. *Born to Run*, p. 33.

51 / What is Przewalski's horse?

This native of Mongolia and northeastern China is the last truly wild horse species. Named for Nikolai Przewalski (1839-1888), the Russian colonel who reported its existence in 1870, Przewalski's horse (*Equus przwalskii*) is a stocky, short-legged animal with a dun-colored coat, a pale muzzle and belly, with dark legs, mane, and tail. Its short mane is bristly and erect. This horse has the unusual feature of possessing 66 chromosomes rather then the customary 64 found in a domestic horse. Last seen in the wild in 1968 and believed to be extinct, there are 1,000 or so in zoos and wildlife parks.

Sources: *The Illustrated Encyclopedia of Wildlife*, vol. 2, p. 347; *National Geographic Magazine*, vol. 181 (March 1992), unnumbered pages; *New Scientist*, vol. 129 (January 12, 1991), p. 26.

Przewalski's horse

52 What is a capybara?

The capybara (*Hydrochaerus hydrochaeris*) is the largest of all living rodents. Also called the water hog, water pig, water cary, or carpincho, it looks like a huge guinea pig. Its body length can be 3¼ to 4½ feet (1 to 1.3 meters), and it usually weighs between 120 to 130 pounds (54.4 to 59 kilograms) or more. A native of northern South America, this rodent leads a semi-aquatic life, feeding on aquatic plants and grasses. A subspecies, native to Panama, is smaller and weighs between 60 to 75 pounds (27.2 to 34 kilograms).

Sources: *The Illustrated Encyclopedia of Wildlife*, vol. 2, p. 279; *Marshall Cavendish International Wildlife Encyclopedia*, vol. 5, pp. 504-505.

53 How long do wombats live and what do they eat?

Native to Australia and Tasmania, the common wombat or coarse-haired wombat (*Vombatus ursinus*) lives between 5 and 26 years (26 years in zoos). It dines mostly on grasses, roots, mushrooms, fresh shoots, and herbaceous plants. A wombat looks like a small bear in appearance, has a thick heavy body ranging from 2.3 to 4 feet (0.7 to 1.2 meters) and weighs 33 to 77 pounds (15 to 35 kilograms). This marsupial resembles a rodent in its manner of feeding and in its tooth structure—all its teeth are rootless and ever growing to compensate for their wear. Shy, it lives in a burrow and is an active digger, and its rough fur ranges from yellowish buff, gray, to dark brown or black.

Sources: *The Complete Encyclopedia of the Animal World*, p. 261; *Grzimek's Encyclopedia of Mammals*, vol. 1, pp. 342-343.

54 / *What is chamois?*

The chamois (of the family Bovidae) is a goat-like animal living in the mountainous areas of Spain, central Europe (the Alps and Apennines), south central Europe, the Balkans, Asia Minor, and the Caucasus. Agile and surefooted, with acute senses, it can jump 6.5 feet (2 meters) in height and 19.69 feet (6 meters) in distance, and run at speeds of 31 miles (50 kilometers) per hour. Its skin has been made into "shammy" leather for cleaning glass and polishing automobiles; although now, many times the shammy or chamois skins sold are specially treated sheepskin.

Sources: Nowak, Robert M. *Walker's Mammals of the World*, vol. 2, p. 1446.

55 / *Is there a cat that lives in the desert?*

The sand cat (*Felis margarita*) is the only member of the cat family tied directly to sand regions. Found in North Africa, the Arabian peninsula, and the deserts of Turkmenistan in Uzbekistan, western Pakistan, the sand cat has adapted to extremely arid desert areas. The padding on the soles of its feet is well-suited to the loose sandy soil, and it can live without drinking "free" water. Having sandy or grayish-ochre dense fur, its body length is 17.6 to 22.2 inches (45 to 57 centimeters). Mainly nocturnal (active at night) the cat feeds on rodents, hares, birds and reptiles.

The Chinese desert cat (*Felis bieti*) does not live in the desert as its name implies, but inhabits the steppe country and mountains. Likewise the Asiatic desert cat (*Felis silvestris ornata*) inhabits the open plains of India, Pakistan, Iran, and Asiatic Russia.

Sources: *Marshall Cavendish International Wildlife Encyclopedia*, vol. 2, p. 206; Nowak, Robert M. *Walker's Mammals of the World*, vol. 2, p. 1193; Taylor, David. *You & Your Cat*, pp. 41-43.

56 / *Are there any mammals that fly?*

Bats (order Chiroptera with 986 species) are the only truly flying mammals, although several gliding mammals are referred to as "flying" (such as flying squirrel and flying lemur). The "wings" of bats are double membranes of skin stretching from the sides of the body to the hind legs and tail, and are actually skin extensions of the back and belly. The wing membranes are supported by the elongated fingers of the forelimbs (or arms). Nocturnal (active at night), ranging in length from 1.5 inches to 1.33 feet (25 to 406 millimeters), living in caves or crevices, bats inhabit most of the temperate and tropical regions of both hemispheres. The majority of species feed on inserts and fruit, and some tropical species eat pollen and nectar of flowers and insects found inside them. Moderate size species usually prey on small mammals, birds, lizards, and frogs, and some eat fish. But true vampire bats (three species) eat the blood of animals by making an incision in the animal's skin—from these bats, animals can contract rabies. Most bats do not find their way around by sight but have evolved a sonar system, called "echolocation," for locating solid objects. Bats emit vocal sounds through the nose or mouth while flying. These sounds, usually above the human hearing range, are reflected back as echoes. This method enables bats, when flying in darkness, to avoid solid objects and to locate the position of flying insects. Bats have the most acute sense of hearing of any land animal, hearing frequencies as high as 120 to 210 kilohertz. The highest frequency humans can hear is 20 kilohertz.

Sources: Burt, William Henry. *A Field Guide to the Mammals*, p. 21; Nowak, Robert M. *Walker's Mammals of the World*, vol. 1, pp. 190-192.

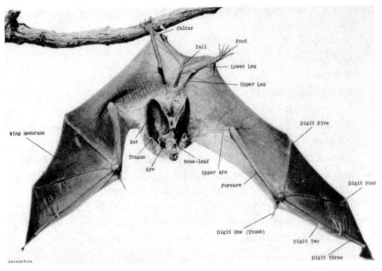

Bat wing structure

▽ **57** ▽
What are some animals that have pouches?

Marsupials (meaning "pouched" animals) differ from all other living mammals in their anatomical and physiological features of reproduction. Most female marsupials—kangaroos, bandicoots, wombats, banded anteaters, koalas, opossums, wallabies, tasmanian devils, etc.—possess an abdominal pouch (called a marsupium), in which their young are carried. However in some small terrestrial marsupials the marsupium is not a true pouch but merely a fold of skin around the mammae (milk nipples).

The short gestation period in marsupials (in comparison to other similarly-sized mammals) allows their young to be born in an "undeveloped" state. Consequently, these animals have been viewed as "primitive" or second-class mammals. However, some now see that the reproductive process of marsupials has an advantage over that of placental mammals. A female marsupial invests relatively few resources during the brief gestation period, more so during the lactation (nursing period) when the young are in the marsupium. If the female marsupial loses its young, it can sooner make a second attempt than a placental mammal in a comparable situation.

Sources: *The Illustrated Encyclopedia of Wildlife*, vol. 4, pp. 607-644; Nowak, Robert M. *Walker's Mammals of the World*, vol. 1, pp. 10-11.

58 ▽ Which bear lives in a tropical rain forest?

The Malayan sun bear (*Ursus malayanus*) is one of the rarest animals in the tropical forests of Sumatra, Malay Peninsula, Borneo, Burma, Thailand, and southern China. The smallest bear, with a length of 3.3 to 4.6 feet (1 to 1.4 meters) and weighing 59.5 to 143.3 pounds (27 to 65 kilograms), it has a black strong, stocky body. With powerful paws having long, curved claws to help it climb trees in the dense forests, it is an expert tree climber. The sun bear tears at tree bark to expose insects, larvae, and the nests of bees and termites. Fruit, coconut palm, and small rodents, too, are part of its diet. Sleeping and sunbathing during the day, it is active at night. Unusually shy and retiring, cautious and intelligent, it is declining in population as the forests are being destroyed.

Sources: *Macmillian Illustrated Animal Encyclopedia*, p. 82; Nowak, Robert M. *Walker's Mammals of the World*, vol. 2, pp. 1084, 1094.

59 ▽ How long does a gray wolf live?

The gray wolf (*Canis lupus*) also known as the timber wolf, is the largest and most widespread species of the family Canidae. It can live in the wild for less than 10 years, but under human care, up to 20 years. However in many areas it has been hunted and killed, because it is believed to pose a threat to humans and domesticated animals (cattle, sheep, and reindeer). It may soon suffer the fate of the red wolf (*Canis rufus*) that once flourished in the southeast and south central United States. Two of the red wolf's three subspecies are extinct with the third found in a small area of the coastal plains and marshes of eastern Texas and Western

Louisiana. The endangered gray wolf is declining faster in the New World than in the Old. In the United States, it is limited to Alaska (10,000), Northern Minnesota (1,200), and Isle Royale, Michigan (20), with perhaps a few packs in Wisconsin, northern Michigan, and the Rocky Mountain area, and in Canada, there are 15,000. Strongly social, living in packs, and weighing between 75-175 pounds (43-80 kilograms), in physical appearance the gray wolf resembles a large domestic dog, such as the Alaskan malamute.

Sources: *Grzimek's Encyclopedia of Mammals*, vol. 4, p. 61; Nowak, Ronald M. *Walker's Mammals of the World*, vol. 2, pp. 1070-1076; *The Official World Wildlife Fund Guide to Endangered Species of North America*, vol. 1, pp. 444-450.

60
What is the chemical composition of a skunk's spray?

The chief odorous components of the spray have been identified as crotyl mercaptan, isopentyl mercaptan, and methyl crotyl disulfide in the ratio of 4:4:3. The liquid is an oily, pale-yellow, foul-smelling spray which can cause severe eye inflammation. This defensive weapon is discharged from two tiny nipples located just inside the skunk's anus—either as a fine spray or a short stream of rain-sized drops. Although the liquid's range is 1.83 to 2.75 yards (2 to 3 meters), its smell can be detected 1.55 miles (2.5 kilometers) downwind.

Sources: *Encyclopedia Americana*, vol. 25, p. 12; Nowak, Ronald M. *Walker's Mammals of the World*, vol. 2, p. 1131.

61
How does a human's heartbeat compare with those of other mammals?

Mammal	Heart beats per minute
Human	70
Elephant	25
Mouse	600–700
Large dog	80
Small dog	120

Sources: Smith, Anthony. *The Body*, p. 492.

62
Which mammals lay eggs and suckle their young?

The duck-billed platypus (*Ornithorhynchus anatinus*) and the echidna or spiny anteater (family Tachyglossidae), indigenous to Australia, Tasmania, and New Guinea, are the only two species of mammals that lay eggs (a non-mammalian feature) but suckle their young (a mammalian feature). These mammals (order Monotremata) resemble reptiles in that they lay rubbery shell-covered eggs that are incubated and hatched outside the mother's body. In addition they resemble reptiles in their digestive, reproductive, and excretory systems, and in a number of anatomical details (eye structure, presence of certain skull bones, pectoral (shoulder) girdle and rib and vertebral structures). However they are classed as mammals because they have fur and a four-chambered heart, nurse their young from gland milk, are warm-blooded, and have some mammalian skeletal features.

Sources: Duplaix, Nicole, and Noel Simon. *World Guide to Mammals*, p. 20; Stimpson, George. *Information Roundup*, p. 114.

63 / What is the name of the seal-like animal in Florida?

The West Indian manatee (*Trichechus manatus*), in the winter, moves to more temperate parts of Florida, such as the warm headwaters of the Crystal and Homosassa rivers in central Florida or the tropical waters of southern Florida. When the air temperature rises to 50°F (10°C), it will wander back along the Gulf coast and up the Atlantic coast as far as Virginia. Long–range offshore migrations to the coast of Guyana and South America have been documented. This large, plant-eating, water mammal may have been the inspiration for the mermaid legend. In 1893, when the population of manatees in Florida was reduced to several thousand, the state gave it legal protection from being hunted or commercially exploited. However, many animals continue to be killed or injured by the encroachment of humans. Entrapment in locks and dams, collisions with barges and power boat propellers, etc., cause at least 30% of the manatee deaths (125 to 130 total annual number of deaths).

Sources: *New York Times* (December 8, 1991), section 5, p. 16; Nowak, Ronald M. *Walker's Mammals of the World*, vol. 2, pp. 1300-1302; *The Universal Almanac 1992*, p. 544.

Manatee

64 ▽ Do camels store water in their humps?

The hump or humps do not store water, since they are fat reservoirs. The ability to go long periods without drinking water, up to ten months if there is plenty of green vegetation and dew to feed on, results from a number of physiological adaptations. One major factor is that camels can lose up to 40% of their body weight with no ill effects. A camel can also withstand a variation of its body temperature by as much as 14°F (-10°C). A camel can drink 30 gallons of water in ten minutes and up to 50 gallons over several hours. A one-humped camel is called a dromedary or Arabian camel; a Bactrian camel has two humps and lives in the wild on the Gobi desert. Today, the Bactrian is confined to Asia, but most of the Arabian camels are on African soil.

Sources: *Marshall Cavendish International Wildlife Encyclopedia*, vol. 5, pp. 484, 488; Voelker, William. *The Natural History of Living Mammals*, pp. 272-273.

65 ▽ How does the breath-holding capability of a human compare with other mammals?

Mammal	Average time in minutes
Human	1
Polar bear	1.5
Pearl diver (human)	2.5
Sea otter	5
Platypus	10
Muskrat	12
Hippopotamus	15
Sea cow	16
Beaver	20
Porpoise	15
Seal	15 to 28
Greenland whale	60
Sperm whale	90
Bottlenose whale	120

Sources: Taber, Robert. *1001 Questions Answered About Oceans and Oceanography*, p. 15.

66 ▽ How deep do marine mammals dive?

Below are listed the maximum depths and the longest durations of time underwater by various aquatic mammals:

Mammal	Maximum depth Feet	Meters	Maximum time underwater
Weddell seal	1,968	600	70 minutes
Porpoise	984	300	6 minutes

Mammal	Maximum depth		Maximum time underwater
	Feet	Meters	
Bottle-nosed whale	1,476	450	120 minutes
Fin whale	1,148	350	20 minutes
Sperm whale	>6,562	>2,000	75–90 minutes

Sources: Duxbury, Alyn C., and Alison Duxbury. *An Introduction to the World's Oceans*, p. 381.

67 / What is the difference between porpoises and dolphins?

Marine dolphins (family Delphinidae) and porpoises (family Phocoenidae) consists of about 40 species. The chief differences between dolphins and porpoises occur in the snout and teeth. True dolphins have a beak-like snout and cone-shaped teeth. True porpoises have a rounded snout and flat or spade-shaped teeth.

Sources: *Macmillan Illustrated Animal Encyclopedia*, pp. 112, 114; Nowak, Ronald M. *Walker's Mammals of the World*, vol. 2, pp. 969-982.

68 / Which mammal has the shortest gestation period? Which one has the longest?

Gestation is the period of time between fertilization and birth in oviparous animals. The shortest gestation period known is 12 to 13 days, shared by three marsupials: the American or Virginian opossum (*Didelphis marsupialis*); the rare water opossum, or yapok (*Chironectes minimus*) of central and northern South America; and the eastern native cat (*Dasyurus viverrinus*) of Australia. The young of each of these marsupials are born while still immature, and complete their development in the ventral pouch of their mother. While 12 to 13 days is the average, the gestation period is sometimes as short as 8 days. The longest gestation period for a mammal is that of the Asiatic elephant (*Elephas maximus*) with an average of 609 days, and a maximum of 760 days.

Mammal	Average gestation period	
	Days	Months
Armadillo	60 to 120	2 to 4
Ass		about 11½
Baboon		5 to 6
Bear, grizzly	210 to 265	7
Bear, polar		about 8
Bear, American black	210 to 215	7
Beaver	about 105	3½
Boar, wild	115	3¼
Bobcat	60 to 63	2
Capybara	150	5
Cat (domestic)	52	2

Mammal	Average gestation period	
	Days	Months
Cattle	about 238	9¼
Cheetah	91 to 95	3
Chimpanzee	230 to 240	7½
Dog	53 to 71	2
Dolphin		10 to 12
Dormouse	21 to 32	
Ermine	about 28	
Fox, red	60 to 63	2
Gerbil	21 to 28	
Gibbon		7 to 8
Giraffe	453 to 464	15¼
Goat	150	5
Gopher	12 to 20	
Guinea pig	63	2
Hamster	from 15 to 37	
Hippopotamus	about 240	8
Horse		about 11½
Kangaroo		6 to 11 (in pouch)
Koala	34 to 36	1¼
Leopard	90 to 105	3¼
Lion	100 to 119	3½
Llama	348 to 368	11½ to 12¼
Mouse	about 20 to 30	
Ocelot	70	2¼
Opossum, American	12 to 13	
Panda, giant	125 to 150	4½
Pig	101 to 129	3¼
Porpoise	183	6
Rabbit	30	
Raccoon	63	2
Reindeer	210 to 240	7 to 8
Rhinoceros, black		15
Sheep	135 to 160	4½ to 5
Shrew	13 to 24	
Skunk, striped	62 to 66	2
Squirrel	about 40	1⅓
Tiger	103	3½
Walrus		15 to 16 including 4 to 5 months of delayed implantation
Weasel	35 to 45	1½
Whale	305 to 365	10 to 12
Whale, beluga		14 to 15
Whale, sperm		14 to 15
Wolf	61 to 63	2
Zebra	340	about 11½

Sources: *The Guinness Book of Answers,* pp. 136-137; *Marshall Cavendish International Wildlife Encyclopedia,* vol. 8, p. 842; *The World Almanac and Book of Facts 1992,* p. 252.

What names are used for groups or companies of mammals?

Mammal	Group name
Antelopes	Herd
Apes	Shrewdness
Asses	Pace, drove, or herd
Baboons	Troop
Bears	Sloth
Beavers	Family or colony
Boars	Sounder
Buffaloes	Troop, herd, or gang
Camels	Flock, train, or caravan
Caribou	Herd
Cattle	Drove or herd
Deer	Herd or leash
Elephants	Herd
Elks	Gang or herd
Foxes	Cloud, skulk, or troop
Giraffes	Herd, corps, or troop
Goats	Flock, trip, herd, or tribe
Gorillas	Band
Horses	Haras, stable, remuda, stud, herd, string, field, set, team, or stable
Jackrabbits	Husk
Kangaroos	Troop, mob, or herd
Leopards	Leap
Lions	Pride, troop, flock, sawt, or souse
Mice	Nest
Monkeys	Troop or cartload
Moose	Herd
Mules	Barren or span
Oxen	Team, yoke, drove, or herd
Porpoises	School, crowd, herd, shoal, or gam
Reindeer	Herd
Rhinoceri	Crash
Seals	Pod, herd, trip, rookery, or harem
Sheep	Flock, hirsel, drove, trip, or pack
Squirrels	Dray
Swine	Sounder, drift, herd, or trip
Walruses	Pod or herd
Weasels	Pack, colony, gam, herd, pod, or school
Whales	School, gam, mob, pod, or herd
Wolves	Rout, route, or pack
Zebras	Herd

Sources: Gray, Peter. *The Encyclopedia of the Biological Sciences*, p. 594; The *Guinness Book of Answers*, pp. 137-138; *The World Almanac and Book of Facts 1992*, p. 573.

How do the great whales compare in weight and length?

Whale	Average weight		Greatest length	
	Tons	**Kilograms**	**Feet**	**Meters**
Sperm	35	31,752	59	18
Blue	84	76,204	98.4	30
Finback	50	45,360	82	25
Humpback	33	29,937	49.2	15
Right	50 (est.)	45,360 (est.)	55.7	17
Sei	17	15,422	49.2	15
Gray	20	18,144	39.3	12
Bowhead	50	45,360	59	18
Bryde's	17	15,422	49.2	15
Minke	10	9,072	29.5	9

Sources: Duxbury, Alyn C., and Alison Duxbury. *An Introduction to the World's Oceans*, pp. 438-439.

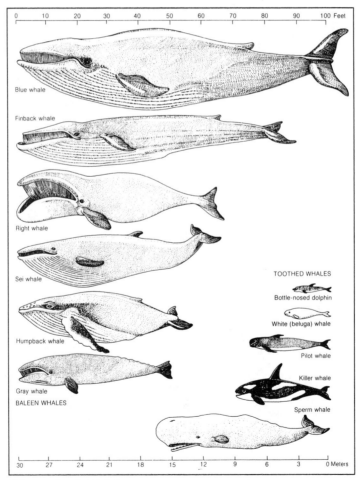

Baleen and toothed whales

NAMES *See also:* THE ANIMAL WORLD—BIRDS; THE ANIMAL WORLD—
   MAMMALS

▽ 71 ▽ *Is Chicago named for an animal?*

Chicago, or *Sikako*, was named for what the Cree Indians called *sikak* or skunk.

Sources: Amory, Cleveland. *Animail*, p. 63.

▽ 72 ▽ *What names are used for groups of animals?*

Animal	Group name
Ants	Nest, army, colony, state or swarm
Bees	Swarm, cluster, nest, hive, or erst
Caterpillars	Army
Eels	Swarm or bed
Fish	School, shoal, haul, draught, run, or catch
Flies	Business, hatch, grist, swarm, or cloud
Frogs	Arm
Gnats	Swarm, cloud, or horde
Goldfish	Troubling
Grasshoppers	Cloud
Hornets	Nest
Jellyfish	Smuck or brood
Lice	Flock
Locusts	Swarm, cloud, or plague
Minnows	Shoal, steam, or swarm
Oysters	Bed
Sardines	Family
Sharks	School or shoal
Snakes	Bed, knot, den, or pit
Termites	Colony, nest, swarm, or brood
Toads	Nest, knot, or knab
Trout	Hover
Turtles	Bale or dole
Wasps	Nest, herd, or pladge

Sources: Gray, Peter. *The Encyclopedia of the Biological Sciences*, p. 594; *The Guinness Book of Answers*, pp. 137-139; *The World Almanac and Book of Facts 1992*, p. 573.

▽ 73 ▽ *What names are used for juvenile animals?*

Animal	Name for young
Ant	Antling
Antelope	Calf, fawn, kid, or yearling
Bear	Cub

Animal	Name for young
Beaver	Kit or kitten
Bird	Nestling
Bobcat	Kitten or cub
Buffalo	Calf, yearling, or spike-bull
Camel	Calf or colt
Canary	Chick
Caribou	Calf or fawn
Cat	Kit, kitling, kitty, or pussy
Cattle	Calf, stot ,or yearling (m. bullcalf or f. heifer)
Chicken	Chick, chicken, poult, cockerel, or pullet
Chimpanzee	Infant
Cicada	Nymph
Clam	Littleneck
Cod	Codling, scrod, or sprag
Condor	Chick
Cougar	Kitten or cub
Cow	Calf (m. bull; f. heifer)
Coyote	Cub, pup, or puppy
Deer	Fawn
Dog	Whelp
Dove	Pigeon or squab
Duck	Duckling or flapper
Eagle	Eaglet
Eel	Fry or elver
Elephant	Calf
Elk	Calf
Fish	Fry, fingerling, minnow, or spawn
Fly	Grub or maggot
Frog	Polliwog or tadpole
Giraffe	Calf
Goat	Kid
Goose	Gosling
Grouse	Chick, poult, squealer, or cheeper
Horse	Colt, foal, stot, stag, filly, hog-colt, youngster, yearling, or hogget
Kangaroo	Joey
Leopard	Cub
Lion	Shelp, cub, or lionet
Louse	Nit
Mink	Kit or cub
Monkey	Suckling, yearling, or infant
Mosquito	Larva, flapper, wriggler, or wiggler
Muskrat	Kit
Ostrich	Chick
Otte	Pup, kitten, whelp, or cub
Owl	Owlet or howlet
Oyster	Set seed, spat, or brood
Partridge	Cheeper
Pelican	Chick or nestling
Penguin	Fledgling or chick
Pheasant	Chick or poult
Pigeon	Squab, nestling, or squealer
Quail	Cheeper, chick, or squealer

Animal	Name for young
Rabbit	Kitten or bunny
Raccoon	Kit or cub
Reindeer	Fawn
Rhinoceros	Calf
Sea Lion	Pup
Seal	Whelp, pup, cub, or bachelor
Shark	Cub
Sheep	Lamb, lambkin, shearling, or yearling
Skunk	Kitten
Squirrel	Dray
Swan	Cygnet
Swine	Shoat, trotter, pig, or piglet
Termite	Nymph
Tiger	Whelp or cub
Toad	Tadpole
Turkey	Chick or poult
Turtle	Chicken
Walrus	Cub
Weasel	Kit
Whale	Calf
Wolf	Cub or pup
Woodchuck	Kit or cub
Zebra	Colt or foal

Sources: Gray, Peter. *The Encyclopedia of the Biological Sciences*, p. 959; *The World Almanac and Book of Facts 1992*, p. 573.

74

What names are used for male and female animals?

Animal	Male name	Female name
Alligator	Bull	
Ant		Queen
Ass	Jack, jackass	Jenny
Bear	Boar or he-bear	Sow or she-bear
Bee	Drone	Queen or queen bee
Camel	Bull	Cow
Caribou	Bull, stag or hart	Cow or doe
Cat	Tom, tomcat, gib, gibeat, boarcat or ramcat	Tabby, grimalkin, malkin, pussy or queen
Chicken	Rooster, cock, stag, or chanticleer	Hen, partlet or biddy
Cougar	Tom or lion	Lioness, she-lion or pantheress
Coyote	Dog	Bitch
Deer	Buck or stag	Doe
Dog	Dog	Bitch
Duck	Drake or stag	Duck
Fox	Fox, dog-fox, stag, reynard or renard	Vixen, bitch or she-fox
Giraffe	Bull	Cow
Goat	Buck, billy, billie, billie-goat or he-goat	She-goat, nanny, nannie, or nannie-goat
Goose	Gander or stag	Goose or dame

Animal	Male name	Female name
Guinea pig	Boar	
Horse	Stallion, stag, horse, stud, stot, stable horse, sire or rig	Mare or dam
Impala	Ram	Ewe
Kangaroo	Buck	Doe
Leopard	Leopard	Leopardess
Lion	Lion or tom	Lioness or she-lion
Lobster	Cock	Hen
Manatee	Bull	Cow
Mink	Boar	Sow
Moose	Bull	Cow
Mule	Stallion or jackass	She-ass or mare
Ostrich	Cock	Hen
Otter	Dog	Bitch
Owl		Jenny or howlet
Ox	Ox, beef, steer, or bullock	Cow or beef
Partridge	Cock	Hen
Peacock	Peacock	Peahen
Pigeon	Cock	Hen
Quail	Cock	Hen
Rabbit	Buck	Doe
Reindeer	Buck	Doe
Robin	Cock	
Seal	Bull	Cow
Sheep	Buck, ram, male-sheep, or mutton	Ewe or dam
Skunk	Boar	
Swan	Cob	Pen
Termite	King	Queen
Tiger	Tiger	Tigress
Turkey	Gobbler or tom	Hen
Walrus	Bull	Cow
Whale	Bull	Cow
Woodchuck	He-chuck	She-chuck
Wren		Jenny or jennywren
Zebra	Stallion	Mare

Sources: Gray, Peter. *The Encyclopedia of the Biological Sciences*, pp. 594-595.

PHYSICAL CHARACTERISTICS, ETC.

▽ **75** ▽
Do animals have blood types?

The number of recognized blood groups varies from species to species:

Animal	Number of blood groups
Pig	16
Cow	12
Chicken	11

Animal	Number of blood groups
Horse	9
Sheep	7
Dog	7
Rhesus monkey	6
Mink	5
Rabbit	5
Mouse	4
Rat	4
Cat	2

Sources: *The Merck Veterinary Manual*, p. 31.

76 Do all animals have red blood?

The color of blood is related to the compounds which transport oxygen. Hemoglobin, containing iron, is red and is found in all vertebrates (animals having a backbone) and a few invertebrates (animals lacking a backbone). Annelids (segmented worms) have either a green pigment chlorocruorin or a red pigment, hemerythrin. Some crustaceans (arthropods having divided bodies and generally having gills) have a blue pigment, hemocyanin, in their blood.

Sources: *New Encyclopedia Britannica*, vol. 2, p. 290.

77 Do any animals snore?

Yes, some dogs definitely snore.

Sources: Walkowicz, Chris. *The Complete Question and Answer Book on Dogs*, p. 178.

78 Besides humans, which animals are the most intelligent?

According to Edward O. Wilson, a behavioral biologist, the ten most intelligent animals are the following:

1. Chimpanzee (two species)
2. Gorilla
3. Orangutan
4. Baboon (seven species, including drill and mandrill)
5. Gibbon (seven species)
6. Monkey (many species, especially macaques, the patas, and the Celebes black ape)
7. Smaller toothed whale (several species, especially killer whale)
8. Dolphin (many of the approximately 80 species)

9. Elephant (two species)
10. Pig

Sources: Wallace, Irving. *The Book of Lists #2*, p. 104.

79 / Do animals have color vision?

Most birds appear to have a well-developed color sense. Most mammals are color-blind. Apes and monkeys have the ability to tell colors apart. Dogs and cats seem to be color-blind and only see shades of black, white, and gray.

Sources: Cruickshank, Allan D. *1001 Questions Answered About Birds*, p. 129; *World Book Encyclopedia*, vol. 13, p. 120.

80 / Which animals spend part of the winter in hibernation?

To avoid the harsh winter conditions, some animals migrate, while others build special nests, hide away in caves and burrows and enter a sleep-like state called hibernation. In hibernation heart rate and respiration slow and body temperature drops. Being so inactive, for several months these animals use little energy and can survive on the fat stores they built up earlier. Hibernators include certain birds, such as nighthawks and swifts; mammals, such as bats, chipmunks, ground squirrels, hamsters, hedgehogs, lemurs, groundhogs, marmots and picket mice; amphibians such as frogs and toads; and reptiles such as lizards, snakes, and turtles. The longest hibernator with 9 months rest is the Burrow ground squirrel (*Spermophilus parryl barrowensis*) of Point Barrow, Alaska.

Some animals like dormice became active at times and feed on nut and seed caches they stored away in the autumn. They are not true hibernators; nor are bears, skunks and opossums. Although these animals lapse into prolonged sleep, they wake from time to time and do not undergo the extreme metabolic changes that true hibernators do.

Some animals go underground and become inactive to avoid the harsh summer conditions; this is called aestivation (estivation). For example, some lungfish will burrow into the dried-up mud of the river or lake bottoms and wait for more water to come.

Sources: Cazeau, Charles J. *Science Trivia*, p. 108; *Guinness Book of Records 1992*, p. 87; Matthews, L. Harrison. *The Life of Mammals*, vol. 1, pp. 97, 293-295, 300; vol. 2, p. 148.

81 / Can animals regenerate parts of their bodies?

Regeneration does occur in animals; however, it progressively declines the more complex the animal species becomes. Among primitive invertebrates (lacking a backbone), regeneration frequently occurs. For example, a planarium (flatworm) can split symmetrically, each being a clone of the other. In higher invertebrates regeneration occurs in echinoderms (such as starfish) and arthropods (such as insects and crustaceans). Regeneration of appendages (limbs, wings, and antennae) occurs in insects (such as cockroaches, fruitflies, and locusts) and in crustaceans (such as lobsters, crabs, and crayfish). For example, regeneration of the crayfish's missing claw occurs its next molt (shedding of its hard cuticle exterior shell/skin in order to grow and the subsequent hardening of a new cuticle exterior). However, sometimes the regenerated claw does not achieve the same size of the missing claw. But after every molt (occurring 2 to 3 times a year) it grows, it eventually will be nearly as large as the original claw. On a very limited basis, some amphibians and reptiles can replace a lost leg or tail.

Sources: Gardiner, Mary S. *The Biology of Invertebrates*, pp. 872-874; Huxley, Thomas H. *The Crayfish*, pp. 38-39; Magill, Frank N. *Magill's Survey of Science: Life Science Series*, vol. 5, p. 2311.

82 / What is a "Skinner box"?

A "Skinner box" is an experimental enclosure for testing animal conditioning in which the subject animal performs a function (e.g., presses a bar or lever) to obtain a reward. By 1938 B.F. Skinner (b. 1904) worked out a technique for measuring the rewarding effect of a stimulus, or the effects of learning on voluntary behavior. The experimental animal (rat) was placed in a special box containing a lever that the animal could manipulate. When it pressed the lever small pellets of food would or would not be released, according to the experimental conditions. This box for animal conditioning has become known as a Skinner box.

Sources: *Dorland's Illustrated Medical Dictionary*, p. 228; Hickman, Cleveland P., Jr., et al. *Integrated Principles of Zoology*, p. A-14.

83 / Which animals can run faster than a human?

The cheetah, the fastest mammal, can accelerate from zero to 45 miles (64.3 kilometers) per hour in two seconds; it has been timed at speeds of 70 miles (112.6 kilometers) per hour over short distances. In most chases, cheetahs average around 40 miles (64.3 kilometers) per hour. Humans can run at almost 28 miles (44.9 kilometers) per hour maximum. Most of the speeds given in the table below are for distances of ¼ mile (0.4 kilometer).

Animal	Maximum speed	
	Miles per hour	Kilometers per hour
Cheetah	70	112.6
Pronghorn antelope	61	98.1
Wildebeest	50	80.5

Animal	Maximum speed	
	Miles per hour	**Kilometers per hour**
Lion	50	80.5
Thomson's gazelle	50	80.5
Quarter horse	47.5	76.4
Elk	45	72.4
Cape hunting dog	45	72.4
Coyote	43	69.2
Gray fox	42	67.6
Hyena	40	64.4
Zebra	40	64.4
Mongolian wild ass	40	64.4
Greyhound	39.4	63.3
Whippet	35.5	57.1
Rabbit (domestic)	35	56.3
Mule deer	35	56.3
Jackal	35	56.3
Reindeer	32	51.3
Giraffe	32	51.3
White-tailed deer	30	48.3
Wart hog	30	48.3
Grizzly bear	30	48.3
Cat (domestic)	30	48.3
Human	27.9	44.9

Sources: *Natural History*, vol. 83 (March 1974), p. 71; Voelker, William. *The Natural History of Living Mammals*, p. 182; *The World Almanac and Book of Facts 1992*, p. 252.

84
What are the largest and smallest living animals?

Largest animals	Name	Length and weight
Mammal	Blue or sulphur-bottom whale (*Balaenoptera musculus*)	100–110 feet (30.5–33.5 meters) long; weighs 135–209 tons (122.4–189.6 tonnes)
Land animal	African bush elephant (*Loxodonta africana*)	Bull is 10½ feet (3.2 meters) tall at shoulder; weighs 5.25–6.2 tons (4.8–5.6 tonnes)
Living bird	North African ostrich (*Struthio C. Camelus*)	8 to 9 feet (2.4–2.7 meters) tall; weighs 345 pounds (156.5 kilograms)
Fish	Whale shark (*Rhincodon typus*)	41 feet (12.5 meters) long; weighs 16½ tons (15 tonnes)
Reptile	Salt-water crocodile (*Crocodylus porosus*)	14–16 feet (4.3–4.9 meters) long; weighs 900–1,500 pounds (408–680 kilograms)
Rodent	Capybara (*Hydrochoerus hydrochaeris*)	3¼–4½ feet (1–1.4 meters) long; weighs 250 pounds (113.4 kilograms)

Smallest animals	Name	Length and weight
Mammal	Commerson's dolphin (*Cephalorhynchus commersonii*)	Weighs 50–70 pounds (236.7–31.8 kilograms)
Land mammal	Bumblebee or Kitti's hog-nosed bat (*Craseonycteris thong longyai*)	1 inch (2.54 centimeters) long; weighs .062 to .07 ounces (1.6–2 grams)

Smallest animals	Name	Length and weight
	or the pygmy shrew (*Suncus erruscus*)	1½–2 inches (3.8–5 centimeters) long; weighs 0.052–.09 ounces (1.5–2.6 grams)
Bird	Bee hummingbird (*Mellisuga helenea*)	2.25 inches (5.7 centimeters) long; weighs 0.056 ounces (1.6 grams)
Fish	Dwarf pygmy goby (*Trimmatam nanus*)	0.35 inches (8.9 millimeters) long
Reptile	Gecko (*Spaerodactylus parthenopion*)	0.67 inches (1.7 centimeters) long
Rodent	Pygmy mouse (*Baiomys taylori*)	4.3 inches (10.9 centimeters) long; weighs 0.24–0.28 ounces (6.8–7.9 grams)

Sources: *Guinness Book of Records 1992*, pp. 54-55, 65, 76-77, 85, 111-112, 117; *Marshall Cavendish International Wildlife Encyclopedia*, vol. 4, p. 384.

85 / How long do animals, in particular mammals, live?

Of the mammals, humans and fin whales live the longest.

Animal	Latin name	Maximum life span in years
Marion's tortoise	*Testudo sumeirii*	152+
Quahog	*Venus mercenaria*	ca. 150
Common box tortoise	*Terrapene carolina*	138
European pond tortoise	*Emys orbicularis*	120+
Spur-thighed tortoise	*Testudo graeca*	116+
Fin whale	*Balaenoptera physalus*	116
Human	*Homo sapiens*	116
Deep-sea clam	*Tindaria callistiformis*	ca. 100
Killer whale	*Orcinus orca*	ca. 90
European eel	*Anguilla anguilla*	88
Lake sturgeon	*Acipenser fulvescens*	82
Freshwater mussel	*Margaritana margaritifera*	80 to 70
Asiatic elephant	*Elephas maximus*	78
Andean condor	*Vultur gryphus*	72+
Whale shark	*Rhiniodon typus*	ca. 70
African elephant	*Loxodonta africana*	ca. 70
Great eagle-owl	*Bubo bubo*	68+
American alligator	*Alligator mississipiensis*	66
Blue macaw	*Ara macao*	64
Ostrich	*Struthio camelus*	62.5
Horse	*Equus caballus*	62
Orangutan	*Pongo pygmaeus*	ca. 59
Bataleur eagle	*Terathopius ecaudatus*	55
Hippopotamus	*Hippopotamus amphibius*	54.5
Chimpanzee	*Pan troglodytes*	51
White pelican	*Pelecanus onocrotalus*	51
Gorilla	*Gorilla gorilla*	50+
Domestic goose	*Anser a. domesticus*	49.75
Grey parrot	*Psittacus erythacus*	49
Indian rhinoceros	*Rhinoceros unicornis*	49

Animal	Latin name	Maximum life span in years
European brown bear	Ursus arctos arctos	47
Grey seal	Halichoerus gryphus	46+
Blue whale	Balaenoptera musculus	ca. 45
Goldfish	Carassius auratus	41
Common toad	Bufo bufo	40
Roundworm	Tylenchus polyhyprus	39
Giraffe	Giraffa camelopardalis	36.25
Bactrian camel	Camelus ferus	35+
Brazilian tapir	Tapirus terrestris	35
Domestic cat	Felis catus	34
Canary	Serinus caneria	34
American bison	Bison bison	33
Bobcat	Felis rufus	32.3
Sperm whale	Physeter macrocephalus	32+
American manatee	Trichechus manatus	30
Red kangaroo	Macropus rufus	ca. 30
American buffalo	Syncerus caffer	29.5
Domestic dog	Canis familiaris	29.5
Lion	Panthera leo	ca. 29
African civet	Viverra civetta	28
Theraphosid spider	Mygalomorphae	ca. 28
Red deer	Cervus elaphus	26.75
Tiger	Panthera tigris	26.25
Giant panda	Ailuropoda melanoleuca	26
American badger	Taxidea taxus	26
Common wombat	Vombatus ursinus	26
Bottle-nosed dolphin	Tursiops truncatus	25
Domestic chicken	Gallus g. domesticus	25
Grey squirrel	Sciurus carolinensis	23.5
Aardvark	Orycteropus afer	23
Domestic duck	Anas platyrhynchos domesticus	23
Coyote	Canis latrans	21+
Canadian otter	Lutra canadensis	21
Domestic goat	Capra hircus domesticus	20.75
Queen ant	Myrmecina graminicola	18+
Common rabbit	Oryctolagus cuniculus	18+
White or beluga whale	Delphinapterus leucuas	17.25
Platypus	Ornithorhynchus anatinus	17
Walrus	Odobenus rosmarus	16.75
Domestic turkey	Melagris gallapave domesticus	16
American beaver	Castor canadensis	15+
Land snail	Helix spiriplana	15
Guinea pig	Cavia porcellus	14.8
Hedgehog	Erinaceus europaeus	14
Burmeister's armadillo	Calyptophractus retusus	12
Capybara	Hydrochoerus hydrochaeris	12
Chinchilla	Chinchilla laniger	11.3
Giant centipede	Scolopendra gigantea	10
Golden hamster	Mesocricetus auratus	10
Segmented worm	Allolobophora longa	10
Purse-web spider	Atypus affinis	9+

Animal	Latin name	Maximum life span in years
Greater Egyptian gerbil	*Gerbillus pyramidum*	8+
Spiny starfish	*Marthasterias glacialis*	7+
Millipede	*Cylindroiulus landinensis*	7
Coypu	*Myocastor coypus*	6+
House mouse	*Mus musculus*	6
Madagascanbrown-tailed mongoose	*Salanoia concolor*	4.75
Cane rat	*Thryonomys swinderianus*	4.3
Siberian flying squirrel	*Pteromys volans*	3.75
Common octopus	*Octopus vulgaris*	2 to 3
Pygmy white-toothed shrew	*Suncus etruscus*	2
Pocket gopher	*Thomomys talpoides*	1.6
Monarch butterfly	*Danaus plexippus*	1.13
Bedbug	*Cimex lectularius*	0.5 or 182 days
Black widow spider	*Latrodectus mactans*	0.27 or 100 days
Common housefly	*Musca domesticus*	0.04 or 17 days

Sources: *The Guinness Book of Answers*, pp. 135-136; Voelker, William. *The Natural History of Living Mammals*, p. 155; Wood, Gerard L. *The Guinness Book of Animal Facts and Feats*, inside cover.

86 / Which animal has the longest gestation period?

The animal with the longest gestation period is not a mammal; it is the viviparous amphibian, the Alpine black salamander, which can have a gestation period of up to 38 months at altitudes above 4,600 feet (1,402 meters) in the Swiss alps; it bears two fully metamorphosed young.

Sources: *The Guinness Book of Records 1992*, p. 91.

PRODUCTS *See:* AROUND THE HOUSE AND GARAGE—CLEANING AND MAINTENANCE; AROUND THE HOUSE AND GARAGE—CLOTHING, JEWELRY, ETC.; METALS AND OTHER MATERIALS—NATURAL SUBSTANCES

REPTILES AND AMPHIBIANS *See also:* AROUND THE HOUSE AND GARAGE—PETS AND PESTS

87 / What are a turtle's upper and lower shell called?

The turtle (order Testudines) uses its shell as a protective device. The upper shell is called the *dorsal carapace* and the lower shell is called the *ventral plastron*. The shell's sections are referred to as *the scutes*. The carapace and the plastron are joined at the sides.

Sources: Conant, Roger. *A Field Guide to Reptiles and Amphibians: Eastern and Central North America*, front end paper.

88 / How fast can a crocodile move?

For short distances a crocodile can raise itself up like a lizard and charge like a galloping horse. A crocodile on land can move at speeds of 15 to 30 miles (24 to 40 kilometers) per hour and in the water, can move at the rate of 20 miles (32 kilometers) per hour.

Sources: *Audubon Magazine*, vol. 86 (May 1984), p. 106.

89 / What is the difference between a reptile and an amphibian?

Reptiles are clad in scales, shields, or plates and their toes have claws. Young reptiles are miniature replicas of their parents in general appearance if not always in coloration and pattern. Reptiles include alligators, crocodiles, turtles, and snakes. Amphibians have moist, glandular skins, and their toes lack claws. Their young pass through a larval, usually aquatic stage before they metamorphose (change in form and structure) into the adult form. Amphibians include salamanders, toads, and frogs.

Sources: Conant, Roger. *A Field Guide to Reptiles and Amphibians: Eastern and Central North America*, facing p. 1.

90 / How is the sex of baby alligators determined?

The sex of an alligator is determined by the temperature at which the eggs are incubated. High temperatures of 90° to 93°F (32° to 34°C) result in males; low temperatures of 82° to 86°F (28° to 30°C) yield females. This determination takes place during the second and third week of the two-month incubation. Further temperature fluctuations before or after this time do not alter the sex of the young. The heat from the decaying matter on top of the nest incubates the eggs.

Sources: *Crocodiles and Alligators*, p. 120; *The New Larousse Encyclopedia of Animal Life*, p. 299; Tyning, Thomas F. *A Guide to Amphibians and Reptiles*, p. 287.

91 / Which poisonous snakes are native to the United States?

Copperheads, rattlesnakes, water moccasins (cottonmouths), and coral snakes are all poisonous. Copperheads (*Agkistrodon contortex*) tend to inhabit areas in the southeastern United States, Texas, and the lower Ohio Valley. Rattlesnakes (genus Crotalus) range throughout the United States, and cottonmouths (*Agkistrodon piscivorus*) range from Texas across to North Carolina. Coral snakes (*Microrus fulvius*), the most venomous snakes in the United States, tend to inhabit the southeastern States and Texas.

Sources: Conant, Roger, and Joseph T. Collins. *A Field Guide to Reptiles and Amphibians*, illus. 188–201.

ZOOS *See: THE ENVIRONMENT—ECOLOGY*

Around the House and Garage

CLEANING AND MAINTENANCE

92

How many square feet does a gallon of paint usually cover?

One gallon of paint usually covers 400 to 500 square feet of surface. This figure can vary depending on the condition of the surface and the various types of paints.

Sources: Libien, Lois. *Paint It Yourself*, p. 21; Nunn, Richard V. *Home Paint Book*, p. 13.

93

How can crayon stains be removed from clothing that has been dried in a clothes dryer?

When a crayon is washed and dried with clothes, use the following method: In a plastic dishpan, add ½ cup of powdered detergent and ½ cup of baking soda. Fill the pan with hot water and put in the clothes. Swish the clothes up and down about every ½ hour for about 8 hours, then wash the stained clothes in the washing machine. Some crayons are made with a permanent dye, and nothing will remove them.

Sources: Chapman, Eugenia. *Clean Your House & Everything in It*, pp. 141–142.

94

Can ivory piano keys be cleaned?

To clean and polish ivory, gently rub the keys in the direction of the grain with a fine abrasive paste. Either lighter fluid or gasoline mixed with finely powdered tailor's chalk or lemon juice combined with whiting are suitable. (CAUTION: Lighter fluid and gasoline are flammable.) Apply the paste to the keys with a damp chamois cloth and let dry. Gently brush the powder off and polish with a soft cloth. White toothpaste may also be used. Never use water on ivory and whenever possible, leave the ivory exposed to sunlight, since ivory yellows in darkness.

Sources: Chapman, Eugenia. *Clean Your House & Everything in It*, p. 100; Wylie, Harriet. *420 Ways to Clean Everything*, p.44.

95 / What is the composition of house dust?

Dust is particles of less than 1 millimeter in diameter, regardless of its content. A significant portion of dust is dead skin fragments from human bodies, tobacco smoke, and airborne particles (pollen or other plant particles, industrial smoke particles, etc.) and clay and other mineral (soil) matter, that is "tracked" into the house. Major volcanic eruptions produce thousands of tons of dust, and dust from outer space can be another source. Over 43 million tons (38.8 metric tons) of dust settle over the United States each year. This includes 31 million tons (28.1 metric tons) from natural sources, with the remainder coming from sources of human activity.

Sources: Cazeau, Charles J. *Science Trivia*, pp. 130–131.

96 / How can jewelry be cleaned?

Dealers in precious stones and collectors use the simplest methods to clean all jewelry, including museum-quality pieces. Everything except wooden beads, beads strung on leather, and pearls can be washed in soap and water. A soft brush can be used to remove dirt from behind and between set stones. A little ammonia added to the water gives crystal and set stones a brilliant shine.

Sources: Levenstein, Mary K. *Caring for Your Cherished Possessions*, p. 22.

97 / Can water spots be removed from wooden furniture?

Water spots appear as white or light-colored rings, usually wherever a glass has been set down. To remove the spots, mix 10 drops of lemon oil into 2 cups of denatured alcohol. Dampen a corner of a soft cloth with this mixture and lightly rub the spots, drying them immediately with the dry end of the cloth.

Another method is to rub the spots lightly with a piece of soft cotton moistened with spirits of camphor. A mixture of one part turpentine (or ammonia) added to one part linseed oil can also be used. (CAUTION: Denatured alcohol and turpentine are flammable.)

Sources: Johnson, Lorraine. *How to Restore and Repair Practically Everything*, p. 22; Stark, Norman. *The Formula Book*, p. 27.

98
How can a quilt be cleaned?

Although many contemporary quilts can be machine-washed and -dried, hand-washing is preferred. If the fabric is colorfast, soak the quilt briefly, using mild soap and tepid water, in a bathtub from which all traces of scouring powder and detergent have been removed. Rinse the quilt well and let it drain in the tub until water no longer drips out of it when it is lifted. Dry flat on a rack. Do not hang the quilt on a line.

A very old or fragile quilt should not be washed. If necessary, gently hand vacuum it, covering the vacuum nozzle with a piece of fine, soft netting held in place with a rubber band.

Sources: Levenstein, Mary K. *Caring for Your Cherished Possessions*, pp. 26-27.

99
Is vinegar a good cleaner?

No, it does not break down dirt and grease like a detergent does. Most common household soils such as oils, fats, and grease are acidic, as is vinegar, and alkaline cleaners work best to remove them. Vinegar is a good rinsing agent; it neutralizes soap residue left by floor cleaners or clothing detergents. It is also good for brightening stained aluminum cookware, removing tarnish from brass and copper, and dissolving hard water spots on dishes and glassware.

Sources: Aslett, Don. *How Do I Clean the Moosehead?*, pp. 137-138.

100
Why should fine textiles not be stored in plastic?

Wedding dresses, table linens, evening gowns, and antique lace should never be stored in plastic. Plastic prevents air circulation, which causes the textile fibers to break down. When the fibers disintegrate, the cotton thread will rot.

Sources: Levenstein, Mary K. *Caring for Your Cherished Possessions*, p. 1.

101
How is common whitewash made?

Although it is not as durable as modern paints, whitewash is still a very practical paint to use on fences, curbs, trellises, and similar outdoor projects. The basic ingredient in it is lime paste which is made by soaking 8 pounds of hydrated "spray lime" in three quarts of water. Dissolve 2 pounds of salt in 2½ gallons of water and add this to 1 gallon of the lime paste. Stir and add enough water to make a thin paint.

An interior whitewash, or calcimine, can be made by dissolving ½ pound of animal glue in 1 quart of water and stirring this mixture into 1 gallon of lime paste. Add enough water to make a thin paint. The mixture may also be tinted with paint pigments.

Sources: Mitchell, Harris. *1200 Household Hints You Wanted to Know*, p. 168.

102
How much humidity should be maintained in the house during the winter?

When the outside temperature is at freezing (32°F or 0°C) or above, an indoor air humidity of 35% to 40% is satisfactory. When the outside temperature decreases to 15°F (9°C), a 30% indoor air moisture content is desirable. At 0°F (-18°C), the humidity should be dropped to about 20%.

Sources: Whitman, Roger C. *More First Aid for the Ailing House*, p. 191.

103
What is the best method for storing silverware and silverplated household items?

Keep silver in an airtight chest or drawer lined with an anti-tarnish cloth. Add a piece of chalk or gum camphor to absorb moisture. Silver can also be stored in felt bags or plastic bags or wrapped in plastic. Plastic wrap or bags may cause moisture to form, but they may be used if the air is carefully squeezed out. The bags should be tied with string rather than with rubber bands. Anti-tarnish strips available from silver dealers can be placed inside the bags.

Sources: *Rodale's Complete Home Products Manual*, p. 278.

104
Why are enzymes no longer used in detergents?

Because enzymes seemed to be an ideal way to remove food, blood, and grass stains without damaging delicate fabrics, they were used in as much as 60% of all detergents until the end of the 1960s. After many consumers complained of skin irritation and rashes from such detergents, in the early 1970s manufacturers ceased to add enzymes to consumer detergents. But detergent makers did not give up on enzymes completely. They developed an encapsulation technique for enzymes that eliminated skin contact. By mid-1988, these encapsulated enzymes were used in 40% of U.S. laundry detergents. Enzymes are used in 80% of detergents in Western Europe and 60% in Japan.

Sources: *The Almanac of Science and Technology*, p. 176.

105

Which Black American invented a dust pan?

On August 3, 1897, Lloyd P. Ray of Seattle, Washington received a patent (U.S. patent no. 587,607) for his long-handled dust-pan. The design would enable the sweeper to perform the debris collection function in an upright standing position. The handle was both removable and adjustable, and the pan portion could be pivoted to allow for convenient hanging storage.

Sources: Burt, McKinley. *Black Inventors of America*, p. 144; *United States Patents*, vol. 1897-16, no. 587,607.

Dust pan patent

106

What can be done to clean the interior of an unanodized aluminum saucepan that has discolored?

Boil 2 teaspoons of cream of tartar (sodium potassium tartrate) in 1 quart of water in the affected saucepan for 10 minutes to lighten the darkened aluminum. The aluminum dark stain is usually aluminum sulfide caused from exposure to an alkaline or mineral-rich food. This oxide can be removed by an acid such as cream of tartar. Acidic foods, as well, can "transfer" the stain, but the food itself becomes discolored. Most contend that this food discoloration is harmless.

Sources: Bear, Marina. *How to Repair Food*, p. 112; Brown, Robert J. *200 Illustrated Science Experiments for Children*, pp. 87-88; Hillman, Howard. *Kitchen Science*, p. 9.

107
How much water does an average person use each day?

64 gallons per day for the average American.

An American Family's Daily Water Consumption

Activity	Water use in gallons
Toilet flushing	100
Shower and bathing	80
Laundry	35
Dishwashing	15
Bathroom sink	8
Utility sink	5
TOTAL	243

Note: Data are for four persons.
Source: U.S. Environmental Protection Agency.

Sources: *The 1992 Information Please Environmental Almanac*, p. 102.

108
Can feathers be cleaned?

Feather trims may be cleaned by sponging them with a cloth dampened in a solution of mild soap and water. Feathers may also be cleaned using a piece of clean flannel moistened with a little paint thinner or turpentine. Dry cleaning solvents containing perchloroethylene or trichloroethylene should not be used. Dyed feathers may be particularly unstable, so first test on a tiny area with a cotton swab. (CAUTION: Except for the soap and water, all of these cleaning solvents are flammable.)

Sources: Jones, Mablen. *Taking Care of Clothes*, p. 256.

109
How is an old piece of chamois revitalized?

Add 1 teaspoon of olive oil to a bucket of warm water. Soak the chamois in this solution for approximately 15 minutes, agitating the water every few minutes. A chamois is a specially treated leather used for cleaning, usually made from sheepskin, and not from the rare Alpine animal of the same name.

Sources: Burnam, Tom. *The Dictionary of Misinformation*, p. 41; *Reader's Digest Practical Problem Solver*, p. 316.

110
Can chewing gum be removed from fabrics, carpets, and hair?

For gum in hair or carpeting, soften the gum with peanut butter and remove. Any remaining stain from the peanut butter can be removed with an oil or grease solvent, or a laundry pre-treat. When using a laundry pre-treat, the item must be rinsed with water. If the gum is on a piece of clothing, try freezing it and carefully scraping the frozen gum. Any remaining gum may be removed with an oil solvent. Oil solvents should not be used on wet fabrics; always wait until the fabric is dry.

Sources: Barndt, Herb. *How to Remove Spots and Stains*, pp. 74–75.

111 ▽ *Is it possible to remove lipstick?*

Apply an oil solvent to the stain. Rinse. Repeat the process on any remaining stain. If the stain persists, treat it with a few drops of liquid soap mixed with ammonia and rinse.

Sources: Barndt, Herb. *How to Remove Spots and Stains*, p. 87.

112 ▽ *How can ballpoint pen ink be removed from clothing?*

For fresh stains, first sponge with water. If color transfers to the sponge cloth, continue this process until no more color comes out. Blot dry. Saturate the stain with acetone, amyl acetate, or hair spray and blot it through onto a fresh cloth or paper towel; continue until no more color is being transferred. Apply laundry pre-treat and launder in warm water; then air dry. If the hairspray leaves a stain, apply alcohol and then wash item in a detergent. Old stains may also need bleaching as well. Washing removes some types of ballpoint ink but sets others. Test first before washing by marking a scrap of similar material with the ink and washing that.

Sources: Aslett, Don. *Stainbusters Bible*, p. 219; Barndt, Herb. *How to Remove Spots and Stains*, pp. 77, 112; *The New York Public Library Desk Reference*, p. 500.

CLOTHING, JEWELRY, ETC. *See also:* AROUND THE HOUSE AND GARAGE—CLEANING AND MAINTENANCE

113 ▽ *What is cordovan leather?*

Cordovan is a very durable, non-porous leather made from the shell of horse butts. Used for the production of fine boots and shoes, the leather is usually finished in black or reddish-brown glossy finishes. The word is derived from the place Cordova, Spain—a center of world-renowned tanners.

Sources: Calasibetta, Charlotte M. *Fairchild's Dictionary of Fashion*, p. 333; Picken, Mary Brooks. *The Fashion Dictionary*, p. 229.

114

What is the name given to the looped shoulder decoration on uniforms that is made of gilt cord, usually braided?

It is known as aiguillette (or aiguilette).

Sources: Picken, Mary Brooks. *The Fashion Dictionary*, pp. 2-3.

115

What is the name of the device used by shoe salespeople to measure feet?

The instrument is known as the Brannock device.

Sources: Berliner, Barbara. *The Book of Answers*, p. 49.

116

How much does a denier weigh?

Denier is the international textile and hosiery unit of measurement to denote the thickness of silk and man-made filament yarns. The low numbers represent finer yarns, the higher numbers the heavier or coarser yarns. Originally the word was taken from the Roman coin *denarius*. The measurement was first established in France in the sixteenth century. One denier equals 0.05 grams. A pound of 1-denier yarn would measure about 4½ million yards in length; a gram of 1-denier yarn would measure 900 meters.

Sources: Calasibetta, Charlotte M. *Fairchild's Dictionary of Fashion*, p. 158; *Encyclopedia of Textiles*, pp. 316–317, 536.

117

How does a gold-filled piece of jewelry differ from one that is gold-plated?

A gold-filled piece of jewelry has a gold coating of 1/20th or more of the total weight of the piece. If the gold coating is less than 1/20th of the weight, the terms "gold-plated," "gold overlay," or "rolled gold" apply.

Sources: Bingham, Anne. *Buying Jewelry*, p. 9.

118

What is the difference between real and artificial pearls?

Imitation pearls, however cleverly made, are not difficult to detect on close inspection. Imitation pearls feel smooth between the teeth; whereas natural oyster pearls are gritty by comparison. The best-quality imitations are pearl-shaped beads coated with a mixture containing fish scale. The surface of imitation pearls does not show the finely spaced parallel serrated markings, which are typical of natural and cultured pearls.

Sources: Anderson, B.W. *Gem Testing*, pp. 352-353.

119

What is the covering on the end of a shoelace called?

Aglet, or aiglet, is the name of the sheath covering the end; its function is to facilitate threading. The term comes from the French word *aiguille*, which means needle.

Sources: Picken, Mary Brooks. *The Fashion Dictionary*, p. 2.

120

What are the equivalent sizes for clothing in the United States, Great Britain, and European countries?

Women's Clothing Sizes

Blouses and sweaters

U.S.	32	34	36	38	40	42	44
British	34	36	38	40	42	44	46
Continental	40	42	44	46	48	50	52

Coats and dresses

U.S.	8	10	12	14	16	18	20
British	30	32	34	36	38	40	42
Continental	36	38	40	42	44	46	48

Shoes

U.S.	5-5½	6-6½	7-7½	8-8½	9
British	3½-4	4½-5	5½-6	6½-7	7½
Continental	36	37	38	39	40

Stockings

U.S. and British	8	8½	9	9½	10	10½
Continental	0	1	2	3	4	5

Men's Clothing Sizes

Suits and coats

U.S. and British	34	36	38	40	42	44	46
Continental	44	46	48	50	52	54	56

Hats

U.S.	6⅝	6¾	6⅞	7	7⅛	7¼	7⅜	7½
British	6½	6⅝	6¾	6⅞	7	7⅛	7¼	7⅜
Continental	53	54	55	56	57	58	59	60

Shirts

U.S. and British	14	14½	15	15½	16	16½	17
Continental	36	37	38	39	41	42	43

Shoes

U.S.	7	7½	8	8½	9	9½	10	10½	11
British	6½	7	7½	8	8½	9	9½	10	10½
Continental	39	40	41	42	43	43	44	44	45

Socks

U.S. and British	9½	10	10½	11	11½	12	12½
Continental	39	40	41	42	43	44	45

Sources: *The New York Public Library Desk Reference*, pp. 530-531.

121 ▽ When was permanent press invented?

"Permanent press" is a term used to describe a fabric or garment which retains its original shape through wear and laundering and needs no ironing. This means it will resist wrinkling and will retain creases and pleats indefinitely. Crease-resistance was first developed for cotton in 1932 by R.S. Willows and his research team for the Manchester textile manufacturer, Tootal Broadhurst Lee. Rather than laminating the fibers with synthetic resin solutions, Willows made the individual cotton fibers absorb the solution—giving the cotton yarn an elasticity that it otherwise lacked. The first products manufactured using this process were Tootal ties (marketed in autumn 1932) and dress fabrics (marketed in April 1933).

In permanent press, the yarn is treated with a cross-linking chemical or reactant which is then cured (or fixed) by the application of heat. For garments needing creases or pleats, a post-cured process is used, giving the fabric "memory." Koret of California developed the process and received a patent in 1961 for its deferred cure process. The first application, in 1964, was for men's and boys' pants introduced to the consumer by Levi Strauss and McCampbell Graniteville Co.

Sources: *Encyclopedia of Textiles*, pp. 187, 408, 411; Robertson, Patrick. *The Book of Firsts*, p. 46.

122 ▽ Who invented velcro®?

Velcro® is the invention of George de Mestral, a Swiss engineer. He got the idea around 1948 when he noticed how burdock burrs often cling to clothing. Granted U.S. patent number 2,717,437 in 1955, de Mestral opened the first factory to produce velcro® in 1957. The name is derived from velvet (vel) and crochet (cro) from the French word for small hook. A velcro® fastener consists of two nylon strips, one containing thousands of tiny hooks, the other, tiny eyes. Pressing the two together locks the hooks in the eyes. To perfect that simple idea required 10 years of continuous effort.

Sources: *Inventive Genius*, p. 72; Panati, Charles. *Panati's Extraordinary Origin of Everyday Things*, pp. 317-318.

▽ **123** ⟋

What is the origin of the fur known as "nutria"?

"Nutria" is the soft brown fur of the coypu (*Myocastor coypus*), a South American aquatic beaver-like rodent about 2 feet (61 centimeters) long with webbed hind feet and a round tail. The fur is often dyed to resemble beaver. Because of the demand for this fur, breeding farms were started in Europe and North America. Some animals escaped or were let go (as during World War II in England) and now there are extensive populations in Louisiana, England, Germany, Holland, etc. The textile industry uses the belly portion of the nutria's hide, which has a beautiful silky undergrowth. The nutria develops fine hair as an insulator against the cold shallow streams it dwells in, where the water temperature is colder than the air.

Sources: *Encyclopedia of Textiles*, p. 122; *Marshall Cavendish International Wildlife Encyclopedia*, vol. 6, pp. 686-687.

▽ **124** ⟋

What is "flax scutching"?

Flax, a bast fiber taken from the stalk of the plant *Linum usitatissimum*, is called linen after it is processed into yarn or fabric. To separate the woody portions from the fiber is the task of "scutching." First, however, the flax is soaked to loosen the gums that bind the fiber to the stem and then dried (a process called "retting"), and then the woody tissue is crushed. A scutching machine removes the broken woody material by means of rotating wooden paddles, separating the flax fibers from the stalk. This scutching operation also can be done by hand. After combing ("hackling") to clean and straighten the fibers, the resulting fibers are ready for spinning.

Sources: *Encyclopedia of Textiles*, pp. 140-142.

▽ **125** ⟋

How do apparel manufacturers know what measurements to use in making different sizes in clothing?

Until 1980, the United States National Bureau of Standards issued voluntary product standards, which included women's, children's, and men's sizing. Some of these measurements were derived from anthropometric data compiled by the United States Department of Agriculture from measurements of military personnel. Women's measurements were re-evaluated in a comprehensive study in 1940.

The American Society for Testing and Materials, in Philadelphia, Pennsylvania in 1993 will be issuing new voluntary guidelines for women's clothing. Clothing officials note that women have changed—that their thighs and arms have become bigger and their shoulders wider.

Sources: *Detroit Free Press* (March 4, 1992), p. 1; Solinger, Jacob. *Apparel Manufacturing Handbook*, pp. 88, 693, 718.

126 *How is silk made?*

Silk fiber is a continuous protein filament produced by a silkworm to form its cocoon. The principal species used in commercial silkmaking is the mulberry silkworm (the larva of the silk moth *Bombyx mori*) belonging to the order Lepidoptera. The raw silk fiber has three elements—two filaments excreted from both of the silkworm's glands and a soluble silk gum called "sericin," which cements the filaments together. It is from these filaments that the caterpiller constructs a cocoon around itself.

The process of silkmaking starts with raising silkworms on diets of mulberry leaves for 5 weeks until they spin their cocoons. Then the cocoons are treated with heat to kill the silkworms inside (otherwise when the moths emerged, they would break the long silk filaments). After the cocoons are soaked in hot water, the filaments of 5 to 10 cocoons are unwound in the reeling process, and twisted into a single thicker filament; still too fine for weaving, these twisted filaments are twisted again into a thread that can be woven. Other insects, of course, produce a kind of silk, e.g. spider's silk, but these generally are too sticky for commercial use.

Sources: Cazeau, Charles J. *Science Trivia*, pp. 105-106; *Encyclopedia of Textiles*, pp. 125-137.

127 *What is polyhexamethyleneadipamide?*

This is another name for nylon, a family of polyamide fibers first developed by Wallace H. Caruthers (1896-1937) of the DuPont Corporation, and patented on February 16, 1937. This new material, called "Fiber 66" by DuPont, had both high strength and flexibility, as well as impact and wear resistance. Nylon's first practical application to a consumer product came in 1938, when the new polymer was introduced to the public in the form of toothbrush bristles. But it was its use in stockings, first sold to consumers on a trial basis in October 1939, that made it an overwhelming commercial success. It was similar to silk in its properties but was far less expensive.

The two most common types of nylon in production are "Nylon 6.6" and "Nylon 6" (the difference is based on the number of carbons in the repeating units of its molecular structure). Nylon 6 accepts dyes better, has great elasticity and softness, but has a lower melting point (420°F or 215.5°C) than Nylon 6.6 (482°F or 250°C).

Sources: *Encyclopedia of Textiles*, pp. 21-24, 568; Snyder, Carl H. *The Extraordinary Chemistry of Ordinary Things*, pp. 536-539; *Van Nostrand Reinhold Encyclopedia of Chemistry*, p. 635.

128 / Who invented the zipper?

On August 29, 1893, a Chicago, Illinois, mechanical engineer, Whitcomb Judson, was awarded a patent for a "clasp-locker." The purpose of this fastener was to close high boots, replacing the long buttonhooked shoelaces of the 1890s. It consisted of a linear sequence of hook-and-eye locks. Judson displayed the invention at the 1893 Chicago World's Fair to drum up business, but it was ignored. Judson and his partner Lewis Walker did receive an order from the United States Postal Service for twenty zippered mail bags, but in use, the zippers jammed up so badly that the bags were discarded. In 1913 Gideon Sundback, a Swedish-American engineer, abandoned Judson's hook-and-eye arrangement to devise a smaller, lighter, more reliable "hookless fastener" with interlocking teeth (similar to today's zippers). The United States Army was Sundback's first customer in 1918. In 1923, when B.F. Goodrich introduced these hookless fasteners on its boots, it called the boots "zippers"—stemming from the "zip" sound the fastener made when closing. The name eventually applied to the hookless fastener itself.

Sources: Panati, Charles. *Panati's Extraordinary Origins of Everyday Things*, pp. 316-317.

129 / What is the origin of blue jeans?

In 1860, Levi Strauss (1829-1902) arrived in San Francisco with bales of canvas cloth he intended to use to make tents and wagon coverings. He used this material to make overalls for the gold rush prospectors who complained that ordinary trousers wore out too quickly in their work environment. In the 1860s, Strauss replaced the canvas with denim, a softer fabric milled in Nimes, France. The indigo dye minimized soil stains. In 1893, copper rivets were added to stress points, such as pockets, which tended to be pulled out by the weight of the tools they held. In 1935, they became a fashion item. The term "jeans" denotes a twilled cotton cloth, similar to denim, that was used in the manufacturing of sturdy work clothing. The textile first was milled in Genoa, Italy, which French weavers called "Genes."

Sources: Panati, Charles. *Panati's Extraordinary Origins of Everyday Things*, pp. 302-303; Robertson, Patrick. *The Book of Firsts*, p. 89.

CRAFTS *See: AROUND THE HOUSE AND GARAGE—HOBBIES, TOYS, ETC.*

ENERGY USE AND CONSERVATION *See: ENERGY—CONSUMPTION AND CONSERVATION*

FABRICS *See: AROUND THE HOUSE AND GARAGE—CLOTHING, JEWELRY, ETC.*

FOOD PRODUCTS, STORAGE, ETC. *See: FOOD AND NUTRITION—FOOD PROCESSING AND STORAGE*

GARDENS AND GARDENING *See: THE PLANT WORLD—GARDENING, FARMING, ETC.*

HOBBIES, TOYS, ETC.

130
How is a salt map made?

    1 part salt
    1 part non-self-rising wheat flour
    ⅔ part water
    food coloring or poster paints
    heavy cardboard

Mix salt and flour in a bowl. Add water until the mixture becomes the consistency of a thick icing. Add food coloring to the mixture, if desired. Draw a map on heavy cardboard. Spread the mixture on the cardboard, shaping hills and valleys. The mixture will dry in 1-2 days. Paint with poster colors when dry.

Sources: Sattler, Helen P. *Recipes for Art and Craft Materials*, p. 50.

131
What are simple recipes for modeling dough?

Modeling Dough
¾ cup flour (not self-rising)
½ cup salt
1½ teaspoons powdered alum
1½ teaspoons vegetable oil
½ cup boiling water
food coloring

Combine flour, salt, and alum in a bowl. Add vegetable oil and boiling water. Stir until dough does not stick to the bowl. Add food coloring. Makes about 1 cup. Will dry overnight or may be kept several months in a jar with a tight lid.

Salt Dough

2 cups salt

⅔ cup water

1 cup cornstarch

½ cup cold water

Stir salt and ⅔ cup water in medium pan and cook until well heated. Remove from heat. Mix cornstarch and ½ cup water. Add to salt water in pan. Stir over medium heat to form a smooth workable mass. Turn out onto a plate and cover with a damp cloth to cool. Shape as desired on a foil-covered baking sheet. Dry in a warm oven for several hours.

Sources: Gardner, Pat. *Dough Creations*, pp. 111–119; Sattler, Helen Roney. *Recipes for Art and Craft Materials*, p. 34.

132 / How are doughballs used as fish bait made?

Cooked Doughball

Bring to a boil 1 cup water and 2 teaspoons sugar in a pan. Add 1 cup cornmeal and ¾ cup plain flour and stir over low heat for 5 minutes. Put mixture in a double boiler, cover it and cook it for ½ hour. Add extra sugar or honey if bait is for carp. Makes enough dough for two anglers for a day.

Uncooked Doughball

Trim the crust from several slices of rye bread, dampen the bread with port or any other sweet wine and knead it into a doughball. Keep it sealed in a plastic bag to prevent it from drying out.

Sources: Evanoff, Vlad. *Fishing with Natural Baits*, pp. 99–100; Oberrecht, Kenn. *The Practical Angler's Guide to Successful Fishing*, p. 101.

133 / How do boats get inside glass bottles?

The ship is built outside the bottle, with the masts laying down on the deck. A long cotton thread is attached to the rigging then the ship is pushed through the neck of the bottle. Pulling on the thread hoists the rigging and up comes the masts and the ship is miraculously in the bottle, in full sail.

Sources: *The Bumper Book of Things a Boy Can Make*, p. 15.

134 How are marbleized end papers, note cards, and fabrics created?

The basic technique of marbling is simple. In a shallow pan of gelatinous medium, paints or inks of different colors are floated to form a surface film of colors that when raked and swirled with combs, picks, or other simple tools form a design. A pretreated piece of paper or cloth is placed gently on the surface of the medium to pick up the medium. A rinse in clear water removes excess paint. Most paints or inks need to be set by resting and heating with an iron. The pan of medium is quickly skimmed with strips of newspaper and is ready for the next print.

A medium made of carrageenan and water, a cloth prepared with an alum solution, and acrylic paints diluted with water and mixed with a surfactant such as oxgall (used by artists to thin paint) are all effective materials.

Sources: Cohen, Daniel. *Marbling on Fabric*, p. 15.

135 What is a bungee cord?

A bungee is an elasticized cord, typically with a hook at each end, used to bind a suitcase to a wheeled carrier. The modern–day craze of bungee jumping had its origins in 1979. Jumpers leap headfirst from bridges, cranes, and hot-air balloons, from 90 to 300 feet (27.4 to 91.4 meters) above the ground, with only a long nylon-cased rubber bungee cord to break their fall. Anchored around the ankles or to a body harness, the wrist-thin cord is long enough to allow a few seconds of free fall before it stretches, dampening the force of the plunge. The diver sometimes hurtles to within a few feet of the ground before rebounding skyward like a yo-yo.

Sources: *Time*, vol. 135 (April 23, 1990), p. 75; *Webster's College Dictionary*, p. 182.

136 How is a pomander made?

Originally small filigree balls of gold, silver, or ivory filled with fragrant spices and an ambergris fixative, pomanders were used as early as the Middle Ages, when they were worn to ward off unpleasant odors. Today's version is made by studding a piece of fruit with cloves and curing it in a mixture of ground spices with orris-root as a fixative.

6 to 8 firm, thin-skinned apples, oranges, lemons, or limes
½ pound whole cloves
¼ cup ground cinnamon
½ cup ground cloves
2 tablespoons ground nutmeg
2 tablespoons ground allspice
¼ cup powdered orrisroot

Insert the cloves at ⅛ to ¼ inch intervals over the entire surface of the fruit. A one-inch wide band without cloves may be left, if a ribbon is to be tied around.

Blend the spices and orrisroot in a small bowl. One at a time, roll each piece of fruit in the mixture, coating it generously. Place the spice-covered fruit in a large bowl, cover with more of the spice mixture, and set in a warm place to dry. Turn the fruit daily; drying can take from two weeks to a month; the pomanders will be hard when they are completely dry. Dust off excess spice mixture. Tie with ribbon if desired.

Fixatives absorb and hold the perfume of flower petals which is often quite fleeting, etc. The most commonly used fixative is ground orrisroot, which has a scent of violets; other fixatives are frankincense, gum benzoin, oakmoss, and tonka beans.

Sources: *A Country Christmas*, p. 47; *The Little Scented Library, vol. 3: Potpourri*, p. 8.

137 ▽ *What is the origin and process of the craft batik?*

Batik is a very ancient craft dating back at least 2,000 years. It is a method of applying a colored design on textiles by waxing those parts that are not to be dyed. White pure cotton or white pure silk are the best materials to use. Hot wax is applied to the fabric in the areas that are to be kept white and then the dye is added. The lightest color is the first dye. The fabric is then waxed in those areas that are to retain the first color and the process is repeated with a second color. Once all the dyeing is completed, the wax is removed, revealing the design.

Sources: Griffin, Heather. *Introduction to Batik*, p. 4.

138 ▽ *Is there a way to keep newspaper clippings from yellowing?*

Drop one Milk of Magnesia® tablet into one quart of warm club soda in a large bowl and dissolve it overnight. On a large cookie sheet with sides pour in ½ inch of the preservative solution. Place the newspaper clippings in the liquid and let it set until thoroughly saturated. Remove the clippings and lay them out to dry on a flat surface. This preservative will prevent clippings from turning yellow as long as the clippings are kept out of direct sunlight.

Sources: Sattler, Helen Roney. *Recipes for Art and Craft Materials*, p. 140.

139 ▽ *What is pysanky?*

Pysanky is the traditional Ukrainian art of decorating and dyeing eggs, using a wax resistant technique similar to batik. The egg is usually divided into geometric areas, then filled with intricate, symmetrical, linear designs that require great skill and control to produce. The dye is applied to uncooked eggs that have been washed in a solution of warm water and vinegar. Wax designs are then added with a brush. The waxed egg is dipped in Easter egg dye or household dye. The egg may be rewaxed and redyed with an additional color. The first coat of dye should be the lightest. Traditionally, the final dye coating is black. After the dyeing is completed the wax is removed by placing the egg, on paper toweling in a pan, into a warm oven for a few minutes. Remove the egg and wipe the wax off with toweling. The process may be repeated until the wax is completely removed.

Sources: *The Encyclopedia of Crafts*, vol. 2, p. 700.

140
What is a thaumatrope?

The thaumatrope (or wonder turner) consists of a piece of cardboard with a picture drawn or pasted on each side, the two pictures being upside-down from one another. Two pieces of string are attached to the cardboard and when they are spun, it produces the visual effect of merging the two pictures into one. Invented by J.A. Paris in 1826, it is believed to be the first cinematographic device.

Sources: Provenzo, Eugene F. *47 Easy-To-Do-Classic Science Experiments*, p. 28.

141
Who invented the Slinky®?

Richard T. James was granted a patent in 1947 for his "toy and process of use." This helical spring toy tumbles smoothly down a flight of stairs when its free end is lifted, rotating about a diameter and quickly dropping to the step below. His invention was the result of a failed attempt to produce an antivibration device for ship instruments. Having built a spring that James hoped would instantaneously counterbalance wave motion on a ship, he accidentally knocked one off a shelf, and saw it literally crawl, coil by coil, to a lower shelf, onto a stack of books, down to the tabletop, and finally come to rest, upright, on the floor. James' wife, Betty, saw its potential as a toy and named it "slinky." Betty James still runs James Industries, the company she founded with her husband in 1946 to market the toy Slinky®.

Sources: *American Scientist*, vol. 75, no. 3 (May/June 1987), p. 289; Harris, Harry. *Good, Old-Fashioned Yankee Ingenuity*, pp. 267-268; Panati, Charles. *Panati's Extraordinary Origins of Everyday Things*, pp. 380-381.

142
Who invented Rubik's cube®?

In 1979, a Hungarian chess fanatic and professor at the University of Budapest, Erno Rubik (b.1944), invented a cube puzzle brain–teaser. This game of logic first interested colleagues and mathematicians, but shortly afterwards, in 1980, Ideal Toys marketed the puzzle.

In May 1982 a patent infringement suit was filed by Arthur S. Obermayer, who claimed his company, Moleculon, held a 1972 United States patent based on the work of its chief research scientist, Larry D. Nichols, for the identical puzzle. In 1984 the court ruled in favor of Moleculon.

Sources: Giscard d'Estaing, Valerie-Anne. *World Almanac Book of Inventions*, p. 132; Harris, Harry. *Good Old-Fashioned Yankee Ingenuity*, p. 256.

143 / What do the different scales in model railroading represent?

These scales are a comparison between the model and a full-sized railroad.

Scale	Inches/Foot	Proportion
O	0.25	1:48
S	0.1875	1:64
OO	0.157 (4 mm)	1:76.2
HO	0.138 (3.5 mm)	1:87.1
TT	0.100	1:120
N	0.075	1:160

Sources: Buehner, Jim. *The Complete Handbook of Model Railroading*, p. 20.

144 / How is papier–mâché made?

Tear or shred cardboard and newspaper into tiny pieces. In a bucket or bowl soak the pieces in water for 1 to 2 days. The pulp then can be combined with paste to make a malleable mixture for modeling.

Although its use is older in the East and in Persia, papier-mâché became popular in Europe in the middle of the 18th century. Made from paper pulp mixed with glue, chalk, and sand, pressed into oiled molds and baked, objects, such as small boxes, trays, picture frames and ornaments, were fashioned and painted. In 1772 Henry Clay of Birmingham, England, patented a method of making the substance so hard that it could be treated as wood—and it was. By using cotton rags in place of paper, the result was stronger than the wood-pulp used today. The Great Exhibition of 1851 showed chairs, couches, tables, and cradles made from papier-mâché.

Sources: DeBono, Edward. *Eureka!*, pp. 87-88; *The Encyclopedia of Crafts*, vol. 2, pp. 621-622.

145 / What is the secret of the "drinking bird," a little glass bird that takes repeated drinks from a container of water?

Utilizing certain physical principles, the drinking bird seems to be in perpetual motion once he is started. However, as the demonstrated in (1), his body (filled with liquid methyl chloride) and head are connected only by a glass tube. When the head is pushed into the water, capillary action wets the entire head-cover. In (2), methyl chloride vapor that rose through the tube to the head has been condensed by the cooling effect of the evaporation of moisture in the cloth head-cover. Condensation lowers the vapor pressure in the head and the greater pressure of the vapor in the body forces liquid up through the tube into the head and beak, which then tips the bird forward to drink. When the bird has tipped all the way over (3), the liquid seal in the tube is broken and the vapor in the body flows into the head to equalize the pressure so that the liquid methyl chloride runs back into the body, pulling the bird's thoroughly soaked head out of the water. The flow of the liquid makes the bird return to the upright position (4). However, the water picked up in his head covering will evaporate again and he will take another drink, repeating the steps again.

Sources: *Popular Science*, vol. 149 (November 1946), p. 74.

Drinking bird mechanism

▽ **146** ▽ *Who invented Monopoly®?*

During the early 1930s in the Great Depression, this high-stakes real estate game called Monopoly® was invented by Charles B. Darrow (1889-1967). In 1934, he approached Parker Brothers, the Massachusetts-based game company, who rejected it as too dull, too complicated, and too time-consuming. When Darrow, after manufacturing five thousand sets, landed a sizable order from the John Wanamaker department store, Parker Brothers reconsidered and offered Darrow a contract, which made him a millionaire, eventually. In 50 years, total sales exceeded 100 million.

Sources: Harris, Harry. *Good Old-Fashioned Yankee Ingenuity*, pp. 193-194; Panati, Charles. *Panati's Extraordinary Origins of Everyday Things*, pp. 379-380.

147 / What were the occupations of Paul Revere?

Paul Revere (1735-1818) was most renowned as a silversmith, but his energy and skill along with the necessity of supporting a growing family led him in many directions. He not only worked in silver but also made surgical instruments, sold spectacles, replaced missing teeth, and engraved copper plates.

Sources: *New Encyclopaedia Britannica*, vol. 10, p. 8.

HOUSEHOLD ITEMS *See also: ENERGY—CONSUMPTION AND CONSERVATION*

148 / Where should a smoke detector be located in the home?

The best location for a home smoke detector is on or near a ceiling close to the sleeping area and away from corners, windows, and doors. Another good location is at the top of a stairwell.

Sources: *Reader's Digest Practical Problem Solver*, p. 189.

149 / When was the lawnmower invented?

The first mechanized mower was patented in 1830 by the English inventor Edwin Budding. This device was a 19-inch (48-centimeter) roller mower that used a set of rotating cutters operating against fixed ones—an adaptation of the rotary shearing machine used to cut nap off of cotton cloth. Previously, grass needed to be dampened first to give it enough "body" before it could be cut with a scythe. Budding's device could be used on dry grass. By the 1880s, the popularity of hand-powered mowers increased as their cost decreased. And in 1919, an American army colonel, Edwin George, produced the first gasoline-powered lawnmower.

Sources: Panati, Charles. *Panati's Extraordinary Origins of Everyday Things*, pp. 162-163; Yarwood, Doreen. *Five Hundred Years of Technology in the Home*, p. 109.

150 ▽ *In electric power receptacles, why do the newer ones have a third hole?*

In 1962 the third hole was added, for grounding purposes. Previously, electrical plugs with two prongs could give the user a shock if a short circuit in the wiring or in the electrical device developed. Since then "double-insulated" pronged plugs were developed (with fatter ends) that do not require a third prong for protection. But these new double–insulated prongs require new wide-slotted receptacles.

Sources: DuVall, Nell. *Domestic Technology*, p. 255; Feldman, David. *Do Penguins Have Knees?*, p. 191; *Technology and Culture*, vol. 27 (July 1986), p. 542.

151 ▽ *What is a Franklin stove?*

The Franklin stove, invented by Benjamin Franklin (1706-1790) in 1740, is a free-standing, enclosed unit usually made from cast iron. Being enclosed with limited air flow gives the stove controllable combustion and greater efficiency. Today, stoves of this type are often called "Franklin stoves."

Franklin's stove was built to heat rooms, and its most important feature was the flue, which doubled back and formed a sort of radiator, around which room air circulated. Since it was free-standing, there was more surface to radiate the heat, and none of its heat was wasted by being held in the fireplace's brickwork.

Sources: Busha, William, and Stephen Morris. *The Book of Heat*, pp. 160-161; Wilson, Mitchell. *American Science and Invention*, p. 22.

Franklin stove

152 ▽ What is a fluting iron?

A household iron with grooves for pressing fabric into pleats or ruffles is called a fluting iron.

Sources: DuVall, Nell. *Domestic Technology*, pp. 201-202.

153 ▽ Who invented the modern flush toilet?

In 1449, Thomas Brightfield of St. Martin's Parish in England, built a water closet flushed by water piped from a cistern. In 1586, English poet Sir John Harington (1561–1612) designed and installed a practical water-flushing system for cleaning toilets. In 1778, Englishman Joseph Bramah (1748-1814) invented the ball-valve and u-bend method. But it was not until 1861 that British sanitary engineer Thomas Crapper (1837-1910) offered his major innovation—a mechanism that shut off the flow of clean water when the tank filled. Previously, the supply of clean water had be turned on and off by the user, and many simply left it running continuously.

Sources: Burnam, Tom. *The Dictionary of Misinformation*, p. 56; DuVall, Nell. *Domestic Technology*, pp. 363, 366; Manchester, Richard B. *Mammoth Book of Fascinating Information*, pp. 452-456.

154 ▽ When was the first microwave oven sold?

Microwave cooking—the first absolutely new method of cooking food since fire was discovered—has no application of fire or a fiery element, direct or indirect, to the food. Pure electromagnetic energy agitates the water molecules in food at almost 2,500 million times a second, producing friction and as a by-product, sufficient heat for cooking. Metal containers reflect microwaves and should not be used in cooking.

The electronic tube that produces microwave energy—a magnetron—was invented by Sir John Randall and H.A. Boot during World War II to be used in radar. In 1946, a Raytheon engineer, Percy Spencer (1894-1970), discovered food was affected by the magnetron. Raytheon developed a very large, commercial microwave oven, the Radar Range, in 1947. Although it was the size of a refrigerator, its cooking space was modest. Before microminiaturization, electronic devices like the microwave oven required much space for vacuum tubes, wiring, etc.—making these devices very bulky. By 1952, the Tappan Company produced an oven with two speeds for the home, but it was still very expensive and large. During the 1960s the ovens became more available, less expensive, and smaller. But it took nearly 20 years for this new way of cooking to become popular. Now 75% of American homes have microwave ovens.

Sources: Cone, Robert J. *How the New Technology Works*; Hillman, Howard. *Kitchen Science*, pp. 19-20; Panati, Charles. *Panati's Extraordinary Origins of Everyday Things*, pp. 125-126.

155
What is the small metal hoop that supports a lampshade called?

The hoop is called a "harp." The finial fastens the shade to the harp.

Sources: Bragonier, Reginald, and David Fisher. *What's What*, p. 234.

Lamp parts

156
How is the proper size of a room air conditioner determined?

Multiply the square footage of the room or rooms to be cooled by 27. This figure will be the size air conditioner needed in British thermal units (a common energy measurement, often abbreviated as "BTU," which represents the energy required to raise the temperature of one pound of water by 1° Fahrenheit).

Sources: Collin, P.H. *Dictionary of Ecology and the Environment*, p. 22; *Rodale's Complete Home Products Manual*, p. 8.

157
When was the window air conditioner invented?

The term "air conditioning" was first used by American physicist Stuart W. Cramer in 1907—years before a practical air-conditioning system was invented. The American inventor Willis Carrier (1876-1950) invented air conditioning in

1911, and commercial air conditioners were produced in 1914. A patent request was filed in 1931 by H.H. Schultz and J.Q. Sherman for an air conditioner to be placed on window ledges. Room air conditioners were available for home use in 1932, but economic constraints arising from the depression and World War II, together with the uneven spread of electrical power distribution systems, prevented widespread use until after the war.

Sources: DuVall, Nell. *Domestic Technology*, p. 247; Giscard d'Estaing, Valerie-Anne. *World Almanac Book of Inventions*, p. 21; Panati, Charles. *Panati's Extraordinary Origins of Everyday Things*, pp. 159-160.

158 / Why are "dummy boards" used?

A "dummy board" is a pasteboard or wooden figure cut and painted to represent any number of things, from people to pets to plants. Dummy boards, used for decoration in today's homes, had a practical use in nineteenth century homes as fireboards. A fireboard (often painted or stained) was needed to stop drafts from whisking down the chimney when a fire was not lit. Today, snug dampers and sophisticated heating systems have made these functional fireboards obsolete in newer or renovated homes.

Sources: *Country Living's Country Look and How to Get It*, pp. 124-125.

159 / When was the electric clothes dryer invented?

The first manual clothes dryer, called a "ventilator" was made by the French inventor Pochan in 1799. It was a crank-operated container pierced with holes, which was suspended above an open fire—clothes would either dry or burn. Spin dryers, first developed around 1909, used centrifugal force to remove moisture, and about 1930 electric clothes dryers first appeared.

Sources: Mount, Ellis, and Barbara A. List. *Milestones in Science and Technology*, p. 26; Panati, Charles. *Panati's Extraordinary Origins of Everyday Things*, p. 146.

160 / What is a Toby jug?

Toby jugs were first made in England by a Staffordshire potter around 1765, and have been popular ever since. Originally these handmade brightly-colored jugs were sold at fairs. Useful as decorations and functional as cream pitchers, these jugs often depicted humorous figures—soldiers, farmers, and famous men. Although sometimes the figures are bust-length, usually the jugs depict full-figures, often seated and holding their own little jugs. These comic portrait jugs, made by machine now in England, are shipped all over the world.

The name "Toby jug" may have derived from "Toby Fillpot," a nickname given to a local heavy drinker who was pictured in an engraving and mentioned in a song called the "Brown Jug." Others suggest the "Toby" character was a Yorkshire farmer who drank huge quantities, and still others think that he is copied from the character Uncle Toby in Laurence Stern's novel *Tristam Shandy*.

Sources: *Ceramics Monthly*, vol. 28, no. 6 (June 1980), pp. 59-60; Fowler, Virginie. *Folk Arts Around the World*, p. 28.

161 ▽ *How do three-way light bulbs work?*

Three-way light bulbs have two filaments, used separately for two of the light levels and together for the third. The highest wattage will always be the sum of the first two wattages (e.g., a 50/100/150 watt three-way bulb). These bulbs only work in specially designed sockets.

　　The principle of the light bulb (more correctly known as the incandescent filament lamp) is that if sufficient electric current can be passed through a conducting filament, the molecules of the filament become excited, the filament gets hot and eventually glows. Discovering the right substance to use for the filament led Thomas Alva Edison (1847-1931) to produce the first practical electric light. For 14 months he tested over 1200 filaments until on October 31, 1879, he discovered that a charred cotton sewing thread was the answer. Simultaneously in England, Joseph Wilson Swan (1824-1914) had designed a thin carbon filament in a vacuum tube, but Swan did not get his light into production until 1881. Filaments today are made from tungsten, which resists melting below 6170°F (3410°C).

Sources: Clarke, Donald. *The Encyclopedia of How It Works*, p. 136; De Bono, Edward. *Eureka!*, p. 146; Harris, Harry. *Good Old Fashioned Yankee Ingenuity*, pp. 172-173.

162 ▽ *What inventions are credited to Benjamin Thompson, Count Rumford?*

Benjamin Thompson (1753-1814), a physicist interested in caloric (heat) theory, invented a kitchen stove in which the fire was enclosed within the range to conserve heat. In the process, he invented the fireplace damper and smoke shelf, and redesigned the fireplace so that smoke went up the chimney rather than into the room. Known for his studies on heat and friction, he argued that heat is a form of motion. He developed a calorimeter and a photometer, a double boiler and a drip coffeepot, as well as a device to measure the power of gunpowder, and devised improvements in lighting. Thompson, a Loyalist born in Massachusetts, became a permanent exile after the American Revolutionary War. The Elector of Bavaria made Thompson a count in 1790 in gratitude for introducing Watt's steam engine, the potato, etc., to Bavaria.

Sources: Asimov, Isaac. *Asimov's Biographical Encyclopedia of Science and Technology*, pp. 242-244; *The Biographical Dictionary of Scientists: Physicists*, pp. 139-140; Elkort, Martin. *The Secret Life of Food*, pp. 164-166.

163 ▽ Who invented the first electric iron?

The electric clothes iron was first invented in France in 1880. On June 6, 1882, Henry W. Seely of New York City received a patent on one which proved impractical, requiring too much reheating. In 1896, heavy electric irons with replaceable heating units were unsuccessfully sold by the Ward Leonard Electric Company. In 1903, Earl H. Richardson devised a small lightweight iron heated by a glowing wire wrapped around a brass core. He later reworked the design so that more heat was directed to the point, facilitating the smoothing of buttonholes and pleats. This iron became so popular that Richardson renamed his company "Hotpoint."

Sources: Cohen, Daniel. *The Last Hundred Years: Household Technology*, pp. 100-102; Desmond, Kevin. *A Timetable of Inventions and Discoveries*, pp. 1880, 1882; Harris, Harry. *Good Old-Fashioned Yankee Ingenuity*, p. 118.

164 ▽ Who invented scissors?

Scissors made of bronze, specially designed for cutting fabric and other materials, were developed by 1500 B.C.E. Early spring-type scissors appeared in 500 B.C.E. Pivoted scissors, similar to those of today, were used in ancient Rome in 100 C.E.; but they were not used domestically in Europe until the 16th century. The modern cast-steel scissors were developed in England by Robert Hinchcliffe in 1761.

Sources: DuVall, Nell. *Domestic Technology*, pp. 164-165.

JEWELRY *See: AROUND THE HOUSE AND GARAGE—CLOTHING, JEWELRY, ETC.*

KITCHEN ITEMS *See: AROUND THE HOUSE AND GARAGE—HOUSEHOLD ITEMS; FOOD AND NUTRITION—KITCHEN EQUIPMENT, MEASUREMENTS, ETC.*

PETS AND PESTS *See also: THE PLANT WORLD—GARDENING, FARMING, ETC.*

165 ▽ Why can a dog or cat hear a whistle which a human cannot?

A dog's sense of hearing is highly developed. Sounds that a human can hear at a hundred yards a dog can hear over ¼ mile (402 meters) away. Moreover, a dog and cat are capable of hearing ultrasonic sounds. In humans, the maximum audible frequency is approximately 20 thousand cycles, or 20 kilohertz, per second, whereas it has been demonstrated that a dog can perceive at least 35 thousand cycles, or 35 kilohertz, per second and a cat can even hear as high as 65 thousand cycles, or 65 kilohertz, per second.

	Hearing Range	
	From (hertz)	To (kilohertz)
Human	20	20
Dog	20	40
Cat	30	60

Sources: Dunbar, Ian. *Dog Behavior: Why Dogs Do What They Do*, p. 45; Fogle, Don. *Know Your Cat*, p. 93.

166 / What are the most popular dog names and breeds?

Dog owners most frequently named their pets: Pepper, Brandy, Lady, Bear, Rocky, Sam or Samantha, Misty, Sheba, Bandit, and Smoky. More than half of the canine owners surveyed gave their dogs human names.

Purebred dogs registered with the American Kennel Club

Breed	Thousands
Cocker spaniels	112
Labrador retrievers	91
Poodles	79
Golden retrievers	64
German shepherds	58
Rottweilers	51
Chow chows	50
Dachshunds	44
Beagles	43
Miniature breeds	42

Note: Data gathered in early 1990s; exact date not given in sources.

Sources: Krantz, Les. *Best and Worst of Everything*, p. 263; Sussman, Lesley. *The Ultimate Dog Catalog*, p. 135.

167 / Do dogs and cats dream?

Dogs have REM (rapid eye movement) sleep periods, which are associated with dreaming in humans. Therefore, it is assumed that dogs dream. Movements of the eyeballs, lip trembling, twitching of the extremities, and vocalizations have been observed during these sleep periods.

Sleeping cats have been studied with electroencephalograms (EEG) which read their brain activity. 30% of the cats' sleep time were deep sleep times or REM (rapid eye movement) sleep periods accompanied by changes in body posture, paw and claw movements, whisker twitching, ear flicking and, in some cases, vocalization. It could be assumed, as well, that cats dream or at least have semi-conscious experiences similar to human dreams.

Sources: Mery, Fernand. *The Life, History, and Magic of the Dog*, p. 181; Moorcroft, William H. *Sleep, Dreaming, and Sleep Disorders*, p. 80; Schneck, Marcus, and Jill Caravan *Cat Facts*, pp. 24–25.

168 / How many dogs and cats are there in the United States?

In 1990 there were 63.2 million cats and 54.5 million dogs in the United States.

Sources: *Pittsburgh Press* (May 26, 1991), Parade supplement, p. 18.

169 / What is the newest method of tagging a dog?

There is now a computer-age dog tag. A microchip is implanted between the dog's shoulder blades. The semiconductor carries a 10–digit code, which can be read by a scanner. When the pet is found, the code can be phoned into Infopet, the inventor of the system. Infopet's computers keep such data as the dog's license number, medical condition, and the owner's phone number.

Sources: *Time*, vol. 135, no. 8 (February 19, 1990), p. 73.

170 / How is the age of a dog or cat computed in human years?

When a cat is one year old, it is about 20 years old in human years. Each additional year is multiplied by four. Another source counts the age of a cat slightly differently. At age one, a cat's age equals 16 human years. At age two, a cat's age is 24 human years. Each additional year is multiplied by four.

When a dog is one year old, it is about 15 years old in human years. At age two it is about 24; after age two, each additional year is multiplied by four.

Sources: *Blair & Ketchum's Country Journal*, vol. 15 (May 1986), p. 5; *Cats Magazine*, vol. 39 (March 1982), p. 20; Taylor, David. *You & Your Cat*, p. 35.

171 / Do dogs and cats have good memories?

Dogs do have long-term memories, especially for those whom they love. Cats have a memory for things that are important to their lives. Some cats seem to have extraordinary "memories" for finding places. Taken away from their homes, they seem able to "remember" where they live. The key to this "homing" ability could be a built-in celestial navigation, similar to that used by birds, or the cats' navigational ability could be attributed to the cats' sensitivity to earth's magnetic fields. When magnets are attached to cats, their normal navigational skills are disrupted.

Sources: Fox, Michael W. *The Animal Doctor's Answer Book*, p. 215; Schneck, Marcus, and Jill Caravan. *Cat Facts*, pp. 20–21.

172
What was the contribution to medical science of a dog named Marjorie?

Marjorie was a diabetic black and white mongrel that was the first creature to be kept alive by insulin (a substance that controls the level of sugar in the blood).

Sources: Eldridge, Wayne B. *The Best Pet Name Book Ever*, p. 31.

173
Which breed of dogs are least likely to bite?

1. Golden retriever
2. Labrador retriever
3. Shetland sheepdog
4. Old English sheepdog
5. Welsh terrier
6. Yorkshire terrier
7. Beagle
8. Dalmatian
9. Pointer

Sources: Wallechinsky, David. *The Book of Lists*, p. 125.

174
What is the rarest breed of dog?

The Tahltan bear dog, of which only a few remain is thought to be the rarest dog; it is in danger of extinction. It was once used by the Tahltan Indians of western Canada to hunt bear, lynx and porcupine.

Sources: Palmer, Joan. *Dog Facts*, p. 16.

175
Which breed is known as the "wrinkled dog"?

The shar-pei, or Chinese fighting dog, is covered with folds of loose skin, as if its coat is in need of ironing. It stands 18 to 20 inches (46 to 51 centimeters) and weighs up to 50 pounds (22.5 kilograms). Its solid-colored coat can be black, red, fawn, or cream. The dog originated in Tibet or the northern provinces of China some 2,000 years ago. The People's Republic of China put such a high tax on shar-peis that few people could afford to keep them and the dog was in danger of extinction. However a few specimens were smuggled out of China and the breed has made a comeback in the United States, Canada, and the United Kingdom. Although bred as a fighting dog, the shar-pei is generally an amiable dog.

Sources: Palmer, Joan. *Dog Facts*, p. 74.

176 / Which breed is known as the "voiceless dog"?

The basenji dog does not bark. When happy, it will make an appealing sound described as something between a chortle and a yodel. It also snarls and growls on occasion. One of the oldest breeds of dogs, and originating in central Africa, the basenji was often given as a present to the Pharaohs of ancient Egypt. Following the decline of the Egyptian civilization, the basenji was still valued in central Africa for its hunting prowess and its silence. The dog was rediscovered by English explorers in the nineteenth century, although it was not widely bred until the 1940s.

The basenji is a small, lightly built dog with a flat skull and a long rounded muzzle. It measures 16 to 17 inches (40 to 43 centimeters) in height at the shoulder and weighs 22 to 24 pounds (10 to 11 kilograms). The coat is short and silky in texture. The feet, chest, and tail tip are white; the rest of the coat is chestnut red, black, or black and tan.

Sources: American Kennel Club. *The Complete Dog Book*, pp. 156-159; *The New Dog Encyclopedia*, pp. 505-506; Palmer, Joan. *Dog Facts*, p. 150.

177 / What are the different classifications of dogs?

Dogs are divided into groups according to the purpose for which they have been bred.

Group	Purpose	Representative breeds
Sporting dogs	Retrieving game birds and water fowl	Cocker spaniel, English setter, English springer spaniel, golden retriever, Irish setter, Labrador retriever, pointer
Hounds	Hunting	Basenji, beagle, dachshund, foxhound, greyhound, saluki, Rhodesian ridgeback
Terriers	Hunting small animals such as rats and foxes	Airedale terrier, Bedlington terrier, bull terrier, fox terrier, miniature schnauzer, Scottish terrier, Skye terrier, West Highland white terrier

Group	Purpose	Representative breeds
Toy dogs	Small companions or lap dogs	Chihuahua, Maltese, Pekingese, Pomeranian, pug, Shih Tzu, Yorkshire terrier
Herding dogs	Protect sheep and other livestock	Australian cattle dog, bouviers des Flandres, collie, German shepherd, Hungarian puli, Old English sheepdog, Welsh corgi
Working dogs	Herding, rescue, and sled dogs	Alaskan malamute, boxer, Doberman pinscher, great Dane, mastiff, St. Bernard, Siberian husky
Non-sporting dogs	No specific purpose, not toys	Boston terrier, bulldog, Dalmatian, Japanese akita, keeshond, Lhasa apso, poodle

Sources: Palmer, Joan. *Dog Facts*, pp. 66-71.

178 ▽ *Where did the pug dog originate?*

The pug's true origin is unknown, but it has existed in China, its earliest known source, for 1,800 years. A popular pet in Buddhist monasteries in Tibet, it next appeared in Japan and then in Europe. It probably was introduced into Holland by the traders of the Dutch East India Company. The name "pug dog" may have come from the dog's facial resemblance to a marmoset monkey. This popular pet in the 1700s was called a "pug." So the term "pug dog" distinguished the dog from the "pug" monkey.

A pug has a square, short compact body, either silver or apricot-fawn in color. Its muzzle is black, short, blunt, and square; its average weight is 14 to 18 pounds (6.4 to 8.2 kilograms). The pug is often described by the motto "Multum in Parvo"—a lot of dog in a small space.

Sources: American Kennel Club. *The Complete Dog Book*, pp. 477-479; Gore, Louise L. *Meet the Pug*, p. 3.

179 ▽ *What is the oldest breed of dog?*

The Saluki is believed to be the oldest pure–bred dog. Sumerian rock carvings in Mesopotamia that date to about 7000 B.C.E. depict dogs bearing a striking resemblance to the Saluki. The dogs are 23 to 28 inches (58.4 to 71 centimeters) tall with a long, narrow head. The coat is smooth and silky and can be white, cream, fawn, gold, red, grizzle (bluish–gray) and tan, black and tan, or tricolor (white, black, and tan). The tail is long and feathered. The Saluki has remarkable sight and tremendous speed which makes him an excellent hunter.

The oldest American pure-bred dog is the American Foxhound. It descends from a pack of foxhounds belonging to an Englishman named Robert Brooke who settled in Maryland in 1650. These dogs were crossed with other strains imported from England, Ireland, and France to develop the American Foxhound. This dog stands 22 to 25 inches (55.9 to 63.5 centimeters) tall. It has a long, slightly domed head, with a straight, squared-out muzzle. The coat is of medium length and can be any color. They are used primarily for hunting.

Sources: American Kennel Club. *The Complete Dog Book*, pp. 210-242, 246-248; Palmer, Joan. *Dog Facts*, pp. 9-11, 144-145.

180 / Which breeds of dogs are the most dangerous?

The dog breeds responsible for the most fatal attacks on people between 1979 and 1989 are

Breed	Known fatal attacks
Pit bull	37
German shepherd	9
Husky	7
Malamute	6
Doberman pinscher	5
Rottweiler	5
Great Dane	4
St. Bernard	4

Sources: Krantz, Les. *The Best and Worst of Everything*, pp. 262–263.

181 / What controls the formation of the color "points" in a Siamese cat?

The color points are due to the presence of a recessive gene, which operates at cooler temperatures, limiting the color to well-defined areas—the mask, ears, tail, lower legs, and paws—the places at the far reaches of the cardiovascular system of the cat.

There are four classic varieties of Siamese cats. Seal-points have a pale fawn to cream colored coat with seal-brown markings. Blue-points are bluish-white with slate blue markings. Chocolate-points are ivory colored with milk-chocolate brown colored markings. Lilac-points have a white coat and pinkish-gray markings. There are also some newer varieties with red, cream, and tabby points.

The Siamese originated in Thailand (once called Siam) and arrived in England in the 1880s. They are medium-sized and have long, slender, lithe bodies, with long heads and long tapering tails. Extroverted and affectionate, Siamese are known for their loud, distinctive voices which are impossible to ignore.

Sources: *Cat Fancy*, vol. 35, no. 4 (April 1991), p. 36; *Standard Guide to Cat Breeds*, p. 41; Taylor, David. *The Ultimate Cat Book*, pp. 108-109.

182

How many bones are in the tail of a cat?

The number of bones can vary a great deal depending on the length of the tail, but usually the minimum number is 20.

Sources: Chiasson, Robert B. *Laboratory Anatomy of the Cat*, p. 12.

SKELETON OF THE CAT

Adapted from *The World Book Encyclopedia.*
© 1992 World Book, Inc. By permission of the publisher.

183

Why and how do cats purr?

Experts cannot agree on how or why cats purr, or on where the sound originates. Some think that the purr is produced by the vibration of blood in a large vein in the chest cavity. Where the vein passes through the diaphragm, the muscles around the vein contract, nipping the blood flow and setting up oscillations. These sounds are magnified by the air in the bronchial tubes and the windpipe. Others think that purring is the vibrations of membranes, called false vocal cords, located near the vocal cords. No one knows for sure why a cat purrs, but many people interpret the sound as one of contentment.

Sources: Schneck, Marcus, and Jill Caravan. *Cat Facts*, p. 26; Taylor, David. *You & Your Cat*, p. 35.

184

What is the original breed of domestic cat in the United States?

The American shorthair is believed by some naturalists to be the original domes-

Europe by the early settlers. The cats readily adapted to their new environment. Selective breeding to enhance the best traits began early in the 20th century.

The American shorthair is a very athletic cat with a lithe, powerful body, excellent for stalking and killing prey. Its legs are long, heavy, and muscular, ideal for leaping and for coping with all kinds of terrain. The fur, in a wide variety of color and coat patterns, is thick enough to protect the animal from moisture and cold but short enough to resist matting and snagging.

Although this cat makes an excellent house pet and companion, it remains very self-sufficient. Its hunting instinct is so strong that it exercises the skill even when well-provided with food. The American shorthair is the only true "working cat" in the United States.

Sources: Müller, Ulrike. *The New Cat Handbook*, p. 125; Taylor, David. *The Ultimate Cat Book*, pp. 102-103; Taylor, David. *You & Your Cat*, pp. 90-91.

185 Why do cats' eyes shine in the dark?

A cat's eyes contain a special light-conserving mechanism called the *tapetum lucidum*, which reflects any light not absorbed as it passes through the retina of each eye. The retina gets a second chance (so to speak) to receive the light, aiding the cat's vision even more. In dim light, when the pupils of the cat's eyes are opened the widest, this glowing or shining effect occurs when light hits them at certain angles. The *tapetum lucidum*, located behind the retina, is a membrane composed of fifteen layers of special glittering cells, that all together act as a mirror. The color of the glow is usually greenish or golden, but the eyes of the Siamese cat reflect a luminous ruby red.

Sources: McNulty, Faith. *Wholly Cats*, p. 17; Schneck, Marcus, and Jill Caravan. *Cat Facts*, p. 40; Taylor, David. *You & Your Cat*, p. 28.

186 Which plants are poisonous to cats?

Certain common houseplants are poisonous to cats and they should not be allowed to eat the following:

Caladium (Elephant's ears)
Dieffenbachia (Dumb cane)
Euphorbia pulcherrima (Poinsettia)
Hedera (True ivy)
Mistletoe
Oleander
Philodendron
Prunus laurocerasus (Common or cherry laurel)

Rhododendron (Azalea)
Solanum capiscastrum (Winter or False Jerusalem cherry)

Sources: Taylor, David, and Daphne Negus. *The Ultimate Cat Book*, p. 141; Taylor, David. *You & Your Cat*, p. 147.

▽ **187** ▽ Why do cats have whiskers?

The function of a cat's whiskers is not fully understood. They are thought to have something to do with the sense of touch. Removing them can disturb a cat for some time. Some people believe that the whiskers act as antennae in the dark, enabling the cat to identify things it cannot see. The whiskers may help the cat to pinpoint the direction from which an odor is coming. In addition, the cat is thought to point some of its whiskers downwards to guide it when jumping or running over uneven terrain at night.

Sources: Fox, Michael W. *Understanding Your Cat*, pp. 182-183; Taylor, David. *You & Your Cat*, p. 35.

▽ **188** ▽ What is a "tabby" cat?

"Tabby," the basic feline coat pattern, goes back to when cats were wild. The tabby coat is an excellent form of camouflage. Each hair has two or three dark and light bands, with the tip always dark. There are four variations on the basic tabby pattern.

The mackerel (also called striped or tiger) tabby has a dark line running down the back from the head to the base of the tail, with several stripes branching down the sides. The legs have stripes and the tail has even rings with a dark tip. There are two rows of dark spots on the stomach. Above the eyes is a mark shaped like an "M" and dark lines run back to the ears. Two dark necklace-like bands appear on the chest.

The blotched, or classic, tabby markings seem to be the closest to those found in the wild. The markings on the head, legs, tail, and stomach are the same as the mackerel tabby. The major difference is that the blotched tabby has dark patches on the shoulder and side rimmed by one or several lines.

The spotted tabby has uniformly-shaped round or oval dark spots all over the body and legs. The forehead has a "M" on it and a narrow dark line runs down the back.

The Abyssinian tabby has almost no dark markings on its body; they appear only on the forelegs, the flanks and the tail. The hairs are banded except on the stomach where they are light and unicolored.

Sources: Müller, Ulrike. *The New Cat Handbook*, pp. 114-115.

189 ⁄ Which types of birds make the best pets?

There are several birds that make good house pets and have a reasonable life expectancy:

Bird	Life expectancy in years	Considerations
Finch	2-3	Easy care
Canary	8-10	Easy care; males sing
Budgerigar	8-15	Easy care
Cockatiel	15-20	Easy care; easy to train
Lovebird	15-20	Cute, but not easy to care for or train
Amazon parrot	50-60	Good talkers, but can be screamers
African grey parrot	50-60	Talkers; never scream

Sources: *Prevention's Giant Book of Health Facts*, p. 463.

190 ⁄ How much does an average adult potbellied pig weigh compared to a regular farm pig?

The adult Chinese potbellied pig usually weighs about 70 to 150 pounds (31.75 to 68 kilograms); adult farm pigs may weigh 1,200 to 1,500 pounds (544.32 to 680.4 kilograms). Chinese potbellies are an ancient breed that stand only 16 inches (40.64 centimeters) tall. They are very intelligent, easily house-broken, and reportedly make affectionate pets.

Sources: Mull, Kayla. *Pot-Bellied Pet Pigs*, pp. 6-7; *People Weekly*, vol. 29 (February 15, 1988), pp. 115-117.

191 ⁄ What kind of care do tadpoles need and what do they eat?

Keep the frog eggs and the tadpoles that hatch from them in water at all times, changing half of the water volume no more than once a week. The best diet is probably baby cereal having a high protein content, fresh greens, and bits of egg yolk. Provide a rock island when the legs of the tadpoles appear. A 5-gallon (18.92-liter) tank is sufficient for a half dozen tadpoles. When they mature (lose their tails and have grown legs) they should be released in a pond or by the lake shore.

Sources: Dolensek, Emil P. *A Practical Guide to Impractical Pets*, pp. 67-68.

192 ⁄ Has there ever been a pet alligator in the White House?

In 1825, the Marquis de Lafayette (1757-1834) toured America and was given an alligator by some grateful citizen. While Lafayette was the guest of President John

Quincy Adams (1767-1848), the alligator took up residence in the East Room of the White House for several months. When Lafayette departed, he took his alligator with him.

Sources: Kelly, Niall. *Presidential Pets*, p. 17; Truman, Margaret. *White House Pets*, pp. 4, 169.

193 / *How can pets be treated to remove skunk odor?*

Purchase one of the products specifically designed to counteract skunk odor from a pet store. Most of these are of the enzyme or bacterial enzyme variety and can be used without washing the pet first.

Sources: Aslett, Don. *Pet Clean-Up Made Easy*, p. 14.

194 / *What is the best way to remove a skunk that stays near the house?*

Skunks are beneficial animals; they are great rodent hunters and eat lots of insects (cutworms, army worms, grubs, and potato beetles). They can be caught in cages, placed inside plastic bags, and baited with cat kibbles—their favorite!

Sources: Klein, Hilary D. *Tiny Game Hunting*, pp. 190–191.

195 / *How can bats be removed from the house?*

To rid an attic or other part of a house of a bat colony, seal, caulk, weatherstrip, or screen all exterior openings greater than ¼ inch wide—except one, through which the bats can escape. Wait 3 days; then, about ½ hour after dark (when the bats are outside feeding), seal the last opening.

If an area can not be sealed, install a light that would shine directly on the roosting area. In confined areas, suspend mothballs in porous bags as a repellent. Avoid exterminating bats; they are desirable as insect eaters.

If a single bat strays into a room, after dark open the windows and any outside doors and turn off all the lights. The bat will fly outside.

CAUTION: Never touch a bat; it can be a carrier of rabies. If bitten by one, go to a hospital immediately.

Sources: *How to Do Just About Anything*, p. 28.

196 / *Any suggestions for keeping mosquitoes away from an outdoor patio in the summertime?*

One idea is to keep a 12-inch oscillating electric fan operating to make it difficult for the mosquitoes to land. Although this is not 100% effective, it helps. Another idea comes from Spain. Plant basil all around the patio; mosquitoes do not like basil.

Sources: *New York Times Book of Indoor and Outdoor Gardening Questions*, p. 84.

STORAGE *See: AROUND THE HOUSE AND GARAGE—CLEANING AND MAINTENANCE; FOOD AND NUTRITION—FOOD PROCESSING AND STORAGE*

TOYS *See: AROUND THE HOUSE AND GARAGE—HOBBIES, TOYS, ETC.*

BIOLOGY

ALGAE See: BIOLOGY—FUNGI, BACTERIA, ALGAE, ETC.

BACTERIA See: BIOLOGY—FUNGI, BACTERIA, ALGAE, ETC.

CLASSIFICATION, MEASUREMENTS, AND TERMS

197 / *What is radiocarbon dating?*

Radiocarbon dating is a process for determining the age of a prehistoric object by measuring its radiocarbon content. The technique was developed by an American chemist, Dr. Willard F. Libby (1908–1980), in the late 1940s. All living things contain radiocarbon (carbon 14), an isotope that occurs in a small percentage of atmospheric carbon dioxide as a result of cosmic ray bombardment. After an animal or plant dies, it no longer absorbs radiocarbon and the radiocarbon present begins to decay (break down by releasing particles) at an exact and uniform rate. Its half-life of 5,730 years made it useful for measuring prehistory and events occurring within the past 35 to 50 thousand years. A recent development, called the Accelerated Mass Spectrometer, which separates and detects atomic particles of different mass, can establish more accurate dates with a smaller sample. The remaining radiocarbon can be measured and compared to that of a living sample. In this way, the age of the 50,000 year old or less animal or plant (or more precisely the elapsed time since its death) can be determined.

Since Libby's work other isotopes having longer half-lives have been used as "geologic clocks" to date very old rocks—at least 100 thousand years old to use the isotope potassium-argon dating. The isotope Uranium-238 (decaying to Lead-206) has a half-life of 4.5 billion years, Uranium-235 (decaying to Lead-207) has a value of 704 million years, Thorium-232 (decaying to Lead-278) has a half–life of 14 billion years, Rubidium-87 (decaying to Stronium-87) has a half-life value of

48.8 billion years, Potassium-40 (decaying to Argon-40) has a value of 1.25 billion years and Samarium-147 (decaying to Neodymium-143) has a value of 106 billion years. These isotopes are used in dating techniques of gas formation light emission (called thermoluminescence). Other ways to date the past is dating by tree rings (counting its annual growth rings), and dating by thermoremanent magnetism (the magnetic field of the rock is compared to a date chart of changes in the earth's magnetic field).

Sources: *The Biographical Dictionary of Scientists: Chemists*, pp. 84-85; Gray, Peter. *The Encyclopedia of the Biological Science*, p. 141.

198

Who coined the term biology?

Biology was first used by Karl Burdach (1776–1847) to denote the study of man. Jean Baptiste Pierre Antoine de Monet Lamarck (1744–1829) gave the term a broader meaning in 1812. He believed in the integral character of science. For the special sciences, chemistry, meteorology, geology, and botany-zoology he coined the term *biology*.

Lamarckism epitomizes the belief that changes acquired during an individual's lifetime as the result of active, quasi-purposive, functional adaptations can somehow be imprinted upon the genes, thereby becoming part of the heritage of succeeding generations. Today, very few professional biologists believe that anything of the kind occurs—or can occur.

Biology is the science that deals with living things (Greek *bios*, "life"). Formerly broadly divided into two areas, zoology (Greek *zoon*, "animal"), the study of animals, and botany (Greek *botanes*, "plant"), the study of plants, biology is now divided and sub-divided into hundreds of special fields involving the structure, function, and classification of the forms of life. These include anatomy, ecology, embryology, evolution, genetics, paleontology, and physiology.

Sources: Bynum, W.F., et al. *Dictionary of the History of Science*, p. 43; Medawar, P.B., and J.S. Medawar. *Aristotle to Zoos*, p. 166; Schneider, Herman, and Leo Schneider. *The Harper Dictionary of Science in Everyday Language*, p. 41.

199

What are the five kingdoms presently used to categorize living things?

Carolus Linnaeus (1707–1778) in 1735 divided all living things into two kingdoms in his classification system that was based on similarities and differences of organisms. However, since then, fungi seemed not to fit nicely into either kingdom. Although fungi were generally considered plants, they had no chlorophyll, roots, stems, or leaves, and hardly resemble any true plant. They also have several features found in the animal kingdom, as well as unique features characteristic of themselves alone. So fungi was considered the third kingdom. In 1959, R.H. Whittaker proposed the current five kingdom system, based on new evidence from biochemical techniques and electron microscope observations that revealed fundamental differences among organisms. Each kingdom is listed below.

Monera—One-celled organisms lacking a membrane around the cell's genetic matter. Prokaryote (or procaryote) is the term used for this condition where the genetic material lies free in the cytoplasm of the cell without a membrane to form the nucleus of the cell. The kingdom consists of bacteria and blue-green algae (also called blue-green bacteria or cyanobacteria). Bacteria do not produce their own food but blue-green algae do. Blue-green algae, the primary form of life 3.5 to 1.5 billion years ago, produce most of the world's oxygen through photosynthesis.

Protista—Mostly single-celled organisms with a membrane around the cell's genetic material called eukaryotes (or eucaryotes) because of the nucleus membrane and other organelles found in the cytoplasm of the cell. Protista consists of true algae, diatoms, slime molds, protozoa, and euglena. Protistans are diverse in their modes of nutrition, etc. They may be living examples of the kinds of ancient single cells that gave rise to the kingdoms of multi–celled eukaryotes (fungi, plants, and animals).

Fungi—One-celled or multicelled eukaryotes (having a nucleus membrane or a membrane around the genetic material). The nuclei stream between cells giving the appearance that cells have multiple nuclei. This unique cellular structure, along with the unique sexual reproduction pattern, distinguish the fungi from all other organisms. Consisting of mushrooms, yeasts, molds, etc., the fungi do not produce their own food.

Plantae—Multicellular organisms having cell nuclei and cell walls, which directly or indirectly nourish all other forms of life. Most use photosynthesis (a process by which green plants, containing chlorophyll, utilize sunlight as an energy source to synthesize complex organic material, especially carbohydrates from carbon dioxide, water, and inorganic salts) and most are autotrophs (produce their own food from inorganic matter).

Animalia—Multicellular organisms whose eukaryotic cells (without cell walls) form tissues (and from tissues form organs). Most get their food by ingestion of other organisms; they are heterotrophs (cannot produce their food from inorganic elements). Most are able to move from place to place (mobile) at least during part of the life cycle.

Sources: Emiliani, Cesare. *The Scientific Companion*, pp. 149-155; Margulis, Lynn. *Five Kingdoms*, p. 3; Starr, Cecie, and Ralph Taggart. *Biology*, pp. 347, 385, 406.

▽ **200** ╱

Who devised the current animal and plant classification system?

The naming and organizing of the millions of species of plants and animals is frequently called taxonomy; such classifications provide a basis for comparisons and generalizations. A common classification format is an hierarchical arrangement in which a group is classified within a group, and the level of the group is denoted in a ranking.

Carolus Linnaeus (1707–1778) composed a hierarchical classification system for plants (1753) and animals (1758) using a system of nomenclature (naming) that continues to be used today. Every plant and animal was given two scientific names (binomial method) in Latin, one for the species and the other for the group or genus within the species. He categorized the organisms by perceived physical differences and similarities. Although Linnaeus started with only two kingdoms, contemporary classifiers have expanded them into five kingdoms. Each kingdom is divided into two or more phyla (major groupings; phylum is the singular form). Members within one phylum are more closely related to one another than they are to members of another phylum. These phyla are subdivided into parts again and again; members of each descending level have a closer relationship to each other than those of the level above. Generally the ranking of the system, going from general to specific, are Kingdom, Phylum (plant world uses the term, division), Class, Order, Family, Genus, and Species. In addition, intermediate taxonomic levels can be created by adding the prefixes "sub" or "super" to the name of the level, for example, "subphylum" or "superfamily." Zoologists working on parts of the animal classification, many times are not uniform in their groupings. The system is still evolving and changing as new information emerges and new interpretations develop. Below is listed a comparison of hierarchy for four of the five kingdoms.

Taxonomic level	Human	Grasshopper	White Pine	Typhoid Bacterium
Kingdom	Animalia	Animalia	Plantae	Protista
Phylum	Chordata	Anthropoda	Tracheophyta	Schizomyco-phyta
Class	Mammalia	Insecta	Gymnospermae	Schizomycetes
Order	Primates	Orthoptera	Coniferales	Eubacteriales
Family	Hominidae	Arcridiidae	Pinaceae	Bacteriaceae
Genus	Homo	Schistocerca	Pinus	Eberthella
Species	sapiens	americana	strobus	typhosa

Sources: Grzimek, Bernhard. *Grzimek's Animal Life Encyclopedia*, vol. 11, p. 545; Otto, James H. *Modern Biology*, p. 179; Zahradnik, Jiri. *A Field Guide in Color to the Animal World*, p. 6.

EVOLUTION AND GENETICS

▽ 201 ▽

What is genetic engineering?

Genetic engineering is the deliberate alteration of the genetic make-up (genome) of an organism by manipulation of its DNA (*deoxyribonucleic acid*) molecule (a double helix chemical structure containing genetic information) to effect a change in heredity traits.

Genetic engineering techniques include cell fusion, and the use of recombinant DNA (RNA) or gene-splicing. In cell fusion, the tough outer membranes of sperm and egg cells are stripped off by enzymes, and then the fragile cells are mixed and combined with the aid of chemicals or viruses. The result may be the creation of a new life form from two species. Recombinant DNA techniques transfer a specific genetic activity from one organism to the next through the use of bacterial plasmids (small circular pieces of DNA lying outside the main bacterial chromosome) and enzymes, such as restriction endonucleases (which cut the DNA strands); reverse transcriptase (which makes a DNA strand from an RNA strand); DNA ligase (which joins DNA strands together); and tag polymerase (which can make a double-stranded DNA molecule from a single stranded "primer" molecule). The process begins with the isolation of suitable DNA strands and fragmenting them. After these fragments are combined with vectors, they are carried into bacterial cells where the DNA fragments are "spliced" on to plasmid DNA that has been opened up. These hybrid plasmids are now mixed with host cells to form transformed cells. Since only some of the transformed cells will exhibit the desired characteristic or gene activity, the transformed cells are separated and grown individually in cultures. This methodology has been successful in producing large quantities of hormones (such as insulin) for the biotechnology industry. However, it is more difficult to transform animal and plant cells. Yet the technique exists to make plants resistant to diseases and to make animals grow larger. Because genetic engineering interferes with the processes of heredity and can alter the genetic structure of our own species, there is much concern over the ethical ramifications of such power, as well as the possible health and ecological consequences of the creation of these bacterial forms. Some applications of genetic engineering in the various fields are listed below:

Agriculture—Crops having larger yields, disease- and drought-resistancy; bacterial sprays to prevent crop damage from freezing temperatures (1987); and livestock improvement through changes in animal traits (1982).

Industry—Use of bacteria to convert old newspaper and wood chips into sugar; oil- and toxin-absorbing bacteria for oil spill or toxic waste clean-ups (1991); and yeasts to accelerate wine fermentation.

Medicine—Alteration of human genes to eliminate disease (1990, experimental stage); faster and more economical production of vital human substances to alleviate deficiency and disease symptoms (but not to cure them) such as insulin (1982), interferon (cancer therapy), vitamins, human growth hormone ADA (1990), antibodies, vaccines, and antibiotics.

Research—Modification of gene structure in medical research (1989), especially cancer research (1990).

Food processing—Rennin (enzyme) in cheese aging (1989).

Sources: Cone, Robert J. *How the New Technology Works*, pp. 34-38; *How in the World?*, p. 190; *Oxford Illustrated Encyclopedia of Invention and Technology*, pp. 153-155.

202 ◹ *What is Batesian mimicry?*

In 1861, Henry Walter Bates (1825–1892), a British naturalist, proposed that a nontoxic species can evolve (especially in color and color pattern) to look like a toxic or unpalatable species, or to act like a toxic species, to avoid being eaten by a predator. The classic example is the viceroy butterfly, which resembles the unpalatable monarch butterfly. This is called Batesian mimicry. Subsequently, Fritz Müller (1821–1897), a German-born zoologist, discovered that all the species of similar appearance become distasteful to predators. This phenomenon is called Müllerian mimicry.

Sources: *The Biographical Dictionary of Scientists: Biologists*, p. 15; Bynum, W.F., et al. *Dictionary of the History of Science*, p. 149; *Discover*, vol. 13 (January 1992), p. 72.

203 ∕ Who coined the phrase "survival of the fittest"?

Although frequently associated with Darwinism, this phrase was coined by Herbert Spencer (1820-1903), an English sociologist. It is the process by which organisms that are less well-adapted to their environment tend to perish and better-adapted organisms tend to survive.

Sources: Asimov, Isaac. *Asimov's Biographical Encyclopedia of Science and Technology*, p. 407; Bowler, Peter J. *Evolution, the History of an Idea*, p. 225; Toothill, Elizabeth. *The Facts On File Dictionary of Biology*, p. 197.

204 ∕ What is meant by Mendelian inheritance?

Mendelian inheritance refers to genetic traits carried through heredity; described by the Austrian monk, Gregor Mendel (1822-1889). Mendel was the first to deduce correctly the basic principles of heredity. Mendelian traits are also called single gene or monogenic traits, because they are controlled by the action of a single gene or gene pair. More than 4,300 human disorders are known or suspected to be inherited as Mendelian traits, encompassing autosomal dominant (e.g., neurofibromatosis), autosomal recessive (e.g., cystic fibrosis), sex-linked dominant and recessive conditions (e.g., color-blindness and hemophilia).

Overall, incidence of Mendelian disorders in the human population is about 1%. Many non-anomalous characteristics that make up human variation are also inherited in Mendelian fashion.

Sources: *The Biographical Dictionary of Scientists: Biologists*, pp. 93–94; Wynbrandt, James. *The Encyclopedia of Genetic Disorders and Birth Defects*, p. 205.

205 ∕ When was the Scopes (monkey) trial?

John T. Scopes (1900-1970), a high-school biology teacher, was brought to trial by the State of Tennessee in 1925 for teaching the theory of evolution. He challenged a recent law passed by the Tennessee legislature that made it unlawful to teach in any public school any theory which denies the divine creation of man. He was convicted and sentenced, but the decision was reversed later and the law repealed in 1967.

At the present, pressure against school boards still affects the teaching of evolution today. Recent drives by anti-evolutionists have tried to either ban the teaching of evolution or have demanded "equal time" for "special creation" as described in the Biblical Book of Genesis. This has raised many questions about the separation of church and state, the teaching of controversial subjects in public schools, and the ability of scientists to communicate with the public. The gradual improvement of the fossil record, the result of comparative anatomy, and many other developments in biological science contributed toward making evolutionary thinking more palatable.

Sources: *Encyclopedia Americana*, vol. 24, p. 397; Magner, Lois N. *A History of the Life Sciences*, p. 395.

206 / How did humans evolve?

Evolution of the *Homo* lineage of modern humans (*Homo sapiens*) began with the hunter of nearly five feet tall, *Homo habilis*, who is widely presumed to have evolved from an australopithecine ancestor. Near the beginning of the Pleistocene (2 million years ago), *Homo habilis* was transformed into *Homo erectus* (Java Man), who used fire and possessed culture. Middle Pleistocene populations of *Homo erectus* are said to show steady evolution toward the anatomy of *Homo sapiens* (Neanderthals, Cro-Magnons, and modern humans), 120 to 40 thousand years ago. *Homo sapiens* built huts and made clothing.

Sources: *Encyclopedia of Human Evolution and Prehistory*, p. 259; Trefil, James. *1001 Things Everyone Should Know About Science*, p. 34; Rovin, Jeff. *Laws of Order*, pp. 120-121.

207 / What is the significance of The Origin of Species?

Charles Darwin (1809-1882) first proposed a theory of evolution based on natural selection in his treatise on *The Origin of Species*. The publication of *The Origin of Species* ushered in a new era in our thinking about the nature of man. The intellectual revolution it caused and the impact it had on man's concept of himself and the world were greater than those caused by the works of Newton and others. The effect was immediate, the first edition being sold out on the day of publication (November 24, 1859). The *Origin* has been referred to as "the book that shook the world." Every modern discussion of man's future, the population explosion, the struggle for existence, the purpose of man and the universe, and man's place in nature rests on Darwin.

The work was a product of his analyses and interpretations of his findings from his voyages on the H.M.S. *Beagle*, as a naturalist. In Darwin's day, the prevailing explanation for organic diversity was the story of creation in the Book of Genesis in the Bible. The *Origin* was the first publication to present scientifically sound, well-organized evidence for evolution. Darwin's theory of evolution was based on natural selection in which the best, the fittest, survive, and if there is a difference in genetic endowment among individuals, the race will, by necessity, steadily improve. It is a two-step process: the first consists of the production of variation, and the second, of the sorting of this variability by natural selection in which the favorable variations tend to be preserved.

Sources: Darwin, Charles. *On the Origin of Species*, pp. vii-xvi.

208
Who is generally known as the father of genetics?

The English biologist, William Bateson (1861-1926), coined the term *genetics* the study of the causes and effects of heritable characteristics, and was a leading proponent of Mendelian views. Although Gregor Mendel (1822-1884) is considered the founder of genetics, there are other contenders, such as William Bateson, to share the honor with Mendel.

Sources: Asimov, Isaac. *Asimov's Chronology of Science and Discovery*, p. 346; Gray, Peter. *The Dictionary of the Biological Sciences*, p. 224; *Encyclopedic Dictionary of Science*, p. 38.

209
Which biological events occurred during the geologic time divisions?

Cenozoic Era
(Age of Mammals)

Period	Epoch	Beginning date in est. millions of years	Plants and microorganisms	Animals
Quaternary	Holocene (Recent)	10,000 years ago	Decline of woody plants and rise of herbaceous plants.	Age of *Homo sapiens*; humans dominate.
	Pleistocene	1.9	Extinction of many species (from 4 ice ages).	Extinction of many large mammals (from 4 ice ages)
Tertiary	Pliocene	6.0	Development of grasslands; decline of forests; flowering plants.	Large carnivores; many grazing mammals; first known human-like primates.
	Miocene	25.0		Many modern mammals evolve.
	Oligocene	38.0	Spread of forests; flowering plants, rise of monocotyledons.	Apes evolve; all present mammal families evolve; saber-toothed cats.

Period	Epoch	Beginning date in est. millions of years	Plants and microorganisms	Animals
Tertiary	Eocene	55.0	Gymnosperms and angiosperms dominant.	Beginning of age of mammals; modern birds.
	Paleocene	65.0		Evolution of primate mammals.

Mesozoic Era
(Age of Reptiles)

Period	Epoch	Beginning date in est. millions of years	Plants and microorganisms	Animals
Cretaceous		135.0	Rise of angiosperms; gymnosperms decline.	Dinosaurs reach peak and then become extinct; toothed birds become extinct; first modern birds; primitive mammals.
Jurassic		200.0	Ferns and gymnosperms common.	Large, specialized dinosaurs; first modern birds; insectivorous marsupials.
Triassic		250.0	Gymnosperms and ferns dominate.	First dinosaurs; egg-laying mammals.

Paleozic Era
(Age of Ancient Life)

Period	Epoch	Beginning date in est. millions of years	Plants and microorganisms	Animals
Permian		285.0	Conifers evolve.	Modern insects appear; mammal-like reptiles; extinction of many Paleozic invertebrates.
Carboniferous (divided into Mississippian and Pennsylvanian periods by some in the U.S.)		350.0	Forests of ferns and gymnosperms; swamps; club mosses and horsetails.	Ancient sharks abundant; many echinoderms, mollusks and insect forms; first reptiles; spread of ancient amphibians.

Period	Epoch	Beginning date in est. millions of years	Plants and microorganisms	Animals
Devonian		410.0	Terrestrial plants est-ablished; first forests; gymnosperms appear.	Age of fish; amphibians; wingless insects and millipedes appear.
Silurian		425.0	Vascular plants appear; algae dominant.	Fish evolve; marine arachnids dominant; first insects; crustaceans.
Ordovician		500.0	Marine algae dominant; terrestrial plants first appear.	Invertebrates dominant; first fish appear.
Cambrian		570.0	Algae dominant.	Age of marine invertebrates.

Precambrian Era

Period	Epoch	Beginning date in est. millions of years	Plants and microorganisms	Animals
Archeozoic and Protozoic Eras		3800.0	Bacterial cells; then primitive algae and fungi; marine protozoans.	Marine invertebrates at end of period.
Azoic		4600.0	Origin of the earth.	

Sources: Rovin, Jeff. *Laws of Order*, pp. 117-119; Villee, Claude. *Biology*, p. 1028.

210 / Can human beings be cloned?

In theory, yes. There are, however, many technical obstacles to human cloning, as well as moral, ethical, philosophical, religious, and economic issues to be resolved before a human being could be cloned.

A clone is a group of cells derived from the original cell by fission (one cell dividing into two cells) or by mitosis (cell nucleus division with each chromosome splitting into two). It perpetuates an existing organism's genetic make-up. Gardeners have been making clones (copies) of plants for centuries by taking cuttings of plants to make genetically identical copies. For plants that refuse to grow from cuttings or for the animal world, modern scientific techniques have greatly extended the range of cloning. The technique for plants starts with taking a cutting of a plant, usually the "best" one in terms of reproductivity or decorativeness or other standard. Since all the plant's cells contain the genetic information from which the entire plant can be reconstructed, the cutting can be taken from any part of the plant. Placed in a culture medium having nutritious chemicals and a growth hormone, the cells in the cutting divide, doubling in size every six weeks until the mass of cells produces small white globular points called embryoids. These embryoids develop roots, or shoots, and begin to look like tiny plants. Transplanted into compost, these plants grow into exact copies of the parent plant. The whole process takes 18 months. This process, called tissue culture, has been used to make clones of oil palm, asparagus, pineapples, strawberries, brussels sprouts, cauliflower, bananas, carnations, ferns, etc. Besides making high productive copies of the best plant available, this method controls viral diseases that are passed through seed generations.

For animals, a technique called nuclear transfer enables up to 32 clones to be produced at one time. An embryo at the 32-cell stage of development is split up using tiny surgical tools. Each of the 32 cells then are combined with single cell embryos (from the same species) from which the nucleus has been removed. This method has been used on mice, frogs, sheep, and cattle. Ultimately, there seems to be no biological reason why human beings could not be cloned, sometime in the future.

Sources: *How in the World?*, p. 188; McKinnell, Robert G. *Cloning of Frogs, Mice, and Other Animals*, pp. 95-107; *Scientific American*, vol. 260 (May 1989), pp. 30-31.

211 / What is the Genome Project?

In late 1990, the biomedical community began work on a 15-year, $3 billion (government-financed) program to map the entire human *genome* (the complete genetic structure of a species), which involves locating 100 genes. So far they have plotted geographic locations on the chromosomes of about 4 human genes (4% of the total). The goal is not only to pinpoint these genes, but also to decode the biochemical information down to the so called "letters" of inheritance, the four basic constituents of all genes called *nucleotides* (A (adenine), C (cytosine), G (guanine), and T (thymine)). Since these letters are linked in pairs of sequences in the double helix of DNA, this means that three billion pairs are involved in this process, of which so far 35 million have been deciphered. The justification of the project is that since genetics is a powerful way to study human disease and other aspects of biology, this project will accelerate biomedical research.

Sources: Brennan, Richard P. *Dictionary of Scientific Literacy*, pp. 122, 124; Cone, Robert J. *How the New Technology Works*, p. 38; *McGraw-Hill Yearbook of Science and Technology 1990*, pp. 204-205.

FUNGI, BACTERIA, ALGAE, ETC.

212 / Are there stone-eating bacteria?

Stone-eating bacteria belong to several families in the genus *Thiobacillus*. They can cause damage to monuments, tombs, buildings, and sculptures by converting marble into plaster. The principal danger seems to come from *Thiobacillus thioparus*. This microbe's metabolic system converts sulfur dioxide gas (found in the air) into sulfuric acid and uses it to transform calcium carbonate (marble) into calcium sulfate (plaster). The bacilli draw their nutrition from carbon dioxide formed in the transformation. *Nitrobacter* and *Nitrosomonas* are other "stone-eating bacteria" that use ammonia from the air to generate nitric and nitrous acid, and there are still other kinds of bacteria and fungi producing organic acids (formic, acetic, and oxalic acids), which can attack the stone as well. The presence of these microbes was first observed by a French scientist, Henri Pochon at Angkor Wat, Cambodia, during the 1950s. The increase of these bacteria and other biological-damaging organisms that threaten tombs and buildings of antiquity are due to the sharp climb in the level of free sulfur dioxide gas in the atmosphere from automotive and industrial emissions.

Sources: *Science 80*, vol. 1 (September-October 1980), p. 58.

213 / What are petite yeasts?

In certain yeasts, abnormally small colonies occur. The cells which form petite colonies are mutants which lack certain components for respiration, therefore they exhibit a reduced growth rate. Mutations occur in nuclear DNA or in mitochondrial DNA. Mitochondria are "organelles" found in most eukanyotic (complex) cells. They are the sites for cellular respiration and other cellular processes. Occurrence is one in 500.

Sources: Singleton, Paul, and Diana Sainsbury. *Dictionary of Microbiology*, p. 300.

214 / Why are Koch's postulates significant?

The German bacteriologist, Robert Koch (1843-1910), developed four rules in his study of disease-producing organisms, which later investigators found useful. The following conditions must be met to prove that a bacterium causes a particular disease:
1. The microorganism must be found in large numbers in all diseased animals but not in healthy ones.
2. The organism must be isolated from a diseased animal and grown outside the body in a pure culture.
3. When the isolated microorganism is injected into other healthy animals, it must produce the same disease.
4. The suspected microorganism must be recovered from the experimental hosts (#3), isolated, compared to the first microorganism, and found to be identical.

Sources: Downs, Robert B. *Landmarks in Science*, p. 67; Johnson, Leland G. *Biology*, p. 948.

215

Who were the founders of modern bacteriology?

The German bacteriologist, Robert Koch (1843-1910), and the French chemist, Louis Pasteur (1822-1895), are considered the founders. Pasteur devised a way to heat food or beverages at a temperature slow enough not to ruin them, but high enough to kill most of the microorganisms that would cause spoilage and disease. This process is called pasteurization. By demonstrating that tuberculosis was an infectious disease caused by a specific bacillus and not by bad heredity, Koch laid the groundwork for public health measures that would significantly reduce such diseases. His working methodologies for isolating microorganisms, his laboratory procedures, and his four postulates for determination of disease agents gave medical investigators valuable insights into the control of bacterial infections.

Sources: *The Biographical Dictionary of Scientists: Biologists*, pp. 77, 103; Starr, Cecie, and Ralph Taggart. *Biology*, p. 680.

216

What are diatoms?

Diatoms are microscopic algae in the phylum *bacillarrophyte* of the protista kingdom. Yellow or brown in color, almost all diatoms are single-celled algae, dwelling in fresh and salt water, especially in the cold waters of the North Pacific Ocean and the Antarctic. Diatoms are an important food source for marine plankton (floating animal and plant life) and many small animals.

Diatoms have hard cell walls; these "shells" are made from silica that they extract from the water. It is unclear how they accomplish this. When they die, their glassy shells, called frustules, sink to the bottom of the sea, which hardens into rock called diatomite. One of the most famous and accessible diatomites is the Monterrey Formation along the coast of central and southern California.

Sources: Campell, Neil A. *Biology*, p. 549; *Encyclopedic Dictionary of Science*, p. 77; Flatow, Ira. *Rainbows, Curve Balls and other Wonders of the Natural World Explained*, pp. 16-17.

217

Who was first to coin the word virus?

The English physician Edward Jenner (1749-1823), founder of virology and a pioneer in vaccination, first coined the word *virus*. Using one virus to immunize against another one was precisely the strategy Jenner used when he inoculated someone with cowpox (a disease that attacks cows) to make them immune to smallpox. This procedure is called vaccination from *vaccine* (the Latin name for cowpox). Vaccines are usually a very mild dose of the disease–causing bacteria or virus (weakened or dead). These vaccines stimulate the creation of antibodies in the body that recognize and attack a particular infection. A virus is a minute parasitic organism that reproduces only inside the cell of its host. Viruses replicate by invading host cells and taking over the cell's "machinery" for DNA replication. Viral particles then can break out of the cells, causing disease.

Sources: *The Biographical Dictionary of Scientists: Biologists*, pp. 73-74; Brennan, Richard P. *Dictionary of Scientific Literacy*, p. 315; *Encyclopedic Dictionary of Science*, p. 247.

218 / *What is a lichen?*

Lichens are organisms which grow on rocks, tree branches, or bare ground. They are composed of a green algae and a colorless fungus living together symbiotically. They do not have roots, stems, flowers, or leaves. The fungus, having no chlorophyll, cannot manufacture its own food, but can absorb food from the algae which it enwraps completely, providing protection from the sun and moisture.

This relationship between the fungus and algae is called symbiosis (a close association of two organisms not necessarily to both their benefits). Lichens were the first recognized and are still the best examples of this phenomenon. An unique feature of lichen symbiosis is that it is so perfectly developed and balanced as to behave as a single organism.

Sources: Hale, Mason E., Jr. *The Biology of Lichens*, p. 69; Lawrence, Eleanor. *Henderson's Dictionary of Biological Terms*, p. 532; *The Wise Garden Encyclopedia*, p. 581.

219 / *How is a fairy ring formed?*

A *fairy ring*, or *fungus ring*, is found frequently in a grassy area. There are three types: those which do not affect the surrounding vegetation, those that cause increased growth, and those that damage the surrounding environment. The ring is started from a mycelium (the underground, food–absorbing part of a fungus). The fungus growth is on the outer edge of the grassy area because the inner band of decaying mycelium "use-up" the resources in the soil at the center. This creates a ring effect. Each suceeding generation is farther out from the center.

Sources: Ainsworth, G.C. *Ainsworth's & Bisby's Dictionary of the Fungi*, p. 148.

220 / *What is the scientific study of fungi called?*

Mycology is the science concerning fungi. In the past, fungi have been classified in other kingdoms, but currently they are recognized as a separate kingdom based on their unique cellular structure and their unique pattern of sexual reproduction.

Fungi are heterotrophs (cannot produce their own food from inorganic matter). They secrete enzymes that digest food outside their bodies and their fungal cells absorb the products. Their activities are essential in the decomposition of organic material and cycling of nutrients in nature.

Some fungi, called saprobes, obtain nutrients from non–living organic matter. Other fungi are parasites; they obtain nutrients from the tissues of living host organisms. The great majority of fungi are multicelled and filamentous. A mushroom is a modified reproductive structure in or upon which spores develop. Each spore dispersed from it may grow into a new mushroom.

Sources: Moore-Landecker, Elizabeth. *Fundamentals of the Fungi*, pp. 1-2; Starr, Cecie, and Ralph Taggert. *Biology*, p. 373.

GENETICS *See: BIOLOGY—EVOLUTION AND GENETICS*

LIFE PROCESSES, STRUCTURES, ETC.

▽ 221 ▽ *What is a biological clock?*

First recognized by the Chinese in the 3rd century B.C.E., the biological clock is an intrinsic mechanism that controls the rhythm of various metabolic activities of plants and animals. Some, such as mating, hibernation, and migration have a yearly cycle; others, such as ovulation and menstrual cycles of women, follow a lunar month. The majority, however, have a 24-hour, day-night cycle called circadian rhythm. This day-night cycle, first recognized in plants over 250 years ago and existing in virtually all species of plants and animals, regulates these organisms' metabolic functions: plants opening and closing their petals or leaves, germination and flowering functions, changes in human body temperature, hormone secretion, blood sugar and blood pressure levels, and sleep cycles.

Research in chronobiology—the study of these daily rhythms—reveals that many accidents occur between 1 and 6 a.m., that most babies are born in the morning hours, that heart attacks tend to occur between 6 and 9 a.m., and that most Olympic records are broken in the late afternoon. The clock regulator may be the pineal gland located in the heads of animals (including man).

Sources: Brennan, Richard P. *Dictionary of Scientific Literacy*, pp. 25, 49; *Encyclopedic Dictionary of Science*, p. 42; Temple, Robert. *The Genius of China*, p. 124.

▽ 222 ▽ *Who was regarded as the father of biochemistry?*

Jan Baptista van Helmont (1577-1644) is called the father of biochemistry because he studied and expressed vital phenomena in chemical terms. The term "biochemistry," coined by F. Hoppe-Seyler in 1877, is the science dealing with the dynamics of living chemical processes or metabolism. It was formed from both the chemists' animal and vegetable chemistry, and the biologists' and doctors' physiological, zoological, or biological chemistry.

Johannes Baptista van Helmont devoted his life to the study of chemistry as the true key to medicine and is considered as one of the founders of modern pathology because he studied the external organs of diseases as well as the anatomical changes caused by diseases.

Sources: Bynum, W.F., et al. *Dictionary of the History of Science*, p. 42; *Encyclopedic Dictionary of Science*, p. 116; *World Who's Who in Science*, p. 782.

223 / What does "ontogeny recapitulates phylogeny" mean?

Ontogeny is the course of development of an organism from fertilized egg to adult; phylogeny is the evolutionary history of a group of organisms. So the phrase, originating in nineteenth century biology means that, as an embryo of an advanced organism grows, it will pass through stages that look very much like the adult phase of less advanced organisms. For example, at one point the human embryo has gills and resembles a tadpole.

Sources: Toothill, Elizabeth. *The Facts On File Dictionary of Biology*, pp. 210, 235; Trefil, James. *1001 Things Everyone Should Know About Science*, p. 23.

224 / Who is considered the "father" of embryology?

Kaspar Friedrich Wolff (1733–1794), a German surgeon, is regarded as the "father" of embryology. Wolff produced his revolutionary work *Theoria generationis* in 1759. Until that time it was generally believed that each living organism developed from an exact miniature of the adult within the seed or sperm. Wolff introduced the idea that cells in a plant or animal embryo are initially unspecified (undefined) but later differentiate to produce the separate organs and systems with distinct types of tissues.

Sources: *The Biographical Dictionary of Scientists: Biologists*, p. 138.

225 / What is the difference between DNA and RNA?

DNA (*deoxyribonucleic acid*) is a nucleic acid formed from a repetition of simple building blocks called nucleotides. The nucleotides consist of phosphate (PO_4), sugar (deoxyribose), and a base which is either adenine (A), thymine (T), guanine (G), or cytosine (C). In a DNA molecule, this basic unit is repeated in a double helix structure made from two chains of nucleotides linked between the bases. The links are either between A and T or between G and C. The structure of the bases does not allow other kinds of links. The famous double helix structure resembles a twisted ladder. The 1962 Nobel Prize in physiology or medicine was awarded to James Watson (b. 1928), Francis Crick (b. 1916), and Maurice Wilkins (b. 1916) for determining the molecular structure of DNA.

RNA (*ribonucleic acid*) is also a nucleic acid, but it consists of a single chain and the sugar is ribose rather than deoxyribose. The bases are the same except that the thymine (T) which appears in DNA is replaced by another base called uracil (U), which links only to adenine (A).

Sources: Magill, Frank N. *Nobel Prize Winners: Physiology or Medicine*, vol. 2, p. 853; Trefil, John. *1001 Things Everyone Should Know About Science*, pp. 260-261.

DNA double helix

226
What is the "Spiegelman monster"?

Sol Spiegelman, an American microbiologist, conducted an experiment to identify the smallest molecule capable of replicating itself. He began with a virus called Q_B, which consisted of a single molecule of ribonucleic acid (RNA) composed of 4,500 nucleotides (units of nucleic acid).

Usually this virus is able to make copies of itself only by invading a living cell because it requires a cellular enzyme called replicase. When Spiegelman added the replicase along with a supply of free nucleotides in a test tube to the virus, the virus replicated itself for several generations when a mutant appeared having fewer than 4,500 nucleotides. Being smaller, this mutant replicated faster than the original virus. Then another mutant appeared and displaced the first, and so it went. Finally, the virus degenerated into a little piece of ribonucleic acid with only 220 nucleotides, the smallest bit necessary for recognizing the replicase. This little test tube monster could continue to replicate at high speed, provided it had a source of building blocks.

Sources: Sherman, Irwin W. *Biology: A Human Approach*, p. 16.

227 / Who originated the idea called panspermia?

Panspermia is the idea that microorganisms, spores, or bacteria attached to tiny particles of matter have travelled through space, eventually landing on a suitable planet and initiated the rise of life there. The word itself means "all-seeding." The British scientist Lord Kelvin (1824-1907) suggested, in the 19th century, that life may have arrived here from outer space, perhaps carried by meteorites. In 1903, the Swedish chemist Svante Arrhenius (1859-1927) put forward the more complex panspermia idea that life on earth was "seeded" by means of extraterrestrial spores, bacteria, and microorganisms coming here on tiny bits of cosmic matter.

Sources: Angelo, Joseph A. *The Extraterrestrial Encyclopedia*, p. 127.

228 / How many mitochondria are there in a cell?

The number of mitochondria varies according to the type of cell, but each cell in the human liver has over one thousand mitochondria. A mitochondrion (singular form) is a self–replicating double-membraned body found in the cytoplasm of all eukaryotic (having a nucleus) cells. The number of mitochondria per cell varies between one and 10,000, and averages about 200. The mitochondria are the sites for much of the metabolism necessary for the production of ATP, lipids, and protein synthesis.

Sources: Bynum, W.F., et al. *Dictionary of the History of Science*, p. 149; Maclean, Norman. *Dictionary of Genetics and Cell Biology*, p. 254; Trefil, James. *1001 Things Everyone Should Know About Science*, p. 101.

229 / What are the stages in the type of cell division called mitosis?

Cell division in eukaryotes (higher organisms) consists of two stages, *mitosis*, the division of the nucleus, and *cytokinesis*, the division of the whole cell.

The first process in the actual division of the cell is mitosis. In mitosis, the replicated chromosomes are maneuvered so that each new cell gets a full complement of chromosomes—one of each. The process is divided into four phases: prophase, metaphase, anaphase, and telophase.

Nuclear division of sex cells is called meiosis. Sexual reproduction generally requires two parents and it always involves two events (meiosis and fertilization).

Sources: Armstrong, Joseph E. *Science in Biology*, p. 64; Curtis, Helena, and N. Sue Barnes. *Invitation to Biology*, pp. 92, 155; Maclean, Norman. *Dictionary of Genetics and Cell Biology*, p. 255.

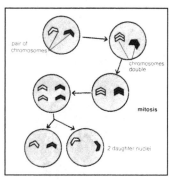

Mitosis process

▽ 230 ▽ *Who is Melvin Calvin?*

Calvin (b. 1911) is an American chemist who won the 1961 Nobel Prize for Chemistry for his achievement of working out the chemical reaction cycles called biosynthetic pathways in photosynthesis. Photosynthesis is the process by which green plants use the energy of the sunlight to convert water and carbon dioxide into carbohydrates and oxygen. This cycle of reactions is now called the Calvin cycle. He also conducted research in organometallic chemistry, the chemical origin of life, and taught chemistry at the University of California at Berkeley.

Sources: *The Biographical Dictionary of Scientists: Chemists*, p. 24; Farber, Edward. *Nobel Prize Winners in Chemistry 1901-1961*, p. 307.

MEASUREMENTS *See: BIOLOGY—CLASSIFICATION, MEASUREMENTS, AND TERMS*

▼ Bridges, Buildings, etc.

BRIDGES, ROADS, AND TUNNELS

231
What unusual method of testing was used to prove the safety of the first bridge across the Mississippi River at St. Louis in 1874?

According to a historical appraisal of the bridge by Howard Miller:
> Progressively heavier trains shuttled back and forth across the bridge as its engineer, James B. Eads, took meticulous measurements. However, the general public was probably more reassured by a nonscientific test. Everyone knew that elephants had canny instincts and would not set foot on an unsound bridge. The crowd cheered as a great beast from a local menagerie mounted the approach without hesitation and lumbered placidly across to the Illinois side.

Sources: *American Scientist*, vol. 80, no. 2 (March/April 1992), p. 123.

232
Which United States place has the most bridges?

With the possible exception of Venice, Italy, Pittsburgh, Pennsylvania is the bridge capital of the world. There are more than 1,700 bridges in Allegheny County's 731 square miles. The county has 2.3 bridges per square mile or one bridge for every mile of highway. Nationwide, there are approximately 564 thousand bridges.

Sources: *American Transportation Builder*, vol. 58, no. 5 (Winter 1981), p. 24.

233
Are there any floating bridges in the United States?

There are three floating pontoon bridges in the United States, all located in the state of Washington. The Evergreen Point Bridge (1963) and the Third Lake Washington Bridge (1989) in Seattle are 7,518 feet (2,291.5 meters) and 6,130 feet

(1868.5 meters) long respectively. The Hood Canal Bridge (1961) in Port Gamble is 6,471 feet (1972.4 meters) long.

Sources: *The World Almanac and Book of Facts 1992*, p. 624.

234 / *Where is the longest suspension bridge in the United States?*

Spanning New York (City) Harbor, the Verrazano-Narrows Bridge is the longest suspension bridge in the United States. With a span of 4,260 feet (1,298.5 meters) from tower to tower, its total length is 7,200 feet (2,194.6 meters). Named after Giovanni da Verrazano (1485-1528), the Italian explorer who discovered New York Harbor in April 1524, it was erected under the direction of Othmar H. Ammann (1879-1965), and completed in 1964. To avoid impeding navigation in and out of the harbor, it provides a clearance of 216 feet (65.8 meters) between the water level and the bottom of the bridge deck. Like other suspension bridges, the bulk of the load of the Verrazano-Narrows Bridge is carried on cables anchored to the banks.

Sources: Jackson, Donald C. *Great American Bridges and Dams*, p. 138; *The World Almanac and Book of Facts 1992*, p. 384.

235 / *What are the world's longest bridge spans?*

The world's longest main-span in a suspension bridge is the Humber Estuary Bridge in England, at 4,626 feet (1,410 meters). Designed by Freeman Fox and Partners and begun on July 27, 1972, it was opened on July 18, 1981. The Humber Bridge is longer than the Verrazano-Narrows Bridge in New York, but it will be surpassed by two bridges now under construction. The double-deck Akashi-Kaikyo Bridge linking Honshū and Shikoku, Japan, begun in 1998 will have a main span of 6,529 feet (1,990 meters). The Messing Bridge that will link Sicily with Calabria on the Italian mainland will become the longest by far when it is completed, with a single span of 10,892 feet (3,320 meters).

The world's longest cable-stayed bridge span is the 1,608-foot (490 meter) Ikuchi Bridge between Honshū and Shikoku, Japan, completed in 1991.

The world's longest cantilever bridge is the Quebec Bridge over the St. Lawrence River in Canada, opened in 1917, which has a span of 1,800 feet (548.6 meters) between piers, with an overall length of 3,239 feet (987.3 meters).

The world's longest steel arch bridge is the New River Gorge Bridge near Fayettesville, West Virginia completed in 1977, with a span of 1,700 feet (518.2 meters).

The world's longest concrete arch bridge is the Jesse H. Jones Memorial Bridge which spans Houston Ship Canal in Texas. Completed in 1982, the bridge measures 1,500 feet (457.2 meters).

The world's longest stone arch bridge is the 3,810-foot–long (1161-meter) Rockville Bridge, completed in 1901, north of Harrisburg, Pennsylvania, with 48 spans containing 216.051 tons (219.52 tonnes) of stone.

Sources: *Guinness Book of Records 1992*, pp. 317-318; Hawkes, Nigel. *Structures*, pp. 170-173.

236 / What are the various types of bridge structures?

There are four basic types of structures that can be used to bridge a stream or similar obstacle: rigid beam, cantilever, arch, and suspension systems.

The rigid beam bridge, the simplest and most common form of bridging, has straight slabs or girders carrying the roadbed. The span is relatively short and its load rests on its supports or piers.

The arch bridge is in compression, and thrusts outward on its bearings at each end.

In the suspension bridge, the roadway hangs on steel cables, with the bulk of the load carried on cables anchored to the banks. It can span a great distance without intermediate piers.

Each arm of a cantilever bridge is, or could be, free-standing, with the load of the short central truss span pushing down through the piers of the outer arms and pulling up at each end. The outer arms are usually anchored at the abutments and project into the central truss.

Sources: *The Illustrated Science and Invention Encyclopedia*, vol. 3, pp. 353, 357-359; *Oxford Illustrated Encyclopedia of Invention and Technology*, pp. 48, 57; Stephens, John H. *The Guinness Book of Structures*, pp. 11-15.

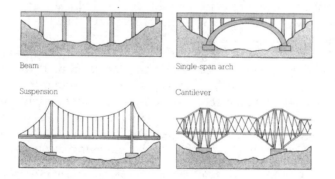

Beam   Single-span arch

Suspension   Cantilever

Bridge types

237 / Who designed the Golden Gate Bridge?

Joseph B. Strauss (1870-1938), formally named chief engineer for the project in 1929, was assisted by Charles Ellis and Leon Moisseiff in the design. An engineering masterpiece that opened to traffic in May 1937, this suspension bridge spans San Francisco Bay, linking San Francisco with Marin County, California. It has a central span of 4,200 feet (1280.2 meters) with towers rising 746 feet (227.4 meters).

Sources: Jackson, Donald C. *Great American Bridges and Dams*, pp. 278-280.

238 / What is a "kissing bridge"?

Covered bridges with roofs and wooden sides are called "kissing bridges," because people inside the bridge could not be seen from outside. Such bridges can be traced back to the early nineteenth century. Contrary to folk wisdom, they were not designed to produce rural "lovers' lanes," but were covered to protect the structures from deterioration.

Sources: Jackson, Donald C. *Great American Bridges and Dams*, p. 20; Maginley, C.J. *Models of America's Past*, p. 100.

239 / Who built the Brooklyn Bridge?

John A. Roebling (1806-1869), a German-born American engineer, constructed the first truly modern suspension bridge in 1855. Towers supporting massive cables, tension anchorage for stays, a roadway suspended from the main cables, and a stiffening deck below or beside the road deck to prevent oscillation are all characteristics of Roebling's suspension bridge. In 1867, Roebling was given the ambitious task of constructing the Brooklyn Bridge. In his design he proposed the revolutionary idea of using steel wire for cables rather than the less-resilient iron. Just as construction began, Roebling died of tetanus when his foot was crushed in an accident and his son, Washington A. Roebling (1837-1926), assumed responsibility for the bridge's construction. Fourteen years later, in 1883, the bridge was completed. At that time, it was the longest suspension bridge in the world, spanning the East River and connecting New York's Manhattan with Brooklyn. The bridge has a central span of 1,595.5 feet (486.3 meters), with its masonry towers rising 276 feet (84.1 meters) above high water. Today, the Brooklyn Bridge is among the best-known of all American civil engineering accomplishments.

Sources: Schodek, Daniel L. *Landmarks in American Civil Engineering*, pp. 132-142; Sutton, Caroline, and Duncan M. Anderson. *How Do They Do That?*, pp. 156-160.

240 / How many covered bridges are there in the United States?

There were more than 10,000 covered bridges built across the United States between 1805 (when the first was erected in Philadelphia) and the early 20th century. As of January 1980, only 893 of these covered bridges remain—231 in Pennsylvania, 157 in Ohio, 103 in Indiana, 100 in Vermont, 54 in Oregon, and 52 in New Hampshire. Three interstate bridges link New Hampshire and Vermont. The remainder are scattered throughout the country. Non-authentic covered bridges—built or covered for visual effect—appear in each state.

Sources: Donovan, Richard T. *World Guide to Covered Bridges*, p. 1.

241 / Where is the longest bridge-tunnel in the world located?

Completed in 1964 after 42 months' work and $200 million and spanning a great distance of open sea, the Chesapeake Bay Bridge-Tunnel is a 17.5 mile (28.2 kilometer) combination of trestles, bridges, and tunnels that connect Norfolk with Cape Charles in Virginia. Its only rival for crossing so much open and deep water is the Zuider Zee Dam in Holland—a road-carrying structure of similar length, but without the same water depth and same length of open sea.

Sources: Hawkes, Nigel. *Structures*, p. 232; Stephens, John H. *The Guinness Book of Structures*, p. 260.

242 / What is the world's longest road tunnel?

The two-lane road tunnel at St. Gotthard Pass between Göschenen, Switzerland and Airolo, Italy, constructed between 1970 and 1980, is 10.1 miles (16.3 kilometers) long. The Mont Blanc tunnel, completed in 1965, is the next longest road tunnel. Its two lanes connect the 7.2 miles (11.6 kilometers) between Chamonix, France, and the Valle d'Aosta, Italy. The longest road tunnel in the United Kingdom is the four-lane Mersey Tunnel, joining Liverpool and Birkenhead, Merseyside. It is 2.13 miles (3.43 kilometers) long. In the United States, the longest road tunnel is the 2.5 mile (4.02 kilometer) Lincoln tunnel, connecting New York City and New Jersey, and constructed under the Hudson River in 1937.

Sources: *Guinness Book of Records 1992*, p. 324; Hawkes, Nigel. *Structures*, p. 160; Stephens, John H. *The Guinness Book of Structures*, p. 113.

243 / How are United States highways numbered?

The main north-south interstate highways always have odd numbers of one or two digits. The system, beginning with Interstate 5 on the West Coast, increases in number as it moves eastward, and ends with Interstate 95 on the East Coast.

The east-west interstate highways have even numbers. The lowest numbered highway begins in Florida with Interstate 4, increasing in number as it moves northward, and ends with Interstate 96. Coast-to-coast east-west interstates, such as Routes 10, 40, and 80, end in zero. An interstate with three digits is either a beltway or a spur route.

U.S. routes follow the same numbering system as the interstate system, but increase in number from east to west and from north to south. U.S. Route 1, for example, runs along the East Coast; U.S. Route 2 runs along the Canadian border. U.S. route numbers may have anywhere from one to three digits.

Sources: Blocksma, Mary. *Reading the Numbers*, pp. 97-98.

244 / Which state has the most miles of roads?

Texas lays claim to the most road mileage with 305,692 miles (491,858 kilometers) lacing through the Lone Star State. California is a distant second with 164,298 miles (264,355 kilometers) of roads. On the other end of the scale, only three states have less than 10,000 miles (16,090 kilometers) of roads: Delaware, Hawaii, and Rhode Island. The total road milage count for the United States is 3,876,501 miles (6,237,290 kilometers).

Sources: *MVMA Motor Vehicle Facts and Figures '91*, p. 84.

245 / When was the first United States coast-to-coast highway built?

Completed in 1923, after ten years of planning and construction, the Lincoln Highway was the first transcontinental highway connecting the Atlantic coast (New York) to the Pacific (California). Its original length of 3,389 miles (5,453 kilometers) was later shortened by relocations and improvements to 3,143 miles (5,057 kilometers). Crossing 13 states—New York, New Jersey, Pennsylvania, Ohio, Indiana, Illinois, Nebraska, Colorado, Wyoming, Utah, Nevada, and California—in 1925, the Lincoln Highway became, for most of its length, U.S. Route 30.

Sources: Kane, Joseph N. *Famous First Facts*, p. 549; Labatut, Jean, and J.L. Wheaton. *Highways in Our National Life*, p. 94.

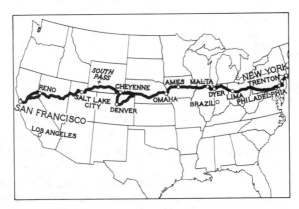

Lincoln Highway

246

Which city had the first traffic light?

On December 10, 1868, the first traffic light was erected on a 22-foot-high (6.7-meter) cast-iron pillar at the corner of Bridge Street and New Palace Yard off Parliament Square in London, England. Invented by J.P. Knight, a railway signalling engineer, the light was a revolving lantern illuminated by gas, with red and green signals. It was turned by hand using a lever at the base of the pole.

Cleveland, Ohio installed an electric traffic signal at Euclid Avenue and 105th Street on August 5, 1914. It had red and green lights with a warning buzzer as the color changed.

Around 1913, Detroit, Michigan used a system of manually-operated semaphores. Eventually the semaphores were fitted with colored lanterns for night traffic. New York City installed the first three-color light signals in 1918; these signals were still operated manually.

Sources: De Bono, Edward. *Eureka!*, P. 39; Robertson, Patrick. *The Book of Firsts*, pp. 191-192; *Traffic Engineering Handbook*, pp. xi-xii.

247

Who is known as the father of civil engineering?

Thomas Telford (1757-1834), the first President of the Institute of Civil Engineers, is the father figure of the British civil engineering profession. Telford established the professional ethos and tradition of the civil engineer—a tradition followed by all engineers today. He built bridges, roads, harbors, and canals. His greatest works include the Menai Strait Suspension Bridge and Pont y Cysyllte Aqueduct, the Gotha Canal in Sweden, the Caledonian Canal, and many Scottish roads. He was the first and greatest master of the iron bridge.

Sources: Ellis, Keith. *Thomas Telford*, p. 15; Stephens, John H. *The Guinness Book of Structures*, pp. 224-225.

248

Why are manhole covers round?

Circular covers are almost universally used on sewer manholes because they cannot drop through the opening. The circular cover rests on a lip that is smaller than the cover. Any other shape—square, rectangle, etc.—could slip into the manhole opening.

Sources: Babbitt, Harold E. *Sewage and Sewage Treatment*, p. 127.

BUILDING MATERIALS See: *METALS AND OTHER MATERIALS—MAN-MADE PRODUCTS*

BUILDINGS AND BUILDING PARTS

249
How much does the leaning tower of Pisa lean?

The leaning tower of Pisa, 184.5 feet (56.23 meters) tall, is about 17 feet (5.18 meters) out of perpendicular, increasing by about 1/20 inch (1.25 millimeters) a year. The Romanesque-style tower was started by Bonanno Pisano in 1173 as a campanile or bell tower for the nearby Baptistry, but was not completed until 1372. Built entirely of white marble, with eight tiers of arched arcades, it began to lean during construction. Although the foundation was dug down to 10 feet (3.04 meters) the builders did not reach bedrock for a firm footing. Ingenious attempts were made to compensate for the tilt by straightening up the subsequent stories and making the pillars higher on the south side than on the north. During the 1960s, cement was added to the foundation to strengthen it, but it is still threatened with collapse.

Sources: Cazeau, Charles J. *Science Trivia*, p. 65; Hawkes, Nigel. *Structures*, p. 228; Stephens, John H. *The Guinness Book of Structures*, p. 104.

250
Which building has the largest non-air-supported clear span roof?

The Suncoast Dome in St. Petersburg, Florida, completed in 1990, has a clear span of 688 feet (209.7 meters). Covering an area of 372 thousand square feet (34,570.1 square meters), the fabric–covered dome is a cable roof structure—the newest structural system for trussed domes. Its structural behavior is exactly opposite that of the traditional dome: the base ring is in compression rather than tension, and the ring at the crown is in tension rather than compression. Also, the space enclosed is not totally free and unobstructed, but includes structural, top-to-bottom members. To retain its characteristic lightness, the roofing surface employs flexible fabric membranes. These must be flexible since cable structures undergo major structural distortions, and they actually change shape under different load conditions. The largest air-supported building is the 80,600-capacity octagonal Pontiac Silverdome Stadium in Pontiac, Michigan. 522 feet (159.1 meters) wide and 722 feet (220.1 meters) long, it is supported by a 10 acre (1.62 hectare) translucent fiberglass roof.

Sources: *Guinness Book of Records 1992*, p. 306; Melaragno, Michele. *An Introduction to Shell Structures*, pp. 322-324; *Yearbook of Science and the Future*, pp. 286-287.

251
Who invented the geodesic dome?

A geodesic line is the shortest distance between two points across a surface. If that surface is curved, a geodesic line across it will usually be curved as well. A geodesic line on the surface of a sphere will be part of a great circle. Buckminster Fuller (1895-1983) realized that the surface of a sphere could be divided into triangles by a network of geodesic lines and that structures could be designed so that their main elements either followed those lines or were joined along them.

This is the basis of his very successful geodesic dome: a structure of generally spherical form, constructed of many light, straight structural elements in tension, arranged in a framework of triangles to reduce stress and weight. These contiguous tetrahedrons are made from lightweight alloys with high tensile strength.

An early example of a geodesic structure is Britain's Dome of Discovery, built in 1951. It was the first dome ever built with principal framing members intentionally aligned along great circle arcs. The ASM (American Society for Metals) dome built east of Cleveland, Ohio, in 1959-1960 is an open lattice-work geodesic dome. Built in 1965, the Houston Astrodome forms a giant geodesic dome.

Sources: *Encyclopedia Americana*, vol. 12, p. 431; Morgan, George W. *Geodesic and Geodetic Domes and Space Structures*, pp. 32-34; Pugh, Anthony. *Polyhedra*, p. 56.

Geodesic dome

▽ **252** /
What is the ground area of the Pentagon Building in Washington, D.C.?

The Pentagon, the headquarters of the United States Department of Defense, is the world's largest office building in terms of ground space. Its construction was completed on January 15, 1943—in just over 17 months. With a gross floor area of over 6.5 million square feet, this five-story, five-sided building has three times the floor space of the Empire State Building and is 1½ times larger than the Sears Tower in Chicago. The World Trade Center complex in New York City, completed in 1973, is larger with over 9 million square feet (836 thousand square meters), but it basically is two structures (towers). Each of the five sides is 921 feet (281 meters) long with a perimeter of 4,610 feet (1,405 meters). The Secretary of Defense, the

Secretaries of the three military departments, and the military heads of the Army, Navy, and Air Force are all located in the Pentagon. The National Military Command Center, which is the nation's military communications hub, is located in the area of the Joint Chiefs of Staff. It is commonly called the "war room."

The largest commercial building in the world under one roof is the flower auction building of the cooperative VBA in Aalsmeer, Netherlands. In 1986, the floor plan was extended to 91.05 acres (36.85 hectares). It measures 2,546 x 2,070 feet (776 x 639 meters). The largest building in the United States, and the largest assembly plant in the world, is the Boeing 747 assembly plant in Everett, Washington, with a capacity of 200 million cubic feet (5.5 million cubic meters) and covering 47 acres (19 hectares).

Sources: *Guinness Book of Records 1992*, pp. 301-302; *New Encyclopedia Brittanica*, vol. 9, p. 267; *The Pentagon: A National Institution*, p. 9, 13, 17.

253 / When was the first skyscraper built?

Designed by William Le Baron Jenney (1832-1907), the first skyscraper, the 10-story Home Insurance Company Building in Chicago, Illinois was completed in 1885. A skyscraper—a very tall building supported by an internal frame (skeleton) of iron and steel rather than by load-bearing walls—maximizes floor space on limited land. Three technological developments made skyscrapers feasible: a better understanding of how materials behave under stress and load (from engineering and bridge design); the use of steel or iron framing to create a structure, with the outer skin "hung" on the frame; and the introduction of the first "safety" passenger elevator, invented by Elisha Otis (1811-1861).

Sources: Bagenal, Philip, and Jonathan Meades. *The Illustrated Atlas of the World's Great Buildings*, p. 161; Berliner, Barbara. *The Book of Answers*, p. 72.

254 / What are the tallest buildings in the United States?

The following buildings are the tallest United States buildings and the tallest in the world:

Building	Location	Year built	Feet	Meters	Stories
Sears Tower	Chicago, Illinois	1974	1454	443	110
plus two T.V. towers			1707	520.9	
World Trade Center	New York, New York				
North tower		1972	1368	417	110
South tower		1973	1362	415	110
Empire State Bldg	New York, New York	1931	1250	381	102
plus two T.V. towers			1414	431	

The tallest self-supporting structure in the World is the CN Tower in Toronto, Canada at 1815 feet (533 meters), built in 1975. The tallest structure in the world has been the Guyed Warszawa Radio mast at Konstantynow, Poland at 2,121 feet (646.4 meters). However, on August 10, 1991, it fell during renovation.

Sources: *Guinness Book of Records 1992*, pp. 302-303, 317, 802; *The World Almanac and Book of Facts 1992*, pp. 616-622.

255 ▽ What is a yurt?

Originally a Mongolian hut, the yurt has been adapted in the United States as a low-cost structure which can be used as a dwelling. The foundation is built of wood on a hexagonal frame. The wooden lattice-work side walls have a tension cable of ³⁄₁₆ inch airplane cable sandwiched between the wall pieces at the top to keep the walls from collapsing. The walls are insulated and covered with boards, log slabs, canvas, or aluminum siding. A shingle roof, electricity, plumbing, and a small heating stove may be installed. The interior can be finished as desired with shelves, room dividers, and interior siding. The yurt is beautiful, practical, and relatively inexpensive to erect.

Sources: Charney, Leonard. *Build a Yurt*, p. 134; Churchill, James E. *Backyard Building Book*, pp. 76-84.

256 ▽ Why are Phillips screws used?

The recessed head and cross-shaped slots are self-centering and allow a closer, tighter fit than conventional screws. Straight slotted screws can cause the screwdriver to slip out of the groove, ruining the wood, etc.

Screws were used in carpentry as far back as the 16th century, but slotted screws with a tapering point were made at the beginning of the 19th century. The great advantage of screws over nails is that they are extremely resistant to longitudinal tension. Larger-size screws that require considerable force to insert have square heads that can be tightened with a wrench. Screws provide more holding power than nails and can be withdrawn without damaging the material. Types of screws include wood screws, lag screws (longer and heavier than wood screws), expansion anchors (for masonry usually), and sheet metal screws. The screw sizes vary from ¼ to 6 inches.

Sources: Feldman, David. *Why Do Clocks Run Clockwise? and Other Imponderables*, pp. 206-207; Guedes, Pedro. *The Macmillan Encyclopedia of Architecture and Technological Change*, p. 209; U.S. Department of the Army. Headquarters. *Carpenter*, pp. 3-6 to 3-8.

257
What is a clerestory?

In architecture, a clerestory is a windowed wall that extends above the ceiling-level of a room or hall, providing lighting and sometimes ventilation. The most highly developed of the many types of clerestories are found in Gothic churches. The earliest known example of the use of the clerestory is at the great Temple of Amon-Ra at Karnak, Egypt in its hypostyle hall (a hall in which the roof is supported only by columns, without vaults or domes).

Sources: Bagenal, Philip, and Jonathan Meades. *The Illustrated Atlas of the World's Great Buildings*, pp. 13, 188; Putnam, R.E., and G.E. Carlson. *Architectural and Building Trades Dictionary*, p. 75; *Encyclopedia Americana*, vol. 7, p. 54.

Clerestory level above roofline

258
What is an STC rating and what does it mean?

The STC (Sound Transmission Class) tells how well a wall or floor assembly inhibits airborne sound. The higher the number, the greater the sound barrier. Typical STC ratings are as follows:

STC number	What it means
25	Normal speech can be easily understood
30	Loud speech can be understood
35	Loud speech is audible, but not intelligible
42	Loud speech audible as a murmur
45	Must strain to hear loud speech
48	Some loud speech barely audible
50	Loud speech not audible

The degree to which sound barriers work depends greatly on the elimination of air gaps under doors, through electrical outlets, and around heating ducts.

Sources: *Progressive Builder*, vol. 12, no. 2 (February 1987), p. 48.

259 / Which part of a door is a doorjamb?

The doorjamb is not part of the door, but is the surrounding case into and out of which the door opens and closes. It consists of two upright pieces, called side jambs, and a horizontal head jamb.

Sources: Sherwood, Gerald E., and Robert C. Stroh. *Wood-Frame House Construction*, pp. 116, 248.

Door parts

260 / Why is the term "penny" used in nail sizes?

The term "penny" is a measurement relating to the length of nails, originating in England. One explanation is that it refers to cost, with the cost of 100 nails of a certain size being 10 pence or 10d ("d" being the British symbol for a penny). Another explanation suggests that the term refers to the weight of 1,000 nails, with "d" at one time being used as an abbreviation for a pound in weight.

Mankind has been using nails for the last 5,000 years or so. Nails are known to have been used in Ur (ancient Iraq) to fasten together sheet metal. Before 1500, nails were made by hand by drawing small pieces of metal through a succession of graded holes in a metal plate. In 1741, 60,000 people were employed in England making nails.

The first nail-making machine was invented by the American Ezekiel Reed. In 1851, Adolphe F. Brown of New York invented a wire-nail-making machine. This enabled nails to be mass-produced cheaply.

Sources: Carter, E.F. *Dictionary of Inventions and Discoveries*, p. 122; Hooper, Meredith. *Everyday Inventions*, pp. 3-5; Putnam, Robert. *Builder's Comprehensive Dictionary*, p. 290.

Size	2d	3d	4d	5d	6d	7d	8d	9d	10d	12d	16d	20d	30d	40d	50d	60d
Diameter (inches)	.072	.08	.098	.098	.113	.112	.131	.131	.148	.148	.162	.192	.207	.225	.244	.262
Steel wire gage	15	14	12	12	11	11	10	10	9	9	8	6	5	4	3	2
Number per pound	900	615	322	254	200	154	106	85	74	57	46	29	23	17	14	11

Sizes of penny-nails

▷ **261** ◁

When was the first shopping center built?

The first shopping center in the world was built in 1896 at Roland Park, Baltimore, Maryland. One of the world's largest shopping centers is the West Edmonton Mall in Alberta, Canada, which covers 5.2 million square feet (480 thousand square meters) on a 121 acre (49 hectare) site. It has 828 stores and services with parking for 20 thousand vehicles.

Sources: *Guinness Book of Records 1992*, p. 305.

▷ **262** ◁

What is the "topping out" party that an iron worker does?

When the last beam is placed on a new bridge, skyscraper, or building, ironworkers hoist up an evergreen tree, attach a flag or a handkerchief, and brightly paint the final beam and autograph it.

This custom of raising an evergreen tree goes back to Scandinavia in 700 C.E. when attaching the tree to the building's ridge pole signalled to all who helped, that the celebration of its completion would begin.

Sources: *Stone*, vol. 91, no. 6 (Winter, 1971), p. 5.

263 / *How does a chimney differ from a flue?*

A chimney is a brick and masonry construction which contains one or more flues. A flue is a passage within a chimney through which smoke, fumes, and gases ascend. A flue is lined with clay or steel to contain the combustion wastes. By channeling the warm, rising gases, a flue creates a draft that pulls the air over the fire and up the flue. Each heat source needs its own flue, but one chimney can have several flues.

Sources: *How Things Work in Your Home*, p. 299.

264 / *What is the BOCA code?*

The Building Officials and Code Administrators (BOCA) International is a service organization, which issues a series of model regulatory construction codes for the protection of public health, safety, and welfare. The codes are published in sections, such as the National Building Code, National Plumbing Code, National Fire Prevention Code, etc. They are designed for adoption by state or local governments and may be amended or modified to accomplish desired local requirements.

Sources: *The BOCA National Building Code 1990*, inside front cover and p. iii.

265 / *What name is given to the the covering at the top of a concrete wall?*

A coping is the protective covering over any vertical construction, such as a wall or chimney. A coping may be made of masonry, metal, or wood and is usually sloping or beveled to shed water in such a way that it does not run down the vertical face of the wall.

Sources: *Means Illustrated Construction Dictionary*, p. 121.

266 / What is crown molding?

Crown molding is a wood, metal, or plaster finishing strip placed on the wall where it intersects the ceiling. If the molding has a concave face, it is called a cove molding. In inside corners it must be cope-jointed to insure a tight joint.

Sources: Putnam, Robert E. *Builder's Comprehensive Dictionary*, p. 103.

267 / How many acres of trees are used in the construction of a single-family home?

On the average, an acre of softwood forest is used for a typical two thousand square-foot house.

Sources: *Harrowsmith Country Life*, vol. 7, no. 37 (January/February 1992), p. 10.

268 / When did Sears, Roebuck and Company sell homes by mail order?

Sears sold mail-order homes from 1909 to 1937. All construction materials and detailed plans for electric wiring were included. The price for the precut homes ranged from $595 to $5,000.

Sources: DuVall, Nell. *Domestic Technology*, p. 251.

269 / What is R-value?

An R-value, or resistance to the flow of heat, is a special measurement of insulation which represents the difficulty with which heat flows through an insulating material. The higher the R-value, the greater the insulating value of the material.

A wall's component R-values can be added up to get its total R-value:

Wall component	R-value
Inside Air Film	0.7
½" Gypsum Wallboard	0.5
R-13 Insulation	13.0
½" Wood Fiber Sheathing	1.3
Wood Siding	0.8
Outside Air Film	0.2
Total R-value	16.5

Building component	R-value
Standard attic ceiling	19
Standard 4-inch-thick insulated wall	11
Typical single-pane glass window	1
Double-glazed window	2
"Superwindows" (inner surface of a pane coated with an infrared radiation reflector material such as tin oxide and argon gas filling in the space between the panes of a double-glazed window.)	4

Sources: Brennan, Richard P. *Dictionary of Scientific Literacy*, pp. 155-156; Nunn, Richard V. *Saving Home Energy*, p. 6; *Thermal and Sound Control*, p. 12.

MISCELLANEOUS STRUCTURES

270 *Will building a seawall protect a beach?*

It may for a while, but during a storm, the sand cannot follow its natural pattern of allowing waves to draw the sand across the lower beach, making the beach flatter. With a seawall, the waves carry off more sand, dropping it into deeper water. A better alternative to the seawall is the revetment. This is a wall of boulders, rubble, or concrete block, tilted back away from the waves. It imitates the way a natural beach flattens out under wave attack.

Sources: Kaufman, Wallace. *The Beaches Are Moving*, pp. 207-208, 211.

271 *How tall is the figure on the Statue of Liberty, and how much does it weigh?*

The Statue of Liberty, conceived and designed by the French sculptor Fréderic-Auguste Bartholdi (1834-1904), was given to the United States to commemorate its first centennial. Called (in his patent, U.S. Design Patent no. 11,023, issued February 18, 1879) "Liberty Enlightening the World," it is 152 feet (46.3 meters) high, weighs 225 tons (204 metric tons), and stands on a pedestal and base that are 151 feet (46 meters) tall. Its flowing robes are made from more than 300 sheets of hand-hammered copper over a steel frame. Constructed and finished in France in 1884, the statue's exterior and interior were taken apart piece-by-piece, packed into 200 mammoth wooden crates, and shipped to the United States in May 1885. The statue was placed by Bartholdi on Bedloe's Island, at the mouth of New York City Harbor. On October 28, 1886, ten years after the centennial had passed, the inauguration celebration was held.

It was not until 1903 that the inscription "Give me your tired, your poor, Your huddled masses yearning to breathe free ..." was added. The verse was taken from *The New Colossus* composed by New York City poet Emma Lazarus in 1883. The statue, the tallest in the United States, and the tallest metal statue in the world, was recently refurbished for its own centennial, at a cost of $698 million, reopening on July 4, 1986. One visible difference is the flame of her torch is now 24-karat gold-leaf, just as in the original design. In 1916 the flame was redone into a lantern of amber glass. Concealed within the rim of her crown is the observation deck that can be reached by climbing 171 steps or by taking the newly installed hydraulic elevator.

Sources: *Guinness Book of Records 1992*, pp. 333-334; *How in the World*, pp. 359-361; Panati, Charles. *Panati's Extraordinary Origins of Everyday Things*, pp. 291-293.

272 / Who invented the Ferris Wheel and when?

Originally called "pleasure wheels," the first such rides were described by English traveller Peter Mundy in 1620. In Turkey he saw a ride for children consisting of two vertical wheels, 20 feet (6.1 meters) across, supported by a large post on each side. Such rides were called "ups-and-downs" at the St. Bartholomew Fair of 1728 in England, and in 1860, a French pleasure wheel was turned by hand and carried sixteen passengers. They were also in use in the United States by then, with a larger, wooden wheel operating at Walton Spring, Georgia.

Wanting a spectacular attraction to rival that of the 1889 Paris Centennial celebration—the Eiffel Tower—the directors of the 1893 Columbian Exposition had a design competition. The prize was won by the American bridge builder George Washington Gale Ferris (1859-1896). In 1893 he designed and erected a gigantic revolving steel wheel whose top reached 264 feet (80.5 meters) above ground. The wheel—825 feet (251.5 meters) in circumference, 250 feet (76.2 meters) in diameter, and 30 feet (9.1 meters) wide—was supported by two 140-foot (42.7 meters) towers. Attached to the wheel were 36 cars, each able to carry 60 passengers. Opening on June 21, 1893 at the exposition in Chicago, Illinois, it was extremely successful. Thousands lined up to pay 50 cents for a 20-minute ride—a large sum in those days, considering that a merry-go-round ride cost only 4 cents. In 1904 it was moved to St. Louis, Missouri, for the Louisiana Purchase Exposition. It was eventually sold for scrap. The largest-diameter wheel currently operating is the Cosmoclock 21 at Yokohama City, Japan. It is 344.48 feet (105 meters) high and 328 feet (100 meters) in diameter.

Sources: Anderson, Norman D., and Walter R. Brown. *Ferris Wheels*, pp. 7-23; *Guinness Book of Records 1992*, p. 313; Harris, Harry. *Good Old-Fashioned Yankee Ingenuity*, pp. 125-126.

273 / How many roller coasters are there in the United States, and how many people ride them?

There were 164 roller coasters in the United States in 1989 and an estimated 214 million people rode them during that year. Roller coasters have had a long history in thrill-giving. During the 15th and 16th centuries, the first known gravity rides were built in St. Petersburg and were called "Russian Mountains." A wheeled roller coaster, called the "Switchback," was used as early as 1784 in Russia. By 1817, the first roller coaster with cars locked to the tracks operated in France. The first United States roller coaster patent was granted to J.G. Taylor in 1872, and LaMarcus Thompson built the first known roller coaster in the United States at Coney Island, Brooklyn, New York in 1884. Until recently, the longest roller coaster in the world was "The Beast" at Kings Island, Ohio, which has a run of 1.4 miles (2.25 kilometers) with 800 feet (243.84 meters) of tunnels. The current record-holder is the "Ultimate" at Lightwater Valley, Ripon, Great Britain, with a run of 1.42 miles (2.28 kilometers).

Sources: Cartnell, Robert. *The Incredible Scream Machine: A History of the Roller Coaster*, pp. 4-5, 19-34; *Guinness Book of Records 1992*, pp. 313-314, 807; *Smithsonian*, vol. 20, no. 5 (August 1989), pp. 82-92.

274 / What are the dimensions of the Eiffel Tower?

Now France's most instantly recognizable landmark, the Eiffel Tower was built in Paris during 1887-1889 by Gustave Eiffel (1832-1923), a French engineer, for the 100th anniversary of the French Revolution. For this momentous occasion, for which a huge exhibition, Exposition Universelle, was planned, Eiffel wished to erect something extraordinary made from modern materials. His design, a structure of wrought iron ribs, held together by rivets and resting on a solid masonry foundation, was initially denounced by some as a "tragic lamppost," but its elegance has come to be recognized by most.

It stands 985 feet 11 inches (300.5 meters) on a square base measuring 328 feet (100 meters) on each side. The first story platform is at 189 feet (57.6 meters), with sides 232 feet (70.7 meters) long. The second story platform is at 379 feet (115.5 meters), with sides 134 feet (40.8 meters) long. There is an intermediate platform at 643 feet (196 meters)—this is only a point to change elevators. The third story platform is at 905 feet (275.8 meters). Later a 66-foot (20-meter) television antenna was attached, which extended the tower's total height to 1052 feet (320.7 meters).

Sources: Poirier, René. *The Fifteen Wonders of the World,* pp. 242, 248.

Eiffel Tower

275 / How big is the Hoover Dam?

Formerly called Boulder Dam, the Hoover Dam is located between Nevada and Arizona on the Colorado River. The highest concrete arch dam in the United States, the dam is 1,244 feet (379 meters) long and 726 feet (221 meters) high. It has a base thickness of 660 feet (201 meters) and a crest thickness of 45 feet (13.7 meters). It stores 21.25 million acre-feet of water in the 115-mile-long (185-kilometer) Lake Mead reservoir.

The dam was built because the Southwest was faced with constantly recurring cycles of flood and drought. Uncontrolled, the Colorado River had limited value, but once regulated, the flow would assure a stabilized, year-round water supply and the low-lying valleys would be protected against floods. On December 21, 1928, the Boulder Canyon Project Act became law, and the project was completed on September 30, 1935—two years ahead of schedule. For 22 years, Hoover Dam was the highest dam in the world.

Sources: U.S. Department of the Interior. *The Story of the Hoover Dam,* pp. 8, 15, 35.

276 / Where is the highest dam in the United States and what is its capacity?

Oroville, the highest dam in the United States, is an earth-fill dam that rises 754 feet (229.8 meters) and extends more than a mile across the Feather River, near Oroville, California. Built in 1968, it forms a reservoir containing about 3.5 million acre-feet (4.3 million cubic meters) of water. The next highest dam in the United States is the Hoover Dam on the Colorado River, on the Nevada-Arizona border. It is 726 feet (221.3 meters) high, and was, for 22 years, the world's highest. Presently, there are nine dams higher than the Oroville—the highest currently is the 984-foot (299.9–meter) Nurek Dam on the Vakhsh River in Tadzhikistan near the Afghanistan border. But it will be surpassed by the 1,098-foot (334.7-meter) Rogun(skaya) earth-fill dam, currently under construction, which also crosses the Vaksh River in Tadzhikistan.

Sources: *Guinness Book of Records 1992*, p. 322; Stephens, John H. *The Guinness Book of Structures*, pp. 61-64; *World Book Encyclopedia*, vol. 14, p. 864.

ROADS *See: BRIDGES, BUILDINGS, ETC.—BRIDGES, ROADS, AND TUNNELS*

TUNNELS *See: BRIDGES, BUILDINGS, ETC.—BRIDGES, ROADS, AND TUNNELS*

CARS, BOATS, PLANES, AND TRAINS

AIRCRAFT See also CARS, BOATS, PLANES, AND TRAINS—MILITARY VEHICLES

277 / Who made the first supersonic flight?

Supersonic flight is flight at or above the speed of sound. The speed of sound is 760 miles (1,223 kilometers) per hour in warm air at sea level. At a height of about 37,000 feet (11,278 kilometers), its speed is only 660 miles (1,062 kilometers) per hour. The first person credited with reaching the speed of sound (Mach 1) was Major Charles E. (Chuck) Yeager (b. 1923), of the United States Air Force. In 1947, he attained Mach 1.45 at 60,000 feet (18,288 meters) while flying the Bell *X-1* rocket research plane. This plane had been carried aloft by a B-29 and released at 30,000 feet (9,144 meters). In 1949, the Douglas *Skyrocket* became the first supersonic jet-powered aircraft to reach Mach 1 when Gene May flew at Mach 1.03 at 26,000 feet (7,925 meters).

Sources: *Encyclopedia of Aviation*, pp. 184-185.

278 / What is the maximum seating capacity in a Boeing 747?

The seating capacity of the 747 and some other jets servicing cities in the United States are given below.

Airplane	Maximum seating capacity
Boeing 707	179
Boeing 707-320, 707-420	189
Boeing 720	149
Boeing 727	125
Boeing 747	498
Boeing 757	196
Boeing 767	289
Boeing 777	375

Airplane	Maximum seating capacity
Concorde (SST)	110
Lockheed L-1011 TriStar	345
McDonnell Douglas DC-8	189
McDonnell Douglas DC-9	
Series 20	119
Series 30 & 40	125
Series 50	139
McDonnell Douglas DC-10	380
Tupolev Tu-144 (Soviet SST)	140

Sources: Rovin, Jeff. *Laws of Order*, pp. 206-207.

▷ 279 ◁ What is avionics?

Avionics, a term derived by combining aviation and electronics, describes all of the electronic navigational, communications, and flight management aids with which airplanes are equipped today. In military aircraft it also covers electronically-controlled weapons, reconnaissance, and detection systems. Until the 1940s, the systems involved in operating aircraft were purely mechanical, electric, or magnetic, with radio apparatus being the most sophisticated instrumentation. The advent of radar and the great advance made in airborne detection during World War II led to the general adoption of electronic distance-measuring and navigational aids. In military aircraft such devices improve weapon delivery accuracy and in commercial aircraft they provide greater safety in operation.

Sources: *Encyclopedia of Aviation*, pp. 20-21; *The Illustrated Science and Invention Encyclopedia*, vol. 2, pp. 206-210; *Science and Technology Illustrated*, vol. 3, p. 328

▷ 280 ◁ Who made the first nonstop transatlantic flight?

The first nonstop flight across the Atlantic Ocean, from Newfoundland, Canada, to Ireland, was made by two British aviators, Capt. John W. Alcock (1892-1919) and Lt. Arthur W. Brown (1886-1948), on June 14-15, 1919. The aircraft, a converted twin-engined Vickers Vimy bomber, took 16 hours 27 minutes to fly 1,890 miles (3,032 kilometers). Later Charles A. Lindbergh (1902-1974) made the first solo crossing flight on May 20-21, 1927 in the single-engined Ryan monoplane *Spirit of St. Louis*. His flight from New York to Paris covered a distance of 3,609 miles (5,089 kilometers) and lasted 33½ hours. The first woman to fly solo across the Atlantic was Amelia Earhart (1897-1937) who flew from Newfoundland to Ireland May 20-21, 1932.

Sources: *Encyclopedia of Aviation*, pp. 12-13, 64, 119.

281

When was the first non-stop, unrefueled airplane flight around the world?

Dick Rutan (b. 1943) and Jeana Yeager (b. 1952) flew the *Voyager*, a trimaran mono-plane, in a closed circuit loop westbound and back to Edwards Air Force Base, California, December 14-23, 1986. The flight lasted 9 days, 3 minutes, 44 seconds and covered 24,986.7 miles (40,203.6 kilometers). The first successful round-the-world flight was made by two Douglas World Cruisers between April 6 and September 28, 1924. Four aircraft originally left Seattle, Washington, and two went down. The two successful planes completed 27,553 miles (44,332.8 kilometers) in 175 days—with 371 hours 11 minutes being their actual flying time. Between June 23 and July 1, 1931, Wiley Post (1900-1935) and Harold Gatty (1903-1957) flew around the world, starting from New York, in their Lockheed Vega, *Winnie Mae*.

Sources: Mondey, David. *The Guinness Book of Aircraft*, pp. 103, 108-109, 126.

282

What is the world record for the fastest kite?

On May 17, 1987, Troy Vickstrom piloted a speeding 10-foot (3.05-meter) Flexifoil across the beach in Lincoln City, Oregon. Documentation of the record came from the local police, who issued Vickstrom a traffic citation for exceeding the beach's posted speed limit of 20 miles (32.18 kilometers) per hour. The fastest speed attained by a flying kite was 120 miles (193 kilometers) per hour on September 22, 1989 by Pete DiGiacomo at Ocean City, Maryland.

Sources: Eden, Mazwell. *Kiteworks*, p. 93; *Guinness Book of Records 1992*, p. 400.

283

Where is the black box carried on an airplane?

Actually painted bright orange to make it more visible in an aircraft's wreckage, the black box is a tough metal and plastic case containing two recorders. Installed in the rear of the aircraft—the area most likely to survive a crash—the case has two shells of stainless steel with a heat-protective material between the shells. The case must be able to withstand a temperature of 2,000°F (1,100°C) for 30 minutes. Inside it, mounted in a shockproof base, is the aircraft's flight data and cockpit voice recorders. The flight data recorder provides information about airspeed, direction, altitude, acceleration, engine thrust, and rudder and spoiler positions from sensors that are located around the aircraft. The data is recorded as electronic pulses on stainless steel tape which is about as thick as aluminum foil. When the tape is played back, it generates a computer printout. The cockpit voice recorder records the previous 30 minutes of the flight crew's conversation and radio transmission on a continuous tape loop. If a crash does not stop the recorder, vital information can be lost.

Sources: Flatow, Ira. *Rainbows, Curve Balls and Other Wonders of the Natural World Explained*, p. 231; *How In the World?*, pp. 99-100; Tunnell, James E. *Latest Intelligence*, p. 27.

284 / When was the first full-scale wind tunnel for testing airplanes used?

It began operations on May 27, 1931, at the Langley Research Center of the National Advisory Committee for Aeronautics, Langley Field, Virginia. This tunnel, still in use, is 30 feet (9.14 meters) high and 60 feet (18.29 meters) wide. A wind tunnel is used to simulate air flow for aerodynamic measurement; it consists essentially of a closed tube, large enough to hold the airplane or other craft being tested, through which air is circulated by powerful fans.

Sources: Kane, Joseph N. *Famous First Facts*, p. 703; Nayler, Joseph L. *Aviation: Its Technical Development*, p. 268.

285 / Why did the dirigible Hindenburg explode?

Despite the official United States and German investigations into the explosion, it still remains a mystery today. The most plausible explanations are structural failure, St. Elmo's Fire, static electricity, or sabotage. The *Hindenburg*, built following the great initial success of the *Graf Zeppelin*, was intended to exceed all other airships in size, speed, safety, comfort, and economy. At 803 feet (244.75 meters) long, it was 80% as long as the liner *Queen Mary*, and it was 135 feet (41.1 meters) in diameter.

In 1935, the German Air Ministry virtually took over the Zeppelin Company to use it to spread Nazi propaganda. After its first flight in 1936, the airship was very popular with the flying public. No other form of transport could carry passengers so swiftly, reliably, and comfortably between continents. During 1936, 1,006 passengers flew over the North Atlantic Ocean in the *Hindenburg*. On May 6, 1937, while landing at Lakehurst, New Jersey, its hydrogen burst into flames, and the airship was completely destroyed. Of the 97 people aboard, 62 survived.

Sources: Gerken, Louis C. *Airships: History and Technology.* pp. 331-353; Stacey, Tom. *The Hindenburg*, pp. 52, 53.

286 / Who designed the Spruce Goose?

Howard Hughes (1905-1976) designed and built the all-wood H-4 Hercules flying boat, nicknamed the *Spruce Goose*. The aircraft had the greatest wingspan ever built and was powered by eight engines. It was only flown once—covering a distance of less than one mile at Los Angeles harbor on November 2, 1947, lifting only 33 feet (10.6 meters) off the surface of the water.

After the attack on Pearl Harbor on December 7, 1941, and the subsequent entry of the United States into World War II, the United States government needed a large, cargo-carrying airplane that could be made from non-critical wartime materials, such as wood. Henry J. Kaiser (1882-1967), whose shipyards were producing Liberty ships at the rate of one per day, hired Howard Hughes to build such a plane. Hughes eventually produced a plane that weighed 400,000 pounds (181,440 kilograms) and had a wingspan of 320 feet (97.5 meters). Unfortunately, the plane was so complicated that it was not finished by the end of the war. In 1947, Hughes flew the boat himself during its only time off the ground—supposedly just to prove that something that big could fly. Today, the plane is on public exhibition in Long Beach, California.

Sources: *Encyclopedia of Aviation*, p. 97; *Milestones of Aviation*, p. 188.

287 / What is the difference between an amphibian plane and a seaplane?

The primary difference is that an amphibian has retractable wheels that enable it to operate from land as well as water, while a seaplane is limited to water take-offs and landings, having only pontoons without wheels. Because its landing gear cannot retract a seaplane is less aerodynamically efficient than an amphibian.

Sources: *Encyclopedia of Aviation*, pp. 172-173; *World and United States Aviation and Space Records*, p. 138.

BOATS AND SHIPS *See also:* CARS, BOATS, PLANES AND TRAINS—MILITARY VEHICLES

288 / Where does the term "mark twain" originate?

Mark twain is a riverboat term meaning 2 fathoms (a depth of 12 feet or 3.65 meters). A hand lead is used for determining the depth of water where there is less than 20 fathoms. The lead consists of a lead weight of 7 to 14 pounds (3.17 to 6.35 kilograms) and a line of hemp or braided cotton, 25 fathoms (150 feet or 46 meters) in length. The line is marked at 2, 3, 5, 7, 10, 15, 17, and 20 fathoms. The soundings are taken by a leadsman who calls out the depths while standing on a platform projecting from the side of the ship, called "the chains." The number of fathoms always forms the last part of the call. When the depth corresponds to any mark on the lead line, it is reported as "By the mark 7," "By the mark 10," etc. When the depth corresponds to a fathom between the marks on the line, it is reported as "By the deep 6," etc. When the line is a fraction greater than a mark, it is reported as "And a half 7," "And a quarter 5"; a fraction less than a mark is "Half less 7," "Quarter less 10," etc. If bottom is not reached, the call is "No bottom at 20 fathoms."

"Mark Twain" was also the pseudonym chosen by American humorist Samuel L. Clemens. Supposedly, he chose the name because of its suggestive meaning, since it was a riverman's term for water that was just barely safe for navigation. One implication of this "barely safe water" meaning was, as his character Huck

Finn would later remark, "Mr. Mark Twain ... he told the truth, mostly." Another implication was that "barely safe water" usually made people nervous, or at least uncomfortable.

Sources: Cornell, Felix. *American Merchant Seaman's Manual*, p. 433; Cox, James M. *Mark Twain: The Fate of Humor*, pp. 22-24; *World Book Encyclopedia*, vol. 19, p. 528.

289 ▽ Why is the right side of a ship called starboard?

In the time of the Vikings, ships were steered by long paddles or boards placed over the right side. They were known in Old English as *steorbords*, evolving into the word *starboard*. The left side of a ship, looking forward, is called *port*. Formerly the left side was called *larboard*, originating perhaps from the fact that early merchant ships were always loaded from the left side. Its etymology is Scandinavian, being *lade* (load) and *bord* (side). The British Admiralty ordered *port* to be used in place of *larboard* to prevent confusion with *starboard*.

Sources: Funk, Charles E. *Horse Feathers and Other Curious Words*, p. 13; McEwan, W.A., and A.H. Lewis. *Encyclopedia of Nautical Knowledge*, p. 278.

290 ▽ What is the world's largest ship?

Among passenger liners, the cruise ship *Sovereign of the Seas*, built in France for the Norwegian Royal Caribbean Cruise Lines, was the world's largest until 1990, with a gross registered tonnage of 73,192. Completed in 1988, its dimensions are 874 x 105 feet (266.4 x 32 meters) with a depth of 49 feet (14.94 meters). It can accommodate 2,763 passengers plus crew (over 3,000 people total). In the autumn of 1990, the cruise ship *Norway*, built in 1961 as the *France* and renamed in 1979, increased her gross registered tonnage from 70,202 to 76,000 to become not only the longest liner at 1035.5 feet (315.66 meters) but also the largest. However, the largest passenger liner ever built was Cunard's *Queen Elizabeth*, which was 1031 feet (314.25 meters) long and 118.5 feet (36.12 meters) wide with a final gross registered tonnage of 82,998 (originally 83,673). With the last voyage ending on November 15, 1968, the *Queen Elizabeth* in 1970 became a floating marine university, *Seaside University*, in Hong Kong, and burned on January 9, 1972.

The largest ship of any kind currently in service is the tanker *Hellas Fos*, owned by the Bilinder Marine Corporation of Athens, Greece. Built in 1979, this steam turbine tanker has a deadweight tonnage of 611,832.5 (gross registered tonnage of 254,583). The largest ship in the world is the *Happy Giant*, formerly the *Seawise Giant*, with a deadweight tonnage of 622,571.5, a length of 1,504 feet (458.42 meters), and a beam of 226 feet (68.88 meters). Attacked and damaged by Iraqi Mirage jets in the Persian Gulf on December 22, 1987 and again on May 14, 1988, the tanker is in the process of being refitted. The steam turbines are being replaced with a diesel engine, which will decrease her deadweight tonnage to about 420,000 long tons.

Sources: Cairis, Nicholas. *Cruise Ships of the World*, p. 121; *The Guinness Book of Answers*, p. 282; *Guinness Book of Records 1992*, p. 348.

291 / What does the term loadline *mean in shipping?*

A loadline or load waterline is an immersion mark on the hull of a merchant ship. This indicates her safe load limit. The lines vary in height for different seasons of the year and areas of the world. Also called the *Plimsoll Line* or *Plimsoll Mark*, it was accepted as law by the British Parliament in the Merchant Shipping Act of 1875, primarily at the instigation of Samuel Plimsoll (1824-1898). This law prevented unscrupulous owners from sending out unseaworthy and overloaded, but heavily insured, vessels (so-called "coffin ships"), which risked the crew's lives.

Sources: *Encyclopedia Americana*, vol. 17, p. 632; Peters, George H. *The Plimsoll Line*, pp. 129, 184.

292 / How is a ship's tonnage calculated?

Tonnage of a ship is not necessarily the number of tons that the ship weighs. There are at least six different methods of rating ships; the most common are:

Displacement tonnage—used especially for warships and U.S. merchant ships—is the weight of the water displaced by a ship. Since a ton of sea water occupies 35 cubic feet, the weight of water displaced by a ship can be determined by dividing the cubic footage of the submerged area of the ship by 35. The result is converted to long tons (2,240 pounds). Loaded displacement tonnage is the weight of the water displaced when a ship is carrying its normal load of fuel, cargo, and crew. Light displacement tonnage is the weight of water displaced by the unloaded ship.

Gross tonnage (GRST) or *gross registered tonnage* (GRT)—used to rate merchant shipping and passenger ships—is a measure of the enclosed capacity of a vessel. It is the sum in cubic feet of the vessel's enclosed space divided by 100 (100 such cubic feet is considered one ton). The result is gross (registered) tonnage. For example, the old *Queen Elizabeth* did not weigh 83,673 tons, but had a capacity of 8,367,300 cubic feet.

Deadweight tonnage (DWT)—used for freighters and tankers—is the total weight in long tons (2,240 pounds) of everything a ship can carry when fully loaded. It represents the amount of cargo, stores, bunkers, and passengers that are required to bring a ship down to her loadline, i.e. the carrying capacity of a ship.

Net registered tonnage (NRT)—used in merchant shipping—is the gross registered tonnage minus the space that cannot be utilized for paying passengers or cargo (crew space, ballast, engine room, etc.).

Sources: Blocksma, Mary. *Reading the Numbers*, p. 160; Burnam, Tom. *The Dictionary of Misinformation*, p. 228; *The Guinness Book of Answers*, p. 281.

293
Which type of wood was used to build Noah's ark?

According to the Bible, Noah's ark was made of gopher wood. This is identified as *Cupressus sempervirens*, one of the most durable woods in the world. Also called the Mediterranean cypress, the tree is native throughout southern Europe and western Asia. It grows up to 80 feet (24.38 meters) tall. Similar to this tree is the Monterey cypress (*Cupressus macrocarpa*), which is restricted to a very small area along the coast of central California. It can become as tall as 90 feet (27.43 meters) with horizontal branches that support a broad, spreading crown. When old, this tree looks very much like the aged cedars of Lebanon.

Sources: Everett, Thomas H. *Living Trees of the World*, p. 35; Platt, Rutherford. *1001 Questions Answered About Trees*, p. 31.

294
What is dead reckoning?

Dead reckoning is the determination of a craft's current latitude and longitude by advancing its previous position to the new one on the basis of assumed distance and direction traveled. The influences of current and wind as well as compass errors are taken into account in this calculation, all done without the aid of any celestial or physical observation. This is a real test of a navigator's skill.

Nuclear-powered submarines, which must retain secrecy of movement and cannot ascend to the surface, use the SINS system (*Ship's Inertial Navigation System*) developed by the United States Navy. It is a fully self-contained system, which requires no receiving or transmitting apparatus and thus involves no detectable signals. It consists of accelerometers, gyroscopes, and a computer. Together they produce inertial navigation, which is a sophisticated form of dead reckoning.

Sources: Bradford, Gershom. *A Glossary of Sea Terms*, pp. 51-52; Kemp, Peter. *Encyclopedia of Ships and Seafaring*, p. 143; *The McGraw-Hill Dictionary of Scientific and Technical Terms*, p. 493

295
When was the first hospital ship built?

It is believed that the Spanish Armada fleet of 1587-1588 included hospital ships. England's first recorded hospital ship was the *Goodwill* in 1608, but it was not until after 1660 that the Royal Navy made it a regular practice to set aside ships for hospital use. The United States Government outfitted six hospital ships, some of which were permanently attached to the fleet, during the Spanish-American War of 1898. Congress authorized the construction of the USS *Relief* in August 1916. It was launched in 1919 and delivered to the Navy in December 1920.

Sources: Kane, Joseph N. *Famous First Facts*, p. 575; Plumridge, John H. *Hospital Ships and Ambulance Trains*, pp. 13, 31.

296 ▽ Who created the Liberty ship?

The Liberty ship of World War II was the brainchild of Henry J. Kaiser (1882-1967), an American industrialist who had never run a shipyard before 1941. The huge loss of merchant tonnage during the war created an urgent need for protection of merchant vessels transporting weapons and supplies, and the Liberty ship was born. It was a standard merchant ship with a deadweight tonnage of 10,500 long tons and a service speed of 11 knots. They were built to spartan standards and production was on a massive scale. Simplicity of construction and operations, rapidity of building, and large cargo carrying capacity were assets. To these, Kaiser added prefabrication, and welding instead of riveting. The ships were a deciding factor on the side of the Allies. In four years 2,770 ships with a deadweight tonnage of 29,292 thousand long tons were produced.

Sources: Kemp, Peter. *The History of Ships*, p. 255; Sawyer, L.A., and W.H. Mitchell. *The Liberty Ships*.

297 ▽ When were the first nuclear-powered vessels launched?

A controlled nuclear reaction generates tremendous heat, which turns water into steam for running turbine engines. The USS *Nautilus* was the first submarine to be propelled by nuclear power, making her first sea run on January 17, 1955. It has been called the first true submarine since it can remain underwater for an indefinite period of time; as a nuclear reactor requires no oxygen to operate, only the crew's need for oxygen limits the submersion time. The *Nautilus*, 324 feet (98.75 meters) long, has a range of 2,500 miles (4,022.5 kilometers) submerged, a diving depth of 700 feet (213.36 meters), and can travel submerged at 20 knots.

The first nuclear warship was the 14,000-ton cruiser USS *Long Beach*, launched on July 14, 1959. The USS *Enterprise* was the first nuclear-powered aircraft carrier. Launched on September 24, 1960, the *Enterprise* was 1,101.5 feet (335.74 meters) long, and designed to carry 100 aircraft.

The first nuclear-powered merchant ship was the *Savannah*, a 20,000-ton vessel, launched in 1962. The United States built it largely as an experiment and it was never operated commercially. In 1969, Germany built the *Otto Hahn*, a nuclear-powered ore carrier. The most successful use of nuclear propulsion in non-naval ships has been as icebreakers. The first nuclear-powered icebreaker was the Soviet Union's *Lenin*, commissioned in 1959.

Sources: Horton, Edward. *The Illustrated History of the Submarine*, pp. 150-160; Kemp, Peter. *Encyclopedia of Ships and Sailing*, p. 99; Robertson, Patrick. *The Book of Firsts*, pp. 13, 171, 243, 244.

298 ▽ Why did the Titanic sink?

On its maiden voyage from Southampton, England, to New York, the British luxury liner, *Titanic*, sideswiped an iceberg at 11:40 p.m. on Sunday, April 14, 1912, and was badly damaged. The 882-foot–long (268.8-meter) liner, whose eight decks rose to the height of an 11–story building, sank two hours and forty minutes later. Of the 2,227 passengers and crew, 705 escaped in 20 lifeboats and rafts; 1,522 drowned.

Famous as the greatest disaster in transatlantic shipping history, circumstances made the loss of life in the sinking of the *Titanic* exceptionally high. Although Capt. E.J. Smith was warned of icebergs in shipping lanes, he maintained his speed of 22 knots, and did not post additional lookouts. Later inquiries revealed that the liner *Californian* was only 20 miles (32.18 kilometers) away and could have helped, had its radio operator been on duty. There was an insufficient number of lifeboats, and those available for use were badly managed, with some leaving only half-full. The only ship responding to distress signals was the ancient *Carpathia*, which saved 705 people.

Contrary to a long-held belief, the Titanic had not been sliced open by the iceberg. When Dr. Robert Ballard (b. 1942) from Woods Hole Oceanographic Institution descended to the site of the sunken vessel in the research vessel *Alvin* in July 1986, he found that the ship's starboard bow plates had buckled under the impact of the collision. This caused the ship to be opened up to the sea.

Ballard found the bow and the stern more than 600 yards apart on the ocean's floor, and speculated on what happened after the collision with the iceberg. "Water entered six forward compartments after the ship struck the iceberg. As the liner nosed down, water flooded compartments one after another, and the ship's stern rose even higher out of the water, until the stress amidships was more then she could bear. She broke apart..." and the stern soon sank by itself.

Sources: Cornell, James. *The Great International Disaster Book*, pp. 403-405; Davie, Michael. *Titanic: The Death and Life of a Legend*, p. xiii; *National Geographic*, vol. 172, no. 4 (October 1987), pp. 454-463.

CARS *See: CARS, BOATS, PLANES, AND TRAINS—MOTOR VEHICLES*

ENERGY CONSUMPTION AND CONSERVATION *See: ENERGY— CONSUMPTION AND CONSERVATION*

MILITARY VEHICLES

▽ **299**

Where did the military tank get its name?

During World War I, when the British were developing the tank, they called these first armored fighting vehicles "water tanks" to keep their real purpose a secret.

This code word has remained in spite of early efforts to call them "combat cars" or "assault carriages."

Sources: Shafritz, Jay M. *The Facts On File Dictionary of Military Science*, p. 454.

300 ▽ Why was the designation MiG chosen for the Soviet fighter plane used in World War II?

The *MiG* designation, formed from the initials of the plane's designers, Artem I. Mikoyan and Mikhail I. Gurevich, sometimes is listed as the Mikoyan-Gurevich MiG. Appearing in 1940 with a maximum speed of 400 miles (643.6 kilometers) per hour, the MiG-3, a piston-engined fighter, was one of the few Soviet planes whose performance was comparable with Western types during World War II. One of the best-known fighters, the MiG-15, first flown in December 1977, was powered by a Soviet version of a Rolls-Royce turbo jet engine. This high performer saw action during the Korean Conflict (1950-1953). In 1955 the MiG-19 became the first Soviet fighter capable of supersonic speed in level flight.

Sources: *Jane's Encyclopedia of Aviation*, p. 655; Taylor, Michael J.H., and John W.R. Taylor. *Encyclopedia of Aircraft*, pp. 164-165; Wragg, David W. *A Dictionary of Aviation*, pp. 192-193.

301 ▽ Who was the "Red Baron"?

Manfred von Richthofen (1892-1918), a German fighter pilot during World War I, was nicknamed "Red Baron" by the Allies because he flew a red-painted Albatros fighter. He was the top ace of the war, shooting down 80 Allied planes. He was killed on April 21, 1918, when he was attacked both by Roy Brown, a Canadian ace, and by Australian ground machine-gunners. Both parties claimed responsibility for his death.

Sources: Shores, Christopher F. *Fighter Aces*, p. 20.

302 ▽ What is a Sopwith Camel and why is it so called?

The most successful British fighter of World War I, the Camel was a development of the earlier Sopwith Pup, with a much larger rotary engine. Its name "camel" was derived from the humped shape of the covering of its twin synchronized machine guns. The highly maneuverable Camel, credited with 1,294 enemy aircraft destroyed, proved far superior to all German types as dogfighters, until the introduction of the Fokker D. VII in 1918. Altogether, 5,490 Camels were built by Sopwith Aircraft. Its top speed was 118 miles (189 kilometers) per hour and it had a ceiling of 24,000 feet (7,300 meters).

Sources: Taylor, Michael J.H., and John W.R. Taylor. *Encyclopedia of Aircraft*, pp. 205-206.

303
Who invented the culin *device on a tank?*

In World War II, American tank man Sergeant Curtis G. Culin devised a crossbar welded across the front of the tank with four protruding metal tusks. This device made it possible to break through the German hedgerow defenses and save thousands of lives. In the hedgerow country of Normandy, France, countless rows or stands of bushes or trees surrounded the fields, limiting tank movement. The culin device, also known as the "Rhinoceros," because its steel angled teeth formed a tusk-like structure, cut into the base of the hedgerow and pushed a complete section ahead of it into the next field, burying any enemy troops dug in on the opposite side.

Sources: Cary, James. *Tanks and Armor in Modern Warfare*, p. 206; Hunnicutt, R.P. *Sherman: A History of the American Medium Tank*, pp. 310-311.

304
Who were the "Flying Tigers"?

They were members of the American Volunteer Group who were recruited early in 1941 by Major General Claire Lee Chennault (1890–1958) to serve in China as mercenaries. Some 90 veteran United States pilots and 150 support personnel served from December 1941 until June 1942 during World War II. The airplanes they flew were P-40 Warhawks, which had the mouths of tiger sharks painted on the planes' noses. These painted-on images gave the group nickname the "Flying Tigers."

Sources: *Encyclopedia of Twentieth Century Warfare*, pp. 159-160.

305
What is the name of the airplane which carried the first atomic bomb?

During World War II, the *Enola Gay*, a modified Boeing B-29 bomber, dropped the first atomic bomb on Hiroshima, Japan at 8:15 a.m. on August 6, 1945. It was piloted by Col. Paul W. Tibbets, Jr. (b.1915) of Miami, Florida. The bombardier was Maj. Thomas W. Ferebee of Mocksville, North Carolina. Bomb designer Capt. William S. Parsons was aboard as an observer.

Three days later, another B-29 called *Bock's Car* dropped a second bomb on Nagasaki, Japan. The Japanese surrendered unconditionally on August 15, which confirmed the American belief that a costly and bloody invasion of Japan could be avoided at Japanese expense.

The *Enola Gay* is currently undergoing restoration at the National Air and Space Museum's Silver Hill, Maryland facility. *Bock's Car* is on display at the U.S. Air Force Museum, Wright-Patterson Air Force Base, Dayton, Ohio.

Sources: *Facts On File News Digest with Index*, vol. 5, no. 250 (August 1-7, 1945), p.244; *Milestones of Aviation*, p. 195; Morlan, Michael. *Kitty Hawk to NASA: A Guide to U.S. Air and Space Museums and Exhibits*, p. 201.

306 *When was the B-17 Flying Fortress introduced?*

A Fortress prototype first flew on July 28, 1935, and the first Y1B-17 was delivered to the Air Corps in March 1937 followed by an experimental Y1B-17A fitted with turbo-supercharged engines in January 1939. An order for 39 planes was placed for this model under the designation B-17B. In addition to its bombing function, the B-17 was used for many experimental duties, including serving as a launching platform in the U.S.A.A.F. guided missile program and in radar and radio-control experiments. It was called a "Flying Fortress" because it was the best defended bomber of World War II. Altogether, it carried thirteen 50 caliber Browning M-2 machine guns, each having about 700 pounds (317.5 kilograms) of armor-piercing ammunition. Ironically, the weight of all its defensive armament and manpower severely restricted the space available for bombs.

Sources: *Jane's All the World's Aircraft*, pp. 192–193; Jerram, Mike. *The World's Classic Aircraft*, pp. 110, 113; Wragg, David A. *A Dictionary of Aviation*, p. 49.

MOTOR VEHICLES *See also: BRIDGES, BUILDINGS, ETC.—BRIDGES, ROADS, AND TUNNELS; ENERGY—CONSUMPTION AND CONSERVATION*

307 *How many new passenger cars in the United States are imported cars?*

In 1990, imported cars accounted for 25.8% of the total passenger car sales. On total sales of 9,301,324 cars, 6,896,888 were domestic and 2,404,416 were imports. Japanese car sales (1,719,839) account for 18.5% of total car sales, and 71.5% of total import car sales. In 1990, the top new passenger car models sold in the United States were the Honda Accord (417,179), Ford Taurus (313,274), Chevrolet Cavalier (295,123), Ford Escort (288,727), and Toyota Camry (264,595).

Sources: *MVMA Motor Vehicle Facts and Figures '91*, p. 7; *The World Almanac and Book of Facts 1992*, p. 679.

308 *How many motor vehicles are registered in the United States?*

Total United States registration of motor vehicles in 1991 is estimated to be 190,741 thousand. Of the total, 145 million were automobiles and 45 million were trucks and buses. Worldwide, there were 556,031,486 total motor vehicles registered in 1989 (approximately 80% of which were passenger cars).

Sources: *MVMA Motor Vehicle Facts and Figures '92*, p. 22; *World Motor Vehicle Data*, p. 38.

309 / Which automobile is the one most often stolen?

The 1986 Camaro was the most stolen car in 1991, followed by the 1984 Oldsmobile Cutlass Supreme and the 1987 Camaro. The three most stolen cars of 1991 were also the top three in 1990. On April 24, 1986, the United States Congress ruled that certain parts on "high-theft" cars must be marked by the manufacturer with an identification number. Cars affected by this were Buick Riviera, Toyota Celica Supra, Cadillac Eldorado, Chevrolet Corvette, Pontiac Firebird, Mazda RX-7, Chevrolet Camaro, Porsche 911, Pontiac Grand Prix, and Oldsmobile Toronado.

Sources: *Edmund's Car Savvy*, pp. 52, 56; *Pittsburgh Press* (January 21, 1992), p. C3.

310 / How many motor vehicles are stolen each year?

Automobile theft has been occurring since an automobile mechanic stole Baron de Zuylen's Peugeot in Paris in June 1896. Motor vehicle thefts in 1989 in the United States totaled 1,564,800—averaging 816 thefts for every 100 thousand vehicles and increasing 9.2% from 1988 figures. Thefts declined in 12 states and the District of Columbia, but were up significantly in several others. Arizona led these states with a 53.5% increase. Other states with a vehicle theft increase of 20% minimum included Arkansas, Delaware, Kansas, Missouri, and Wisconsin.

Sources: *MVMA Motor Vehicle Facts & Figures '91*, p. 93; Robertson, Patrick. *The Book of Firsts*, p. 105.

311 / Who invented the automobile?

Although the idea of self-propelled road transportation originated long before, Karl Benz (1844-1929) and Gottlieb Daimler (1834–1900), are both credited with the invention of the gasoline-powered automobile, because they were the first to make their automotive machines commercially practicable. Benz and Daimler worked independently, unaware of each other's endeavors. Both built compact, internal-combustion engines to power their vehicles. Benz built his three-wheeler in 1885; it was steered by a tiller. Daimler's four-wheeled vehicle was produced in 1887.

Earlier self-propelled road vehicles include a steam-driven contraption invented by Nicolas-Joseph Cugnot (1725-1804), who rode the Paris streets at 2.5 miles (4 kilometers) per hour in 1769. Richard Trevithick (1771-1833) also produced a steam-driven vehicle that could carry eight passengers. It first ran on December 24, 1801 in Camborne, England. Londoner Samuel Brown built the first practical four-horsepower gasoline-powered vehicle in 1826. The Belgian engineer J.J. Etienne Lenoir (1822-1900) built a vehicle with an internal combustion engine that ran on liquid hydrocarbon fuel in 1862, but he did not test it on the road until September 1863 when it travelled a distance of 12 miles (19.3 kilometers) in 3 hours. The Austrian inventor Siegfried Marcus (1831–1898) invented a four-wheeled gasoline-powered handcart in 1864 and a full-size car in 1875; the Viennese police objected to the noise that the car made, and Marcus did not continue its development. Edouard Delamare-Deboutteville invented an eight-horsepower vehicle in 1883, which was not durable enough for road conditions.

Sources: Day, John, and C. Eng. *The Bosch Book of the Motor Car*, pp. 14-15; *Guinness Book of Records 1992*, p. 357; Robertson, Patrick. *The Book of Firsts*, pp. 99-100.

▽312▽ Who started the first American automobile company?

Charles Duryea (1861-1938), a cycle manufacturer from Peoria, Illinois, and his brother, Frank (1869-1967), founded America's first auto-manufacturing firm and became the first to build cars for sale in the United States. The Duryea Motor Wagon company, set up in Springfield, Massachusetts in 1895, built gasoline-powered, horseless carriages similar to those built by Benz in Germany.

However, the Duryea brothers did not build the first automobile factory in the United States. Ransom Eli Olds (1864-1950) built it in 1899 in Detroit, Michigan, to manufacture his Oldsmobile. Over 10 vehicles a week were produced there by April 1901 (433 in 1901; 2,500 in 1902; and 5,508 in 1904). In 1906, there were 125 companies making automobiles in the United States. In 1908, the American engineer Henry Ford (1863-1947) initiated assembly line techniques that made automotive manufacture quick and cheap; his company sold 10,660 vehicles that year.

Sources: Harris, Harry. *Good Old-Fashioned Yankee Ingenuity*, pp. 36-37, 183; Robertson, Patrick. *The Book of Firsts*, p. 104; Wise, David Burgess. *The Motor Car*, p. 74.

▽313▽ Has there ever been a nuclear-powered automobile?

In the 1950s, Ford automotive designers envisioned the Ford Nucleon, which was to be propelled by a small atomic reactor core, located under a circular cover at the rear of the car. It was to be recharged with nuclear fuel. The car was never built.

Sources: *Yearbook of Science and the Future 1991*, p. 22.

▽314▽ Is the electric automobile an idea of the last twenty years or so?

During the last decade of the 19th century, the electric vehicle became especially popular in the cities. People had grown familiar with electric trolleys and railways, and technology had produced motors and batteries in a wide variety of sizes. The Edison Cell, a nickel-iron battery, became the leader in electric vehicle use. By 1900, electric vehicles nearly dominated the pleasure car field. In that year, 4,200 automobiles were sold in the United States. Of these 38% were powered by electricity, 22% by gasoline, and 40% by steam. By 1911, the automobile starter motor did away with hand-cranking gasoline cars, and Henry Ford had just begun to mass-produce his Model T's. By 1924, not a single electric vehicle was exhibited at the National Automobile Show, and the Stanley Steamer was scrapped the same year.

Because of the energy crises of the 1970s and the 1990s concern for the environment (as well as "Clean Air" legislation), the large American auto manufacturers, along with the Japanese, are working toward producing a practical electric automobile. Chrysler and Ford have built electric versions of their minivans, and Nissan, Mitsubishi, and Daihatsu are competing, having announced aggressive programs of their own. In Europe, where electric vehicles (EVs) such as the General Motors one-ton van the Griffon have been used for years for local deliveries and in service fleets, VW, BMW, Audi, Fiat, and Peugeot are also developing products. But the odds-on favorite for the first electric car to win a mass market is General Motors' Impact, due out in 1995.

The basic problem, a technological one, still remains—the limitations of battery technology. The electric car remains a poor performer in speed and range, and the cost of battery pack replacement is expensive. Two new types of batteries are under development: the lithium-sulfide battery and the sodium-sulfur unit. Also, manufacturers are experimenting with an alternating-current (AC) engine to replace the present heavier direct-current (DC) power trains.

Sources: *Discover*, vol. 13, no. 3 (March 1992), p. 97; Shacket, Sheldon R. *The Complete Book of Electric Vehicles*, pp. 1-5; *Smithsonian*, vol. 23, no. 1 (April 1992), p. 35.

315 / Which vehicle had the first modern automobile air conditioner?

The first air-conditioned automobile was manufactured by the Packard Motor Car Company in Detroit, Michigan, and was exhibited publicly November 4–12, 1939 at the 40th Automobile Show in Chicago, Illinois. Air in the car was cooled to the temperature desired, dehumidified, filtered, and circulated. The first fully automatic air conditioning system was Cadillac's "Climate Control," introduced in 1964.

Sources: Harding, Anthony. *The Guinness Book of the Car*, p. 28; Kane, Joseph N. *Famous First Facts*, p. 51.

316 / What was the first car manufactured with an automatic transmission?

The first of the modern generation of automatic transmissions was General Motors' Hydramatic, first offered as an option on the Oldsmobile during the 1940 season. Between 1934 and 1936, a handful of 18 horsepower Austins were fitted with the American-designed Hayes infinitely variable gear. The direct ancestor of the modern automatic gearbox was patented in 1898.

Sources: Harding, Anthony. *The Guinness Book of Car Facts and Feats*, pp. 32, 252.

317 / *What is the difference between a medium truck and a heavy truck?*

Medium trucks weigh 14,001 to 33,000 pounds (6,351 to 14,969 kilograms). They span a wide range of sizes and uses from step-van route trucks to truck tractors. Common examples include beverage trucks, city cargo vans, and garbage trucks. Heavy trucks weigh 33,001 pounds (14,969 kilograms) or greater. Heavy trucks include over-the-road 18-wheelers, dump trucks, concrete mixers, and fire trucks. These trucks have come a long way from the first carrying truck, built in 1870 by John Yule, which moved at a rate of 0.75 miles (12 kilometers) per hour.

Sources: Robertson, Patrick. *The Book of Firsts*, p. 108; *Versatility of Trucks*, p. 8.

318 / *When did the first used-car dealership open?*

The first used-car dealership in Britain, the Motor Car Company of London, England, opened in September 1897 with 17 second-hand vehicles.

Sources: Robertson, Patrick. *The Book of Firsts*, p. 162.

319 / *Where was the first automobile license plate issued?*

Leon Serpollet of Paris, France obtained the first license plate in 1889. They were first required in the United States by New York State in 1901. Registration was required within 30 days. Owners had to provide their names and addresses as well as a description of their vehicles. The fee was one dollar. The plates bore the owner's initials and were required to be over 3 inches high. Permanent plates made of aluminum were first issued in Connecticut in 1937.

Sources: *Automotive Industries* (October 4, 1930), p. 494; Kane, Joseph N. *Famous First Facts*, p. 56.

320 / *When was the Michelin tire introduced?*

The first pneumatic (air-filled) tire for automobiles was produced in France by André (1853-1931) and Edouard Michelin (1859-1940) in 1885. The first radial-ply tire, the Michelin X, was made and sold in 1948. In radial construction, layers of cord materials called plies are laid across the circumference of the tire from bead to bead (perpendicular to the direction of the tread centerline). The plies can be made of steel wires or belts which circle the tire. Radial tires are said to give longer tread life, better handling, and a softer ride at medium and high speeds than bias or belted bias tires (both of which have plies laid diagonally). Radials give a firm, almost hard, ride at low speeds.

Sources: *Automotive Encyclopedia*, pp. 589-590; Harding, Anthony. *The Guinness Book of the Car*, p. 11.

321 ▽ When were tubeless automobile tires first manufactured?

In Akron, Ohio, the B.F. Goodrich Company announced the manufacture of tubeless tires on May 11, 1947. Dunlop was the first British firm to make tubeless tires in 1953.

Sources: Harding, Anthony. *Car Facts and Feats*, p. 206; Robertson, Patrick. *The Book of Firsts*, p. 133.

322 ▽ What do the numbers mean on automobile tires?

The numbers and letters associated with tire sizes and types are complicated and confusing. The "Metric P" system of numbering is probably the most useful method of indicating tire sizes. For example, if the tire had P185/75R-14, then "P" means the tire is for a passenger car. The number 185 is the width of the tire in millimeters. 75 is the aspect ratio, i.e., the height of the tire from the rim to the road is 75% of the width. R indicates that it is a radial tire. 14 is the wheel diameter. 13 and 15 inches are also common sizes.

Sources: Hopkins, Nigel J., John W. Mayne, and John R. Hudson. *The Numbers You Need*, p. 206.

323 ▽ How does police radar work?

The Austrian physicist, Christian Doppler (1803-1853), discovered that the reflected radio waves bouncing off a moving object are returned at a different frequency (shorter or longer waves, cycles, or vibrations). This phenomenon, called the *Doppler effect*, is the basis of police radar. Directional radio waves are transmitted from the radar device. The waves bounce off the targeted vehicle and are received by a recorder. The recorder compares the difference between the sent and received waves, and translates the information into miles per hour and displays the speed on a dial.

Sources: Bogner, Bruce F. *Vehicular Traffic Radar*, p. 4; Schneider, Herman. *The Harper Dictionary of Science in Everyday Language*, p. 90.

324

Which states do not allow radar detection devices in motor vehicles?

Only Connecticut, Virginia, and the District of Columbia do not allow drivers to install radar detectors.

Sources: Boyer, Rick. *Places Rated Almanac*, p. 169

325

How does VASCAR work?

Invented in 1965, VASCAR (*V*isual *A*verage *S*peed *C*omputer *a*nd *R*ecorder) is a calculator that determines a car's speed from two simple measurements of time and distance. No radar is involved. VASCAR can be used at rest or while moving to clock traffic in both directions. The patrol car can be behind, ahead of, or even perpendicular to the target vehicle. The device measures the length of a speed trap and then determines how long it takes the target car to cover that distance. An internal calculator does the math and displays the average speed on an LED readout. Most police departments now use several forms of moving radar which are less detectable and more accurate.

Sources: *Popular Science*, vol. 237 (September 1990), p. 68.

326

When was a speed trap first employed to apprehend speeding automobile drivers?

In 1905, William McAdoo, police commissioner of New York City, was stopped for travelling at 12 miles per hour in an 8 mile per hour zone in rural New England. The speed detection device consisted of two lookout posts, camouflaged as dead tree trunks, spaced one mile apart. A deputy with a stopwatch and a telephone kept watch for speeders. When a car appeared to be travelling too fast, the deputy pressed his stopwatch and telephoned ahead to his confederate who immediately consulted a speed-mileage chart and phoned ahead to another constable manning a road block to apprehend the speeder. McAdoo invited the New England constable to set up a similar device in New York City.

One of the most famous speed traps was in the Alabama town of Fruithurst, on the Alabama-Georgia border. In one year, this town of 250 people collected over $200,000 in fines and forfeitures from unwary "speeders."

Sources: *The Inventive Yankee*, pp. 186-187; Power, Rex. *How to Beat Police Radar*, p. 13

327

Which states require regular automobile safety inspections?

Twenty-six states require regular inspection of automobiles to rid the highways of dangerous vehicles with bald tires, wobbly suspensions, smoke-filled exhausts, and defective brakes and lights. These states are

Arkansas	New York
California	North Carolina
Delaware	Ohio
District of Columbia	Oklahoma
Hawaii	Pennsylvania
Louisiana	Rhode Island
Maine	South Carolina
Maryland	Tennessee
Massachusetts	Texas
Michigan	Utah
Missouri	Vermont
New Hampshire	Virginia
New Jersey	West Virginia

Sources: Boyer, Rick. *Places Rated Almanac*, p. 169.

328
How does an air bag work to prevent injury in an automobile crash?

An air bag, stored in the steering wheel or dashboard, is activated during a frontal collision involving a force roughly equivalent to hitting a solid barrier head-on at about 12 miles (19.31 kilometers) per hour. Sensors trigger the release of chemicals (sodium azide and nitrogen) that inflate the bag fully in about 1/30 of a second after impact to create a protective cushion. Immediately the airbag begins to deflate and the harmless nitrogen gas escapes through holes in the back, so that it is out of the way almost at once.

Sources: Brennan, Richard P. *Dictionary of Scientific Literacy*, pp. 4-5; *Consumers' Research Magazine*, vol. 74, no. 1 (January 1991), p. 11.

329
When was the automobile safety device, the air bag, invented?

Patented ideas on air bag safety devices began appearing in the early 1950s. U.S. patent 2,649,311 was granted on August 18, 1953, to John W. Hetrick for an inflated safety cushion to be used in automotive vehicles. The Ford Motor Company studied the use of air bags around 1957, and other undocumented work was carried out by Mr. Assen Jordanoff before 1956. There are other earlier uses of an air bag concept, including a rumored method of some World War II pilots inflating their life vests before a crash.

In the mid-70s, General Motors geared up to sell 100 thousand air bag–equipped cars a year in a pilot program to offer them as a discounted option on luxury models. GM dropped the option after only eight thousand buyers ordered air bags in three years. As of September 1, 1989, all new passenger cars produced for sale in the United States are required to be equipped with passive restraints (either automatic seatbelts or air bags). Currently, a major consumer shift toward air bags is taking place, and in 1991, air bags were in 5 million United States cars.

Sources: *Motor Trend*, vol. 22 (August 1970), p. 9; *MVMA Motor Vehicle Facts and Figures '91*, p. 3, *Status Report*, October 6, 1990.

330
What is the origin of the term taxicab?

The term *taxicab* is derived from two words—*taximeter* and *cabriolet*. The taximeter, an instrument invented by Wilhelm Bruhn in 1891, automatically recorded the distance traveled and/or the time consumed. This enabled the fare to be accurately measured. The cabriolet is a two-wheeled, one-horse carriage which was often rented.

The first taxicabs for hire were two Benz-Kraftdroschkes operated by "Droschkenbesitzer" Dütz in the spring of 1896 in Stuttgart, Germany. In May 1897 a rival service was started by Friedrich Greiner. In a literal sense, Greiner's cabs were the first "true" taxis because they were the first motor cabs fitted with taximeters.

Sources: *New Encyclopaedia Britannica*, vol. 11, p. 585; Robertson, Patrick. *The Book of Firsts*, p. 173.

331
How much does it cost to operate an automobile in the United States?

Below is listed the average cost per mile to operate an automobile in the United States in cents per mile. Figures are given for suburban driving conditions:

	Large	Intermediate	Compact	Sub-compact	Passenger van
Depreciation	9.6	8.6	7.3	5.9	10.7
Maintenance	6.0	5.2	4.6	5.1	6.9
Gas and oil	7.0	5.7	4.6	4.4	9.1
Parking and tolls	0.9	0.9	0.9	0.9	0.9
Insurance	4.9	5.6	4.3	5.0	8.9
Taxes	2.2	1.8	1.6	1.4	2.7
Total costs	30.6	27.8	23.3	22.7	39.2

Large (weight not more than 3,500 pounds)
Intermediate (weight less than 3,500 pounds)
Compact (weight less than 3,000 pounds)
Subcompact (weight less than 2,500 pounds)
Passenger van (weight less than 5,000 pounds)

Sources: *Edmund's Car Savvy*, p. 23

332
What information is available from the vehicle identification number (VIN), body number plate, and engine on a car?

These coded numbers reveal the model and make, model year, type of transmission, plant of manufacture and sometimes even the date and day of the week a car was made. The form and content of these codes is not standardized and often changes from one year to the next for the same manufacturer. Various components of a car may be made in different plants so a location listed on a VIN may differ from one on the engine number. The official shop manual lists the codes for a particular make of car.

Sources: *Catalog of American Car ID Numbers, 1970-79*, pp. 1-1, 2-1.

333
What is the braking distance for an automobile at different speeds?

Average stopping distance is directly related to vehicle speed. On a dry, level concrete surface, the minimum stopping distances are as follows (including driver reaction time to apply brakes):

Speed		Reaction time distance		Braking distance		Total distance	
Mph	Kph	Feet	Meters	Feet	Meters	Feet	Meters
10	16	11	3.4	9	2.7	20	6.1
20	32	22	6.7	23	7.0	45	13.7
30	48	33	10.1	45	13.7	78	23.8
40	64	44	13.4	81	24.7	125	38.1
50	80	55	16.8	133	40.5	188	57.3
60	97	66	20.1	206	62.8	272	82.9
70	113	77	23.5	304	92.7	381	116.1

Sources: *Automotive Encyclopedia*, p. 637.

334
Which colors of cars are the safest?

Tests at the University of California concluded that either blue or yellow is the best color for car safety. Blue shows up best during daylight and fog; yellow is best at night. The worst color from the visibility standpoint is gray. In another study by Mercedes-Benz in Germany, white ranked the highest in all-around visibility, except in situations of completely snow-covered roads or white sand. In such extremes, bright yellow and bright orange ranked second and third respectively in visibility. The least visible car color in the Mercedes-Benz test was dark green.

Sources: National Safety Council. *Accident Facts 1989*, p. 73; Carper, Jean. *Stay Alive*, p. 282.

335
How did the term horsepower originate?

Horsepower is the unit of energy needed to lift 550 pounds the distance of 1 foot in 1 second. Near the end of the 18th century the Scottish engineer James Watt (1736-1819) made improvements in the steam engine and wished to determine how its rate of pumping water out of coal mines compared with that of horses, which had previously been used to operate the pumps. In order to define a horsepower, he tested horses and concluded that a strong horse could lift 150 pounds 220 feet in 1 minute. Therefore, 1 horsepower was equal to 150 x 220/1 or 33,000 foot pounds per minute (also expressed as 745.2 joules per second, 7,452 million ergs per second, or 745.2 watts).

The term horsepower was frequently used in the early days of the automobile because the "horseless carriage" was generally compared to the horse-drawn carriage. Today this inconvenient unit is still used routinely to express the power of motors and engines, particularly of automobiles and aircraft. A typical automobile requires about 20 horsepower to propel it at 50 miles (80.5 kilometers) per hour.

Sources: Asimov, Isaac. *Understanding Physics*, vol. 1, p. 93; Brennan, Richard P. *Dictionary of Scientific Literacy*, p. 144; Davis, G.J. *Automotive Reference*, pp. 209-210.

RAILROADS, *See: CARS BOATS, PLANES, AND TRAINS—TRAINS AND TROLLEYS*

TRAINS AND TROLLEYS

336
What is a Johnson bar on a railroad?

It is the reverse lever of a locomotive. "To put the Johnson bar against the running board" means to get speed.

Sources: Adams, Ramon F. *The Language of the Railroader*, pp. 87, 120.

337
What was the route of the Orient Express?

This luxury train service was inaugurated in June 1883 to provide through connection between France and Turkey. It was not until 1889 that the complete journey could be made by train. The route left Paris and went via Chalons, Nancy, and Strasbourg into Germany (via Karlsruhe, Stuttgart, and Munich), then into Austria (via Salzburg, Linz, and Vienna), into Hungary (through Gyor and Budapest), south to Belgrade, Yugoslavia, through Sofia, Bulgaria and finally to Istanbul

(Constantinople), Turkey. It ceased operation in May 1977. In 1982, part of the line, the *Venice-Simplon-Orient Express* went into operation.

Sources: Nock, O.S. *Encyclopedia of Railways*, p. 239.

▽ **338**
What does the term gandy dancer mean?

A track laborer. The name derived from the special tools used for track work made by the Gandy Manufacturing Company of Chicago, Illinois. These tools were used during the 19th century almost universally by section gangs.

Sources: Beck, James H. *Rail Talk*, unnumbered alphabetical entry.

▽ **339**
When was the first railroad in the United States chartered?

The first American railroad charter was obtained on February 6, 1815 by Colonel John Stevens (1749-1838) of Hoboken, New Jersey, to build and operate a railroad between the Delaware and Raritan rivers near Trenton and New Brunswick. However, lack of financial backing prevented its construction. The Granite Railway built by Gridley Bryant was chartered on October 7, 1826. It ran from Quincy, Massachusetts to the Neponset River—a distance of three miles (4.8 kilometers). The main cargo was granite blocks used in building the Bunker Hill Monument.

Sources: Berliner, Barbara. *The Book of Answers*, p. 8; Marshall, John. *Rail: The Records*, p. 16; Nock, O.S. *Encyclopedia of Railways*, p. 52.

▽ **340**
What is the fastest train?

The world's fastest train is France's TGV (Train à Grande Vitesse or train of great speed) with a top speed of 186 miles (299 kilometers) per hour and an average speed of 132 miles (212.5 kilometers) per hour. However, soon it will be outdistanced by the MAGLEV (*magnetic levitation*) trains of Japan and Germany, which can travel between 250 to 300 miles (402 to 483 kilometers) per hour. These trains run on a bed of air produced from the repulsion or attraction of powerful magnetic fields (based on the principle that like poles of magnets repel and unlike poles (north and south) attract). The German Transrapid uses conventional magnets to levitate the train. The principle of attraction in magnetism, the employment of wing-like flaps extending under the train to fold under a T-shaped guideway, and the use of electromagnets on board (that are attracted to the non–energized magnetic surface) are the guiding components. Interaction between the train's electromagnets and those built on top of the T-shaped track lift the vehicle ⅜ inch (1 centimeter) off the guideway. Another set of magnets along the rail sides provides lateral guidance. The train rides on electromagnetic waves. Alternating current in the magnet sets in the guideway changes their polarity to alternately push and pull the train along. Braking is done by reversing the direction of the magnetic field (caused by reversing the magnetic poles). To increase train speed, the frequency of current is raised.

The Japanese MLV002 uses the same propulsion system, the difference is in the levitation design in which the train rests on wheels until it reaches a speed of 100 miles (161 kilometers) per hour. Then it levitates 4 inches (10 centimeters) above the guideway. The levitation depends on superconducting magnets and a repulsion system (rather than the attraction system that the German system uses).

Sources: Brennan, Richard P. *Levitating Trains and Kamikaze Genes*, pp. 218-220.

341 / What is the world's longest railway?

The Trans-Siberian Railway, from Moscow to Vladivostok, is 5,777 miles (9,297 kilometers) long. If the spur to Nakhodka is included, the distance becomes 5,865 miles (9,436 kilometers). It was opened in sections and the first goods train reached Irkutsk on August 27, 1898. The Baikal-Amur Northern Main Line, begun in 1938, shortens the distance by about 310 miles (500 kilometers). The journey takes approximately seven days, two hours and crosses seven time zones. There are nine tunnels, 139 large bridges or viaducts, and 3,762 smaller bridges or culverts on the whole route. Nearly the entire line is electrified.

In comparison, the first American transcontinental railroad, completed on May 10, 1869, is 1,780 miles (2,864 kilometers) long. The Central Pacific Railroad built eastward from Sacramento, California and the Union Pacific Railroad built westward to Promontory Point, Utah, where the two lines met to connect the line.

Sources: *Guinness Book of Records 1992*, p. 378; Marshall, John. *The Guinness Railway Book*, pp. 62-64; McNeil, Ian. *An Encyclopedia of the History of Technology*, p. 574.

342 / What is a standard gauge railroad?

The first successful railroads in England used steam locomotives built by George Stephenson (1781-1848) to operate on tracks with a gauge of 4 feet 8.5 inches (1.41 meters) probably because that was the wheel spacing common on the wagons and tramways of the time. Stephenson, a self-taught inventor and engineer, had developed in 1814 the steam-blast engine that made steam locomotives practical. His railroad rival, Isambard K. Brunel (1806-1859), laid out the line for the Great Western Railway at 7 feet 0.25 inches (2.14 meters), and the famous "Battle of the gauges" began. A commission appointed by the British Parliament decided in favor of Stephenson's narrower gauge, and the Gauge Act of 1846 prohibited using other gauges. This width eventually became accepted by the rest of the world. The distance is measured between the inner sides of the heads of the two rails of the track at a distance of ⅝ inch (1.58 centimeters) below the top of the rails.

Sources: Nock, O.S. *Encyclopedia of Railways*, pp. 11–12; Panati, Charles. *Panati's Browser's Book of Beginnings*, p. 311; *Standard Handbook for Civil Engineers*, p. 19–15.

What was the railroad velocipede?

In the nineteenth century, railroad track maintenance workers used a three-wheeled handcar to speed their way along the track. The handcar was used for interstation express and package deliveries and for delivery of urgent messages between stations that could not wait until the next train. Also called an "Irish Mail," this 150-pound (68 kilogram) three-wheeler resembled a bicycle with a sidecar. The operator sat in the middle of the two-wheel section and pushed a crank back and forth, which propelled the triangle-shaped vehicle down the tracks. This manually-powered handcar was replaced after World War I by a gasoline-powered track vehicle. This, in turn, was replaced by a conventional pickup truck fitted with an auxiliary set of flanged wheels.

Sources: *Technology and Culture*, vol. 17, no. 3 (July 1976), pp. 515–517.

Velocipede

How does a cable car, like those in San Francisco, move?

A cable runs continuously in a channel, between the tracks located just below the street. The cable is controlled from a central station, and usually moves about 9 miles (14.5 kilometers) per hour. Each cable car has an attachment, on the underside of the car, called a grip. When the car operator pulls the lever, the grip latches onto the moving cable and is pulled along by the moving cable. When the operator releases the lever, the grip disconnects from the cable and comes to a halt when the operator applies the brakes. Also called an endless ropeway, it was invented by Andrew S. Hallide (1836-1900) who first operated his system in San Francisco in 1873.

Sources: Smallwood, Charles A., et al. *The Cable Car Book*, pp. 6–7; *World Book Encyclopedia*, vol. 3, p. 7.

CHEMISTRY AND PHYSICS

CHEMICAL ELEMENTS, ETC. *See also: METALS AND OTHER MATERIALS*

345 / What is meant by Group I elements?

These are the elements at the left of the periodic table: lithium (Li, element 3), potassium (K, element 19), rubidium (Rb, element 37), cesium (Cs, element 55), francium (Fr, element 87), and sodium (Na, element 11). They are alkali metals, sometimes called the sodium family of elements. Because of their great chemical reactivity (easily form positive ions), none exist in nature in the elemental state.

Sources: *The Facts On File Dictionary of Chemistry*, p. 7; Tzimopoulos, Nicholas D., et al. *Modern Chemistry*, p. 713.

346 / Which elements are called Group II elements?

These are beryllium (Be, element 4); magnesium (Mg, element 12); calcium (Ca, element 20); strontium (Sr, element 38); barium (Ba, element 56); and radium (Ra, element 88). They are known as the alkaline-earth metals. Like the alkali metals, they are never found as free elements in nature and are moderately reactive metals. Harder and less volatile than the alkali metals (Group I), these elements all burn in air.

Sources: *The Facts On File Dictionary of Chemistry*, p. 7; Tzimpoulos, Nicholas D., et al. *Modern Chemistry*, p. 717.

347 / What are the transition elements?

The transition elements are the 10 subgroups of elements between group 2 and group 13, starting with period 4. They include gold (Au, element 79), silver (Ag, element 97), platinum (Pt, element 78), iron (Fe, element 26), copper (Cu, element 29), and other metals. The transition elements or metals are so named because they represent a gradual shift from the strongly electropositive elements of Group I and II to the electronegative elements of Groups VI and VII.

Sources: Hampel, Clifford A. *Glossary of Chemical Terms*, pp. 98, 288; Tzimopoulos, Nicholas D., et al. *Modern Chemistry*, p. 729.

348 / Which elements are liquid at room temperature?

Mercury ("liquid silver", Hg, element 80) and bromine (Br, element 35) are liquid at room temperature 68° to 70°F (20° to 25°C). Gallium (Ga, element 31) with a melting point of 85.6°F (29.8°C) and cesium (Cs, element 55) with a melting point of 83°F (28.4°C), are liquids at slightly above room temperature.

Sources: Brown, Theodore L. *Chemistry*, pp. 140, 143; Hackh, Ingo W.D. *Grant & Hackh's Chemical Dictionary*; pp. 96, 362; *Science and Technology Illustrated*, vol. 6, p. 658.

349 / What compound contains the most elements in it?

Chemical Abstracts Service searched its database to find the compound with the most elements—ten.

Sources: Nickon, Alex, and Ernest F. Silversmith. *Organic Chemistry: The Name Game*, p. 149.

THE COMPOUND WITH MOST ELEMENTS (10)

Tungsten, dichloro(diiodocadmium)bis[μ-[3-(diphenylphosphino)-1-propanaminato-*N*:*P*]]
dinitrosylbis[tris(3,5-dimethyl-1*H*-pyrazolato-*N*¹)hydroborato(1-)-*N*², *N*², *N*²] di-

350 / What is Harkin's rule?

Atoms having even atomic numbers are more abundant in the universe than are atoms having odd atomic numbers. Chemical properties of an element are determined by its atomic number, which is the number of protons in the atom's nucleus or the number of electrons revolving around this nucleus.

Sources: Fisher, David J. *Rules of Thumb for Engineers and Scientists*, p. 85; *A New Dictionary of Physics*, p. 33.

351
What are the names and numbers of the transuranic chemical elements?

These elements include 17 established artificial elements. They have atomic numbers higher than that of uranium (U, element 92), which is the last stable element of the periodic system.

Element Number	Name	Symbol
93	Neptunium	Np
94	Plutonium	Pu
95	Americium	Am
96	Curium	Cm
97	Berkelium	Bk
98	Californium	Cf
99	Einsteinium	Es
100	Fermium	Fm
101	Mendelevium	Md
102	Nobelium	No
103	Lawrencium	Lr
104	Unnilquadium	Unq
105	Unnilpentium	Unp
106	Unnilhexium	Unh
107	Unnilseptium	Uns
108	Unniloctium	Uno
109	Unnilennium	Une

Sources: *CRC Handbook of Physics and Chemistry*, pp. 1-7 – 1-8; Hackh, Ingo W.D. *Grant & Hackh's Chemical Dictionary*, pp. 599, 611.

352
Which chemical element is the most abundant on earth?

Oxygen (O, element 8) is the most abundant element in the earth's crust, waters, and atmosphere. It composes 49.5% of the total mass of these compounds.

Sources: Tzimopoulos, Nicholas D., et al. *Modern Chemistry*, p. 289.

353
What is the second most abundant chemical element on earth?

Silicon (Si, element 14) is the second most abundant element. Silicon dioxide and silicates make up about 87% of the materials in the earth's crust.

Sources: Tzimopoulos, Nicholas D., et al. *Modern Chemistry*, p. 758.

354
Which elements have the most isotopes?

The elements with the most isotopes, with 36 each, are xenon (Xe) with 9 stable isotopes (identified from 1920 to 1922) and 27 radioactive isotopes (identified from 1939 to 1981), and cesium (Cs) with one stable isotope (identified in 1921) and 35 radioactive isotopes (identified from 1935 to 1983).

The element with the least number of isotopes is hydrogen (H), with three isotopes, including two stable ones—protium (identified in 1920) and deuterium (identified in 1931)—and one radioactive isotope—tritium (first identified in 1934, but later considered a radioactive isotope in 1939).

Sources: *Guinness Book of Records 1992*, p. 262.

▽ 355 ▽ Why is Berzelius considered one of the founders of science in its modern form?

Swedish chemist Jöns Jakob Berzelius (1779–1848) devised chemical symbols, determined atomic weights, contributed to the atomic theory, and discovered several new elements. Between 1810 and 1816, he described the preparation, purification, and analysis of 2,000 chemical compounds. Then he determined atomic weights for 40 elements. He simplified chemical symbols, introducing a notation (still used today)—letters with numbers—that replaced the pictorial symbols his predecessors used. He discovered cerium (in 1803, with Wilhelm Hisinger), selenium (1818), silicon (1824), and thorium (1829).

Sources: *The Biographical Dictionary of Scientists: Chemists*, p. 17.

▽ 356 ▽ Why are the rare earth elements "rare"?

Elements numbered 58 through 71 in the periodic table plus yttrium (Y, element 39) and thorium (Th, element 90) are called "rare earths" because they are difficult to extract from monazite ore, where they occur. The term has nothing to do with scarcity or rarity in nature.

Sources: Brady, George S. *Materials Handbook*, pp. 661–662.

▽ 357 ▽ Who developed the periodic table?

Dmitri Ivanovich Mendeleyev (1834-1907) was a Russian chemist whose name will always be linked with his outstanding achievement, the development of the periodic table. He was the first chemist really to understand that all elements are related members of a single ordered system. He changed what had been a highly fragmented and speculative branch of chemistry into a true, logical science. His nomination for the 1906 Nobel Prize for Chemistry failed by one vote, but his name became recorded in perpetuity 50 years later when element 101 was called mendelevium.

According to Mendeleyev, the properties of the elements, as well as those of their compounds, are periodic functions of their atomic weights (in the 1920s, it was discovered that atomic number was the key rather than weight). Mendeleyev compiled the first true periodic table listing all the 63 (then-known) elements. In order to make the table work, Mendeleyev had to leave gaps, and he predicted that further elements would eventually be discovered to fill them. Three were discovered in Mendeleyev's lifetime: gallium, scandium, and germanium.

There are 94 naturally-occurring elements; of the 15 remaining elements (elements 95 to 109), 10 are undisputed. By 1984, over 6.8 million chemical compounds had been produced from these elements; 65 thousand of them in common use.

Sources: *The Biographical Dictionary of Scientists: Chemists*, p. 91; *Guinness Book of Records 1992*, p. 260; *The Universal Almanac 1992*, p. 505.

THE PERIODIC TABLE OF THE ELEMENTS

358
Which elements are the "noble metals"?

The noble metals are gold (Au, element 79), silver (Ag, element 47), mercury (Hg, element 80), and the platinum (Pt, element 78) group (including palladium (Pd, element 46), iridium (Ir, element 77), rhodium (Rh, element 45), ruthenium (Ru, element 44), and osmium (Os, element 76)). The term refers to those metals highly resistant to chemical reaction or oxidation (resistant to corrosion) and is contrasted to "base" metals which are not so resistant. The term has its origins in ancient alchemy whose goals of transformation and perfection were pursued through the different properties of metals and chemicals. The term is not synonymous with "precious metals," although a metal, like platinum, may be both.

Sources: Brady, George S. *Materials Handbook*, p. 644; Thrush, Paul W. *A Dictionary of Mining, Mineral, and Related Terms*, p. 753.

359
Which chemical element is the most abundant one in the universe?

Hydrogen (H, element 1) makes up about 75% of the mass of the universe. It is estimated that more than 90% of all atoms in the universe are hydrogen atoms. Most of the rest are helium (He, element 2) atoms.

Sources: *Science and Technology Illustrated*, vol. 6, p. 659.

360
What were the proposed names for elements 104 and 105 which were in use before the official names were chosen by the International Union of Pure and Applied Chemistry?

Rutherfordium (Rf) and kurchatovium (Ku) were the names proposed by rival groups for element 104 (unnilquadium (Unq)). Hafnium (Ha) and nielsbohrium (Ns) were proposed for element 105 (unnilpentium (Unp)).

Sources: *American Scientist*, vol. 68 (May/June 1980), p. 285; Rovin, Jeff. *Laws of Order*, p. 181; *Science 80*, vol. 1, no. 2 (January/February 1980), p. 77.

ENERGY, MOTION, FORCE, AND HEAT *See also: ENERGY*

361
What is the Kelvin temperature scale?

Temperature is the level of heat in a gas, liquid, or solid. The freezing and boiling points of water are used as standard reference levels in both the metric (centigrade or Celsius) and the English system (Fahrenheit). In the metric system, the difference between freezing and boiling is divided into 100 equal intervals called degree Celsius or degree centigrade (°C). In the English system, the intervals are divided into 180 units, with 1 unit called degree Fahrenheit (°F). But temperature can be measured from absolute zero (no heat, no motion); this principle defines thermodynamic temperature and establishes a method to measure it upward. This scale of temperature is called the Kelvin temperature scale, after its inventor, William Thomson, Lord Kelvin (1824–1907), who devised it in 1848. The Kelvin (symbol K) has the same magnitude as the degree Celsius (the difference between freezing and boiling water is 100 degrees), but the two temperatures differ by 273.15 degrees (absolute zero, which is -273.15°C on the Celsius scale). Below is a comparison of the three temperatures:

Characteristic	K°	C°	F°
Absolute zero	0	-273.15	-459.7
Freezing point of water	273.15	0	32
Normal human body temperature	310.15	37	98.6
Boiling point of water	373.15	100	212

To convert Celsius to Kelvin: Add 273.15 to the temperature (K = C + 273.15). To convert Fahrenheit to Celsius: Subtract 32 from the temperature and multiply the difference by 5; then divide the product by 9 (C = ⁵⁄₉ [F - 32]). To convert Celsius to Fahrenheit: Multiply the temperature by 1.8, then add 32 (F = ⁹⁄₅ C + 32 or F = 1.8C + 32).

Sources: Emiliani, Cesare. *The Scientific Companion*, pp. 34–35; Giscard d'Estaing. *World Almanac Book of Inventors*, p. 246; *The Universal Almanac 1992*, p. 530.

▽ **362** ◿
Who invented the thermometer?

The Greeks of Alexandria knew that air expanded as it was heated, and it is known that Hero of Alexandria (1st century C.E.) and Philo of Byzantium made simple thermometers or "thermoscopes," but they were not real thermometers. In 1592, Galileo (1564–1642) made a kind of thermometer which also functioned as a barometer, but in 1612, his friend Santorio Santorio (1561–1636) first adapted the air thermometer (a device in which a colored liquid was driven down by the expansion of air) to measure the body's temperature change during illness and recovery. Still, it was not until 1713 that Daniel Fahrenheit (1686–1736) began developing a thermometer having a fixed scale. He worked out his scale from two "fixed" points: the melting point of ice and the heat of the healthy human body. He realized that the melting point of ice was a constant temperature, whereas the freezing point of water varied. Fahrenheit put his thermometer into a mixture of ice, water, and salt (which he marked off as 0°) and using this as a starting point, marked off melting ice at 32° and blood heat at 96°. In 1835, it was discovered that normal blood measured 98.6°F. Sometimes, Fahrenheit used spirit of wine as the liquid in the thermometer tube, but more often he used specially-purified mercury. Later, the boiling point of water (212°F) became the upper fixed point.

Sources: De Bono, Edward. *Eureka!*, p. 194; Hooper, Meredith. *More Everyday Inventions*, pp. 112–116.

▽ **363** ◿
What was unusual about the original Celsius temperature scale?

In 1742, the Swedish astronomer Anders Celsius (1701–1744) set the freezing point of water at 100°C and the boiling point of water at 0°C. It was Carolus

Linnaeus (1707–1778), who reversed the scale, but a later textbook attributed the modified scale to Celsius and the name has remained.

Sources: *American Scientist*, vol. 80, no. 2 (March/April 1992), p. 127.

364 / How are Celsius temperatures converted into Fahrenheit temperatures?

The formulas for converting Celsius or Centigrade temperatures into Fahrenheit (and the reverse) are as follows:

$$F = (C \times \tfrac{9}{5}) + 32$$
$$C = (F - 32) \text{ times } \tfrac{5}{9}$$

Some useful comparisons of the two scales:

Temperature	Fahrenheit	Celsius or Centigrade
Absolute Zero	-459.67	-273.15
Point of Equality	-40.0	-40.0
Zero Fahrenheit	0.0	-17.8
Freezing Point of Water	32.0	0.0
Normal Human Blood Temperature	98.4	36.9
100 Degrees F	100.0	37.8
Boiling Point of Water (at standard pressure)	212.0	100.0

Sources: Dean, John A. *Lange's Handbook of Chemistry*, pp. 2–48; McWhirter, Norris. *The Book of Answers*, p. 52.

Fahrenheit and Celsius scales

365 ╱ How is "absolute zero" defined?

Absolute zero is the theoretical temperature at which all substances have zero thermal energy. Originally conceived as the temperature at which an ideal gas at constant pressure would contract to zero volume, absolute zero is of great significance in thermodynamics and is used as the fixed point for absolute temperature scales. Absolute zero is equivalent to 0°K, -459.69°F, or -273.16°C.

The velocity of molecules of substances determines their temperature; the faster they move, the more volume they require, and the higher the temperature becomes. The lowest actual temperature ever reached was two-billionth of a degree above absolute zero (2×10^9 Kelvin) by a team at the Low Temperature Laboratory in the Helsinki University of Technology, Finland, in October, 1989.

Sources: *Encyclopedic Dictionary of Science*, p. 10.

366 ╱ Does hot water freeze faster than cold?

A bucket of hot water will not freeze faster than a bucket of cold water. However, a bucket of water that *has been* heated or boiled, then allowed to cool to the same temperature as the bucket of cold water, may freeze faster. Heating or boiling drives out some of the air bubbles in water; because air bubbles cut down thermal conductivity, they can inhibit freezing. For the same reason, previously heated water forms denser ice than unheated water, which is why hot-water pipes tend to burst before cold-water pipes.

Sources: Burnam, Tom. *The Dictionary of Misinformation*, p. 112.

367 ╱ What is superconductivity?

Superconductivity is a condition occurring in many metals, alloys, etc., usually at low temperatures, involving zero electrical resistance and perfect diamagnetism. In such a material an electric current will persist indefinitely without any driving voltage, and applied magnetic fields are exactly cancelled out by the magnetization they produce. Superconductivity was discovered by Heike Kamerlingh Onnes (1853-1926) in 1911.

Sources: *Encyclopedic Dictionary of Science*, p. 231.

368 ╱ What is inertia?

Inertia is a tendency of all objects and matter in the universe to stay still, or if moving, to continue moving in the same direction, unless acted on by some outside force. This forms the first law of motion formulated by Isaac Newton (1642–1727). To move a body at rest, enough external force must be used to overcome the object's inertia; the larger the object is, the more force is required to move it. In his *Philosophae Naturalis Principia Mathematica*, published in 1687, Newton sets forth all three laws of motion. Newton's second law is that the force to move a body is equal to its mass times its acceleration (F = MA), and the third law states that for every action there is an equal and opposite reaction.

Sources: Barnhart, Robert K. *The American Heritage Dictionary of Science*, p. 316; *The Biographical Dictionary of Scientists: Physicists*, pp. 119–121; Trefil, James. *1001 Things Everyone Should Know About Science*, pp. 146–147.

369
Why do golf balls have dimples?

The dimples minimize the drag (a force that makes a body lose energy as it moves through a fluid), allowing the ball to travel further than a smooth ball would. In a dimpled ball the air, as it passes a ball, tends to cling to the ball longer, reducing the eddies or wake effect that drain the ball's energy. The dimpled ball can travel up to 300 yards (275 meters) but a smooth ball only goes 70 yards (65 meters). A ball can have 300 to 500 dimples which can be 0.01 inch (0.25 millimeters) deep. Another effect to get distance is to give the ball a backspin. With a backspin there is less air pressure on the top of the ball, so the ball stays aloft longer (much like an airplane).

Sources: Flatow, Ira. *Rainbows, Curve Balls and Other Wonders of the Natural World Explained*, pp. 80–82; *How in the World?*, 1990, p. 432.

370
Who is the father of the science of magnetism?

The English scientist William Gilbert (1544–1603) regarded the earth as a giant magnet, and investigated its field in magnetism terms of dip and variation. He explored many other magnetic and electrostatic phenomena. The Gilbert (symbol Gb), a unit of magnetism, is named for him.

Sources: *Encyclopedic Dictionary of Science*, p. 108.

371
What is Maxwell's demon?

An imaginary creature who, by opening and shutting a tiny door between two volumes of gases, could, in principle, concentrate slower molecules in one (making it colder) and faster molecules in the other (making it hotter), thus breaking the second law of thermodynamics. This law states that heat cannot by its own accord flow from a colder to a hotter body. This hypothesis was formulated in

1871 by James C. Maxwell (1831-1879), who is considered to be the greatest theoretical physicist of the 19th century. Maxwell discovered that light consists of electro-magnetic waves, proved the nature of Saturn's rings, set forth the principles of color vision and established the kinetic theory of gases, wherein the nature of heat resides in the motion of the molecules.

Sources: *The Biographical Dictionary of Scientists: Physicists,* pp. 109-111.

372 / *When was spontaneous combustion first recognized?*

Spontaneous combustion is the ignition of materials stored in bulk. This is due to internal heat build-up caused by oxidation (generally a reaction in which electrons are lost, specifically when oxygen is combined with a substance, or when hydrogen is removed from a compound). Because this oxidation heat cannot be dissipated into the surrounding air, the temperature of the material rises until the material reaches its ignition point and bursts into flame.

A Chinese text written before 290 C.E. recognized this phenomenon in a description of the ignition of stored oiled cloth. The first Western recognition of spontaneous combustion was by J.P.F. Duhamel in 1757, when he discussed the gigantic conflagration of a stack of oil-soaked canvas sails drying in the July sun. Before spontaneous combustion was recognized, such events were usually blamed on arsonists.

Sources: Parker, Sybil P. *McGraw-Hill Concise Encyclopedia of Science & Technology,* p. 1765; Temple, Robert K.G. *The Genius of China,* pp. 166-167.

373 / *What is phlogiston?*

Phlogiston was a name used in the 18th century to identify a supposed substance given off during the process of combustion. The phlogiston theory was developed in the early 1700s by the German chemist and physicist, Georg Ernst Stahl (1660-1734).

In essence, Stahl held that combustible material such as coal or wood was rich in a material substance called "phlogiston." What remained after combustion was without phlogiston and could no longer burn. The rusting of metals also involved a transfer of phlogiston. This accepted theory explained a great deal previously unknown to chemists. For instance, metal smelting was consistent with the phlogiston theory. Charcoal in burning lost weight. Thus the loss of phlogiston either decreased or increased weight.

The French chemist Antoine Lavoisier (1743-1794) demonstrated that the gain of weight when a metal turned to a calx was just equal to the loss of weight of the air in the vessel. Lavoisier also showed that part of the air (oxygen) was indispensable to combustion, and that no material would burn in the absence of oxygen. The transition from Stahl's phlogiston theory to Lavoisier's oxygen theory marks the birth of modern chemistry at the end of the 18th century.

Sources: *The Biographical Dictionary of Scientists: Chemists*, pp. 131–132; Bynum, W.F. *Dictionary of the History of Science*, pp. 322–323; Downs, Robert B. *Landmarks in Science*, pp. 170–172.

374 / Who is generally regarded as the father of quantum mechanics?

The German mathematical physicist, Werner Karl Heisenberg (1901-1976), is regarded as the father of quantum mechanics (theory of small-scale physical phenomena). His theory of uncertainty in 1927 overturned traditional classical mechanics and electromagnetic theory regarding energy and motion when applied to sub-atomic particles such as electrons and parts of atomic nuclei. The theory states that it is impossible to specify precisely both the position and the simultaneous momentum (mass x volume) of a particle, but they could only be predicted. This meant that a result of an action can only be expressed in terms of probability that a certain effect will occur, not certainty.

Sources: *The Biological Dictionary of Scientists: Physicists*, pp. 79–81; *Encyclopedic Dictionary of Science*, p. 116.

375 / Does water running down a drain rotate in a different direction in the Northern and Southern Hemispheres?

It water runs out from a perfectly symmetrical bathtub, basin, or toilet bowl, in the Northern Hemisphere it would swirl counterclockwise; in the Southern Hemisphere, the water would run out clockwise. This is due to the Coriolis effect (the earth's rotation influencing any moving body of air or water). However, some scientists think that the effect does not work on small bodies of water. Exactly on the equator, the water would run straight down.

Sources: Burnam, Tom. *The Dictionary of Misinformation*, p. 27; Flatow, Ira. *Rainbows, Curve Balls and Other Wonders of the Natural World Explained*, pp. 151–153; Schweighauser, Charles A. *Astronomy from A to Z*, p. 27

376 / What is a Leyden jar?

A Leyden jar, the earliest form of capacitor, is a device for storing an electrical charge. First described in 1745 by E. Georg van Kleist (c.1700–1748), it was also used by Pieter van Musschenbroek (1692–1761), a professor of physics at the University of Leyden. The device came to be known as a Leyden jar and was the first device that could store large amounts of electric charge. The jars contained an inner wire electrode in contact with water, mercury, or wire. The outer electrode was a human hand holding the jar. An improved version coated the jar inside and outside with separate metal foils with the inner foil connected to a conducting rod and terminated in a conducting sphere. This eliminated the need for the liquid electrolyte. In use, the jar was normally charged from an electrostatic generator. The Leyden jar is still used for classroom demonstrations of static electricity.

Sources: *Encyclopedia Americana*, vol. 17, p. 280; *New Encyclopedia Brittanica*, vol. 18, p. 189.

Leyden jar

377
Why is Van Vleck considered to be one of the fathers of modern magnetic theory?

Awarded the Nobel Prize in 1977, John H. Van Vleck (1899–1980), an American physicist, made significant contributions in the field of magnetism. He explained the magnetic, electrical, and optical properties of many elements and compounds with the ligand field theory, demonstrated the effect of temperature on paramagnetic materials (called Van Vleck paramagneticism), and developed a theory on the magnetic properties of atoms and their components.

Sources: *The Biographical Dictionary of Scientists: Physicists*, p. 164.

FORCE *See:* CHEMISTRY AND PHYSICS—ENERGY, MOTION, FORCE, AND HEAT

HEAT *See:* CHEMISTRY AND PHYSICS—ENERGY, MOTION, FORCE, AND HEAT

INERTIA *See:* CHEMISTRY AND PHYSICS—ENERGY, MOTION, FORCE, AND HEAT

MATTER *See also:* CHEMISTRY AND PHYSICS—CHEMICAL ELEMENTS, ETC.

378
How did the quark get its name?

This mathematical particle, considered to be the fundamental unit of matter, was named by Murray Gell-Mann (b. 1929) an American theoretical physicist and Nobel Prize winner. Its name was initially a playful tag that Gell-Mann invented, sounding something like "kwork". Later Gell-Mann came across the line "Three quarks for Master Marks" in James Joyce's *Finnegan's Wake*, and the tag became known as a quark. There are six kinds or "flavors" (up, down, strange, charm, bottom, and top) of quarks, and each "flavor" has 3 varieties or "colors" (red, blue, and green). All eighteen types have different electric charges (a basic characteristic of all elementary particles). Three quarks form a proton (having one unit of positive electric charge) or a neutron (zero charge), and two quarks (a quark and an antiquark) form a meson. Like all known particles, a quark has its anti-matter opposite, known as an antiquark (having the same mass but opposite charge).

Sources: Asimov, Isaac. *Asimov's Chronology of Science and Discovery*, p. 606; *Biographical Encyclopedia of Scientists*, vol. 1, p. 310; Emiliani, Cesare. *The Scientific Companion*, pp. 11-13; *Science News*, vol. 142 (September 12, 1992) p. 174.

379
What is the fourth state of matter?

Plasma, a mixture of free electrons and ions or atomic nuclei, is sometimes referred to as a "fourth state of matter." Plasmas occur in thermonuclear reactions as in the sun, in fluorescent lights, and in stars. When gas temperature is raised high enough the collision of atoms become so violent that electrons are knocked loose from their nuclei. The result of a gas having loose negatively charged electrons and heavier positively-charged nuclei is called a plasma.

All matter is made up of atoms. Animals and plants are organic matter; minerals and water are inorganic matter. Whether matter appears as a solid, liquid, or gas depends on how the molecules are held together in their chemical bonds. Solids have a rigid structure in the atoms of the molecules; in liquids the molecules are close together but not packed; in a gas, the molecules are widely spaced and move around, occasionally colliding but usually not interacting. These states—solid, liquid, and gas—are the first three states of matter.

Sources: Barnhart, Robert K. *American Heritage Dictionary of Science*, p. 385; Daintith, John. *The Facts On File Dictionary of Physics*, p. 154; Trefil, James. *1001 Things Everyone Should Know About Science*, p. 201.

380
Who is generally regarded as the discoverer of the electron?

The British physicist, Sir Joseph John Thomson (1856–1940), in 1897 researched electrical conduction in gases, which led to the important discovery that cathode rays consisted of negatively charged particles called electrons. The discovery of the electron inaugurated the electrical theory of the atom, and this with other work entitled Thomson to be regarded as the founder of modern atomic physics.

Sources: *Encyclopedic Dictionary of Science*, p. 238; Hechtlinger, A. *Modern Science Dictionary*, pp. 124, 308; Howard, A.V. *Chamber's Dictionary of Scientists*, p. 435.

381
What substance, other than water, is less dense as a solid than as a liquid?

Only bismuth and water share this characteristic. Density (the mass per unit volume or mass/volume) refers to how compact or crowded a substance is. For instance the density of water is 1 gram/cm³ or 1 kg/l (kilogram/per liter); the density of a rock is 3.3 g/cm³; pure iron is 7.9 g/cm³; and the earth (as a whole) 5.5 g/cm³ (average). Water as a solid (i.e., ice) floats; which is a good thing, otherwise ice would sink to the bottom of every lake or stream, which would result in great flooding.

Sources: Burnam, Tom. *The Dictionary of Misinformation*, p. 281; Shipman, James T., and Jerry D. Wilson. *An Introduction to Physical Science*, p. 10.

382
Why is liquid water more dense than ice?

Pure liquid water is most dense at 39.2°F (3.98°C) and decreases in density as it freezes. The water molecules in ice are held in a relatively rigid geometric pattern by their hydrogen bonds, producing an open, porous structure. Liquid water has fewer bonds, therefore more molecules can occupy the same space, making liquid water more dense than ice.

Sources: Leopold, Luna B. *Water*, p. 30.

383
Who is known as the father of crystallography?

The French priest and mineralogist, Rene-Just Haüy (1743-1822), is called the father of crystallography. In 1781 Haüy had a fortunate accident when he dropped a piece of calcite and it broke into small fragments. He noticed that the fragments broke along straight planes that met at constant angles. He hypothesized that each crystal was built up of successive additions of what is now called a unit cell to form a simple geometric shape with constant angles. An identity or difference in crystalline form implied an identity or difference in chemical composition. This was the beginning of the science of crystallography.

Sources: Asimov, Isaac. *Asimov's Biographical Encyclopedia of Science and Technology*, p. 221; Berry, James. *Exploring Crystals*, p. 4.

384 / What does half-life mean?

Half-life is the time it takes for the number of radioactive nuclei originally present in a sample to decrease to one half of their original number. Thus, if a sample has a half-life of one year, its radioactivity will be reduced to half its original amount at the end of a year and to one quarter at the end of two years. The half-life of a particular radionuclide is always the same, independent of temperature, chemical combination, or any other condition.

Sources: Barnhart, Robert K. *The American Heritage Dictionary of Science*, p. 273; *Encyclopedia Americana*, vol. 13, p. 712.

385 / Who made the first organic compound to be synthesized from inorganic ingredients?

Urea was synthesized by Friedrich Wohler (1800-1882) in 1828 from ammonia and cyanic acid. This synthesis dealt a deathblow to the vital-force theory. This theory assumed the existence of a mysterious "Vital Force," which intervened in the formation of compounds so that their preparation in the laboratory could hardly be expected.

Sources: *Encyclopedia of Chemical Technology*, vol. 23, p. 548; Partington, J.R. *A Short History of Chemistry*, p. 221.

386 / What is a chemical garden and how is one made?

Mix 4 tablespoons of bluing, 4 tablespoons of salt, and 1 tablespoon household ammonia. Pour this mixture over pieces of coal or brick in a suitable dish or bowl. Put several drops of red or green ink or mercurochrome on various parts of the coal and leave undisturbed for several days.

A crystal garden—a dishful of crystals that grow like plants and look like coral—will begin to appear. How soon the crystals will begin to appear depends on the temperatures and humidity in the room. Before long, crystals will be growing all over the briquettes, on the side of the dish, and down onto the plate. The crystals will be pure white with a snow-like texture.

Sources: Herbert, Don. *Mr. Wizard's Experiments for Young Scientists*, pp. 55-57.

MEASUREMENT, METHODOLOGY, ETC.

▽ **387** ▽
Who is considered the father of modern chemistry?

Several contenders share this honor. Robert Boyle (1627–1691), a British natural philosopher, is considered one of the founders of modern chemistry. Best known for his discovery of Boyle's Law (volume of a gas is inversely proportional to its pressure at constant temperature), he was a pioneer in the use of experiments and the scientific method. A founder of the Royal Society, he worked to remove the mystique of alchemy from chemistry to make it a pure science.

The French chemist, Antoine-Laurent Lavoisier (1743–1794) is regarded as another founder of modern chemistry. His wide-ranging contributions include the discrediting of the phlogiston theory of combustion, which had been for so long a stumbling block to a true understanding of chemistry. He established modern terminology for chemical substances and did the first experiments in quantitative organic analysis. He is sometimes credited with having discovered or established the law of conservation of mass in chemical reactions.

Sources: *Encyclopedic Dictionary of Science*, p. 50; *The Great Scientists*, vol. 7, p. 84.

▽ **388** ▽
Who was the father of analytical chemistry?

René Descartes' (1596–1650) reform of science was the establishment of a mechanical philosophy, which sought to explain the properties and actions of bodies in terms of the parts of which they are composed. He proposed the use of mathematical methods to investigate scientific problems. Descartes rejected all untested ancient and medieval authority as opinion that must be subjected to objective scientific analysis to test its validity. Honored in physics and mathematics, he is also considered one of the foremost philosophers of the modern era. Descartes believed in a firm unity of all knowledge, scientific and philosophical, which he symbolized in the metaphor of a tree whose roots are metaphysics, whose trunk is physics, and whose branches are specific topics (such as medicine, mechanics, and morality).

Sources: Cohen, I. Bernard. *Revolution in Science*, p. 146; Downs, Robert B. *Landmarks in Science*, pp. 117–119; *The Great Scientists*, vol. 4, p. 13.

▽ **389** ▽
What is STP?

The abbreviation *STP* is often used for *s*tandard *t*emperature and *p*ressure. As a matter of convenience, scientists have chosen a specific temperature and pressure as standards for comparing gas volumes. The standard temperature is 0°C (273°K) and the standard pressure is 760 torr (1 atmosphere).

Sources: Kroschwitz, Jacqueline I. *Chemistry*, p. 212.

390 / How did the electrical term "ampere" originate?

It was named for Andre Marie Ampere (1775-1836), the physicist who formulated the basic laws of the science of electrodynamics. The *ampere* (symbol A), often abbreviated as "amp," is the unit of electric current, defined at the constant current, that, maintained in two straight parallel infinite conductors placed one meter apart in a vacuum, would produce a force between the conductors of 2 x 10^{-7} Newton per meter. For example, the amount of current flowing through a 100-watt light bulb is 1 amp; through a toaster, 10 amps; a TV set, 3 amps; a car battery, 50 amps (while cranking). Newton (symbol N) is defined as a unit of force needed to accelerate one kilogram by one meter second^{-2}, or $1N = 1Kg^{MS-2}$.

Sources: *Facts On File Dictionary of Physics*, pp. 5, 136; *Great Engineers and Pioneers in Technology*, vol. 1, p. 378; Trefil, James. *1001 Things Everyone Should Know About Science*, pp. 133-134.

391 / What is a mole in chemistry?

A mole (symbol Mol), a fundamental measuring unit for the amount of a substance, refers to either a gram atomic weight or a gram molecular weight of a substance. It is the quantity of a substance that contains 6.02 x 10^{23} atoms, molecules, or formula units of that substance. This number is called Avogadro's number or constant after Amedeo Avogadro (1776-1856) who is considered to be one of the founders of physical science.

Sources: *The Biographical Dictionary of Scientists: Chemists*, pp. 12-13; Henrickson, Charles H. *Chemistry for the Health Professions*, pp. 66-67.

392 / How does a gram atomic weight differ from a gram formula weight?

A gram atomic weight is the amount of an *element* (substance made up of atoms having the same atomic number) equal to its atomic weight (the number of protons) in grams. A gram formula weight is an amount of a *compound* (a combination of elements) equal to its formula weight in grams.

Sources: Henrickson, Charles H. *Chemistry for the Health Professions*, p. 65; *McGraw-Hill Dictionary of Physics and Mathematics*, pp. 59, 316.

393

How did the electrical term volt originate?

The unit of voltage is the volt, named after Alessandro Volta (1745-1827), the Italian scientist who built the first battery. Voltage measures the force or "oomph" with which electrical charges are pushed through a material. Some common voltages are 1.5 volts for a flashlight battery; 12 volts for a car battery; 115 volts for ordinary household receptacles, and 230 volts for a heavy-duty household receptacle.

Sources: Trefil, James. *1001 Things Everyone Should Know About Science*, p. 134.

MOTION See: CHEMISTRY AND PHYSICS—ENERGY, MOTION, FORCE, AND HEAT

NOBEL PRIZES See: GENERAL SCIENCE AND TECHNOLOGY—AWARDS, PRIZES AND FAIRS

SOUND, LIGHT, AND OTHER WAVES

394

What is the speed of light?

In round numbers, the speed of light in a vacuum is 186 thousand miles (300 thousand kilometers) per second. The exact figure is 186,282 miles (299,792.458 kilometers) per second.

Sources: *McGraw-Hill Encyclopedia of Science and Technology*, vol. 10, p. 51.

395

Where is Albert Einstein's brain?

Einstein was cremated after his death on April 18, 1955. The doctor who performed the autopsy removed Einstein's brain, part of which now rests in a bottle somewhere in Weston, Missouri. Albert Einstein (1879-1955) was the principal founder of modern theoretical physics; his theory of relativity (speed of light is a constant and not relative to the observer or source of light), and the relationship of mass and energy ($E=MC^2$), fundamentally changed human understanding of the physical world. His stature as a scientist, together with his strong humanitarian stance on major political and social issues, made him one of the outstanding men of the twentieth century.

Sources: *The Great Scientists*, vol. 4, pp. 109–111; Pais, Abraham. *'Subtle is the Lord—'*, p. 477.

396
What are the primary colors in light?

Color is determined by the wavelength of its light (the distance between one crest of the light wave and the next). Those colors that blend to form "white light" are from shortest wave length to longest: red, orange, yellow, green, blue, indigo, and violet. All these monochromatic colors, except indigo, occupy large areas of the spectrum (entire range of wavelengths produced when a beam of electromagnetic radiation is broken up). These colors can be seen when a light beam is refracted through a prism. Some consider the primary colors to be six monochromatic colors that occupy large areas of the spectrum: red, orange, yellow, green, blue, and violet. Many physicists recognize three primary colors: red, yellow, and blue; or red, green, and blue; or red, green, and violet. All other colors can be made from these by adding two primary colors in various proportions. Within the spectrum, scientists have discovered 55 distinct hues. Infra-red and infra-violet rays at each end of the spectrum are invisible to the human eye.

Sources: Field, Gary C. *Color and Its Reproduction*, pp. 8–9; *The World Almanac and Book of Facts 1992*, p. 305.

397
Why does a double sonic boom occur when the space shuttle enters the atmosphere?

As long as an airborne object, such as a plane, is moving below the speed of sound (called Mach 1), the disturbed air remains well in front of the craft. But as the craft passes Mach 1 and is flying at supersonic speeds, a sharp air pressure rise occurs in front of the craft. In a sense the air molecules are crowded together and collectively impact. What is heard is a claplike thunder called a sonic boom or a supersonic bang. There are many shocks coming from a supersonic aircraft but these shocks usually combine to form two main shocks, one coming from the nose of the aircraft and the other from the aft. Each of the shocks moves at different velocities. If the time difference between the two shock waves is greater than 0.10 seconds apart, two sonic booms will be heard. This usually occurs when the aircraft ascends quickly. If the aircraft ascends more slowly, the two booms will sound like only one boom to the observer.

Sources: Cazeau, Charles J. *Science Trivia*, pp. 74–75; *Van Nostrand's Scientific Encyclopedia*, vol. 2, p. 2744.

398
Why does the color of clothing appear different in direct light than it does in a store under fluorescent light?

White light is a blend of all the colors, and each color has a different wavelength. Although sunlight and fluorescent light both appear as "white light," they each contain slightly different mixtures of these varying wavelengths. When sunlight and fluorescent light (white light) are absorbed by the piece of clothing, only some of the wavelengths (composing white light) reeflect from the clothing. When the retina of the eye perceives the "color" of the clothing, it is really perceiving these reflected wavelengths. The mixture of wavelengths determines the color perceived. This is why an article of clothing sometimes appears to be a different color in the store than it does on the street.

Sources: Trefil, James. *1001 Things Everyone Should Know About Science*, pp. 111-113.

399 ▽ *What causes the sounds that are heard in a seashell?*

When a seashell is held to an an ear, the sounds heard are ambient, soft sounds that have been resonated and thereby amplified by the seashell's cavity. The extreme sensitivity of the human ear to sound is illustrated by the seashell resonance effect.

Sources: Berg, Richard E. *The Physics of Sound*, p. 145.

400 ▽ *What is the Doppler effect?*

The Austrian physicist, Christian Doppler (1803-1853) in 1842 explained the phenomenon of the apparent change in wavelength of radiation—such as sound or light—emitted either by a moving body (source) or by the moving receiver. The frequency of the wave-lengths increases and the wavelength becomes shorter as the moving source approaches, producing high-pitched sounds and bluish light (called blue shift). Likewise as the source recedes from the receiver the frequency of the wavelengths decreases, the sound is pitched lower and light appears reddish (called red shift). This Doppler effect is commonly demonstrated by the train whistle of an approaching train or jet aircraft.

There are three differences between acoustical (sound) and optical (light) Doppler effects: The optical frequency change is not dependent on which is moving—the source or observer—nor is it affected by the medium through which the waves are moving, but acoustical frequency is affected by such conditions. Optical frequency changes are affected if the source or observer moves at right angles to the line connecting the source and observer. Observed acoustical changes are not affected in such a situation. Applications of the Doppler phenomenon include the Doppler radar and the measurement by astronomers of the motion and direction of celestial bodies.

Sources: Brennan, Richard P. *Dictionary of Scientific Literacy*, pp. 79-80; Parker, Sybil P. *McGraw-Hill Concise Encyclopedia of Science and Technology*, p. 616; Rickard, Teresa. *Barnes & Noble Thesaurus of Physics*, p. 110.

401
What are the characteristics of alpha, beta, and gamma radiation?

Radiation is a term that describes all the ways energy is emitted by the atom as x-rays, gamma rays, neutrons, or as charged particles. Most atoms, being stable, are nonradioactive; but some are unstable and give off either particles or gamma radiation. Substances bombarded by radioactive particles can become radioactive and yield alpha particles, beta particles, and gamma rays.

Alpha particles, first identified by Antoine Henri Becquerel (1852–1908), have a positive electrical charge and consist of two protons and two neutrons. Because of their great mass, alpha particles can travel only a short distance, around 5 centimeters in air, and can be stopped by a sheet of paper.

Beta particles, identified by Ernest Rutherford (1871–1937), are extremely high-speed electrons or protons that move at the speed of light. They can travel far in air and can pass through solid matter several millimeters thick.

Gamma rays, identified by Marie (1867–1934) and Pierre Curie (1859–1906), are similar to x-rays, but usually have a shorter wave length. These rays, which are bursts of protons, or very short–wave electromagnetic radiation, travel at the speed of light. They are much more penetrating than either the alpha or beta particles and can go through 7 inches of lead.

Sources: Brennan, Richard P. *Dictionary of Scientific Literacy*, pp. 253–255; *World Book Encyclopedia*, vol. 2, p. 274; vol. 16, p. 97.

402
What were Anders Angstrom's contributions to the development of spectroscopy?

Swedish physicist and astronomer, Anders Jonas Angstrom (1814–1874), was one of the founders of spectroscopy. His early work provided the foundation for spectrum analysis (analysis of the ranges of electromagnetic radiation emitted or absorbed). He investigated the sun spectra as well as that of the Aurora Borealis. In 1868, he established measurements for wavelengths of greater than 100 Frauenhofer. In 1907, the angstrom (Å, equal to 10^{-10m}), a unit of wavelength measurement, was officially adopted.

Sources: *The Biographical Dictionary of Scientists: Physicists*, p. 11; *Encyclopedic Dictionary of Science*, p. 23.

403
What is the sound frequency of the musical scale?

EQUAL TEMPERED SCALE

Note	Frequency	Note	Frequency
C♭	261.63	G	392.00
C#	277.18	G#	415.31
D	293.67	A	440.00

Note	Frequency	Note	Frequency
D#	311.13	A#	466.16
E	329.63	B	493.88
F	349.23	C♮	523.25
F#	369.99		

Notes: ♭ indicates flat; # indicates sharp; ♮ indicates return to natural.

The lowest frequency distinguishable as a note is about 30 hertz. The highest audible frequency is about 40 thousand hertz. A hertz (symbol Hz) is a unit of frequency that measures the number of the wave cycles per second frequency of a periodic phenomenon whose periodic time is one second (cycles per second).

Sources: Childs, W.H.J. *Physical Constants*, p. 90; *Sound*, p. 241.

404 / What is the speed of sound?

The speed of sound is not a constant; it varies depending on the medium in which it travels. The measurement of sound velocity in the medium of air must take into account many factors, including air temperature, pressure, and purity. At sea level and 32°F (0°C), scientists do not agree on a standard figure; estimates range from 740 to 741.5 miles (1191.6 to 1193.22 kilometers) per hour. As air temperature rises, sound velocity increases. Sound travels faster in water than in air and even faster in iron and steel. Sounds traveling a mile in 5 seconds in air, will travel the same distance in 1 second underwater and travel ⅓ of a second in steel.

Sources: Brennan, Richard B. *The Dictionary of Scientific Literacy*, p. 276; *The New York Times*, (May 27, 1986), pp. C1, C5; *The World Almanac and Book of Facts 1992*, p. 35.

405 / Why was the Michelson-Morley experiment important?

This experiment on light waves, first carried out in 1881 by physicists Albert A. Michelson (1852–1931) and E.W. Morley (1838–1923) in the United States, is one of the historically significant experiments in physics and led to the development of Einsteins's theory of relativity. The original experiment, using the Michelson interferometer, attempted to detect the velocity of the earth with respect to the hypothetical "luminiferous ether," a medium in space proposed to carry light waves. The procedure measured the speed of light in the direction of the earth and the speed of light at right angles to the earth's motion. No difference was found. This result discredited the ether theory and ultimately led to the proposal by Albert Einstein (1879–1955) that the speed of light is a universal constant.

Sources: Bolton, W.C. *Physics Experiments and Projects*, vol. 2, pp. 29-30; *New Encyclopedia Brittanica*, vol. 8, p. 98.

▼ COMMUNICATION

CODES, SYMBOLS, AND WRITING

406 ▽ *How much does a cubic foot of books weigh?*

A cubic foot of books weighs 65 pounds (29.48 kilograms). This does not take into consideration any type of variation in paper quality, binding, etc.

Sources: Ballast, David Kent. *Architects' Handbook of Formulas, Tables and Mathematical Calculations*, p. 461.

407 ▽ *What is the international phonetic alphabet?*

Letter	Phonetic equivalent
A	Alpha
B	Bravo
C	Charlie
D	Delta
E	Echo
F	Fox Trot
G	Gold
H	Hotel
I	India
J	Juliett
K	Kilo
L	Lima
M	Mike
N	November
O	Oscar
P	Papa
Q	Quebec
R	Romeo
S	Sierra
T	Tango
U	Uniform
V	Victor

Letter	Phonetic equivalent
W	Whiskey
X	X-ray
Y	Yankee

Sources: Helms, Harry L. *Shortwave Listening Handbook*, p. 234.

408

What is the fog index formula?

This is a system that has been developed to judge the reading level of a text. It is calculated in three steps:

1. From several randomly-selected samples of about 100 words divide the number of words in the total sample by the number of complete thought sentences to get average sentence average. Count every independent clause (separated by colons, commas, and semicolons) as a thought sentence.
2. Count the number of "hard" words having three of more syllables. Do not count as "hard" words the following: proper nouns, compound words made from shorter words (e.g., basketball), or verb forms whose third syllable ends in "ed" or "es" (e.g., "decided"). Count the syllables as they are generally pronounced, not as the dictionary divides them. If the same word is repeated, count it again. Divide the number of hard words by the total number of words in the sample. The figure should be computed to four decimal places. (In order to avoid a distorted result, terms such as "nuclear magnetic resonance" should be counted as "NMR.")
3. Add the end number from 1 and 2 together and multiply by 0.4. This will give the grade level of the reading sample.

Sources: Fisher, David J. *Rules of Thumb for Engineers and Scientists*, pp. 79-90.

409

What are the "10-codes"?

There are almost as many different codes as there are agencies using codes in radio transmission. The following are officially suggested by the Associated Public Safety Communications Officers (APCO):

Ten-1   Cannot understand your message.
Ten-2   Your signal is good.
Ten-3   Stop transmitting.
Ten-4   Message received ("O.K.").
Ten-5   Relay information to _____.
Ten-6   Station is busy.
Ten-7   Out of service.
Ten-8   In service.
Ten-9   Repeat last message.
Ten-10   Negative ("no").

Ten-11	_____ in service.
Ten-12	Stand by.
Ten-13	Report _____ conditions.
Ten-14	Information.
Ten-15	Message delivered.
Ten-16	Reply to message.
Ten-17	Enroute.
Ten-18	Urgent.
Ten-19	Contact _____.
Ten-20	Unit location.
Ten-21	Call _____ by telephone.
Ten-22	Cancel last message.
Ten-23	Arrived at scene.
Ten-24	Assignment completed.
Ten-25	Meet _____.
Ten-26	Estimated time of arrival is _____.
Ten-27	Request for information on license.
Ten-28	Request vehicle registration information.
Ten-29	Check records.
Ten-30	Use caution.
Ten-31	Pick up.
Ten-32	Units requested.
Ten-33	Emergency! Officer needs help.
Ten-34	Correct time.

Sources: _Gene Hughes' Police Call Radio Guide_, back cover.

▽ **410** /
Which animals other than horses have been used to deliver the mail?

During the 19th century, cows hauled mail wagons in some German towns. In Texas, New Mexico, and Arizona, camels were used. In Russia and Scandinavia, reindeer pulled mail sleighs. The Belgian city of Liége even tried cats, but they proved to be unreliable.

Sources: _Reader's Digest Practical Problem Solver_, p. 161.

▽ **411** /
What were Enigma and Purple in World War II?

Enigma and _Purple_ were the electric rotor cipher machines of the Germans and Japanese, respectively. The Enigma machine, used by the Nazis, was invented in the 1920s. The Japanese machine, called Purple by the Americans, was adapted from Enigma.

Cryptography—the art of sending messages in such a way that the real meaning is hidden from everyone but the sender and the recipient—is done in two ways: code and cipher. A code is like a dictionary in which all the words and phrases are replaced by codewords or codenumbers. A codebook is used to read the code. A cipher works with single letters, rather than complete words or phrases. There are two kinds of ciphers: transposition and substitution. In a transposition cipher, the letters of the ordinary message (or plain text) are jumbled to form the cipher text. In substitution, the plain text letters can be replaced by other letters, numbers, or symbols.

Sources: Norman, Bruce. *Secret Warfare: The Battle of Codes and Ciphers*, pp. 13-14; Yost, Graham. *Spy Tech*, p. 216.

412 / What is the Morse Code?

The success of any electrical communication system lies in its coding interpretation, for only series of electric impulses can be transmitted from one end to the other. These impulses must be "translated" from/into words, numbers, etc. This problem plagued early telegraphy until American painter-turned-scientist Samuel F.B. Morse (1791-1872) with the help of Alfred Vail (1807-1859) devised in 1835 a code composed of dots and dashes to represent letters, numbers, and punctuation. Telegraphy uses an electromagnet—a device that becomes magnetic when activated and raps against a metal contact. A series of short electrical impulses repeatedly can make and break this magnetism, resulting in a tapped-out message.

Having secured a patent on the code in 1837, Morse and Vail established a communications company on May 24, 1844. The first long-distance telegraphed message was sent by Morse in Washington, D.C. to Vail in Baltimore. This was the same year that Morse took out a patent on telegraphy; Morse never acknowledged the unpatented contributions of Joseph Henry (1797-1878), who invented the first electric motor and working electromagnet in 1829 and the electric telegraph in 1831.

The International Morse Code (shown below) uses sound or a flashing light to send messages. The dot is a very short sound or flash; a dash equals three dots. The pauses between sounds or flashes should equal one dot. An interval of the length of one dash is left between letters; an interval of two dashes is left between words.

A .-	J .---	S ...
B -...	K -.-	T -
C -.-.	L .-..	U ..-
D -..	M --	V ...-
E .	N -.	W .--
F ..-.	O ---	X -..-
G --.	P .--.	Y -.--
H	Q --.-	Z --..
I ..	R .-.	
1 .----	6 -....	Period .-.-.-
2 ..---	7 --...	Comma --..--
3 ...--	8 ---..	
4 ....-	9 ----.	
5	0 -----	

Sources: *How To Do Just About Anything*, p. 259; Panati, Charles. *Panati's Browser's Book of Beginnings*, pp. 212-214.

Who invented the Braille alphabet?

The Braille system, used by the blind to read and write, are combinations of raised dots that form characters corresponding to the letters of the alphabet, punctuation marks, and common words such as "and" and "the." Louis Braille (1809-1852), blind himself since the age of three, began working on developing a practical alphabet for the blind shortly after he started a school for the blind in Paris. He experimented with a communication method called night-writing, which the French army used for nighttime battlefield missives. With the assistance of an army officer, Captain Charles Barbier, Braille pared the method's 12-dot configurations to a 6-dot one and devised a code of 63 characters. The system was not widely accepted for several years; even Braille's own Paris school did not adopt the system until 1854 (two years after his death). In 1916, the United States sanctioned Louis Braille's original system of raised dots, and in 1932 a modification called "Standard English Braille, Grade 2" was adopted throughout the English-speaking world. The revised version changed the letter-by-letter codes into common letter combinations, such as "ow," "ing," and "ment," making reading and writing a faster activity.

Before Braille's system, one of the few effective alphabets for the blind was devised by another Frenchman, Valentin Haüy (1745-1822), who was the first to emboss paper to help the blind read. Haüy's letters in relief were actually a punched alphabet, and imitators immediately began to copy and improve on his system. Another letter-by-letter system of nine basic characters was devised by Dr. William Moon (1818-1894) in 1847, but it is less versatile in its applications.

Sources: *How In the World?*, p. 280; Panati, Charles. *Panati's Browser's Book of Beginnings*, pp. 84-85.

What does the code that follows the letters ISBN mean?

ISBN, or International Standard Book Number, is an ordering and identifying code for book products. It forms an unique number to identify that particular item. The first number of the series relates to the language the book is published in, for example, the zero is designated for the English language. The second set of numbers identifies the publisher and the last set of numbers are unique numbers that are used only to identify the particular item. The very last number is a "check number," which mathematically makes certain that the previous numbers have been entered correctly.

Sources: Rovin, Jeff. *Laws of Order*, p. 6.

415 / *What do the lines in a Universal Product Code mean?*

Also called a bar code, a Universal Product Code (UPC) is a product description code designed to be read by a computerized scanner or cash register. It consists of 11 numbers in groups of "0"s (dark strips) and "1"s (white strips). A bar will be thin if it has only one strip or thicker if there are two or more strips side by side.

The first number describes the type of product. Most products begin with a "0"; exceptions are variable weight products such as meat and vegetables (2), health-care products (3), bulk-discounted goods (4), and coupons (5). Since it might be misread as a bar, the number 1 is not used.

The next five numbers describe the product's manufacturer. The following five numbers after that describe the product itself including its color, weight, size, and other distinguishing characteristics. The price of the item is not included; it is coded into the cash register separately.

The last number is a check digit which is used to tell the scanner if there is an error in the other numbers. The preceding numbers, when added, multiplied, and subtracted in a certain way will equal this number. If they do not, there is a mistake somewhere.

Sources: Cone, Robert E. *How the New Technology Works*, p. 107; Hopkins, Nigel J., John W. Mayne, and John R. Hudson. *The Numbers You Need*, pp. 173-174; Rovin, Jeff. *Laws of Order*, p. 48.

FACSIMILE (FAX) *See: COMMUNICATION—TELECOMMUNICATIONS, RECORDING, ETC.*

RADAR *See: CARS, BOATS, PLANES—MOTOR VEHICLES*

RADIO AND TELEVISION

▽ **416** ▽
Who invented radio?

Guglielmo Marconi (1874-1937), of Bologna, Italy, was the first to prove that radio signals could be sent over long distances. Radio is the radiation and detection of signals propagated through space as electromagnetic waves to convey information. It was first called wireless telegraphy because it duplicated the effect of telegraphy without using wires. On December 12, 1901, Marconi successfully sent Morse code signals from Newfoundland to England.

In 1906, the American inventor Lee DeForest (1873-1961) built what he called "the Audion," which became the basis for the radio amplifying vacuum tube. This device made voice radio practical, because it magnified the weak signals without distorting them. The next year, DeForest began regular radio broadcasts from Manhattan, New York. As there were still no home radio receivers, DeForest's only audience was ship wireless operators in New York City Harbor.

Sources: Asimov, Isaac. *Asimov's Biographical Encyclopedia of Science and Technology*, pp. 650-651; Panati, Charles. *Panati's Browser's Book of Beginnings*, pp. 217-219.

▽ **417** ▽
Why do AM stations have a wider broadcast range at night?

This variation is caused by the nature of the ionosphere of the earth. The ionosphere consists of several different layers of rarefied gases in the upper atmosphere that have become conductive through the bombardment of the atoms of the atmosphere by solar radiation, by electrons and protons emitted by the sun, and by cosmic rays. These layers, sometimes called the Kennelly-Heaviside layer, reflect AM radio signals, enabling AM broadcasts to be received by radios that are great distances from the transmitting antenna. With the coming of night the ionosphere layers partially dissipate and become an excellent reflector of the short waveband AM radio waves. This causes distant AM stations to be heard more clearly at night.

Sources: Asimov, Isaac. *Understanding Physics*, p. 45; Churchman, Lee W. *Survey of Electronics*, pp. 64–66; *World Book Encyclopedia*, vol. 16, p. 85.

418

How do submerged submarines communicate?

Using frequencies from very high to extremely low, submarines can communicate by radio when submerged if certain conditions are met, and depending on whether or not detection is important. Submarines seldom transmit on long-range high radio frequencies if detection is important, as in war. However, *Super* (SHF), *Ultra* (UHF), or *Very High Frequency* (VHF) two-way links with cooperating aircraft, surface ships, via satellite, or with the shore are fairly safe with high data rate, though they all require that the boat show an antenna or send a buoy to the surface.

Sources: Moore, John E. *Submarine Warfare*, pp. 49–50.

419

How are the call letters beginning with K or W assigned to radio stations?

These beginning call letters are assigned on a geographical basis. For the majority of radio stations located east of the Mississippi River, their call letters begin with the letter *W*; if the stations are west of the Mississippi, their first call letter is the letter *K*. However there are exceptions to this rule. Stations founded before this rule went into effect kept their old letters. So, for example, KDKA in Pittsburgh has retained the first letter *K*; likewise some western pioneer stations have retained the letter *W*. Since many AM licensees also operate FM and TV stations, a common practice is to use the AM call letters followed by *-FM* or *-TV*.

Sources: *Broadcasting Yearbook 1991*, p. A2.

420

Why do FM radio stations have a limited broadcast range?

Usually radio waves higher in frequency than approximately 50 to 60 megahertz are not reflected by the earth's ionosphere, but are lost in space. Television, FM radio, and high frequency communications systems are therefore limited to approximately line-of-sight ranges. The line-of-sight distance depends on the terrain and antenna height, but is usually limited to from 50 to 100 miles (80 to 161 kilometers). FM (frequency-modulation) radio uses a wider band than AM (amplitude-modulation) radio to give broadcasts high fidelity, especially noticeable in music—crystal clarity to high frequencies and rich resonance to base notes, all with a minimum of static and distortion. Invented by Edwin Howard Armstrong (1891–1954) in 1933, FM receivers became available in 1939.

Sources: Churchman, Lee W. *Survey of Electronics*, p. 66; Harris, Harry. *Good Old-Fashioned Yankee Ingenuity*, pp. 128-129; Panati, Charles. *Panati's Browser's Book of Beginnings*, pp. 220-222.

421
What was the first radio broadcasting station?

Credit for the first radio broadcasting station has customarily gone to Westinghouse station KDKA in Pittsburgh for its broadcast of the Harding-Cox presidential election returns on November 2, 1920. Although there were other earlier radio transmissions, KDKA was one of the first stations to use electron tube technology to generate the transmitted signal and hence to have what could be described as broadcast quality. It was the first to have a well-defined commercial purpose—it was not a hobby or a publicity stunt. It was the first broadcast station to be licensed on a frequency outside the amateur bands. Altogether, it was the direct ancestor of modern broadcasting.

Sources: Inglis, Andrew F. *Behind the Tube: A History of Broadcasting Technology and Business*, p. 61.

422
Can radio transmissions between space shuttles and ground control be picked up by shortwave radio?

Amateur radio operators at Goddard Space Flight Center, Greenbelt, Maryland, retransmit shuttle space-to-ground radio conversations on shortwave frequencies. These retransmissions can be heard freely around the world. To hear astronauts talking with ground controllers during liftoff, flight, and landing, a shortwave radio capable of receiving single-sideband signals should be tuned to frequencies of 3.860, 7.185, 14.295, and 21.395 megahertz. British physics teacher Geoffrey Perry, at the Kettering Boys School, has taught his students how to obtain telemetry from orbiting Russian satellites. Since the early 1960s Perry's students have been monitoring Russian space signals using a simple taxicab radio and using the data to calculate position and orbits of the spacecraft.

Sources: Curtis, Anthony R. *Space Almanac*, p. 278; *Science Digest*, vol. 92, no. 1, (January, 1984), pp. 32, 36, 81.

423
Who was the father of television?

The idea of television (or "seeing by electricity" as it was called in 1880) was offered by several people over the years, and several individuals contributed a multiplicity of partial inventions. For example, in 1897 Ferdinand Braun (1850-1918) constructed the first cathode ray oscilloscope, a fundamental component to all television receivers. In 1907, Boris Rosing proposed using Braun's tube to receive images, and in the following year Alan Campbell-Swinton likewise suggested using the tube, now called the cathode-ray tube, for both transmission and receiving. However, the key figure, frequently called the father of television was the Russian-born American, Vladimir K. Zworykin (1889-1982). A former pupil of Rosing, he produced a practical method of amplifying the electron beam so that the light/dark pattern would produce a good image. In 1923, he patented the iconoscope (which would become the television camera) and in 1924 he patented the kinoscope (television tube). Both inventions rely on streams of electrons for both scanning and creating the image on a fluorescent screen. By 1938, after adding new and more sensitive photo cells, Zworykin demonstrated his first practical model.

During the earlier twentieth century others worked on different approaches to television. The best-known is John L. Baird (1888-1946) who in 1936, used a mechanized scanning device, to transmit the first recognizable picture of a human face. Limitations in his designs made any further improvements in the picture quality impossible.

Sources: Asimov, Isaac. *Understanding Physics*, vol. 3, pp. 48-51; Cipolla, Carlo M., and Derek Birdsall. *The Technology of Man*, p. 236; McNeil, Ian. *An Encyclopedia of the History of Technology*, pp. 744-746.

424 / *How does rain affect television reception from a satellite?*

The incoming microwave signals are absorbed by rain and moisture, and severe rainstorms can reduce signals by as much as 10 decibels (reduction by a factor of 10). If the installation cannot cope with this level of signal reduction, the picture may be momentarily lost. Even quite moderate rainfall can reduce signals enough to give noisy reception on some receivers. Another problem associated with rain is an increase in noise due to its inherent noise temperature. Any body above the temperature of absolute zero ($0°K$ or $-459°F$ or $-273°C$) has an inherent noise temperature generated by the release of wave packets from the body's molecular agitation (heat). These wave packets have a wide range of frequencies, some of which will be within the required bandwidth for satellite reception. The warm earth has a high noise temperature, and consequently rain does as well.

Sources: Stephenson, D.J. *Newnes Guide to Satellite TV*, p. 20.

425 / *What name is used for a satellite dish that picks up TV broadcasts?*

Earth station is the term used for the complete satellite receiving or transmitting station. It includes the antenna, the electronics, and all associated equipment necessary to receive or transmit satellite signals. It can range from a simple inexpensive receive-only earth station that can be purchased by the individual consumer, to elaborate, two-way communications stations that offer commercial access to the satellite's capacity. Signals are captured and focused by the antenna into a feedhorn and low noise amplifier. These are relayed by cable to a down converter and then into the satellite receiver/modulator.

Sources: Baylin, Frank, and Brent Gale. *The Home Satellite TV Installation and Troubleshooting Manual*, pp. 3, 5, 307; Green, James Harry. *The Dow Jones-Irwin Handbook of Telecommunications*, p. 448.

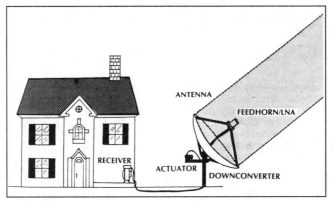

Earth station

426 What is high definition television?

The amount of detail shown in a television picture is limited by the number of lines that make it up and by the number of picture elements on each line. The latter is mostly determined by the width of the electron beam. To obtain pictures closer to the quality associated with 35 millimeter photography, a new television system HDTV (*High Definition Television*) has been developed which will have more than twice the number of scan lines with a much smaller picture element. Currently American and Japanese television has 525 scanning lines, while Europe uses 625 scanning lines. HDTV has received wide publicity in recent years, but it is currently in an engineering phase, and not yet commercially available.

The Japanese are generally given credit for being pioneers in HDTV, ever since NHK, the Japanese broadcasting company, began research in 1968. In fact, the original pioneer was RCA's Otto Schade who began his research after the end of World War II. Schade was ahead of his time, and decades passed before television pickup tubes and other components became available to take full advantage of his research.

HDTV cannot be used in the commercial broadcast bands until technical standards are approved by the United States Federal Communications Commission (FCC) or the various foreign regulating agencies. The more immediate problem however, is a technological one—HDTV needs to transmit five times more data than is currently assigned to each television channel. One approach is signal compression—squeezing the 30 megahertz bandwidth signal that HDTV requires into the 6 megahertz bandwidth currently used for television broadcasting. The Japanese and Europeans are exploring analog systems that use wavelike transmission, while the Americans are basing their HDTV development on digital transmission systems.

Sources: Brennan, Richard P. *Dictionary of Scientific Literacy*, pp. 140-141; Inglis, Andrew F. *Behind the Tube: A History of Broadcasting Technology and Business*,pp. 473-487; *The New Book of Popular Science*, vol. 6, p. 259.

TELECOMMUNICATIONS, RECORDING, ETC.

427
How does a car phone work?

There are three parts to a cellular phone system: the installation in the car; the cell site; and the mobile telephone switching office. Each cell has at its center a cell site where the fixed radio receiver and transmitter are located. All the cell sites belonging to a particular system are connected together at a mobile telephone switching office, which ties them to the local phone system. As a call passes from one cell site and enters into another, the call is transferred or "handed off" to an adjoining cell without any noticeable interruption.

Sources: Bernard, Josef. *The Cellular Connection*, p. 7.

428
What is the Dolby noise reduction system?

The magnetic action of the tape produces a background hiss—a drawback in sound reproduction on tape. A noise reduction system known as *Dolby* (named after R.M. Dolby (b. 1933), its American inventor) is widely used to deal with the hiss. In quiet passages, electronic circuits automatically boost the signals before they reach the recording head, drowning out the hiss. On playback, the signals are reduced to their correct levels. The hiss is reduced at the same time, becoming inaudible.

Sources: *How in the World?*, p. 221.

429
What is digital audio tape (DAT)?

DAT, a new concept in magnetic recording, produces a mathematical value for each sound based on the binary code. When the values are reconstructed during playback, the reconstructed sound is so much like the original that the human ear cannot distinguish the difference. The reproduction is so good that American record companies have lobbied lawmakers to prevent DAT from being sold in the United States, claiming that it could encourage illegal CD copying.

Sources: *The Almanac of Science and Technology*, p. 247.

How are compact discs (CDs) made?

The master disc for a CD is an optically flat glass disc coated with a resist. The resist is a chemical that is impervious to an etchant that dissolves glass. The master is placed on a turntable. The digital signal to be recorded is fed to the laser, turning the laser off and on in response to the binary on-off signal. When the laser is on, it burns away a small amount of the resist on the disc. While the disc turns, the recording head moves across the disc, leaving a spiral track of elongated "burns" in the resist surface. After the recording is complete, the glass master is placed in the chemical etchant bath. This developing removes the glass only where the resist is burned away. The spiral track now contains a series of small pits of varying length and constant depth. To play a recorded CD, a laser beam scans the three miles (five kilometers) of playing track and converts the "pits" and "lands" of the CD into binary codes. A photodiode converts these into a coded string of electrical impulses. In October 1982, the first CDs were marketed; they were invented by Phillips (Netherlands) Company and Sony in Japan in 1978.

Sources: Flatow, Ira. *Rainbows, Curve Balls, and Other Wonders of the Natural World*, pp. 217-221; *The New Book of Popular Science*, vol. 6, p. 268.

Compact disk player

What is the lifespan of a CD-ROM disc?

Although manufacturers claim that a CD-ROM disc will last twenty years, recent statements by the United States National Archives and Records Administration suggest that a lifespan of three to five years is more accurate. The main problem is that the aluminum substratum on which the data is recorded is vulnerable to oxidation.

Sources: *PRLC Technical Bulletin*, new series no. 33 (January/February 1992), item 16.

432 ▽ What is a Clarke belt?

Back in 1945, Arthur C. Clarke (b. 1917), the famous scientist and science fiction writer, predicted that an artificial satellite placed at a height of 22,248 miles (35,803 kilometers) directly above the equator would orbit the globe at the same speed with which the earth was rotating. As a result, the satellite would remain stationary with respect to any point on the earth's surface. This equatorial belt, rather like one of Saturn's rings, is affectionately known as the Clarke belt.

Sources: Stephenson, D.J. *Newnes Guide to Satellite TV*, pp. 10–11.

433 ▽ When was the first commercial communications satellite used?

In 1960 *ECHO 1*, the first communications satellite, was launched. Two years later, on July 10, 1962, the first commercially-funded satellite, *Telstar 1*, (paid for by American Telephone and Telegraph) was launched into low earth orbit. It was also the first true communications satellite, being able to relay not only data and voice, but television as well. The first broadcast, which was relayed from the United States to England, showed an American flag flapping in the breeze. The first commercial satellite (in which its operations are conducted like a business) was *Early Bird* which went into regular service on June 10, 1965 with 240 telephone circuits. *Early Bird* was the first satellite launched for Intelsat (International Telecommunications Satellite Organization). Still in existence, the system is owned by member nations—each nation's contribution to the operating funds are based on its share of the system's annual traffic.

Sources: Gatland, Kenneth. *The Illustrated Encyclopedia of Space Technology*, pp. 90-92; McNeil, Ian. *An Encyclopedia of the History of Technology*, pp. 747-748.

434 ▽ How does a telefax work?

Telefacsimile (also telefax or facsimile or fax) transmits graphic and textual information from one location to another through telephone lines. A transmitting machine uses either a digital or analog scanner to convert the black and white representations of the image into electrical signals that are transmitted through the telephone lines to a designated receiving machine. The receiving unit converts the transmission back to an image of the original and prints it. In its broadest definition, a facsimile terminal is simply a copier equipped to transmit and receive graphics images.

Sources: Green, James Harry. *The Dow Jones-Irwin Handbook of Telecommunications*, pp. 561-577.

▽435▽ What is a fiber optic cable?

A fiber optic cable is composed of many very thin strands of coated glass fibers. It can transmit messages or images by directing beams of light inside itself over very short or very long distances (thousands of miles). The pattern of light waves forms a code that carries a message. At the receiving end the light beams are converted back into electric current and decoded. Since light beams are immune to electrical noise and can be carried greater distances before fading, this technology is used heavily in telecommunications. Other applications include using medical fiber optic viewers, such as endoscopes and fiberscopes, to see internal organs; fiber optic message devices in aircraft and space vehicles; and fiber optic connections in automotive lighting systems.

Sources: Cone, Richard J. *How the New Technology Works*, pp. 30-33.

TELEGRAPH *See: COMMUNICATION—CODES, SYMBOLS, AND WRITING*

TELEPHONE *See: COMMUNICATION—TELECOMMUNICATIONS, RECORDING, ETC.*

TELEVISION *See: COMMUNICATION—RADIO AND TELEVISION*

WRITING *See: COMMUNICATION—CODES, SYMBOLS, AND WRITING*

THE EARTH

AGE *See: SPACE—PLANETS AND MOONS*

AIR *See also: WEATHER AND CLIMATE*

436

What is the weight of one cubic foot of dry air at 1 atmosphere of barometric pressure?

Temperature (Fahrenheit)	Weight per cubic foot (pounds)
50°	0.07788
60°	0.07640
70°	0.07495

Sources: Kidder, Frank. *Architects' and Builders' Handbook*, p. 1569

437

Why is the sky blue?

The sunlight interacting with the earth's atmosphere makes the sky blue. In outer space the astronauts see blackness because outer space has no atmosphere. Sunlight consists of light waves of varying wavelengths, each of which is seen as a different color. The minute particles of matter and molecules of air in the atmosphere intercept and scatter the white light of the sun. A larger portion of the blue color in white light is scattered, more so than any other color because the blue wavelengths are the shortest. When the size of atmospheric particles are smaller than the wavelengths of the colors, selective scattering occurs—the particles only scatter one color and the atmosphere will appear to be that color. Blue wavelengths especially are affected, bouncing off the air particles to become visible. This is why the sun looks yellow (yellow equals white minus blue). At sunset, the sky changes color because as the sun drops to the horizon, sunlight has more atmosphere to pass through and loses more of its blue wavelengths (the shortest of all the colors). The orange and red, having the longer wavelengths and making up more of sunlight at this distance, are most likely to be scattered by the air particles.

Sources: Flatow, Ira. *Rainbows, Curve Balls, and Other Wonders of the Natural World Explained*, pp. 33-35; Forrester, Frank H. *1001 Questions Answered About the Weather*, p. 167; Trefil, James. *1001 Things Everyone Should Know About Science*, p. 113.

438
What is the composition of the earth's atmosphere?

The earth's atmosphere, apart from water vapor and pollutants, is composed of 78% nitrogen, 20% oxygen, and less than 1% each of argon and carbon dioxide. There are also traces of hydrogen, neon, helium, krypton, xenon, methane, and ozone. The earth's original atmosphere was probably composed of ammonia and methane; 20 million years ago the air started to contain a broader variety of elements.

Sources: Brennan, Richard P. *Dictionary of Scientific Literacy*, p. 18.

439
How many layers are there in the earth's atmosphere?

The atmosphere, the "skin" of gas that surrounds the earth, consists of five layers that are differentiated by temperature:

The troposphere—is the lowest level; it averages about 7 miles (11 kilometers) in thickness, varying from 5 miles (8 kilometers) at the poles to 10 miles (16 kilometers) at the equator. Most clouds and weather form in this layer. Temperature decreases with altitude in the troposphere.

The stratosphere—ranges between 7 miles (11 kilometers) to 30 miles (48 kilometers) above the earth's surface. The ozone layer, important because it absorbs most of the sun's harmful ultraviolet radiation, is located in this band. Temperatures rise slightly with altitude to a maximum of about 32°F (0°C).

The mesosphere—(above the stratosphere) extends from 30 miles (48 kilometers) to 55 miles (85 kilometers) above the earth. Temperatures decrease with altitude to -130°F (-90°C).

The thermosphere—(also known as the hetereosphere) is between 55 miles (85 kilometers) to 435 miles (700 kilometers). Temperatures in this layer range to 2696°F (1475°C).

The exosphere—beyond the thermosphere, applies to anything above 435 miles (700 kilometers). In this layer, temperature no longer has any meaning.

The ionosphere—is a region of the atmosphere that overlaps the others, reaching from 30 miles (48 kilometers) to 250 miles (402 kilometers). In this region, the air becomes ionized (electrified) from the sun's ultraviolet rays, etc. This area affects the transmission and reflection of radio waves. It is divided into three regions: the D region (at 35 to 55 miles, or 56 to 88 kilometers), the E Region

(Heaviside-Kennelly Layer, 55 to 95 miles, or 56 to 153 kilometers), and the F Region (Appleton Layer, at 95 to 250 miles, or 153 to 402 kilometers).

Sources: *The Guinness Book of Answers*, p. 96; Rovin, Jeff. *Laws of Order*, p. 91; *The Universal Almanac 1992*, p. 494.

440 / What is the density of air?

The density of dry air is 1.29 grams per liter at 32°F (0°C) at average sea level. Average sea level, measured by the height of a column of mercury in a barometer, is 29.92 inches (760 millimeters).

Sources: *Hawley's Condensed Chemical Dictionary*, p. 29.

441 / What are the Van Allen belts?

The Van Allen belts (or zones) are two regions of highly–charged particles above the earth's equator trapped by the magnetic field, which surrounds the earth. Also called the magnetosphere, the first belt extends from a few hundred to about 2,000 miles (3,200 kilometers) above the earth's surface and the second is between 9,000 and 12,000 miles (14,500 to 19,000 kilometers). The particles, mainly protons and electrons, come from the solar wind and cosmic rays. The belts are named in honor of James Alfred Van Allen (b.1914), the American physicist who discovered them in 1958 and 1959, with the aid of radiation counters carried aboard the artificial satellites, *Explorer I* (1958) and *Pioneer 3* (1959).

Sources: *The Biographical Dictionary of Scientists: Astronomers*, pp. 154-155; Moore, Patrick, and Garry Hunt. *The Atlas of the Solar System*, pp. 140-141; Room, Adrian. *Dictionary of Astronomical Names*, p. 160.

DIMENSIONS See: THE EARTH—PHYSICAL CHARACTERISTICS, ETC.; SPACE—PLANETS AND MOONS

EARTHQUAKES See: THE EARTH—VOLCANOES AND EARTHQUAKES

LAND See also: THE EARTH—VOLCANOES AND EARTHQUAKES; SPACE—PLANETS AND MOONS

442 / What is a tufa?

It is a general name for calcium carbonate ($CaCO_3$) deposits or spongy porous limestone found at springs in limestone areas, or in caves as massive stalactite or stalagmite deposits. Tufa, derived from the Italian word for "soft rock," is formed by the precipitation of calcite from the water of streams and springs.

Sources: *Encyclopaedic Dictionary of Physical Geography*, p. 445; Leet, L. Don and Sheldon Judson. *Physical Geography*, p. 143.

443

How does a stalactite differ from a stalagmite?

A stalactite is a conical or cylindrical calcite ($CaCO_3$) formation hanging from a cave roof. It forms from the centuries-long buildup of mineral deposits resulting from the seepage of water from the limestone rock above the cave. This water containing calcium bicarbonate evaporates, losing some carbon dioxide, to deposit small quantities of calcium carbonate (carbonate of lime), which eventually forms a stalactite.

A stalagmite is a stone formation which develops upward from the cave floor and resembles an icicle upside down. Formed from water containing calcite which drips from the limestone walls and roof of the cave, it sometimes joins a stalactite to form a column.

Sources: Bates, Robert L., and Julia A. Jackson. *Glossary of Geology*, p. 640; Leet, L. Don, and Sheldon Judson. *Physical Geology*, pp. 269-270.

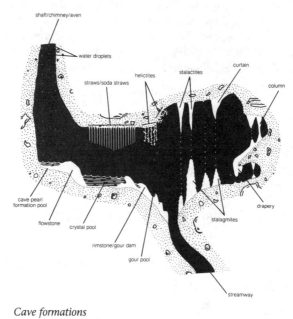

Cave formations

444
What is the difference between spelunking and speleology?

Spelunking, or sport caving, is exploring caves as a hobby or for recreation. Speleology is the scientific study of caves and related phenomena, such as the world's deepest cave, Réseau Jean Bernard, Haute Savoie, France with a depth of 5,256 feet (1,602 meters), or the world's longest cave system, Mammoth Cave in Kentucky, with a length of 348 miles (560 kilometers).

Sources: Halliday, William R. *Depths of the Earth*, p. 407.

445
How is "speleothem" defined?

Speleothem is a term given to those cave features that form after a cave itself has formed. They are secondary mineral deposits that are created by the solidification of fluids or from chemical solutions. These mineral deposits usually contain calcium carbonate ($CaCO_3$) or limestone, but gypsum or silica may also be found. Stalactites, stalagmites, soda straws, cave coral, boxwork and cave pearls are all types of speleothems.

Sources: Mayhew, Susan, and Anne Penny. *The Concise Oxford Dictionary of Geography*, p. 28; Silverman, Sharon H. *Going Underground*, pp. 14-25.

446
Which natural attractions in the United States are the most popular?

1. The Grand Canyon, Arizona
2. Yellowstone National Park, Wyoming
3. Niagara Falls, New York
4. Mount McKinley, Alaska
5. California's "Big Trees": the sequoias and redwoods
6. Hawaii's volcanoes
7. Florida's Everglades

Sources: Wallechinsky, David. *The Book of Lists*, p. 25.

447
What are the LaBrea tar pits?

The tar pits are located in an area of Los Angeles, California formerly known as Rancho LaBrea. Heavy, sticky tar oozed out of the earth, the scum from great petroleum reservoirs far underground. The pools were cruel traps for uncounted numbers of animals. Today the tar pits are a part of Hancock Park. In the park, many fossil remains are displayed along with life-sized reconstructions of these prehistoric species.

Sources: *Cities of the United States: The West*, p. 66; *Natural History*, vol. 46 (December 1940), p. 284.

448
Where are the world's largest deserts?

A desert is an area that receives little precipitation and has little plant cover. Many deserts form a band north and south of the equator at about 20 degrees latitude because moisture-bearing winds do not release their rain over these areas. As the moisture-bearing winds from the higher latitudes approach the equator, their temperatures increase and they rise higher and higher in the atmosphere. When the winds arrive over the equatorial areas and come in contact with the colder parts of the earth's atmosphere, they cool down, and release all their water to create the tropical rain forests near the equator.

Desert	Location	Area	
		Square miles	**Square kilometers**
Sahara	North Africa	3,500,000	9,065,000
Gobi	Mongolia-China	500,000	1,295,000
Kalahari	Southern Africa	225,000	582,800
Great Sandy	Australia	150,000	338,500
Great Victoria	Australia	150,000	338,500

The Sahara desert is three times the size of the Mediterranean Sea. In the United States, the largest desert is the Mojave Desert in southern California with an area of 15 thousand square miles (38,900 square kilometers).

Sources: Cazeau, Charles J. *Science Trivia*, p. 64; Diagram Group. *Comparisons*, p. 82; *The Universal Almanac 1992*, p. 303.

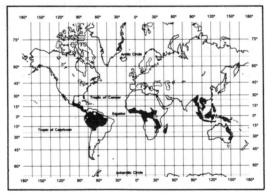

Rain forest locations

449
How tall is a sand dune?

Dunes can range from 3 feet (1 meter) to greater than 650 feet (200 meters) in height. Star dunes (having one central peaked mound with radiating ridges) can grow as high as 1,000 feet (300 meters).

Sources: *McGraw-Hill Encyclopedia of Science and Technology*, vol. 5, pp. 433-434.

450
How long is the Grand Canyon?

The Grand Canyon, cut out by the Colorado River over a period of 15 million years in the northwest corner of Arizona, is the largest land gorge in the world. It is 4 to 13 miles (6.4 to 21 kilometers) wide at its brim, 4,000 to 5,500 feet (1,219 to 1,676 meters) deep, and 217 miles (349 kilometers) long, extending from the mouth of the Little Colorado River to Grand Wash Cliffs (and 277 miles, 600 feet or 445.88 kilometers if Marble Canyon is included).

However, it is not the deepest canyon in the United States; that distinction belongs to Kings Canyon, which runs through the Sierra and Sequoia National Forests near East Fresno, California, with its deepest point being 8,200 feet (2,500 meters). Hell's Canyon of the Snake River between Idaho and Oregon is the deepest United States canyon in low-relief territory. Also called the Grand Canyon of the Snake, it plunges 7,900 feet (2,408 meters) down from Devil Mountain to the Snake River.

Sources: *Guinness Book of Records 1992*, p. 35; *Webster's New Geographical Dictionary*, p. 455; *The World Almanac and Book of Facts 1992*, p. 627.

451
What is the composition of the Rock of Gibraltar?

It is composed of gray limestone, with a dark shale overlay on parts of the western slopes. Located on a peninsula at the southern extremity of Spain, the Rock of Gibraltar is a mountain at the east end of the Strait of Gibraltar, the narrow passage between the Atlantic Ocean and the Mediterranean Sea. "The Rock" is 1,398 feet (426 meters) tall at its highest point.

Sources: *Encyclopedia Americana*, vol. 12, p. 734; *Webster's New Geographical Dictionary*, p. 442.

452
From what type of stone was Mount Rushmore National Monument carved?

Granite. The monument, in the Black Hills of southwestern South Dakota, depicts the 60-foot-high (18-meter-high) faces of four United States presidents: George Washington, Thomas Jefferson, Abraham Lincoln, and Theodore Roosevelt. Sculptor Gutzon Borglum (1867-1941) designed the monument, but died before the completion of the project; his son, Lincoln, finished it. From 1927 to 1941, 360 people, mostly construction workers, drillers, and miners, "carved" the figures using dynamite.

Sources: *How In the World*, pp. 362-364; Panati, Charles. *Panati's Extraordinary Origins of Everyday Things*, pp. 282-284; *The World Almanac and Book of Facts 1992*, p. 646.

453
Where is the southernmost city in the world?

Punta Arenas (formerly Magallanes), Chile, is farther south than any other settlement of sufficient size and having enough commercial importance to deserve being called a city. It lies on the extreme southern end of Patagonia on the Strait of Magellan at 53 degrees 10 minutes south latitude. The world's southernmost village is Puerto Williams (population about 350), Tierra del Fuego, Chile, being 680 miles (1,094 kilometers) north of Antarctica.

Sources: Stimpson, George. *Information Roundup*, p. 285; *Webster's New Geographical Dictionary*, p. 988; *The World Almanac and Book of Facts 1992*, p. 203.

454
What and where is the continental divide of North America?

The continental divide is a continuous line that runs north and south the length of North America. About 50 miles west of Denver, U.S. Highway Route 40 crosses Berthoud Pass at an altitude of 11,314 feet (3,450 meters) above sea level. At the crest of this pass is the continental divide. West of this divide, all water flows eventually to the Pacific Ocean. East of this line, the waters will flow toward the Atlantic Ocean.

Sources: Press, Frank. *Earth*, p. 180.

455
What is a hoodoo?

A hoodoo is a fanciful name for a grotesque rock pinnacle or pedestal, usually of sandstone, that is the result of weathering in a semi-arid region. An outstanding example of hoodoos occurs in the Wasatch Formation at Bryce Canyon, Utah.

Sources: Goulty, George A. *A Dictionary of Landscape*, p. 135.

456
Where are the northernmost and southernmost points of land?

The most northern point of land is Cape Morris K. Jesup on the northeastern extremity of Greenland. It is at 83 degrees, 39 minutes north latitude and is 440 miles (708 kilometers) from the North Pole. However, the *Guinness Book of Records 1992* reports that, an islet of 100 feet across, called Oodaq, is more northerly at 83 degrees, 40 minutes north latitude and 438.9 miles (706 kilometers) from the North Pole. The southernmost point of land is the South Pole (since the South Pole, unlike the North Pole, is on land).

In the United States, the northernmost point of land is Point Barrow, Alaska (71 degrees, 23 minutes north latitude), and the southernmost point of land is Ka Lae or South Cape (18 degrees, 55 minutes north latitude) on the island of Hawaii.

In the 48 contiguous states, the northernmost point is Northwest Angle, Minnesota (49 degrees, 23 minutes north latitude); the southernmost point is Key West, Florida (24 degrees, 33 minutes north latitude).

Sources: *Guinness Book of Records 1992*, p. 25; Stimpson, George. *Information Roundup*, p. 416; *The World Almanac and Book of Facts 1992*, p. 384.

457
Do the continents move?

In 1912, a German geologist, Alfred Lothar Wegener (1880-1930), theorized that the continents had drifted or floated apart to their present locations and that once all the continents had been a single land mass near Antarctica, which is called *Pangaea* (from the Greek word meaning all-earth). Pangaea then broke apart some 200 million years ago into two major continents called Laurasia and Gondwanaland. These two continents continued drifting and separating until the continents evolved their present shapes and positions. Wegener's theory was discounted but it has since been found that the continents do move sideways (not drift) at an estimated 0.75 inch (1.9 centimeters) annually because of the action of plate tectonics. American geologist, William Maurice Ewing (1906-1974) and Harry Hammond Hess (1906-1969) proposed that the earth's crust is not a solid mass, but composed of eight major and seven minor plates that can move apart, slide by each other, collide, or override each other. Where these plates meet are major areas of mountain-building, earthquakes, and volcanoes.

Sources: Asimov, Isaac. *Isaac Asimov's Guide to Earth and Space*, pp. 42-43; Byrum, W.F., et al. *Dictionary of the History of Science*, p. 78; Emiliani, Cesare. *Scientific Companion*, pp. 177–193, 203-205.

Lithosphere plates of the earth

458 / Are there tides in the solid part of the earth as well as in its waters?

The solid earth is distorted about 4.5 to 14 inches (11.4 to 35.6 centimeters) by the gravitational pull of the sun and moon. It is the same gravitational pull that creates the tides of the waters. When the moon's gravity pulls water on the side of the earth near to it, it pulls the solid body of the earth on the opposite side away from the water to create bulges on both sides, and causing high tides. These occur every 12½ hours. Low tides occur in those places from which the water is drained to flow into the two high–tide bulges. The sun causes tides on the earth that are about 33 to 46% as high as those due to the moon. During a new moon or a full moon when the sun and moon are in a straight line, the tides of the moon and the sun reinforce each other, to make high tides higher; these are called spring tides. At the quarter moons, the sun and moon are out of step (at right angles), the tides are less extreme than usual; these are called neap tides. Smaller bodies of water, such as lakes, have no tides because the whole body of water is raised all at once, along with the land beneath it.

Source: Asimov, Isaac. *Isaac Asimov's Guide to Earth and Space*, pp. 97–98; Feldman, David. *Do Penguins Have Knees?*, pp. 138–139; Pickering, James S. *1001 Questions Answered About Astronomy*, p. 41.

459 / How thick is the ice that covers Antarctica?

The ice that covers Antarctica is 15,700 feet (4,785 meters) in depth at its thickest point. This is about ten times taller than the Sears Tower in Chicago, the world's tallest building. However, the average thickness is 7,100 feet (2,164 meters).

Sources: *World Book Encyclopedia*, vol. 1, pp. 530, 532.

460 / When was the Ice Age?

Ice ages, or glacial periods, have occurred at irregular intervals for over 2.3 billion years. During an ice age, sheets of ice cover large portions of the continents. The exact reasons for the changes in the earth's climate are not known, although some think they are caused by changes in the earth's orbit around the sun.

The Great Ice Age occurred during the Pleistocene Epoch which began about 2 million years ago and lasted until 11 thousand years ago. At its height, about 27% of the world's present land area was covered by ice. In North America, the ice covered Canada and moved southward to New Jersey; in the Midwest, it reached as far south as St. Louis. Small glaciers and ice caps also covered the western mountains. Greenland was covered in ice as it is today. In Europe, ice moved down from Scandinavia into Germany and Poland; the British Isles and the Alps also had ice caps. Glaciers also covered the northern plains of Russia, the plateaus of Central Asia, Siberia, and the Kamchetka Peninsula.

The glaciers' effect on the United States can still be seen. The drainage of the Ohio River and the position of the Great Lakes were influenced by the glaciers. The rich soil of the Midwest is mostly glacial in origin. Rainfall in areas south of the glaciers formed large lakes in Utah, Nevada, and California. The Great Salt Lake in Utah is a remnant of one of these lakes. The large ice sheets locked up a lot of water; sea level fell about 450 feet (137 meters) below what it is today. As a result, some states, such as Florida, were much larger during the ice age.

The glaciers of the last ice age retreated about 11 thousand years ago. Some believe that the ice age is not over yet; the glaciers follow a cycle of advance and retreat many times. There are still areas of the earth covered by ice and this may be a time in between glacial advances.

Sources: Leet, L. Don, and Sheldon Judson. *Physical Geology*, pp. 304–305; U.S. Geological Survey. *Our Changing Continent*, p. 11

Great ice age

▽ **461** ▽

Who was the first person on Antarctica?

Historians are unsure who first set foot on Antarctica, the fifth largest continent covering 10% of the earth's surface with its area of 5.4 million square miles (14 million square kilometers). In 1771-1775 British Captain James Cook (1728-1779) circumnavigated the continent. American explorer Nathaniel Palmer (1799-1877) discovered Palmer Peninsula in 1820, without realizing that this was a continent.

In this the same year Fabian Gottlieb von Bellingshausen (1779-1852) sighted the Antarctic continent. American sealer John Davis went ashore at Hughes Bay on February 7, 1821. In 1823, sealer James Weddell (1787-1834) travelled the farthest south (74 degrees south) that anyone had until that time and entered what is now called the Weddell Sea. In 1840, American Charles Wilkes (1798-1877), who followed the coast for 1,500 miles, announced the existence of Antarctica as a continent. In 1841, Sir James Clark Ross (1800-1862) discovered Victoria Land, Ross Island, Mount Erebus, and the Ross Ice Shelf. In 1895, the whaler Henryk Bull landed on the Antarctic continent. Norwegian explorer Roald Amundsen (1872-1928) was the first to reach the South Pole on December 14, 1911.

Sources: May, John. *The Greenpeace Book of Antarctica*, pp. 16, 110–113; *The World Almanac and Book of Facts 1992*, pp. 528-529; *World Book Encyclopedia*, vol. 1, pp. 530, 536.

462 / What is a moraine?

A moraine is a mound, ridge, or any other distinct accumulation of unsorted, unstratified material or drift, deposited chiefly by direct action of glacier ice.

Sources: Bates, Robert L., and Julia A. Jackson. *Glossary of Geology*, pp. 432-433.

463 / How much of the earth's surface is covered with ice?

About 10.4% of the world's land surface is glaciated, or permanently covered with ice. Approximately 6,020 thousand square miles (15,600 thousand square kilometers) are covered by ice in the form of ice sheets, ice caps, or glaciers. An ice sheet is a body of ice that blankets an area of land, completely covering its mountains and valleys. Ice sheets have an area of over 19 thousand square miles (50 thousand square kilometers); ice caps are smaller. Glaciers are larger masses of ice that flow, under the force of gravity, at a rate of between ten and one thousand feet (three and 300 meters) per year. Glaciers on steep slopes flow faster, for example, the Quarayoq Glacier in Greenland averages 65 to 80 feet (20 to 24 meters) per day. The area of glaciation in parts of the world are:

Place	Area	
	Square miles	Square kilometers
Antarctia	5,250,000	12,588,000
North Polar Regions (Greenland, Northern Canada, Arctic Ocean islands)	799,000	2,070,000
Asia	44,000	115,800
Alaska & Rocky Mountains	29,700	76,900
South America	10,200	26,500
Iceland	4,699	12,170
Alpine Europe	3,580	9,280

Place	Area	
	Square miles	**Square kilometers**
New Zealand	391	1,015
Africa	5	12

Sources: *The Guinness Book of Answers*, pp. 92-93; Van der Leeden, Frits. *The Water Encyclopedia*, p. 203.

464 ▽ How much of the earth's surface is permanently frozen?

About one fifth of the earth's land is permafrost, or ground that is permanently frozen. This classification is based entirely on temperature and disregards the composition of the land. It can include bedrock, sod, ice, sand, gravel, or any other type of material in which the temperature has been below freezing for over 2 years. Nearly all permafrost is thousands of years old.

Sources: Dyson, James L. *The World of Ice*, p. 149; *Van Nostrand's Scientific Encyclopedia*, vol. 1, p. 363; *World Book Encyclopedia*, vol. 15, p. 289.

465 ▽ Are all craters part of a volcano?

No, not all craters are of volcanic origin. A crater is a nearly circular area of deformed sedimentary rocks, with a central vent–like depression. Some craters are caused by the collapse of the surface when underground salt or limestone dissolves. The withdrawal of ground-water and the melting of glacial ice can also cause the surface to collapse, forming a crater.

Craters are also caused by large meteorites, comets, and asteroids which hit the earth. A notable impact crater is Meteor Crater near Winslow, Arizona. It is 4,000 feet (1,219 meters) in diameter, six hundred feet (183 meters) deep and is estimated to have been formed 30 thousand to 50 thousand years ago.

Sources: Tilling, Robert I. *Volcanoes*, p. 21.

Meteor Crater, Arizona

OBSERVATION AND MEASUREMENT

466
What is a Foucault pendulum?

An instrument devised by Jean Foucault (1819-1868) in 1851 to prove that the earth rotates on an axis, the pendulum consisted of a heavy ball suspended by a very long fine wire. Sand beneath the pendulum recorded the plane of rotation of the pendulum over time.

A reconstruction of Foucault's experiment is located in Portland, Oregon at its Convention Center. It swings from a cable 90 feet (27.4 meters) long, making it the longest pendulum in the world.

Sources: *The Cambridge Dictionary of Science and Technology*, pp. 362, 812; Daintith, John. *The Facts On File Dictionary of Physics*, p. 78; *Guinness Book of Records 1992*, p. 806.

467
Who is regarded as the father of American geology?

Born in Scotland, the American William Maclure (1763-1840) was a member of a commission set up to settle claims between the United States and France during 1803-1807. In 1809 he made a geographical chart of the United States in which the land areas were divided by rock types. In 1817 he revised and enlarged this map. Maclure wrote the first English language articles and books on United States geology.

Sources: *Dictionary of Scientific Biography*, vol. 8, pp. 615–617; *Encyclopedic Dictionary of Science*, p. 155.

468
When were relief maps first used?

The Chinese were the first to use relief maps where the contours of the terrain were represented in models. Relief maps in China go back at least to the third century B.C.E. Some early maps were modeled in rice or carved in wood. It is likely that the idea of making relief maps was transmitted from the Chinese to the Arabs and then to Europe. The earliest known relief map in Europe was a map showing part of Austria, made in 1510 by Paul Dox.

Sources: Temple, Robert K.G. *The Genius of China*, pp. 179-181.

469
What is the prime meridian?

The north-south lines on a map run from the North Pole to the South Pole and are called "meridians," a word that meant "noon," for when it is noon on one place on the line, it is noon at any other point as well. The lines are used to measure longitudes, or how far east or west a particular place might be, and they are 69 miles (111 kilometers) apart at the equator. The east-west lines are called parallels, and unlike meridians, are all parallel to each other. They measure latitude, or how far north or south a particular place might be. There are 180 lines circling the earth, one for each degree of latitude. The degrees of both latitude and longitude are divided into 60 minutes, further divided into 60 seconds each.

The prime meridian is the meridian of 0 degrees longitude, used as the origin for measurement of longitude. The meridian of Greenwich, England, is used almost universally for this purpose.

Sources: Blocksma, Mary. *Reading the Numbers*, p. 107; *The McGraw-Hill Dictionary of Scientific and Technical Terms*, p. 1492.

470 Who was the first person to map the Gulf Stream?

In his travels to and from France as a diplomat, Benjamin Franklin (1706-1790) noticed a difference in speed in the two directions of travel between France and America. He was the first to study ships' reports seriously to determine the cause of the speed variation. As a result, he found that there was a current of warm water coming from the Gulf of Mexico that crossed the North Atlantic Ocean in the direction of Europe. In 1770, Franklin mapped it.

Sources: Asimov, Isaac. *Asimov's Chronology of Science and Discovery*, pp. 215-216.

471 What is magnetic declination?

It is the angle between magnetic north and true north at a given point on the earth's surface. It varies at different points on the earth's surface and at different times of the year.

Sources: *The Chambers Science and Technology Dictionary*, p. 543.

472 What is the Piri Re'is map?

In 1929, a map was found in Constantinople that caused great excitement. Painted on parchment and dated in the Moslem year 919 (1513 according to the Christian calendar), it was signed by an admiral of the Turkish navy known as Piri Re'is. This map appears to be one of the earliest maps of America and it shows South America and Africa in there correct relative longitudes. The mapmaker also indicated that he had used a map drawn by Columbus for the western part. It was an exciting statement because for several centuries geographers had been trying to find a "lost map of Columbus" supposedly drawn by him in the West Indies.

Sources: Hapgood, Charles H. *Maps of the Ancient Sea Kings*, p. 1.

473 / *Which direction does a compass needle point at the north magnetic pole?*

The compass needle would be attracted by the ground and point straight down.

Sources: *Science and Technology Illustrated*, vol. 7, p. 791.

474 / *What are Landsat maps?*

They are images of the earth taken at an altitude of 567 miles (912.3 kilometers) by an orbiting Landsat satellite, or ERTS (*Earth Resources Technology Satellite*). The Landsats were originally launched in the 1970s. Rather than cameras, the Landsats use multispectral scanners, which detect visible green and blue wavelengths, and four infrared and near-infrared wavelengths. These scanners can detect differences between soil, rock, water, and vegetation; types of vegetation; states of vegetation (e.g., healthy/unhealthy or underwatered/well-watered); and mineral content. The differences are especially accurate when multiple wavelengths are compared using multispectral scanners. Even visible light images have proved useful—some of the earliest Landsat images showed that some small Pacific islands were up to 10 miles (16 kilometers) away from their charted positions.

The results are displayed in "false-color" maps, where the scanner data is represented in shades of easily distinguishable colors—usually, infrared is shown as red, red as green, and green as blue. The maps are used by farmers, oil companies, geologists, foresters, foreign governments, and others interested in land management. Each image covers an area approximately 115 miles (185 kilometers) square. Maps are offered for sale by the United States Geological Survey.

Other systems that produce similar images include the French SPOT satellites, the Russian Salyut and Mir manned space stations, and NASA's Airborne Imaging Spectrometer, which senses 128 infrared bands. NASA's Jet Propulsion Laboratories are developing instruments which will sense 224 bands in infrared, which will be able to detect specific minerals absorbed by plants.

Sources: *How In the World?*, pp. 174-175; *The Map Catalog*, pp. 123, 125; Sutton, Caroline, and Duncan M. Anderson. *How Do They Do That?*, pp. 43-47.

475 / *What are the major eras, periods, and epochs in geologic time?*

Modern dating techniques have given a range of dates as to when the various geologic time periods have started, as they are listed below:

Era	Period	Epoch	Beginning date (Est. millions of years)
Cenozoic	Quaternary	Holocene	10,000 years ago
		Pleistocene	1.9
	Tertiary	Pliocene	6
		Miocene	25
		Oligocene	38
		Eocene	55
		Paleocene	65
Mesozoic	Cretaceous		135
	Jurassic		200
	Triassic		250
Paleozoic	Permian		285
	Carboniferous (divided into Mississippian and Pennsylvanian periods by some in the U.S.)		350
	Devonian		410
	Silurian		425
	Ordovician		500
	Cambrian		570
Precambrian	Proterozoic		2500
	Archeozoic		3800
	Azoic		4600

Sources: *The Guinness Book of Answers*, p. 61; Lambert, David. *The Field Guide to Early Man*, p. 16.

PHYSICAL CHARACTERISTICS, ETC. *See also: SPACE—PLANETS AND MOONS*

476
How much of the earth's surface is land and how much is water?

Approximately 30% of the earth's surface is land. This is about 57,259 thousand square miles (148,300 thousand square kilometers). The area of the earth's water surface is approximately 139,692 thousand square miles (361,800 thousand square kilometers), or 70% of the total surface area.

Sources: Ojakargas, Richard. *Schaum's Outline of Theory and Problems of Introductory Geology*, p. 2.; *World Book Encyclopedia*, vol. 6, p. 17.

477
How much does the earth weigh?

It is estimated to weigh 6 sextillion, 588 quintillion short tons (6.6 sextillion short tons) or 5.97×10^{24} kilograms with the earth's mean density being 5.515 times that of water (the standard). This is calculated from using the parameters of an ellipsoid adopted by the International Astronomical Union in 1964 and recognized by the International Union of Geodesty and Geophysics in 1967 in which the area of the ellipsoid is 196,938,800 square miles.

Sources: Asimov, Issac. *Asimov's Guide to the Earth and Space*, pp. 34-35; Emiliani, Cesare. *The Scientific Companion*, p. 167; *The World Almanac and Book of Facts 1992*, p. 275.

478
What are the highest and lowest points on earth?

The highest point on land is the top of Mt. Everest (in the Himalayas on the Nepal-Tibet border) at 29,028 feet (8,848 meters) above sea level, plus or minus 10 feet (3 meters) because of snow. This height was established by the Surveyor General of India in 1954 and accepted by the National Geographic Society. Prior to that the height was taken to be 29,002 feet (8,840 meters). Satellite measurements taken in 1987 indicate that Mt. Everest is 29,864 feet (9,102 meters) high but this measurement has not been adopted by the National Geographic Society.

The lowest point on land is the Dead Sea between Israel and Jordan which is 1,312 feet (399 meters) below sea level. The lowest point on the earth's surface is thought to be in the Marianas Trench in the western Pacific Ocean extending from southeast of Guam to northwest of the Marianas Islands. It has been measured as 36,198 feet (11,034 meters) below sea level.

Sources: Ojakangas, Richard. *Schaum's Outline of Theory and Problems of Introductory Geology*, p. 5; *The World Almanac and Book of Facts 1992*, p. 530.

479
What is the highest and lowest elevation in the United States?

Mt. McKinley, Alaska, at 20,320 feet (6,194 meters), is the highest point in the United States and North America. Mt. Whitney, California, at 14,494 feet (4,421 meters), is the highest point in the continental United States. Death Valley, California, at 282 feet (86 meters) below sea level, is the lowest in the United States and in the western hemisphere.

Sources: *The World Almanac and Book of Facts 1992*, pp. 531, 536-537.

480
How does the temperature of the earth change as one goes deeper underground?

The earth's temperature increases with depth. Measurements made in deep mines and drill–holes indicate that the rate of temperature increase varies from place to place in the world, ranging from 59° to 167°F (15 to 75°C) per kilometer in depth. Actual temperature measurements cannot be made beyond the deepest drill–holes which are a little more than 6.2 miles (10 kilometers) deep. Estimates show that the temperatures at the earth's center can reach values of 5,000°F (2,760°C) or higher.

Sources: Skinner, Brian J. *The Dynamic Earth*, p. 5.

481 / Which elements are in the earth's crust?

The most abundant elements in the earth's crust are listed in the table below. In addition, nickel, copper, lead, zinc, tin, and silver account for less than 0.02% with all other elements comprising 0.48%.

Element	Percentage
Oxygen	47.0
Silicon	28.0
Aluminum	8.0
Iron	4.5
Calcium	3.5
Magnesium	2.5
Sodium	2.5
Potassium	2.5
Titanium	0.4
Hydrogen	0.2
Carbon	0.2
Phosphorous	0.1
Sulfur	0.1

Sources: *Van Nostrand's Scientific Encyclopedia*, vol. 1, p. 962.

482 / What is the interior of the earth like?

The earth is divided into a number of layers. The topmost layer is the crust, which contains about 0.6% of the earth's volume. The depth of the crust varies from 3.5 to 5 miles (5 to 9 kilometers) beneath the oceans to 50 miles (80 kilometers) beneath some mountain ranges. The crust is formed primarily of rocks such as granite and basalt.

Between the crust and the mantle is a boundary known as the Mohorovičić discontinuity (or Moho for short), named for the Croatian seismologist, Andrija Mohorovičić (1857-1936), who discovered it in 1909. Below the Moho is the mantle, extending down about 1,800 miles (2,900 kilometers). The mantle is composed mostly of oxygen, iron, silicon, and magnesium, and accounts for about 82% of the earth's volume. Although mostly solid, the upper part of the mantle, called the asthenosphere, is partially liquid.

The core-mantle boundary, also called the Gutenberg discontinuity for the German-American seismologist, Beno Gutenberg (1889-1960), separates the mantle from the core. Made up primarily of nickel and iron, the core contains about

17% of the earth's volume. The outer core is liquid and extends from the base of the mantle to a depth of about 3,200 miles (5,155 kilometers). The solid inner core reaches from the bottom of the outer core to the center of the earth, about 3,956 miles (6,371 kilometers) deep. The temperature of the inner core is estimated to be about 7,000°F (3,850°C).

Sources: *The Guinness Book of Answers*, p. 54; *How in the World?*, p. 210.

VOLCANOES AND EARTHQUAKES

483
What is the Richter scale?

On a machine called a seismograph, the Richter scale measures the magnitude of an earthquake, i.e., the size of the ground waves generated at the earthquake's source. The scale was devised by American geologist Charles W. Richter (1900-1985) in 1935. Every increase of one number means a tenfold increase in magnitude.

Richter Scale

Magnitude	Possible effects
1	Detectable only by instruments
2	Barely detectable, even near the epicenter
3	Felt indoors
4	Felt by most people; slight damage
5	Felt by all; damage minor to to moderate
6	Moderately destructive
7	Major damage
8	Total and major damage

There is another widely-used system of measurement developed by Guiseppe Mercalli (1850-1914) in 1902 and modified by Harry Wood and Frank Neumann in the 1930s. This scale, the Mercalli scale, measures the effects of the shock.

Sources: *California Geology*, vol. 37 (September 1984), p. 198; *The Guinness Book of Answers*, p. 121; *The Universal Almanac 1992*, p. 496.

484
Who invented the ancient Chinese earthquake detector?

The detector, invented by Zhang Heng (78-139 C.E.) around 132 C.E., was a copper domed urn with dragons' heads circling the outside, each containing a bronze ball. Inside the dome was suspended a pendulum that would swing when the earth shook, and knock a ball from the mouth of a dragon into the waiting open mouth of a bronze toad below. The ball made a loud noise and signaled the occurrence of an earthquake. Knowing which ball had been released, one could determine the direction of the earthquake's epicenter (the point on the earth's surface directly above the quake's point of origin).

Sources: Temple, Robert K.G. *The Genius of China*, pp. 162-163; Tufty, Barbara. *1001 Questions About Earthquakes, Avalanches, Floods, and Other Natural Disasters*, pp. 44-45.

Chinese quake detector

485

Of what magnitude was the earthquake that hit San Francisco on April 18, 1906 on the Richter scale?

Causing 700 deaths and damage estimated at 500 million dollars, the earthquake measured 8.3 on the Richter scale. The October 17, 1989 earthquake in San Francisco measured 7.1 on the Richter scale, killed 67 people, and caused billions of dollars worth of damage.

Sources: *The Public Health Consequences of Disasters 1989*, p. 18; *The Universal Almanac 1992*, pp. 497-498.

486

Where is the "Circle of Fire"?

The belt of volcanoes bordering the Pacific Ocean is often called the "Circle of Fire" or the "Ring of Fire". The earth's crust is composed of 15 pieces, called plates, which "float" on the partially molten layer below them. Most volcanoes, earthquakes, and mountain building occur along the plate boundaries. The Circle of Fire marks the boundary between the plate underlying the Pacific Ocean and the surrounding plates. It runs up the west coast of the Americas from Chile to Alaska (through the Andes Mountains, Central America, Mexico, California, the Cascade Mountains, and the Aleutian Islands) then down the east coast of Asia from Siberia to New Zealand (through Kamchatka, the Kurile Islands, Japan, the Philippines, Celebes, New Guinea, the Solomon Islands, New Caledonia, and New

Zealand). Of the 850 active volcanoes in the world, over 75% of them are part of the Circle of Fire.

Sources: Pearl, Richard M. *1001 Questions Answered About the Mineral Kingdom*, pp. 221-222.

Circle or Ring of Fire

487
What is a tsunami?

A tsunami is a giant wave set in motion by a large earthquake occurring under or near the ocean which causes the ocean floor to shift vertically. This vertical shift pushes the water ahead of it, starting a tsunami. Ocean earthquakes below a magnitude of 6.5 on the Richter scale, and those that shift the sea floor only horizontally, do not produce these destructive waves.

Sources: Lee, Sally. *Predicting Violent Storms*, pp. 99-100.

488
When did the most severe earthquake in American history occur?

The New Madrid earthquakes (a series of quakes starting on December 16, 1811 and lasting until March 1812) is considered to be the most severe earthquake event in United States history. It shook more than two-thirds of the United States and was felt in Canada. It changed the level of land by as much as 20 feet, altered the course of the Mississippi River, and created new lakes, such as Lake St. Francis west of the Mississippi and Reelfoot Lake in Tennessee. Because the area was so sparsely populated, there was no known loss of life.

Sources: Tufty, Barbara J. *1001 Questions Answered About Earthquakes, Avalanches, Floods, and Other Natural Disasters*, pp. 9-10.

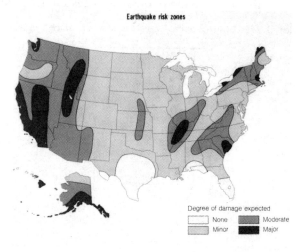

Earthquake risk zones

Degree of damage expected
| | None | | Moderate |
| | Minor | | Major |

489 / *Which volcanoes have been the most destructive?*

The five most destructive eruptions from volcanoes since 1700 are as follows:

Volcano	Date of eruption	Number killed	Lethal agent
Mt. Tambora, Indonesia	April 5, 1815	92,000	2,000 directly by the volcano, 80,000 from starvation afterwards
Karkatoa, Indonesia	Aug. 26, 1883	36,417	90% killed by a tsunami
Mt. Pelee, Martinque	Aug. 30, 1902	29,025	Pyroclastic flows
Nevada del Ruiz, Colombia	Nov. 13, 1985	23,000	Mud flow
Unzen, Japan	1792	14,300	70% killed by cone collapse; 30% by a tsunami

Sources: *The Guinness Book of Answers*, p. 72; *The Public Health Consequences of Disasters 1989*, p. 26.

490 / *When did Mount St. Helens erupt?*

Mount St. Helens, located in southwestern Washington state in the Cascades mountain range, erupted on May 18, 1980. Sixty one people died as a result of the eruption. This was the first known eruption in the 48 contiguous United States to claim a human life. Geologists call Mount St. Helens a composite volcano (a steep-sided, often symmetrical cone constructed of alternating layers of lava, flows, ash, and other volcanic debris). Composite volcanoes tend to erupt explosively. Mount St. Helens and the other active volcanoes in the Cascade Mountains are a part of the "Ring of Fire" (Pacific zone having frequent and destructive volcanic activity).

Sources: *Great Disasters*, p. 272; Tilling, Robert I. *Eruptions of Mount St. Helens*, p. 3.

Cascade Range volcanoes

▽ **491** / *How does a seismograph work?*

A seismograph records earthquake waves. When an earthquake occurs, three types of waves are generated. The vibrations are detected by a delicately balanced beam; the movements of the beam are recorded on a moving tape. The three types of waves arrive at different times after the quake. By comparing the arrival times of the waves at different seismographs throughout the world, it is possible to determine exactly where the earthquake took place.

Sources: Rhodes, Frank H.T. *Geology*, pp. 126-127.

Seismograph device

492

How many kinds of volcanoes are there?

Volcanoes are usually cone-shaped hills or mountains built around a vent connecting to reservoirs of molten rock, or magma, below the surface of the earth. At times the molten rock is forced upwards by gas pressure until it breaks through weak spots in the earth's crust. The magma erupts forth as lava flows or shoots into the air as clouds of lava fragments, ash, and dust. The accumulation of debris from eruptions cause the volcano to grow in size. There are four kinds of volcanoes:

Cinder cones are built of lava fragments. They have slopes of 30 degrees to 40 degrees and seldom exceed 1,640 feet (500 meters) in height. Sunset Crater in Arizona and Paricutin in Mexico are examples of cinder cones.

Composite cones are made of alternating layers of lava and ash. They are characterized by slopes of up to 30 degrees at the summit, tapering off to 5 degrees at the base. Mount Fuji in Japan and Mount St. Helens in Washington are composite cone volcanoes.

Shield volcanoes are built primarily of lava flows. Their slopes are seldom more than 10 degrees at the summit and 2 degrees at the base. The Hawaiian Islands are clusters of shield volcanoes. Mauna Loa is the world's largest active volcano, rising 13,653 feet (4,161 meters) above sea level.

Lava domes are made of viscous, pasty lava squeezed like toothpaste from a tube. Examples of lava domes are Lassen Peak and Mono Dome in California.

Sources: Leet, L. Don, and Sheldon Judson. *Physical Geology*, pp. 74-75; Bair, Frank E. *The Weather Almanac*, pp. 200-204.

WATER

493

What is the chemical composition of the ocean?

The ocean contains every known naturally occurring element plus various gases, chemical compounds, and minerals.

Constituent	Concentration (parts per million)
Chloride	18,980
Sodium	10,560
Sulfate	2,560
Magnesium	1,272
Calcium	400
Potassium	380
Bicarbonate	142
Bromide	65
Strontium	13

Constituent	Concentration (parts per million)
Boron	4.6
Fluoride	1.4

Sources: Groves, Don. *The Ocean Book*, pp. 45-46; Van der Leeden, Frits. *The Water Encyclopedia*, p. 283.

494
Is the Dead Sea really dead?

The Dead Sea, on the boundary between Israel and Jordan, is called "dead" because nothing can live in it. It has a salt content of 25%. Because it is the lowest body of water on the earth's surface, any water that flows into the Dead Sea has no outflow. As the water evaporates, dissolved minerals are left in the sea.

Sources: Cazeau, Charles J. *Science Trivia*, pp. 58-59.

495
Why is the sea blue?

There is no single cause for the colors of the sea. What is seen depends in part on when and from where the sea is observed. Eminent authority can be found to support almost any explanation. Some explanations include absorption and scattering of light by pure water; suspended matter in sea water; the atmosphere; and color and brightness variations of the sky. For example, one theory is that when sunlight hits seawater, part of the white light, composed of different wavelengths of various colors, is absorbed, and some of the wavelengths are scattered after colliding with the water molecules. In clear water, red and infrared light are greatly absorbed but blue is least absorbed, so that the blue wavelengths are reflected out of the water. The blue effect requires a minimum depth of 10 feet (3 meters) of water.

Sources: Bohren, Craig F. *Clouds in a Glass of Beer*, p. 155; Feldman, David. *Why Do Clocks Run Clockwise and Other Imponderables*, p. 213.

496
What is an aquifer?

Some rocks of the upper part of the earth's crust contain many small holes, or pores. When these holes are large or are joined together so that water can flow through them easily, the rock is considered to be permeable. A large body of permeable rock in which water is stored and flows through is called an aquifer (from the Latin for "water" and "to bear"). Sandstones and gravels are excellent examples of permeable rock.

As water reservoirs, aquifers provide about 60% of American drinking water. The huge Ogallala Aquifer, underlying about two million acres of the Great Plains, is a major source of water for the central United States. It has been estimated that after oceans (containing 1,370 million cubic kilometers of water), aquifers, with an estimated 50 million cubic kilometers, are the second largest store of water. Water is purified as it is filtered through the rock, but it can be polluted by spills, dumps, acid rain, and other causes. In addition, recharging of water by rainfall often cannot keep up with the volume removed by heavy pumping. The Ogallala Aquifer's supply of water could be depleted by 25% by the year 2020.

Sources: Hopkins, Nigel J., John W. Mayne, and John R. Hudson. *The Numbers You Need*, p. 112; *The 1992 Information Please Environmental Almanac*, p. 92; *New Scientist*, vol. 129, no. 1756 (February 16, 1991), pp. 1-4.

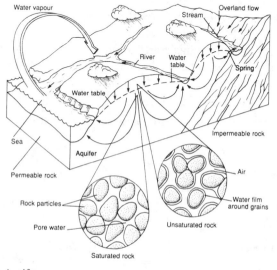

Aquifer

What causes waves in the ocean?

The most common cause of surface waves is air movement (the wind). Waves within the ocean can be caused by tides, interactions among waves, submarine earthquakes or volcanic activity, and atmospheric disturbances. Wave size depends on wind speed, wind duration, and the distance of water over which the wind blows. The longer the distance the wind travels over water, or the harder it blows, the higher the waves. As the wind blows over water it tries to drag the surface of the water with it. The surface cannot move as fast as air, so it rises. When it rises, gravity pulls the water back, carrying the falling water's momentum below the surface. Water pressure from below pushes this swell back up again. The tug of war between gravity and water pressure constitutes wave motion. Capillary waves are caused by breezes of less than two knots. At 13 knots the waves grow taller and faster than they grow longer, and their steepness cause them to break, forming whitecaps. For a whitecap to form, the wave height must be ½ the distance between wave crests. Tidal waves are really seismic waves or tsunamis, and are caused by earthquakes or volcanoes. These are very long waves—100 to 200 miles (161 to 322 kilometers) with high speeds (500 miles, or 805 kilometers, per hour)—that, when approaching shallow water, can grow into a 100-foot–high (30.5–meter) wave as its wavelength is reduced abruptly.

Sources: Flatow, Ira. *Rainbows, Curve Balls, and Other Wonders of the Natural World Explained*, pp. 28-32; Groves, Don. *The Ocean Book*, p. 29.

498

How far can sunlight penetrate into the ocean?

Because seawater is relatively transparent, approximately 5% of sunlight can penetrate clean ocean water to a depth of 262 feet (80 meters). When the water is turbid (cloudy) due to currents, mixing of silt, increased growth of algae, or other factors, the depth of penetration is reduced to less than 164 feet (50 meters).

Sources: Duxbury, Alyn C., and Alison Duxbury. *An Introduction to the World's Oceans*, pp. 382-383.

499

How deep is the ocean?

The average depth of the ocean floor is 13,124 feet (4,000 meters). The average depth of the four major oceans is given below:

Ocean	Average Depth	
	Feet	Meters
Pacific	13,740	4,188
Atlantic	12,254	3,735
Indian	12,740	3,872
Arctic	3,407	1,038
Average overall	13,124	4,000

There are great variations in depth because the ocean floor is often very rugged. The greatest depth variations occur in deep, narrow depressions known as trenches along the margins of the continental plates. The deepest measurements made—36,000 feet (11,033 meters), deeper than the height of the world's tallest mountains—was taken in Mariana Trench east of the Mariana Islands. In January 1960, the French oceanographer Jacques Piccard, together with the United States Navy Lieutenant David Walsh, took the bathyscaphe *Trieste* to the bottom of the Mariana Trench.

Ocean	Deepest Point	Depth	
		Feet	Meters
Pacific	Mariana Trench	36,200	11,033
Atlantic	Puerto Rico Trench	28,374	8,648

Ocean	Deepest Point	Depth	
		Feet	Meters
Indian	Java Trench	25,344	7,725
Arctic	Eurasia Basin	17,881	5,450

Sources: *The New Book of Popular Science*, vol. 2, p. 268; *Van Nostrand's Scientific Encyclopedia*, vol. 2, pp. 2044-2045.

500
What is the difference between an ocean and a sea?

There is no neatly-defined distinction between ocean and sea. One definition terms the ocean as a great body of interconnecting salt water that covers 71% of the earth's surface. There are four major oceans—the Arctic, Atlantic, Indian, and Pacific—but some sources do not include the Arctic Ocean, calling it a marginal sea. The terms "ocean" and "sea" are often used interchangeably but a sea is generally considered to be smaller than an ocean. The name is often given to salt-water areas on the margins of an ocean, such as the Mediterranean Sea.

Sources: Mayhew, Susan and Anne Perry. *The Concise Oxford Dictionary of Geography*, p. 164; *The Guinness Book of Answers*, pp. 62-65; *Universal Almanac*, p. 296.

501
How salty is seawater?

Sea water is, on average, 3.3 to 3.7% salt. The amount of salt varies from place to place. In areas where large quantities of fresh water are supplied by melting ice, rivers, or rainfall, such as the Arctic or Antarctic, the level of salinity is lower. Areas such as the Persian Gulf and the Red Sea have salt contents over 4.2%. If all the salt in the ocean were dried, it would form a mass of solid salt the size of Africa. Most of the ocean salt comes from processes of dissolving and leaking from the solid earth over hundreds of millions of years. Some is the result of salty volcanic rock that flows up from a giant rift that runs through all the ocean's basins.

Sources: Feldman, David. *Do Penguins Have Knees?*, p. 149; Groves, Don. *The Ocean Book*, pp. 46.48; *Van Nostrand's Scientific Encyclopedia*, vol. 2, p. 2047.

502
What is the world's highest waterfall?

Angel Falls, named after the explorer and bush pilot, Jimmy Angel, on the Carrao tributary in Venezuela is the highest waterfall in the world. It has a total height of 3,212 feet (979 meters) with its longest unbroken drop being 2,648 feet (807 meters).

It is difficult to determine the height of a waterfall because many are composed of several sections rather than one straight drop. The highest waterfall in the United States is Yosemite Falls on a tributary of the Merced River in Yosemite National Park, California, with a total drop of 2425 feet (739 meters). There are three sections to the Yosemite Falls: Upper Yosemite is 1,430 feet (435 meters); Cascades (middle portion), 675 feet (205 meters); and Lower Yosemite, 320 feet (97 meters).

Sources: Burnam, Tom. *Dictionary of Misinformation*, p. 8; *The Guinness Book of Answers*, p. 83.

503 / When will Niagara Falls disappear?

The water dropping over Niagara Falls digs great plunge pools at the base, undermining the shale cliff and causing the hard limestone cap to cave in. Niagara has eaten itself seven miles (11 kilometers) upstream since it was formed 10 thousand years ago. At this rate, it will disappear into Lake Erie in 22,800 years. The Niagara River connects Lake Erie with Lake Ontario, and marks the U.S.-Canada boundary (New York-Ontario).

Sources: Leopold, Luna B. *Water*, p. 94.

504 / What is a tidal bore?

It is a wave which moves inland or upriver as the incoming tidal current surges against the flow of river water. A tidal bore occurs in a long narrow estuary (the area of a river mouth in which tides have an effect) or in a river that flows into an estuary.

Sources: Barnhart, Robert K. *The American Heritage Dictionary of Science*, p. 670; Mayhew, Susan, and Anne Penny. *The Concise Oxford Dictionary of Geography*, p. 78.

505 / Where are the world's highest tides?

The Bay of Fundy (New Brunswick, Canada) has the world's highest tides. They average about 45 feet (14 meters) high in the northern part of the bay, far surpassing the world average of 2.5 feet (0.8 meter).

Sources: *World Book Encyclopedia*, vol. 14, p. 181.

506 / Where is the world's deepest lake?

Lake Baikal, located in southeast Siberia, is 5,314 feet (1,620 meters) deep at its maximum depth, making it the deepest lake in the world.

Sources: Matthews, Rupert O. *The Atlas of Natural Wonders*, p. 96.

507
Where are the five largest lakes in the world located?

Location	Area		Length		Depth	
	Square miles	Square km[a]	Miles	Km[a]	Feet	Meters
Caspian Sea[b], Asia-Europe	143,244	370,922	760	1,225	3,363	1,025
Superior, North America	31,700	82,103	350	560	1,330	406
Victoria, Africa	26,828	69,464	250	360	270	85
Aral[b], Asia	24,904	64,501	280	450	220	67
Huron, North America	23,010	59,600	206	330	750	229

[a]-Kilometers
[b]-Salt water lake

Sources: *The Guinness Book of Answers*, p. 89; *The Universal Almanac 1992*, p. 296; *The World Almanac and Book of Facts 1992*, p. 535.

508
What is the bearing capacity of ice on a lake?

The following chart indicates the maximum safe load. It applies only to clear lake ice that has not been heavily travelled. For early winter slush ice, ice thickness should be doubled for safety.

Ice thickness		Examples	Maximum safe load	
Inches	Centimeters		Tons	Kilograms
2	5	One person on foot		
3	7.6	Group in single file		
7.5	19	Car or snowmobiles	2	907.2
8	20.3	Light truck	2½	1,361
10	25.4	Medium truck	3½	1,814.4
12	30.5	Heavy truck	9	7,257.6
15	38		10	9,072
20	50.8		25	22,680

Sources: *National Wildlife*, vol. 16, no. 1 (December/January 1978), p. 27.

509
Which of the Great Lakes is the largest?

Lake	Surface area		Maximum Depth	
	Square miles	Square kilometers	Feet	Meters
Superior	31,700	82,103	1,333	406
Huron	23,010	59,600	750	229
Michigan	22,300	57,757	923	281
Erie	9,910	25,667	210	64
Ontario	7,540	19,529	802	244

Sources: *The Universal Almanac 1992*, p. 296; Van der Leeden, Frits. *The Water Encyclopedia*, pp. 188-189.

510 / What is an oxbow lake?

An oxbow or ox-bow lake is a crescent-shaped lake lying alongside a winding river, formed when a meander (a bend or loop in the river) becomes separated from the main stream. Erosion and deposition tend to accentuate the meander, with the faster-flowing water on the outside edges of the curves eroding the banks, while the slower-moving water on the inside edges deposits silt on the opposite banks. Over time, this process widens the loop of the meander, while narrowing the neck of land dividing the straight path of the river, until the neck vanishes and the river runs past the isolated meander. Without a current to keep it clear, the narrow horseshoe-shaped lake silts up. During the comparatively brief period after the meander is cut off and before it fills with silt, it is an oxbow lake.

A different explanation for the mechanics involved in cutting off a meander has been suggested by laboratory experiments. A stream needs a minimum slope or gradient (horizontal distance divided by height change) to be able to flow and to transport sediment; a meander decreases the gradient of a stream by increasing the distance that it must cover before dropping to a lower height. If a stream's gradient falls below the minimum required, it will tend to cut across the meanders in order to continue flowing.

Sources: The American Geological Institute. *Dictionary of Geological Terms*, pp. 190, 191, 271-272, 311; Mayhew, Susan, and Anne Perry. *The Concise Oxford Dictionary of Geography*, pp. 98, 144, 169; Stiegeler, Stella E. *A Dictionary of Earth Sciences*, p. 199.

Fig. 1: Pattern of erosion and deposition in a meandering reach

Fig. 2: As a result of the pattern of erosion and deposition, meander migrates as shown

Fig. 3: In a storm, the narrow neck is breached leaving the oxbow

Oxbow lake formation

511 / How much salt is in brackish water?

Brackish water has a saline (salt) content between that of fresh water and sea water. It is neither fresh nor salty, but somewhere in between. Brackish waters are usually regarded as those containing 0.5 to 30 parts per thousand salt, while the average saltiness of seawater is 35 parts per thousand.

Sources: Allaby, Michael. *Dictionary of the Environment*, p. 57; American Geological Institute. *Dictionary of Geological Terms*, p. 54.

512 ⟍ What is a yazoo?

A yazoo is a tributary of a river that runs parallel to the river, being prevented from joining the river because the river has built up high banks. The name is derived from the Yazoo River, a tributary of the Mississippi River, which demonstrates this effect.

Sources: Goulty, George A. *A Dictionary of Landscape*, p. 306.

513 ⟍ Which rivers flow northward?

The major rivers that flow north are the Ob, Yenesei, and Lena in Asia; the Nile in Africa; the Rhine in Europe; the Amazon in South America; the Mackenzie and Yukon Rivers in North America.

In the United States, the St. John's River in Florida flows north and the Willamette River in Oregon also flows northward for almost all its length. The Monongahela River, originating in West Virginia, winds northeast into Pennsylvania and then flows northward to Pittsburgh.

Sources: Burnam, Tom. *The Dictionary of Misinformation*, p. 216; *Encyclopedia Americana*, vol. 23, p. 468; Lane, Ferdinand C. *Earth's Grandest Rivers*, pp. 36, 46, 141, 184, 219, 236.

514 ⟍ What are the longest rivers in the world?

River	Length	
	Miles	**Kilometers**
Nile (Egypt)	4,145	6,670
Amazon (Brazil)	4,000	6,404
Chang jiang–Yangtze (China)	3,964	6,378
Mississippi-Missouri (river system) (U.S.)	3,740	6,021
Yenisei-Angara (river system) (U.S.S.R.)	3,442	5,540

Sources: *The Guinness Book of Answers*, p. 76; *The Universal Almanac 1992*, p. 295; *The World Almanac and Book of Facts 1992*, pp. 533-534.

ENERGY

ALTERNATIVE FUELS *See: ENERGY—NON-NUCLEAR FUELS*

BIOMASS FUELS *See: ENERGY—NON-NUCLEAR FUELS*

COAL *See: ENERGY—NON-NUCLEAR FUELS*

CONSUMPTION AND CONSERVATION

515

How much energy is saved by recycling one aluminum can?

Some sources indicate that one recycled aluminum can saves as much energy as it takes to run a TV set for four hours or the energy equivalent of ½ gallon (1.9 liters) of gasoline.

Sources: *Heloise: Hints for a Healthy Planet*, p. 56.

516

When is it more economical to restart an automobile rather than let it idle?

Tests by the Environmental Protection Agency have shown that is it more economical to turn the engine off rather than let it idle if the idle time would exceed 60 seconds.

Sources: Sikorsky, Robert. *How to Get More Miles Per Gallon in the 1990's*, p. 15.

517

How much energy is saved by raising the setting for a house air conditioner?

For every 1°F the inside temperature is increased the energy needed for air conditioning is reduced by 3%. If all consumers raised the settings on their air conditioners by 6°F, for example, 190 thousand barrels of oil could be saved each day.

Sources: *50 Simple Things You Can Do to Save the Earth*, p. 30; Luetzelschwab, John. *Household Energy Use & Conservation*, p. 168.

518

How are fireplace logs made from newspaper?

In a large tub, make a neat stack of old newspapers. Cover the pile with water and let them soak until all the layers are saturated. Pick up the layers of the paper and roll them tightly into logs. Make thin ones for kindling. Stand on end to dry.

Sources: *Pittsburgh Press* (November 12, 1973), p. 15.

519

How much money can be saved by lowering the setting on a home furnace thermostat?

Tests have shown that a 5°F reduction in the home thermostat setting for approximately eight hours will save up to 10% in fuel costs.

Sources: Dalley, Robert L. *Are You Burning Money?*, p. 121.

520

How much gasoline do underinflated tires waste?

Underinflated tires waste as much as 1 gallon out of every 20 gallons of gasoline consumed by an automobile.

Sources: *Reader's Digest Consumer Adviser*, p. 144.

521

What is the advantage of switching from incandescent to fluorescent light bulbs?

One 18-watt fluorescent bulb provides the light of a 75-watt incandescent bulb and lasts ten times as long. Even though the purchase price is higher, over its useful life, an 18-watt fluorescent light bulb saves 80 pounds of coal used to produce electricity. This translates into 250 fewer pounds of carbon dioxide released into the earth's atmosphere.

Sources: MacEachern, Diane. *Save Our Planet*, p. 34.

522

Does the United States currently produce enough energy to meet its con-sumption needs?

No. From 1958 forward, the United States consumed more energy than it pro-duced, and the difference was met by energy imports. In 1990, 67.59 quadrillion BTUs was the total United States energy production with coal being 22.6 quadrillion BTUs; natural gas being 18 quadrillion BTUs;, crude oil and natural plant liquids, 17.6; hydroelectric, 2.9; nuclear, 6.2; and geothermal, biomass, pho-tovoltaic, wind, and solar energy, 0.2. In that year the United States imported 13.83 quadrillion BTUs in petroleum to make up the deficit since its 1990 energy consumption was 81.44 quadrillion BTUs (29.2 in residential and commercial; 30.2 in industrial; and 22 in transportation).

Sources: The Council on Environmental Quality. *Environmental Quality*, pp. 239, 245-246; *Annual Energy Review 1990*, pp. 5,6,7.

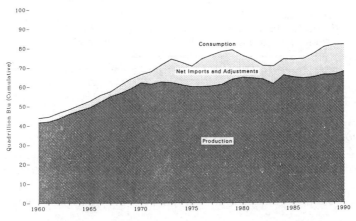

U.S. energy consumption, production, and imports 1960-1990

523

How much energy is required to use various electrical appliances?

The table below indicates the annual estimated energy consumption for various household electrical products.

Appliance	Estimated kilowatt-hours
Air conditioner (room)	1,389
Blender	15
Broiler	100
Clock	17
Clothes dryer	993
Clothes washer	103
Coffee maker	106

Appliance	Estimated kilowatt-hours
Computer	25–400
Dehumidifier	377
Dishwasher	165–363
Fan (circulating)	43
Fan (attic)	291
Food mixer	13
Freezer (frost-free)	1,820
Frying pan	186
Garbage disposal	30
Hair dryer	14
Iron	144
Microwave oven	300
Radio	86
Range (self-cleaning oven)	1,205
Refrigerator-freezer (frost-free)	1,591–1,829
Television (black-and-white)	362
Television (color)	502
Toaster	39
Vacuum cleaner	46
Video cassette recorder (VCR)	10–70
Water heater (standard)	4,219

Sources: *How Things Work in Your Home*, p. 131; *The 1992 Information Please Environmental Almanac*, p. 64.

524
When should a fluorescent light be turned off to save energy?

Fluorescent lights use much electric current getting started, and frequent switching the light on and off will shorten the lamp's life and efficiency. It is energy-efficient to turn off a fluorescent light only if it will not be used again within an hour or more.

Sources: Hand, A.J. *Home Energy How-To*, p. 2.

525
Is it more economical to run an automobile with its windows open rather than using its air conditioner?

At speeds greater than 40 miles (64.4 kilometers) per hour, less fuel is used in driving an automobile with the air conditioner on and the windows up than with the windows rolled down. This is due to the air drag effect—the resistance that a vehicle encounters as it moves through a fluid medium, such as air. In automobiles, the amount of engine power required to overcome this drag force increases with the cube of the vehicle's speed—twice the speed requires eight times the power. For example, it takes 5 horsepower for the engine to overcome the air resistance at 40 miles (64.4 kilometers) per hour; but at 60 miles (96.5 kilometers) per hour, it takes 18 horsepower; at 80 miles (128 kilometers) per hour, it takes 42 horsepower improved aerodynamics in which the drag coefficient (measure of air drag effect) is reduced, significantly increases fuel efficiency. The average automobile in 1990 has a drag coefficient of about 0.4. In the early 1960s it was 0.5, on the average, to 0.47 in the 1970s. The lowest maximum level possible for wheeled vehicles is 0.15.

Sources: Brennan, Richard P. *Levitating Trains and Kamikaze Genes*, pp. 206-208; Doo, Jack. *The Ultimate Owner's Manual*, p. 61.

▽ **526** ╱
What is the current per capita energy consumption in the United States?

In 1990 in the United States, it was about 260 million/BTUs (British Thermal Units) per year. In France and Japan it is about 100 million BTUs per year. A BTU, a common energy measurement, is defined as the amount of energy required to raise 1 pound of water by 1°F.

End-use is the total energy consumption minus losses from generation, transmission, and distribution of power. Below is listed per capita consumption for representative years:

Year	Total (Per capita) (million BTU)	End-use (Per capita) (million BTU)
1950	219	194
1955	235	206
1960	244	212
1965	272	232
1970	327	270
1975	327	261
1980	335	259
1985	310	232
1989	328	246
1990	327	260

Sources: *Changing by Degrees: Steps to Reduce Greenhouse Gases*, pp. 32-33; The Council on Environmental Quality. *Environmental Quality*, p. 247; *The Universal Almanac* 1992, p. 280.

▽ **527** ╱
Which countries consume the most energy?

Top Energy-Consuming Countries (1987)

Country	World total %	Oil %	Coal %	Gas %
United States	24.6	41.1	23.4	22.4
U.S.S.R.	18.0	32.3	22.6	38.3
China	7.8	18.4	75.5	2.3
Japan	4.9	55.9	18.0	9.9
West Germany	3.6	42.2	27.5	17.1

Country	World total %	Oil %	Coal %	Gas %
Canada	3.2	31.5	10.7	19.8
United Kingdom	2.7	36.2	32.9	23.8
France	2.7	42.6	9.6	12.4
India	2.0	32.1	55.8	4.2
Italy	2.0	59.3	9.9	21.9

Sources: *Yearbook of Science and the Future, 1991*, p. 343.

528

How long, at the present rate of consumption, will U.S. current major energy reserves last?

Assuming 1986 production rates, U.S. estimated energy reserves are 32.5 years for oil, 58.7 years for gas, and 226 years for coal.

Sources: Zakrzewski, Sigmund F. *Principles of Environmental Toxicology*, p. 8.

529

What is a pedicar?

As a response to the concern of energy conservation, the pedicar was introduced in 1973. It was a pedal-powered, all-weather one passenger vehicle with straight-line pedal action, disc brakes, five forward speeds plus neutral and reverse. Costing about $550 in 1973, the vehicle was conceived as an alternative to the automobile. It had a speed of 8 to 15 miles (12.9-24.1 kilometers) per hour and was developed mainly for fun use around parks, resorts, college campuses, country clubs and similar protected areas.

Sources: *Popular Mechanics*, vol. 141 (May 1974), pp. 98-99, 166.

530

How does driving speed affect gas mileage for most automobiles?

Most automobiles get about 28% more miles per gallon of fuel at 50 miles (80.5 kilometers) per hour than at 70 miles (112.6 kilometers) per hour, and about 21% more at 55 miles (88.5 kilometers) per hour than at 70 miles per hour.

Sources: Dalley, Robert L. *Are You Burning Money?*, p. 126.

ELECTRICITY *See: ENERGY—NON-NUCLEAR FUELS*

GAS *See: ENERGY—NON-NUCLEAR FUELS*

GEOTHERMAL ENERGY *See: ENERGY—NON-NUCLEAR FUELS*

MEASURES AND MEASUREMENT

▽ **531** ▽ *What is the equivalent of 1 ton of coal in terms of other heating fuels?*

Below is listed some comparisons (approximate equivalents) for the energy sources as of 1990:

Energy unit	Equivalent
1 BTU of energy	1 match tip 250 calories (International Steam Table) 0.25 kilocalories (food calories)
1,000 BTU of energy	2 5-ounce glasses of wine 250 kilocalories (food calories) 0.8 peanut butter and jelly sandwiches
1 million BTU of energy	90 pounds of coal 120 pounds of oven-dried hardwood 8 gallons of motor gasoline 10 therms of dry natural gas 11 gallons of propane 2 months of the dietary intake of a laborer
1 quadrillion BTU of energy	45 million short tons of coal 60 million short tons of oven-dried hardwood 1 trillion cubic feet of dry natural gas 170 million barrels of crude oil 470 thousand barrels of crude oil per day for 1 year 28 days of U.S. petroleum imports 26 days of U.S. motor gasoline 26 hours of world energy use (1989)
1 barrel of crude oil	5.6 thousand cubic feet of dry natural gas 0.26 short tons (520 pounds) of coal 1,700 kilowatthours of electricity
1 short ton of coal	3.8 barrels of crude oil 21 thousand cubic feet of dry natural gas 6,500 kilowatt hours of electricity
1,000 cubic feet of natural gas	0.18 barrels (7.4 gallons) of crude oil 0.05 short tons (93 pounds) of coal 300 kilowatthours of electricity
1,000 kilowatthours of electricity	0.59 barrels of crude oil 0.15 short tons (310 pounds) of coal 3,300 cubic feet of dry natural gas

Notes: One quadrillion equals 1,000,000,000,000,000.
Because of energy losses associated with the generation of electricity, about three

times as much fossil fuel is required to generate 1,000 kilowatthours: 1.8 barrels of oil, 0.47 short tons of coal, or 10,000 cubic feet of dry natural gas.

Sources: The Council on Environmental Quality. *Environmental Quality*, p. 238; *Annual Energy Review 1990*, p. 306.

532 How is a heating degree day defined?

Early this century engineers developed the concept of heating degree days as a useful index of heating fuel requirements. They found that when the daily mean temperature is lower than 65°F, most buildings require heat to maintain a 70°F temperature. Each degree of mean temperature below 65°F is counted as "one heating degree day." For every additional heating degree day, more fuel is needed to maintain a 70°F indoor temperature. For example, a day with a mean tempera-ture of 35°F would be rated as 30 heating degree days and would require twice as much fuel as a day with a mean temperature of 50°F (15 heating degree days). The heating degree concept has become a valuable tool for fuel companies for evalua-tion of fuel use rates and efficient scheduling of deliveries. Detailed daily, monthly, and seasonal totals are routinely computed for the stations of the National Weather Service.

Sources: Bair, Frank E. *The Weather Almanac*, p. 148.

533 What does the term cooling degree day mean?

It is a unit for estimating the energy needed for cooling a building. One unit is given for each degree Fahrenheit above the daily mean temperature when the mean temperature exceeds 75°F.

Sources: *The McGraw-Hill Dictionary of Scientific and Technical Terms*, p. 431.

534 What are the fuel equivalents to produce one quad of energy?

One quad (meaning 1 quadrillion) is equivalent to:
    1×10^{15} BTU
    252×10^{15} calories or 252×10^{12} K calories.
In fossil fuels, one quad is equivalent to
    180 million gallons of crude oil
    0.98 trillion cubic feet of natural gas
    37.88 million tons of anthracite coal
    38.46 million tons of bituminous coal

In nuclear fuels, one quad is equivalent to 2500 tons of U_3O_8 if only U_{235} is used. In electrical output, one quad is equivalent to 2.93×10^{11} kilowatt-hours electric.

Sources: Cuff, David J. *The United States Energy Atlas*, p. 372.

535 / *How many gallons are in a barrel of oil?*

The barrel, a common measure of crude oil, contains 42 U.S. gallons and 34.97 imperial gallons.

Sources: Hyne, Norman J. *Dictionary of Petroleum Exploration, Drilling and Production*, p. 33.

536 / *How much heat will 100 cubic feet of natural gas provide?*

One hundred cubic feet of natural gas can provide about 75,000 BTUs (British *thermal units*) of heat. A British thermal unit, a common energy measurement, is defined as the amount of energy required to raise 1 pound of water by 1°F.

Sources: *Warm House, Cool House*, p. 85.

537 / *What are the approximate heating values of fuels?*

Fuel	BTU	Unit of measure
Oil	141,000	gallon
Coal	31,000	pound
Natural gas	1,000	cubic feet
Steam	1,000	cubic feet
Electricity	3,413	kilowatt hour
Gasoline	124,000	gallon

A BTU (British *thermal unit*), a common energy measurement, is defined as the amount of energy required to raise the temperature of one pound of water by 1°F.

Sources: Stamper, Eugene. *Handbook of Air Conditioning, Heating and Ventilating*, pp. 6-50.

538 / *How much does a barrel of oil weigh?*

A barrel of oil weighs about 306 pounds (139 kilograms).

Sources: Hyne, Norman J. *Dictionary of Petroleum Exploration, Drilling & Production*, p. 33.

539
What is the weight per gallon of common fuels?

One gallon of fuel	Weight in pounds
Butane	4.86
Propane	4.23
Kerosene	6.75
Gasoline	6.00
Aviation gasoline	6.46–6.99

Sources: Considine, Douglas M. *Energy Technology Handbook*, pp. 3-180 to 3-181.

540
How are utility meters read?

Older electric and gas meters have a series of four or five dials, which indicate the amount of energy being consumed. The dials are read from left to right, and if the pointer falls between two numbers, the lower number is recorded. In the electric meter depicted below, the dial registers 13021 kilowatt hours. Gas meters are set to read hundred of cubic feet. New meter models have digital displays.

Sources: Nunn, Richard V. *Saving Home Energy*, p. 5.

Electric meter

Gas meter

541
How much wood is in a cord?

A cord of wood is a pile of logs 4 feet (1.2 meters) wide and 4 feet (1.2 meters) high and 8 feet (2.4 meters) long. It may contain from 77 to 96 cubic feet of wood. The larger the unsplit logs the larger the gaps, with fewer cubic feet of wood actually in the cord.

Sources: *Pennsylvania Forests*, vol. 76 (November–December, 1986), p. 6; Self, Charles. *Wood Heating Handbook*, p. 118.

542 / *How many BTUs are equivalent to 1 ton of cooling capacity?*

1 ton = 288,000 BTUs/24 hours or 12,000 BTUs/hour.

Sources: Zimmerman, O.T. *Conversion Factors and Tables*, p. 257.

NON-NUCLEAR FUELS *See also:* ENERGY—CONSUMPTION AND CONSERVATION; ENERGY—MEASURES AND MEASUREMENT

543 / *What kinds of additives are in gasoline and why?*

Additive	Function
Antiknock compounds	Increase octane number
Scavengers	Remove combustion products of antiknock compounds
Combustion chamber	Suppress surface ignition and spark plug deposit modifiers fouling
Antioxidants	Provide storage stability
Metal deactivators	Supplement storage stability
Antirust agents	Prevent rusting in gasoline-handling systems
Anti-icing agents	Suppress carburetor and fuel system freezing
Detergents	Control carburetor and induction system cleanliness
Upper cylinder lubricants	Lubricate upper cylinder areas and control intake system deposits
Dyes	Indicate presence of antiknock compounds and identify makes and grades of gasoline

Sources: Considine, Douglas M. *Energy Technology Handbook*, pp. 3–27.

544 / *When the price of fuel oil skyrocketed in the 1970s, how many people changed from using fuel oil to using gas?*

High prices for fuel oil and promotion by the gas industry convinced almost a million homeowners to convert their heating systems from oil to gas. Another 1,847 thousand conversions took place between 1980 and 1989.

Sources: *Warm House, Cool House*, p. 85.

545 / *Why is Pennsylvania crude oil so highly valued?*

The waxy, sweet paraffinic oils found in Pennsylvania first became prominent because high quality lubricating oils and greases could be made from them.

Sources: Considine, Douglas M. *Energy Technology Handbook*, pp. 3-13.

546 / What is a reformulated gasoline?

Oil companies are being required to offer new gasolines that burn more cleanly and have less impact on the environment. Typically, reformulated gasolines contain lower concentrations of benzene, aromatics, and olefins; less sulfur; a lower *Reid vapor pressure* (RVP); and some percentage of an oxygenate (non-aromatic component) such as *methyl tertiary butyl ether* (MTBE). MTBE is a high-octane gasoline blending components produced by the reaction of isobutylene and methanol.

Sources: *The Motor Gasoline Industry: Past, Present, and Future*, pp. 37-38; *Road & Track*, vol. 42, no. 12 (August 1991), pp. 104, 107.

547 / When was the first oil well in the United States drilled?

The Drake well at Titusville, Pennsylvania was completed on August 28, 1859 (some sources list the date as August 27). The driller, William "Uncle Billy" Smith, went down 69.5 feet (21.18 meters) to find oil for Edwin L. Drake (1819-1880), the well's operator. Within 15 years, Pennsylvania oil field production reached over 10 million 360-pound (163.3-kilograms) barrels a year.

Sources: Hyne, Norman J. *Dictionary of Petroleum Exploration, Drilling & Production*, p. 585; Kane, Joseph N. *Famous First Facts*, p. 439; McNeil, Ian. *An Encyclopedia of the History of Technology*, p. 211.

548 / What is the process known as hydrocarbon cracking?

Cracking is a process that uses heat to decompose complex substances. Hydrocarbon cracking is the decomposition by heat, with or without catalysts, of petroleum or heavy petroleum fractions (groupings) to give materials of lower boiling points. Thermal cracking, developed by William Burton in 1913, uses heat and pressure to break some of the large heavy hydrocarbon molecules into smaller gasoline-grade ones. The cracked hydrocarbons are then sent to a flash chamber where the various fractions (groupings) are separated. Thermal cracking not only doubles the gasoline yield, but has improved gasoline quality, producing gasoline components with good anti-knock characteristics.

Sources: Allaby, Michael. *Dictionary of the Environment*, pp. 103, 204; *The Motor Gasoline Industry: Past, Present, and Future*, p. 49.

549 / How is gasohol made?

Gasohol, a mixture of 90% unleaded gasoline and 10% ethyl alcohol (ethanol), has gained some acceptance as a fuel for motor vehicles. It is comparable in performance to 100% unleaded gasoline with the added benefit of superior anti-knock properties (no premature fuel ignition). No engine modifications are needed for the use of gasohol.

Since corn is the most abundant United States grain crop, it is predominately used in producing ethanol. However, the fuel can be made from other organic raw materials, such as oats, barley, wheat, milo, sugar beets, or sugar cane. Potatoes, and cassava (a starchy plant) and cellulose (if broken up into fermentable sugars) are possible other sources. The corn starch is processed through grinding and cooking. The process requires the conversion of a starch into a sugar, which in turn is converted into alcohol by reaction with yeast. The alcohol is distilled and any water is removed until it is 200 proof (100% alcohol).

One acre of corn yields 250 gallons (946 liters) of ethanol; an acre of sugar beets yields 350 gallons (1,325 liters), while an acre of sugar can produce 630 gallons (2,385 liters). In the future motor fuel could conceivably be produced almost exclusively from garbage, but currently its conversion remains an expensive process.

Sources: Hunt, V. Daniel. *The Gasohol Handbook*, pp. 1, 4; Lincoln, John W. *Driving Without Gas*, p. 7; Sutton, Caroline, and Duncan M. Anderson. *How Do They Do That?*, pp. 210-211.

550 / *What do the octane numbers of gasoline mean?*

The octane number is a measure of the gasoline's ability to resist engine knock (pinging caused by premature ignition). Two test fuels, normal heptane and isooctane, are blended for test results to determine octane number. Normal heptane has an octane number of zero and isooctane a value of 100. Gasolines are then compared with these test blends to find one that makes the same knock as the test fuel. The octane rating of the gasoline under testing is the percentage by volume of isooctane required to produce the same knock. For example, if the test blend has 85% isooctane, the gasoline has an octane rating of 85. The octane rating that appears on gasoline pumps is an average of research octane determined in laboratory tests with engines running at low speeds, and motor octane, determined at higher speeds.

Sources: *Chambers Science and Technology Dictionary*, p. 622; *Reader's Digest Consumer Adviser*, p. 143.

551 / *Where are the largest oil and gas fields in the world and in the United States?*

The Ghawar field, discovered in 1948 in Saudi Arabia, is the largest in the world; it measures 150 x 22 miles (241 x 35 kilometers). In the continental United States, the Hugoton field in Kansas is second in size only to the Prudhoe field in Alaska.

Sources: *Guinness Book of Records 1992*, p. 271; Hyne, Norman J. *Dictionary of Petroleum Exploration, Drilling & Production*, pp. 591, 593.

552 ▽ When did gasoline stations open?

The first service station (or garage) was opened in Bordeaux, France in December 1895 by A. Barol. It provided overnight parking, repair service, and refills of oil and "motor spirit." In April 1897 a parking and refueling establishment—Brighton Cycle and Motor Co.—opened in Brighton, England.

The pump which would be used to eventually dispense gasoline was devised by Sylanus Bowser of Fort Wayne, Indiana, but in September 1885, it dispensed kerosene. Twenty years later Bowser manufactured the first self-regulating gasoline pump. In 1912, a Standard Oil of Louisiana superstation opened in Memphis, Tennessee, featuring 13 pumps, a ladies' rest room, and a maid who served ice water to waiting customers. On December 1, 1913 in Pittsburgh, Pennsylvania the Gulf Refining Company opened the first drive-in station as a 24 hour-a-day operation. Only 30 gallons (114 liters) of gasoline were sold the first day.

Sources: Harris, Harry. *Good Old-Fashioned Yankee Ingenuity*, pp. 136-137; Kane, Joseph N. *Famous First Facts*, p. 598.

553 ▽ When was offshore drilling for oil first done?

The first successful offshore oil well was built off the coast at Summerland, Santa Barbara County, California, in 1896.

Sources: Kane, Joseph N. *Famous First Facts*, p. 438.

554 ▽ Why is lead added to gasoline and why is lead-free gasoline used in new cars?

Tetraethyl lead has been used for more than 40 years to improve the combustion characteristics of gasoline. It reduces or eliminates "knocking" (pinging caused by premature ignition) in high performance large engines and in smaller high compression engines. It provides lubrication to the extremely close fitting engine parts where oil has a tendency to wash away or burn off. However, lead will ruin and effectively destroy the catalyst presently used in emission control devices installed in new cars. So lead-free gasoline must be used.

Sources: *Mark's Standard Handbook for Mechanical Engineers*, pp. 7-16.

555 ▽ What is cogeneration?

Cogeneration is an energy production process involving the simultaneous generation of thermal (steam or hot water) and electric energy by using a single primary heat source. By producing two kinds of useful fuels in the same facility the net energy yield from the primary fuel increases from 30–35% to 80–90%.

Sources: *Oxford Illustrated Encyclopedia of Invention and Technology*, p. 81.

556 / Why are coal, oil, and natural gas called fossil fuels?

They are composed of the remains of organisms that lived as long ago as 500 million years. These organisms (such as *phytoplankton*) became incorporated into the bottom sediments and then were converted, with time, to oil and gas. Coal is the remains of plants and trees that were buried and subjected to pressure, temperature, and chemical processes (changing into peat and then lignite) for millions of years.

Sources: Ojakangas, Richard. *Schuman's Outline of Theory and Problems of Introductory Geology*, p. 78.

557 / What are the advantages and disadvantages of the alternatives to gasoline to power automobiles?

Because the emissions of gasoline is a major air pollution problem in most U.S. urban areas, alternatives are being worked on. Currently none of the alternatives deliver as much energy content as gasoline so more of each of these fuels must be consumed to equal the distance that the energy of gasoline propels the automobile. The most viable alternative is flexible fuel, a combination of methanol and gasoline, which would add at least $300.00 to car prices for an expensive fuel sensor and a longer fuel tank.

Alternative	Advantages	Disadvantages
Electricity from batteries	No vehicle emissions, good for stop and go driving	Short-lived bulky batteries; limited trip range
Ethanol from corn, biomass, etc.	Relatively clean fuel	Costs, corrosive damage
Hydrogen from electrolysis; etc.	Plentiful supply; non-toxic emissions	High cost; highly flammable
Methanol from methanol gas, coal, biomass, wood	Cleaner combustion; less volatile	Corrosive; some irritant emissions
Natural gas from hydrocarbons and petroleum deposits	Cheaper on energy basis; relatively clean	Cost to adapt vehicle bulky storage; sluggish performance

Sources: Brennan, Richard. *Dictionary of Scientific Literacy*, pp. 113-114.

558 ▽ *How and when was coal formed?*

Coal is formed from the remains of plants that have undergone a series of far-reaching changes, turning into a substance called peat, which subsequently was buried. Through million of years, the earth's crust buckled and folded, subjecting the peat deposits to very high pressure and changing the deposits into coal. The Carboniferous, or coal-bearing period, occurred about 250 million years ago. Geologists in the United States sometimes divide this period into the Mississippian and the Pennsylvanian periods. Most of the high-grade coal deposits are to be found in the strata of the Pennsylvanian period.

Sources: *The New Book of Popular Science*, vol. 2, pp. 320-321.

559 ▽ *How is underground coal mined?*

There are two basic types of underground mining methods: 1) room and pillar and 2) longwall. In room and pillar mines, coal is removed by cutting rooms, or large tunnels, in the solid coal, leaving pillars of coal for roof support. Longwall mining takes successive slices over the entire length of a long working face. In the United States, almost all of the coal recovered by underground mining is by room and pillar method. In Great Britain, however, most of the coal is mined by long-wall methods.

Sources: Cuff, David J. *The United States Energy Atlas*, p. 21.

560 ▽ *What is a "miner's canary?"*

Both mice and birds have been used by miners to test the purity of the air in the mines. A canary or two carried in a small cage will show signs of distress in the presence of carbon monoxide more quickly thans humans will, thus alerting the miner to toxic air. This method of safety was used prior to the more sophisticated equipment used today.

Sources: U.S. Bureau of Mines. *Bulletin No. 42, 1913*, p. 66.

561 ▽ *How thick is the coal in a coal seam?*

Coal seams in the United States range in thickness from a thin film to 50 feet (15 meters) or more. The thickest coalbeds are in the western states, ranging from 10 feet (3 meters) in Utah and New Mexico to 50 feet (15 meters) in Wyoming.

Sources: Cuff, David J. *The United States Energy Atlas*, p. 19.

562

What is the difference between passive solar energy systems and active solar energy systems?

Passive solar energy systems use the architectural design, the natural materials or absorptive structures of the building as an energy saving system. The building itself serves as a solar collector and storage device. An example would be thick-walled stone and adobe dwellings that slowly collect heat during the day and gradually release it at night. Passive systems require little or no investment of external equipment.

Active solar energy systems require a separate collector, a storage device, and controls linked to pumps or fans that draw heat from storage when it is available. Active solar systems generally pump a heat-absorbing fluid medium (air, water, or an antifreeze solution) through a collector. Collectors, such as insulated water tanks, vary in size, depending on the number of sunless days in a locale. Another heat storage system uses eutectic (phase-changing) chemicals to store a large amount of energy in a small volume.

Sources: Cunningham, William P., and Barbara Woodhouse Saigo. *Environmental Science: A Global Concern*, p. 377.

563

Which plant has been investigated as a source of petroleum?

The shrub called the gopher plant (*Euphorbia lathyrus*) produces significant quantities of a milk-like sap—called latex—that is an emulsion of hydrocarbons in water.

Sources: *Science*, vol. 194 (October 1, 1976), p. 46.

564

What is biomass energy?

The catch-all term *biomass* includes all the living organisms in an area. Wood, crops and crop waste, and wastes of plant, mineral, and animal matter are part of the biomass. Much of it is in garbage, which can be burned for heat energy, or allowed to decay and produce methane gas. It has been estimated that 90% of United States waste products could be burned to provide energy as 100 million tons of coal (20% will not burn, but can be recycled). In Western Europe, there are over 200 power plants that burn rubbish to produce electricity. Biomass can be converted into biofuels such as biogas or methane, methanol, ethanol, etc. However, the process has been more costly than the conventional fossil fuel processes. Rubbish buried in the ground can provide methane gas through an aerobic decomposition. One ton of refuse can produce eight thousand cubic feet (227 cubic meters) of methane. Worldwide there are 140 such schemes that tap into the underground rubbish "tips."

Sources: Cunningham, William, and Barbara Woodhouse Saigo. *Environmental Science: A Global Concern*, pp. 384-386; *How in the World?*, pp. 118-119; Wright, R. Thomas. *Understanding Technology*, p. 72.

565

What are some examples of biomass crops that could be used for energy production?

Some crops grown specifically for energy include sugar cane, sorghum, ocean kelp, water hyacinth, and various species of trees.

Sources: Cuff, David J. *United States Energy Atlas*, p. 321.

566

Which woods have the best heating quality in a wood-burning stove?

The woods with the best heating quality are white ash, beech, yellow birch, shag-bark hickory, hophornbeam, black locust, sugar maple, red oak, and white oak. Hardwood will cost more than softwood, but will deliver more heat for the money.

Sources: Harrington, Geri. *The Wood-Burning Stove Book*, pp. 54, 172.

567

There are three types of primary energy that flow continuously on or to the surface of the earth. What are they?

Geothermal energy is heat contained beneath the earth's crust, and brought to the surface in the form of steam or hot water. The five main sources of this geothermal reservoir are dry, super–heated steam from steam fields below the earth's surface, mixed hot water, wet steam, etc., from geysers, etc., dry rocks (into which cold water is pumped to create steam), pressurized water fields of hot water and natural gas beneath ocean beds, and magma (molten rock in or near volcanoes and 5 to 30 miles below the earth's crust). Most Iceland buildings are heated by geothermal energy, a few communities in the United States, such as Boise, Idaho, use geothermal home heating. Electric power production, industrial processing, space heating, etc., are fed from geothermal sources. The California Geysers project is the world's largest geothermal electric generating complex with 200 steam wells that provide some 1,300 megawatts of power. The first geothermal power station was built in 1904 at Larderello, Italy.

Solar radiation utilization depends on the weather, number of cloudy days, and the ability to store energy for night use. The   process of collecting and storing is difficult and expensive. A solar thermal facility (LUZ International Solar Thermal Plant), in the Mojave Desert, currently produces 274 megawatts and is used to supplement power needs of the Los Angeles utilities companies. Japan has 4 million solar panels on roofs and ⅔ of the houses in Israel have them; 90% of Cyprus homes do as well. Solar photo voltaic cells can generate electric current when exposed to the sun. Virtually every spacecraft and satellite since 1958 utilizes this kind of resource.

Tidal and wave energy contain enormous amounts of energy to be harnessed. The first tidal-powered mill was built in England in 1100; another in Woodbridge, England, built in 1170, has functioned for over 800 years. The Rance River Power Station in France, in operation since 1966, was the first large tidal electric generator plant, producing 160 megawatts. A tidal station works like a hydropower dam, with its turbines spinning as the tide flows through them. Unfortunately the tidal period of 13½ hours causes problems of integrating the peak use with the peak generation ability. Ocean wave energy can also be made to drive electrical generators.

Sources: Cone, Robert J. *How the New Technology Works*, pp. 23-23; Cunningham, William P., and Barbara Woodhouse Saigo. *Environmental Science: A Global Concern*, pp. 392–395. *How in the World?*, pp. 125-128.

NUCLEAR POWER

▽ **568** ▽
Where did the expression "China Syndrome" originate?

In a discussion on the theoretical problems of a nuclear reactor if it were to undergo a meltdown, a scientist commented that the molten core could bore a hole through the earth, coming out—if one happened to be standing in America—in China. Ever since then, this scenario has been nicknamed the "China Syndrome." Although the scientist was grossly exaggerating, some people took him seriously. In fact, the core would bore a hole about 30 feet into the earth, but this would have grave repercussions for the small distance.

Sources: Blair, Ian. *Taming the Atom*, p. 171.

▽ **569** ▽
What is the Rasmussen report?

Professor Rasmussen of the Massachusetts Institute of Technology (MIT) conducted a study of nuclear reactor safety for the United States Atomic Energy Commission. The study cost 4 million dollars and took 3 years to complete. It concluded that the odds against a worst-case accident occurring were astronomically large—ten million to one. The worse case accident projected about three thousand early deaths and 14 billion dollars in property damage due to contamination. Cancers occurring later due to the event might number 1,500 per year. The study concluded that the safety features engineered into a plant are very likely to prevent serious consequences from a meltdown. Other groups criticized the Rasmussen report and in particular declared that the estimates of risk were too low. After the Chernobyl disaster in 1986, some scientists estimated that a major nuclear accident might in fact happen every decade.

Sources: Lean, Geoffrey, et al. *WWF Atlas of the Environment*, p. 119; ReVelle, Penelope. *The Environment*, pp. 317-318.

570 / What is a meltdown?

A meltdown is a type of accident in a nuclear reactor in which the fuel core melts, resulting in the release of dangerous amounts of radiation. In most cases the large containment structure that houses a reactor would prevent the radioactivity from escaping. However, there is a small possibility that the molten core could become hot enough to burn through the floor of the containment structure and go deep into the earth. Nuclear engineers call this type of situation the "China Syndrome." All reactors are equipped with emergency systems to prevent such an accident from occurring.

Sources: Blair, Ian. *Taming the Atom*, pp. 170-171; *World Book Encyclopedia*, vol. 14, p. 593.

571 / What is the life of a nuclear power plant?

The working life of a nuclear power plant is approximately 40 years, which is about the same as that of other types of power stations.

Sources: Blair, Ian. *Taming the Atom*, p. 78.

572 / What caused the Chernobyl accident?

Site of the worst nuclear power accident in history, the Chernobyl nuclear power plant in Byelorussia will affect, in one form or another, 20% of the republic's population (2.2 million people). On April 26, 1986 at 1:23:40 a.m., during unauthorized experiments by the operators, in which safety systems were deliberately circumvented in order to learn more about the plant's operation, one of the four reactors rapidly overheated and its water coolant "flashed" into steam. The hydrogen formed from the steam reacted with the graphite moderator to cause two major explosions and a fire. The explosions blew apart the thousand-ton (907 metric ton) lid of the reactor, and released radioactive debris high into the atmosphere. It is estimated that 3.5% of the reactor's fuel and 10% of the graphite reactor itself was emitted into the atmosphere. Human error and design features (positive void coefficient type of reactor, use of graphite in construction, and lack of a containment building) are generally cited as the causes of the accident. Thirty-one people initially died from trying to stop the fires. More than 240 others sustained severe radiation sickness. Eventually 150 thousand people living near the reactor were relocated; some of whom may never be allowed to return home. Fallout from the explosions, containing radioactive isotope cesium—137, was carried by the winds westward across Europe.

Sources: Magill, Frank N. *Great Events from History II: Science and Technology series*, vol. 5, pp. 2321-2323; May, John. *The Greenpeace Book of the Nuclear Age*, pp. 215-228; *Oxford Illustrated Encyclopedia of Invention and Technology*, p. 72.

573 / Where is the oldest operational nuclear power plant in the United States?

The Yankee Plant at Rowe, Massachusetts, which was constructed in 1960, is the oldest United States power plant.

Sources: *Nuclear News*, vol. 35 (February 1992), p. 66.

574 / How many nuclear power plants are there worldwide?

As of 1990 there were 423 reactors operational with 83 power plants under construction and three planned reactors cancelled.

Country	Number of units
Argentina	2
Belgium	7
Brazil	1
Bulgaria	5
Canada	20
Czechoslovakia	8
Finland	4
France	56
Germany	26
Hungary	4
India	7
Japan	41
Korea, South	9
Mexico	1
Netherlandsa	2
Pakistan	1
South Africa	2
Spain	9
Sweden	12

Country	Number of units
Switzerland	5
United Kingdom	37
United States	112
U.S.S.R.	45
Yugoslavia	1
Total	423

As of August 1991, the United States had 111 commercial plants in operation, generating 576,784 million net kilowatt hours or 20.6% of domestic electricity generation with eight more under construction.

Sources: Cuff, David J. *United States Energy Atlas*, p. 143; *Nuclear News*, vol. 34 (August 1991), p. 80; *The World Almanac and Book of Facts 1992*, p. 196.

575 / What actually happened at Three Mile Island?

The Three Mile Island nuclear power plant in Pennsylvania experienced a partial meltdown of its reactor core and radiation leakage. On March 28, 1979, just after 4:00 a.m., a water pump in the secondary cooling system of the Unit 2 pressurized water reactor failed. A relief valve jammed open, flooding the containment vessel with radioactive water. A backup system for pumping water was down for maintenance. Temperatures inside the reactor core rose, fuel rods ruptured, and a partial (52%) meltdown occurred, because the radioactive uranium core was almost entirely uncovered by coolant for 40 minutes. The thick steel-reinforced containment building prevented nearly all the radiation from escaping—the amount of radiation released into the atmosphere was one-millionth of that at Chernobyl. However, if the coolant had not been replaced, the molten fuel would have penetrated the reactor containment vessel, where it would have come into contact with the water, causing a steam explosion, breaching the reactor dome, and leading to radioactive contamination of the area similar to the Chernobyl accident.

Sources: Brennan, Richard P. *Dictionary of Scientific Literacy*, pp. 297-298; May, John. *The Greenpeace Book of the Nuclear Age*, pp. 215-228.

576 / Which nuclear reactors have had accidents?

Incidents with core damage in nuclear reactors

Description of incident	Site	Date	Adult thyroid dose (in rems)
Minor core damage (no release of radiologic material)	Chalk River, Ontario, Canada	1952	not applicable
	Breeder Reactor Idaho	1955	not applicable

Description of incident	Site	Date	Adult thyroid dose (in rems)
	Westinghouse Test Reactor	1960	not applicable
	Detroit Edison Fermi, Michigan	1966	not applicable
Major core damage (radioiodine released)			
Noncommercial	Windscale, England	1957	16
	Idaho Falls SL-1, Idaho	1961	0.035
Commercial	Three Mile Island, Pennsylvania	1979	0.005
	Chernobyl, Soviet Union	1986	100 (estimated)

Sources: *The Public Health Consequences of Disasters, 1989*, p. 122

OIL *See: ENERGY—NON-NUCLEAR FUELS*

SOLAR ENERGY *See: ENERGY—NON-NUCLEAR FUELS*

WIND ENERGY *See: ENERGY—NON-NUCLEAR FUELS*

THE ENVIRONMENT

ANIMALS *See: THE ANIMAL WORLD; AROUND THE HOUSE AND GARAGE—PETS AND PESTS; THE ENVIRONMENT—EXTINCT AND ENDANGERED PLANTS AND ANIMALS*

CONSERVATION *See: THE ENVIRONMENT—RECYCLING, CONSERVATION, AND GARBAGE*

ECOLOGY, RESOURCES, ETC. *See also: THE ENVIRONMENT—POLLUTION*

577 / What is a "green product"?

Green products are environmentally safe products that contain no chlorofluorocarbons, are degradable (can decompose), and are made from recycled materials. Deep-green products are those from small suppliers who build their identities around their claimed environmental virtues. Greened-up products come from the industry giants and are environmentally-improved versions of established brands.

Sources: *Buzzworm: The Environmental Journal*, vol. 3, no. 6 (November/December 1991), p. 37; *Environment*, vol. 33, no. 9 (November 1991), p. 10.

578 / How does the process work in a food chain?

A food chain is the transfer of food energy from the source in plants through a series of organisms with repeated eating and being eaten. The number of steps or "links" in a sequence is usually four to five. The first trophic (group of organisms that get their energy the same way) is plants; the animals that eat plants (called herbivores) form the second trophic level. The third level consists of primary carnivores (animal-eating animals like wolves) who eat herbivores, and the fourth level are animals (like killer whales) that eat primary carnivores. Food chains overlap because many organisms eat more than one type of food, so that these chains can look more like food webs. In 1891 German zoologist Karl Semper introduced the food chain concept.

Sources: Odum, Eugene. *Fundamentals of Ecology*, p. 63; Otto, James H. *Modern Biology*, p. 666; Trefil, James. *1001 Things Everyone Should Know About Science*, p. 43.

Marsh food chain

579

Why is El Niño harmful?

Along the west coast of South America, near the end of each calendar year, a warm current of nutrient-poor tropical water moves southward, replacing the cold, nutrient-rich surface water. Because this condition frequently occurs around Christmas, local residents call it *El Niño* (Spanish for child), referring to the Christ child. In most years the warming lasts for only a few weeks. However, when *El Niño* conditions last for many months, the economic results can be catastrophic. It is this extended episode of extremely warm water that scientists now refer to as *El Niño*. During a severe *El Niño*, large numbers of fish and marine plants may die. Decomposition of the dead material depletes the water's oxygen supply which leads to the bacterial production of huge amounts of smelly hydrogen sulfide. A greatly reduced fish (especially anchovy) harvest affects the world's fishmeal supply, leading to higher prices for poultry and other animals which normally are fed fishmeal.

Studies reveal that *El Niño* is not an isolated occurrence, but is instead part of a pattern of change in the global circulation of the oceans and atmosphere. The 1982-1983 *El Niño* was one of the most severe climate events of the twentieth century in both its geographical extent as well as in the degree of warming (14°F or 8°C). The 1986-1987 *El Niño* might have been responsible in part for record global warmth in 1987—the warmest year in the last 100 years.

Sources: Ahrens, C. Donald. *Meteorology Today*, p. 311; *The Almanac of Science and Technology*, pp. 296-302.

580 ▽ Where is Hawk Mountain Sanctuary?

Hawk Mountain Sanctuary, founded in 1934 as the first sanctuary in the world to offer protection to migrating hawks and eagles, is near Harrisburg, Pennsylvania, on the Kittatinny Ridge. Each year between the months of August and December, over 15 thousand migrating birds pass by. Rare species such as Golden Eagles may be seen there.

Sources: Heintzelman, Donald S. *A Guide to Eastern Hawk Watching*, pp. 57, 82-83.

581 ▽ What is red tide and what causes it?

Red tide is a term used for a brownish or reddish discoloration occurring in ocean, river, or lake water. It is caused by the rapid reproduction of a variety of toxic organisms, especially the toxic red dinoflagellates that are members of two genera Gymnodidium and Gonyaulax. Some red tides are harmless, but millions of fish may be killed during a "bloom," as the build-up is called. Other red tides can poison shellfish and the birds or humans who eat the contaminated food. Scientists do not fully understand why the "bloom" occurs.

Sources: Ashworth, William. *The Encyclopedia of Environmental Studies*, pp. 324-325; *Science Digest*, vol. 75 (February 1974), pp. 64-69.

582 ▽ When did the symbol of Smokey the Bear begin to be used for forest fire prevention?

A nationwide effort in wildfire prevention was begun in 1942. A poster featuring Smokey the Bear was circulated in 1945. The symbol soon became very popular. The slogan "Only you can prevent forest fires" was coined in 1947.

Sources: *Fire Management Notes*, vol. 43, no. 4 (Fall 1982), p. 30.

583 / What is a biome?

It is a plant and animal community that covers a large geographical area. Complex interactions of climate, geology, soil types, water resources, and latitude all determine the kinds of plants and animals that thrive in different places. Fourteen major ecological zones, called "biomes", exist over five major climatic regions and eight zoo geographical regions. Three types of biomes are desert, tropical rain forest, and tundra.

Sources: Lean, Geoffrey et al. *WWF Atlas of the Environment*, p. 11; *World Book Encyclopedia*, vol. 2, pp. 322-323.

584 / Who owns America's forest lands?

More than 282 million acres, or 39% of the total amount of United States forest land, are owned by private individuals. Government owns 137 million acres, or 19%, and the forest industry owns 68 million acres—just 9% of the total United States forest land. The remaining 33% is the 242 million acres of park lands, wilderness reserves, and non-working forest lands.

Sources: *Managing the Future of America's Forests*, p. 4.

585 / What causes the most forest fires in the western United States?

Lightning is the single largest cause of forest fires in the western states.

Sources: *The Odds on Virtually Everything*, p. 181.

586 / How much of the earth is protected as national parks and similar sites?

Below is a table of protected areas by country for the year 1990.

Country	Percent of total land area
Venezuela	22.2
Bhutan	19.8
Chile	18.2
Botswana	17.4
Panama	16.9
Czechoslovakia	15.4
Namibia	12.7
United States	10.5
Indonesia	9.3
Australia	5.9
Canada	5.0

Country	Percent of total land area
Mexico	4.8
Brazil	2.4
Madagascar	1.8
Former Soviet Union	1.1
WORLD	4.9

Sources: *American Forests*, vol. 98, no. 3 & 4 (March/April 1992), p. 42.

587
How many acres in the United States are infested by the gypsy moth?

In 1990, the gypsy–moth infested area in the United States was 7,374,816 acres, of which 4,357,700 were in Pennsylvania. More defoliation has occurred in Pennsylvania than in any other state.

Sources: *American Forests*, vol. 98, nos. 3 & 4 (March/April 1992), p. 24.

588
What environmental landmarks soon followed the celebration of the first Earth Day?

Important official actions that began in 1970 or shortly thereafter were: the birth of the EPA (Environmental Protection Agency), the enactment of the National Environmental Policy Act, the creation of the President's Council on Environmental Quality, and the passage of the new Clean Air Act establishing national air quality standards for the first time.

Sources: *EPA Journal*, vol. 16, no. 1 (January/February 1990), p. 14.

589
How does ozone benefit life on earth?

Ozone in the upper atmosphere (stratosphere) is a major factor in making life on earth possible. The ozone belt shields the earth from excessive ultraviolet radiation generated by the sun. Scientists predict that depletion of this layer could lead to increased health problems for humans and disruption of sensitive terrestrial and aquatic ecosystems. Ozone, a form of oxygen with three atoms instead of the normal two, is highly toxic; less than one part per million of this blue-tinged gas is poisonous to humans. While beneficial in the stratosphere, near ground level it is a pollutant that helps form photochemical smog and acid rain.

Sources: Lean, Geoffrey, et al. *WWF Atlas of the Environment*, pp. 97–98; Schweitzer, Glenn E. *Borrowed Earth, Borrowed Time*, pp. 214, 218.

590 Who started Earth Day?

The first Earth Day was April 22, 1970. It was coordinated by Denis Hayes at the request of United States Senator from Wisconsin, Gaylord Nelson, who is sometimes referred to as the "father" of Earth Day. His main objective was to organize a nationwide public demonstration so large it would get the attention of politicians and force the environmental issue into the political dialogue of the nation.

Sources: *EPA Journal*, vol. 16, no. 1 (January/February 1990), pp. 10-11; *Smithsonian*, vol. 21, no. 1 (April 1990), p. 47.

591 How long will it be before all tropical forests have been destroyed, if present rates of destruction continue?

If present deforestation rates continue, all tropical forests will be cleared in 177 years. These forests contain 155 thousand of the 250 thousand known plant species and innumerable insect and animal species. Half of all medicines prescribed worldwide are originally derived from wild products and the United States National Cancer Institute has identified more than two thousand tropical rainforest plants with the potential to fight cancer. Rubber, timber, gums, resins and waxes, pesticides, lubricants, nuts and fruits, flavorings and dyestuffs, steroids, latexes, essential and edible oils, and bamboo are among the forest's products that would be drastically affected by the depletion of the tropical forests.

Sources: Lean, Geoffrey, et al. *WWF Atlas of the Environment*, pp. 65–68; *National Wildlife*, vol. 30, no. 3 (April/May 1991), p. 16.

592 How rapidly is deforestation occurring?

In the 1990 figures given below, the annual deforestation rate is given in per cent of forests eradication as well as amount in square kilometers.

Country	Deforestation (square kilometers)	Percent
Brazil	13,820	0.4
Colombia	6,000	1.3
Mexico	7,000	1.5
Indonesia	10,000	0.9
Peru	2,700	0.4
Malaysia	3,100	1.5
Ecuador	3,400	2.4
India	10,000	2.7
Zaire	4,000	0.4
Madagascar	1,500	1.5

Sources: *American Forests*, vol. 98, nos. 3 & 4 (March/April 1992), p. 38.

▽ 593 ⟋
How many acres of tropical forest does the world lose annually?

Using data from satellite observations, it is estimated that between 16.4 to 20.4 million hectares are being destroyed each year (a hectare equals 107,639.1 square feet or 10 thousand square meters). Only 50% of the mature tropical forests remain with 750 to 800 million hectares of the original 1.5 to 1.6 billion hectares destroyed. There are two types of tropical forests: wet and dry; the wet ("rain") forests have been incurring the most losses. Latin America has lost 37% of them; Asia, 42%; and Africa, 52%. Logging, fuelwood gathering, and conversion of forests to agriculture are the main causes. Yet tropical forests have been called "green deserts," because their soils are poor in nutrients. The forest vegetation, seemingly lush, has survived through ingenious life–support systems. When the trees are stripped away, the exposed soil deteriorates rapidly, eroded by torrential rains. After the rain ceases, the sun bakes the earth into a hard mass, rendering the soil incapable of vegetative growth.

Sources: Lean, Geoffrey, et al. *WWF Atlas of the Environment*, pp. 65-68; *National Wildlife*, vol. 30, no. 3 (April/May 1992), p. 16.

▽ 594 ⟋
When was the first zoo in the United States established?

The Philadelphia Zoological Garden, chartered in 1859, was the first zoo in the United States. The zoo was delayed by the Civil War, financial difficulties, and restrictions on transporting wild animals. It opened in 1874 on 33 acres, and 282 animals were exhibited.

Sources: Danilov, Victor J. *America's Science Museums*, p. 422.

▽ 595 ⟋
What is the greenhouse effect?

The greenhouse effect is a warming near the earth's surface that results when the earth's atmosphere traps the sun's heat. The atmosphere acts much like the glass walls and roof of a greenhouse. The effect was described by John Tyndall (1820-1893) in 1861. It was given the greenhouse analogy much later in 1896 by the Swedish chemist Svante Arrhenius (1859-1927). The greenhouse effect is what makes the earth habitable. Without the presence of water vapor, carbon dioxide, and other gases in the atmosphere, too much heat would escape and the earth would be too cold to sustain life. Carbon dioxide, methane, nitrous oxide, and other "greenhouse gases" absord the infrared radiation rising from the earth and holds this heat in the atmosphere instead of reflecting it back into space.

In the twentieth century, the increased build-up of carbon dioxide, caused by the burning of fossil fuels has been a matter of concern. There is some controversy concerning whether the increase noted in the earth's average temperature is due to the increased amount of carbon dioxide and other gases, or is due to other causes. Volcanic activity, destruction of the rain forests, use of aerosols, and increased agricultural activity may also be contributing factors.

Sources: Gay, Kathlyn. *The Greenhouse Effect*, pp. 2, 122-125; *McGraw-Hill Encyclopedia of Science and Technology*, vol. 8, pp. 225-226.

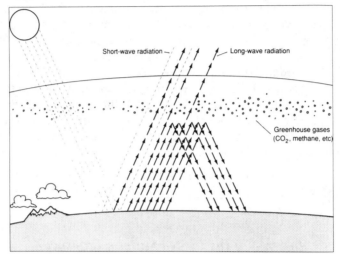

Greenhouse effect

596

How large is Dinosaur National Monument?

Dinosaur National Monument consists of 211,272 acres (85,499 hectares) on the border of northeast Utah and northwest Colorado. It contains the largest known concentration of fossilized bones. Most of the monument is a scenic wilderness area of canyons formed by the Green River and its tributary the Yampa. For a comparison of its size with other parks the ten largest national parks in the 48 contiguous states are listed below:

Park	Location	Acreage (Federal and Non-Federal)
Yellowstone National Park	Idaho, Montana, Wyoming	2,219,785
Death Valley National Monument	California, Nevada	2,067,628
Lake Mead National Recreation Area	Arizona, Nevada	1,496,600
Everglades National Park	Florida	1,398,938
Grand Canyon National Park	Arizona	1,218,375
Glacier National Park	Montana	1,013,572
Olympic National Park	Washington	914,818
Yosemite National Park	California	761,170
Big Bend National Park	Texas	735,416

Park	Location	Acreage (Federal and Non-Federal)
Isle Royale National Park	Michigan	571,790
Big Cypress National Preserve	Florida	570,000
Joshua Tree National Monument	California	559,954

Sources: *The Universal Almanac 1992*, p. 40; *World Book Encyclopedia*, vol. 14, p. 51.

597 / Who was the father of conservation?

American naturalist John Muir (1838-1914) was the father of conservation and the founder of the Sierra Club. He fought for the preservation of the Sierra Nevada Mountains in California, and the creation of Yosemite National Park. He directed most of the Sierra Club's conservation efforts and was a lobbyist for the Antiquities Act.

Sources: Ashworth, John. *The Encyclopedia of Environmental Studies*, p. 248; *Buzzworm: The Environmental Journal*, vol. 4, no. 2 (March/April 1992), p. 26.

598 / What is eutrophication?

A process in which the supply of plant nutrients in a lake or pond is increased. In time, the result of natural eutrophication may be dry land where water once flowed, caused by plant overgrowth.

Natural fertilizers, washed from the soil, result in an accelerated growth of plants, producing overcrowding. As the plants die off, the dead and decaying vegetation depletes the lake's oxygen supply causing fish to die. The accumulated dead plant and animal material eventually changes a deep lake to a shallow one, then to a swamp, and finally it becomes dry land.

While the process of eutrophication is a natural one, it has been accelerated enormously by human activities. Fertilizers from farms, sewage, and industrial wastes and some detergents all contribute to the problem.

Sources: Schneider, Herman. *The Harper Dictionary of Science in Everyday Language*, p. 113.

599 / Who coined the term "Spaceship Earth?"

American inventor and environmentalist, Buckminster Fuller (1895-1983), coined the term "Spaceship Earth" as an analogy of the need for technology to be self-contained and to avoid waste.

Sources: Ashworth, William. *The Encyclopedia of Environmental Studies*, p. 155; Nebel, Bernard J. *Environmental Science*, p. 11.

EXTINCT AND ENDANGERED PLANTS AND ANIMALS

600

What was the first animal on the United States endangered species list?

The peregrine falcon was the first to be listed in the late 1970s.

Sources: Berliner, Barbara. *The Book of Answers*, p. 32.

601

Are turtles an endangered group?

Of the world's 270 species of turtles, 42% are rare or threatened with extinction.

Sources: *American Forests*, vol. 98, no. 3 and 4 (March/April 1992), p. 40.

602

Which animal species has become extinct since 1973 when the Endangered Species Act was signed?

Seven domestic species have been declared extinct: Florida's dusky seaside sparrow, the Santa Barbara song sparrow, the blue pike, the Tecopa pupfish, the Sampson's pearly mussel, and the fishes longjaw cisco and the Amistad gambusic.

Sources: *Buzzworm: The Environmental Journal*, vol. 3, no. 6 (November/December 1991), p. 33.

603

When did the last passenger pigeon die?

At one time, 200 years ago, the passenger pigeon (*Ectopistes migratrius*) was the world's most abundant bird. Although the species was found only in eastern North America, it had a population of 3 to 5 billion birds (25% of the North American land bird population). Overhunting caused a chain of events that reduced their numbers below minimum threshold for viability. In the 1890s several states passed laws to protect the pigeon, but it was too late. The last known wild bird was shot in 1900. The last passenger pigeon, named Martha, died on September 1, 1914 in the Cincinnati Zoo.

Sources: Cunningham, William P., and Barbara Woodhouse Saigo. *Environmental Science: A Global Concern*, p. 260; Schorger, A.W. *The Passenger Pigeon*, p. 29.

604

What is a quagga?

The quagga, a native of South Africa, was basically a brown, rather than striped, zebra with white legs and tail. In the early 19th century, it lived in the wild in great herds and was tamed to become a harness animal or was killed for its skin. The species became extinct in 1883.

Sources: Day, David. *The Doomsday Book of Animals*, pp. 207-209.

605 ▽ How did the dodo become extinct?

The dodo became extinct around 1800. Although thousands were slaughtered for meat, pigs and monkeys, which destroyed dodo eggs, were probably most responsible for the dodo's extinction. They were native to the Mascarene Islands in the Central Indian Ocean. They became extinct on Mauritius soon after 1680 and on Réunion about 1750. They remained on Rodriguez until 1800.

Sources: *Encyclopedia Americana*, vol. 9, pp. 232-233.

606 ▽ Are there any endangered species in the United States that can be legally hunted?

The grizzly bear is the only one of more than 600 threatened species in the United States that can be legally hunted. However, this situation is currently being challenged in the courts under the terms of the Endangered Species Act.

Sources: *Buzzworm: The Environmental Journal*, vol. 4, no. 2 (March/April 1992), p. 11.

607 ▽ How many species have been removed from the United States' endangered and threatened species list, since 1973, the year the Endangered Species Act was passed by Congress?

Six species have been removed from the federal endangered and threatened species list since 1973 because they have recovered. Seven species have been removed from the federal list because they have become extinct.

Sources: *National Wildlife*, vol. 30, no. 3 (April/May 1992), p. 16.

608 ▽ What is the status of an elephant in Africa?

From 1979 to 1989, Africa lost half of its elephants from poaching and illegal ivory trade, with the population decreasing from an estimated 1.3 million to 600 thousand. This led to the transfer of the African elephant from threatened to endangered status in October 1989 by CITES (the Convention in International

Trade in Endangered Species). An ivory ban took effect on January 18, 1990; but six African countries (South Africa, Zimbabwe, Botswana, Namibia, Malawi, and Zambia) are trying to downlist this mammal to a threatened status to allow them to trade in ivory again.

Altogether there are 35 African nations that have elephants, and all want to conserve this resource for their countries' benefit. Kenya now values a living elephant at $14,375 in tourism income for every year of its life giving it a potential lifetime value of $900,000. The ivory from an average elephant killed for its tusk would only be worth $1,000 (the price paid before the ban on ivory as of October 1989).

Sources: *New Scientist*, vol. 133 (February 29, 1992), pp. 49-50 and vol. 133 (March 14, 1992), p. 11.

609 ▽ *What is the current population and status of the great whales?*

Species	Latin Name	Original Population (thousands)	Current Population (thousands)	Status
Sperm	*Physeter macro cephalas*	2400	1974	Insufficient
Blue	*Balenoptera*	226	13	Endangered
Finback	*Balenoptera physalus*	543	123	Vulnerable
Humpback	*Megaptera novaeangliae*	146	4	Vulnerable
Right	*Eubalaena glacialis*	120	3	Endangered
	Eubalaena australis			Vulnerable
Sei	*Balaenoptera*	254	51	Vulnerable
Gray	*Eschrichtius*	20	11	Unlisted
Bowhead	*Balaena mysticetus*	20	2	Vulnerable
Bryde's	*Balaenoptera edeni*	92	92	Insufficient
Minke	*Balaenoptera acutorostrata*	295	280	Insufficient

Sources: Duxbury, Alyn C., and Alison Duxbury. *An Introduction to the World's Oceans*, p. 441; Klinowska, Margaret. *Dolphins, Porpoises and Whales of the World*, p. 9.

610 ▽ *How many species of plants and animals are threatened in the United States?*

There are 157 threatened species (96 animals and 61 plants) in the United States; and 504 endangered species (275 animals and 229 plants).

Category	Threatened	Endangered	Threatened & Endangered
Mammals	8	55	63
Birds	12	73	85
Reptiles	18	16	34
Amphibians	5	6	11
Fishes	34	54	88

Category	Threatened	Endangered	Threatened & Endangered
Snails	6	7	13
Clams	2	40	42
Crustaceans	2	8	10
Insects	9	13	22
Arachnids	0	3	3
Plants	61	229	290
TOTAL	157	504	661

Sources: Council on Environmental Quality. *Environmental Quality*, p. 323; *Endangered Species Technical Bulletin Reprint*, vol. 16, nos. 7 & 8 (July/August 1991), p. 16.

U.S. threatened and endangered species

Why did dinosaurs become extinct?

There are many theories as to why dinosaurs disappeared from the earth about 65 million years ago. Scientists argue over whether the dinosaurs became extinct gradually or all at once. The gradualists believe that the dinosaur population steadily declined at the end of Cretaceous Period. Numerous reasons have been proposed for this. Some claim the dinosaurs' extinction was caused by biological changes which made them less competitive with other organisms, especially the mammals who were just beginning to appear. Overpopulation has been argued, as has the theory that mammals ate too many dinosaur eggs for the animals to reproduce themselves. Others believe that disease—everything from rickets to constipation—wiped them out. Changes in climate, continental drift, volcanic eruptions, and shifts in the earth's axis, orbit, and/or magnetic field have also been held responsible.

The catastrophists argue that a single disasterous event caused the extinction, not only of the dinosaurs, but also of a large number of other species that coexisted with them. In 1980, American physicist Luis Alvarez (1911-1988) and his geologist son, Walter Alvarez (b. 1940), proposed that a large comet or meteoroid struck the earth 65 million years ago. They pointed out that there is a high concentration of the element iridium in the sediments at the boundary between the Cretaceous and Tertiary Periods. Iridium is rare on earth, so the only source of such a large amount of it had to be outer space. This iridium anamoly has since been discovered at over 50 sites around the world. In 1990, tiny glass fragments, which could have been caused by the extreme heat of an impact, were identified in Haiti. A 110-mile-wide (177-kilometer) crater in the Yucatan Peninsula, long covered by sediments, has been dated to 64.98 million years ago, making it a leading candidate for the site of this impact.

A hit by a large extraterrestrial object, perhaps as much as 6 miles (9.3 kilometers) wide, would have had a catastrophic effect upon the world's climate. Huge amounts of dust and debris would have been thrown into the atmosphere, reducing the amount of sunlight reaching the surface. Heat from the blast may also have caused large forest fires which would have added smoke and ash to the air. Lack of sunlight would kill off plants and have a domino—like effect on other organisms in the food chain, including the dinosaurs.

It is possible that the reason for the dinosaurs' extinction may have been a combination of both theories. The dinosaurs may have been gradually declining, for whatever reason. The impact of a large object from space merely delivered the coup de grâce.

The fact that dinosaurs became extinct has been cited as proof of their inferiority and that they were evolutionary failures. However, these animals flourished for 150 million years. By comparison, the earliest ancestors of humanity appeared only about 3 million years ago. Humans have a long way to go before they can claim the same sort of success as the dinosaurs.

Sources: *The Almanac of Science and Technology*, pp. 73-84; Michard, Jean-Guy. *The Reign of the Dinosaurs*, pp. 88-94; Norman, David. *Dinosaur!*, pp. 144-159.

▽ **612** /

Did dinosaurs and humans live at the same time?

No. Dinosaurs first appeared in the Triassic Period (about 220 million years ago) and disappeared at the end of the Cretaceous Period (about 65 million years ago). Modern humans (homo sapiens) appeared only about 25 thousand years ago.

Movies which show humans and dinosaurs existing together are only Hollywood fantasies.

Sources: Norman, David. *Dinosaur!*, p. 36; *The Universal Almanac 1992*, p. 512.

▽ **613** ▽
What is a dolphin-safe tuna?

The order Cetacea, composed of whales, dolphins, and porpoises, were spared from the extinction of large mammals at the end of the Pleistocene about 10 thousand years ago. But from 1,000 C.E. on they, especially the whale, have been relentlessly hunted by man for their valuable products. The 20th century, with its many technological improvements, has become the most destructive period for the cetacea. In 1972, the United States Congress passed the Marine Mammal Protection Act; one of its goals was to reduce the number of small cetaceans (notably *Stenalla* and *Delphinus*) killed and injured during commercial fishing operations, such as the incidental catch of dolphins in tuna purse-seines (nets that close up to from a huge ball to be hoisted aboard ship). Dolphins are often found swimming with schools of yellowfin tuna and are caught along with the tuna by fisherman who use purse-seine nets. The dolphins are drowned by this fishing method because they must be able to breathe air to survive. The number of incidental deaths and injuries in 1972 was estimated at 368 thousand for United States fishing vessels and 55,078 for non-United States vessels. In 1979 the figures were reduced to 17,938 and 6,837 respectively. But in the 1980s, the dolphins killed by foreign vessels rose dramatically to over 100,000 a year. Most of the slaughter occurs in the eastern Pacific Ocean from Chile to Southern California.

To further reduce the numbers of dolphins killed during tuna catches, the three largest sellers of canned tuna in the United States, spearheaded by the Starkist company, decided that they would not sell tuna that has been caught by these methods harmful to dolphins. Only dolphin-safe tuna would be sold by them.

Sources: *Newsweek*, vol. 115 (April 23, 1990), p. 76; Nowak, Ronald M. *Walker's Mammals of the World*, vol. 1, pp. 971-972; *Time*, vol. 135 (April 23, 1990), p. 63.

▽ **614** ▽
What were the smallest and largest dinosaurs?

Compsognathus, a carnivore from the late Jurassic period (131 million years ago) was about the size of a chicken and measured, at most, 35 inches (89 centimeters) from the tip of its snout to the tip of its tail. It probably weighed up to 15 pounds (6.8 kilograms).

The largest species for which a whole skeleton is known is *Brachiosaurus*. A specimen in the Humboldt Museum in Berlin measures 72.75 feet (22.2 meters) long and 46 feet (14 meters) high. It weighed an estimated 34.7 tons (31,480 kilograms). *Brachiosaurus* was a four-footed plant-eating dinosaur with a long neck and a long tail and lived from about 155 to 121 million years ago.

Sources: *Guinness Book of Records 1992*, pp. 110-112; Michard, Jean-Guy. *The Reign of the Dinosaur*, p. 14.

615 / How long did dinosaurs live?

The lifespan has been estimated at 75 to 300 years. Such estimates are educated guesses. From examination of the microstructure of dinosaur bones, scientists have inferred that they matured slowly and probably had proportionately long lifespans.

Sources: Wilford, John N. *The Riddle of the Dinosaur*, p. 249.

616 / How does a mastodon differ from a mammoth?

Although the words are sometimes used interchangeably, the mammoth and the mastodon were two different animals. The mastodon seems to have appeared first and a side branch may have led to the mammoth.

The *mastodon* lived in Africa, Europe, Asia, and North and South America. It appears in the Oligocene (25 to 38 million years ago) and survived until less than 1 million years ago. It stood a maximum of 10 feet (3 meters) tall and was covered with dense woolly hair. Its tusks were straight forward and nearly parallel to each other.

The *mammoth* evolved less than 2 million years ago and died out about 10 thousand years ago. It lived in North America, Europe, and Asia. Like the mastodon, the mammoth was covered with dense, woolly hair, with a long, coarse layer of outer hair to protect it from the cold. It was somewhat larger than the mastodon, standing 9 to 15 feet (2.7 to 4.5 meters). The mammoth's tusks tended to spiral outward, then up.

The gradual warming of the earth's climate and the change in environment were probably primary factors in the animals' extinction. But early man killed many of them as well, perhaps hastening the process.

Sources: Nowak, Ronald M. *Walker's Mammals of the World*, vol. 2, pp. 1277-1278; Schneck, Marcus. *Elephants: Gentle Giants of Africa and Asia*, pp. 91-94.

617 / Under what conditions is a species considered "endangered"?

This determination is a complex process that has no set of fixed criteria that can be applied consistently to all species. The known number of living members in a species is not the sole factor. A species with a million members known to be alive but living in only one small area could be considered endangered, whereas another species having a smaller number of members, but spread out in a broad area, would not be considered so threatened. Reproduction data—the frequency of reproduction, the average number of offspring born, the survival rate, etc.—enter into such determinations. In the United States, the director of the Fish and Wildlife Service (within the Department of the Interior) determines which species are to be considered endangered, based on research and field data from specialists, biologists, botanists, and naturalists.

According to the Endangered Species Act of 1973, a species can be listed if it is threatened by any of the following:

1. The present or threatened destruction, modification, or curtailment of its habitat or range.

2. Utilization for commercial, sporting, scientific, or educational purposes at levels that detrimentally affect it.

3. Disease or predation.

4. Absence of regulatory mechanisms adequate to prevent the decline of a species or degradation of its habitat.

5. Other natural or man–made factors affecting its continued existence.

If the species is so threatened, the director then determines the "critical habitat," that is the species' inhabitation areas that contain the essential physical or biological features necessary for the species' preservation. The critical habitat can include non–habitation areas, which are deemed necessary for the protection of the species.

Sources: Sutton, Caroline, and Duncan M. Anderson. *How Do They Do That?*, pp. 92-93.

FORESTS *See: THE ENVIRONMENT—ECOLOGY, RESOURCES, ETC.*

GARBAGE *See: RECYCLING, CONSERVATION, AND GARBAGE*

HEALTH OF HUMANS, *See: HEALTH AND MEDICINE—HEALTH HAZARDS, RISKS, ETC.*

PARKS *See: THE ENVIRONMENT—ECOLOGY, RESOURCES, ETC.*

PLANTS *See: THE ENVIRONMENT—EXTINCT AND ENDANGERED PLANTS AND ANIMALS; THE PLANT WORLD*

POLLUTION *See also:* HEALTH AND MEDICINE—HEALTH HAZARDS, RISKS, ETC.

618

Why are the "greenhouse gases" important?

These gases influence how much of the sun's energy is absorbed on earth and how much is radiated back into space. The popular conception is that these gases act like a giant greenhouse, trapping energy emitted from ground, etc., to make earth inhabitable. But too much of these gases in the atmosphere has caused global warming. Greenhouse gases include a variety of man-made air pollutants. Carbon dioxide is the most troublesome gas. Others are chlorofluorocarbons (CFCs), methane, and nitrogen oxides.

Sources: Schweitzer, Glenn E. *Borrowed Earth, Borrowed Time*, p. 221.

619

Does the United States government or any of the states regulate the pollutant emissions of gasoline lawn mowers, hedge trimmers, leaf blowers, snow blowers, or chain saws?

The United States government does not regulate these everyday tools of American living. The State of California, through its Air Resources Board, wants these emissions reduced in two stages in 1994 and 1999.

Sources: *The New York Times*, (December 14, 1990), p. A1.

620

What is the Toxic Release Inventory (TRI)?

TRI is a government mandated, publicly available compilation of information on the release of over 300 individual toxic chemicals and 20 categories of chemical compounds by manufacturing facilities in the United States. The law requires manufacturers to state the amounts of chemicals they release directly to air, land, or water, or that they transfer to off-site facilities that treat or dispose of wastes. The U.S. Environmental Protection Agency compiles these reports into an annual inventory and makes the information available in a computerized database. In 1989, 22,560 facilities released 5.7 billion pounds of toxic chemicals into the environment. Over 189 million pounds of this total were released into the water; 2.4 billion pounds emitted into the air; 445 million pounds into landfills; 1.2 billion pounds into underground wells; 551 million pounds to municipal wastewater treatment plants; and 916 million pounds to treatment and disposal facilities (offsite).

Sources: Council on Environmental Quality. *Environmental Quality*, pp. 152-153; *Toxics in the Community, 1988*, pp. xxi, 14-15.

621 ⟋ How harmful are balloon releases?

Both latex and metallic balloons can be harmful. A latex balloon can land in water, lose its color, and resemble a jellyfish, which if eaten by sea animals can cause their death because they cannot digest it. A metal balloon can get caught in electric wires and cause power outages.

Sources: *50 Simple Things You Can Do to Save the Earth,* p. 58.

622 ⟋ How is asbestos removed from existing sites?

There are three acceptable procedures to use in removing asbestos from existing sites: removal, encapsulation, and enclosure. In the removal process, asbestos is stripped from the underlying surface, collected, and placed in containers for burial in an approved site. Encapsulation, the quickest and least expensive method, coats the asbestos material with a sealant. This is often used when the asbestos is not readily accessible. In the enclosure process, a barrier is placed between the asbestos and the building environment. This is rarely used because the asbestos remains in place and has the potential for exposure during routine maintenance.

Sources: Miller, E. Willard. *Environmental Hazards: Toxic Waste and Hazardous Material,* p. 84.

623 ⟋ What are Operation Ranch Hand and Agent Orange?

Operation Ranch Hand was the tactical military project for the aerial spraying of herbicides in South Vietnam during the Vietnam Conflict (1961-1975). In these operations Agent Orange, the collective name for the herbicides 2,4-D and 2,4,5-T, was used for the defoliation. The name derives from the color-coded drums in which the herbicides were stored.

Sources: *Agent Orange and Its Associated Dioxin: Assessment of a Controversy,* p. 10; Miller, E. Willard. *Environmental Hazards: Toxic Waste and Hazardous Material,* p. 13.

624 ⟋ How much oil is dumped into the oceans?

Every year well over three million metric tons of oil contaminate the sea. A half comes from ships, but the rest come from land–based pollution, with only 33% of it spilled by accident. More than 1.1 million metric tons of oil are deliberately discharged from tankers washing out their tanks.

Source of oil	% of total
Tankers operational discharge	22%
Municipal wastes	22%

Source of oil	% of total
Tanker accidents	12.5%
Atmospheric rainout (oil released by industry and cars)	9.5%
Bilge and fuel oils	9%
Natural seeps	7.5%
Non-refining industrial waste	6%
Urban runoff	3.5%
Coastal oil refineries	3%
Offshore production	1.5%
River runoff	1%
Others	2.5%

Sources: Lean, Geoffrey, et al. *WWF Atlas of the Environment*, pp. 175–176.

625
What are the components of smog?

Photochemical air pollution, commonly known as smog, is the result of a number of complex chemical reactions. The hydrocarbons, hydrocarbon derivations, and nitric oxides emitted from such sources as automobiles are the raw materials for photochemical reactions. In the presence of oxygen and sunlight, the nitric oxides combine with organic compounds, such as the hydrocarbons from unburned gasoline, to produce a whitish haze, sometimes tinged with a yellow-brown color. In the process, a large number of new hydrocarbons and oxyhydro-carbons are produced. These secondary hydrocarbon products may compose as much as 95% of the total organics in a severe smog episode.

Sources: Miller, E. Willard. *Environmental Hazards: Air Pollution*, p. 77.

626
How do chlorofluorocarbons affect the earth's ozone layer?

Chlorofluorocarbons (CFCs) are hydrocarbons, such as freon, in which part or all of the hydrogen atoms have been replaced by fluorine atoms. These can be liquids or gases, are non-flammable and heat-stable, and are used as refrigerants, aerosol propellants, and solvents. When released into the air, they slowly rise into the earth's atmosphere where they are broken apart by ultraviolet rays from the sun. Some of the resultant molecular fragments react with the ozone in the atmosphere reducing the amount of ozone. The CFC molecules' chlorine atoms act as catalysts in a complex set of reactions that convert two molecules of ozone into three molecules of ordinary hydrogen. This is depleting the beneficial ozone layer faster than it can be recharged by natural processes. In 1978, the United States government banned the use of fluorocarbon aerosols. Chemical industries must cut CFC manufacture to 50% by the year 2000; they have mounted a massive research effort to find a safer chemical substitute.

Sources: Gay, Kathlyn. *Ozone*, pp. 50, 56-57; Hazen, Robert, and James Trefil. *Science Matters*, pp. 268-269.

▽ **627** ▽
What is acid rain?

The term "acid rain" was coined by British chemist Robert Angus Smith (1817-1884) who, in 1872, published *Air & Rain; The Beginnings of a Chemical Climatology*. Since then, acid rain has unfortunately become an increasingly-used term for rain, snow, sleet, or other precipitation that has been polluted by acids such as sulfuric and nitric acids.

When gasoline, coal, or oil are burned, their waste products of sulfur dioxide and nitrogen dioxide combine in complex chemical reactions with water vapor in clouds to form acids. The United States alone discharges 40 million metric tons of sulfur and nitrogen oxides into the atmosphere. This, combined with natural emissions of sulfur and nitrogen compounds, has resulted in severe ecological damage. Hundreds of lakes in North America (especially northeastern Canada and United States) and in Scandinavia are so acidic that they cannot support fish life. Crops, forests, and building materials, such as marble, limestone, sandstone, and bronze, have been affected as well, but the extent is not as well documented. However, in Europe, where so many living trees are stunted or killed, a new word *Waldsterben* (forest death) has been coined to describe this new phenomenon.

In 1990, amendments to the (United States) Clean Act contained provisions to control emissions that cause acid rain. It included the reductions of sulfur dioxide emissions from 19 million tons to 9.1 million tons annually and the reduction of industrial nitrogen oxide emissions from six to four million tons annually, both by the year 2000. Also the elimination of 90% of industrial benzine, mercury, and dioxin emissions, the reduction of automotive nitrogen oxide by 60%, and hydrocarbons by 40% by year 1997 were specified.

Sources: Adler, Bill. *The Whole Earth Quiz Book*, p. 135; Bair, Frank E. *The Weather Almanac*, pp. 264-267; Hazen, Robert, and James Trefil. *Science Matters*, pp. 269-270.

▽ **628** ▽
How acidic is acid rain?

Acidity or alkalinity is measured by a scale known as the pH (potential for Hydrogen) scale. It runs from zero to 14. Since it is logarithmic, a change in one unit equals a tenfold increase or decrease. So a solution at pH 2 is 10 times more acidic than one at pH 3 and 100 times as acidic as a solution at pH 4. Zero is extremely acid, 7 is neutral, and 14 is very alkaline. Any rain below 5.0 is considered acid rain; some scientists use the value of 5.6 or less. Normal rain and snow containing dissolved carbon dioxide (a weak acid) measure about pH 5.6. Actual values range according to geographical area. Eastern Europe and parts of Scandinavia have 4.3 to 4.5; the rest of Europe is 4.5 to 5.1; eastern United States and Canada ranges from 4.2 to 4.6, and Mississippi Valley has a range of 4.6 to 4.8. The worst North American area, having 4.2, is centered around Lake Erie and Lake Ontario. For comparison, some common items and their pH values are listed below:

Concentrated sulfuric acid	1.0
Lemon juice	2.3
Vinegar	3.3
Acid rain	4.3
Normal rain	5.0 to 5.6
Normal lakes and rivers	5.6 to 8.0
Distilled water	7.0
Human blood	7.35 to 7.45
Seawater	7.6 to 8.4

Sources: *How in the World?*, p. 126; Lean, Geoffrey, et al. *WWF Atlas of the World*, p. 86; Wolke, Robert L. *Chemistry Explained*, pp. 386–393.

629
What are PCBs?

Polychlorinated biphenyls (PCBs) are a group of chemicals that were widely used before 1970 in the electrical industry, as a coolant for transformers and in capacitors and other electrical devices. They caused environmental problems because they do not break down, and can spread through the water, soil, and air. They have been linked by some scientists to cancer and reproductive disorders and have been shown to cause liver function abnormalities. Government action has resulted in the control of the use, disposal, and production of PCBs in nearly all areas of the world including the United States.

Sources: *Encyclopedia of Chemical Technology*, vol. 5, p. 844; Schweitzer, Glenn E. *Borrowed Earth, Borrowed Time*, pp. 9–10.

630
What causes formaldehyde contamination in homes?

Formaldehyde contamination is related to the widespread construction use of wood products bonded with urea-formaldehyde resins and products containing formaldehyde. Major formaldehyde sources include subflooring of particle board; wall paneling made from hardwood plywood or particle board; and cabinets and furniture made from particle board, medium density fiberboard, hardwood plywood, or solid wood. *Urea-formaldehyde foam insulation* (UFFI) has received the most media notoriety and regulatory attention. Formaldehyde is also used in drapes, upholstery, carpeting, and wallpaper adhesives, milk cartons, car bodies, household disinfectants, permanent-press clothing, and paper towels. In particular, mobile homes seem to have higher formaldehyde levels than houses do. Six billion pounds of formaldehyde are used in the United States each year.

The release of formaldehyde into the air by these products (called outgassing) can develop poisoning symptoms in humans. The EPA classifies formaldehyde as a potential human carcinogen (cancer–causing agent).

Sources: Godish, Thad. *Indoor Air Pollution Control*, pp. 37-39, 336-337; Hunter, Linda Mason. *The Healthy Home*, pp. 75-80.

631

How do spray cans of deodorant, paint, air fresheners, whipped cream, etc., affect the atmosphere?

Every time a button is pushed on a spray can using fluorocarbon propellant, a tiny bit of the propellant is released into the air. The fluorocarbon drifts hundreds of miles into the earth's upper atmosphere where it changes some ozone into oxygen creating a "hole" that lets through more ultraviolet light to the earth's surface. This creates health problems for humans such as cataracts and skin cancer, and disturbs delicate ecosystems (for example, plants produce less seed). Currently propellants have been changed to hydrocarbons, such as butane. Chlorofluorocarbons (organic compounds containing chlorine and fluorine) have been banned in aerosols and are being phased out elsewhere.

Sources: Woods, Geraldine. *Pollution*, p. 26.

632

What was the distribution of radioactive fallout after the Chernobyl accident in Byelorussia?

Radioactive fallout, containing the isotope cesium 137, and nuclear contamination covered an enormous area including Byelorussia, Latvia, Lithuania, the central portion of the then Soviet Union, the Scandinavian countries, the Ukraine, Poland, Austria, Czechoslovakia, Germany, Switzerland, northern Italy, eastern France, Romania, Bulgaria, Greece, Yugoslavia, the Netherlands, and the United Kingdom. The fallout, extremely uneven because of the shifting wind patterns, extended 1,200 to 1,300 miles (1,930 to 2,090 kilometers) from the point of the accident. Roughly 5% of the reactor fuel or seven tons of fuel containing 50 to 100 million curies were released. Estimates of the effects of this fallout range from 28 thousand to 100 thousand deaths from cancer and genetic defects within the next 50 years. In particular, livestock in high rainfall areas received unacceptable dosages of radiation.

Sources: Cunningham, William P., and Barbara Woodhouse Saigo. *Environmental Science: A Global Concern*, pp. 354-355; Miller, E. Willard. *Environmental Hazards: Air Pollution*, pp. 67-68; *Oxford Illustrated Encyclopedia of Invention and Technology*, p. 72.

633

Where did the first major oil spill occur?

The first major commercial oil spill occurred on March 18, 1967, when the tanker *Torrey Canyon* grounded on the Seven Stones Shoal off the coast of Cornwall, England, spilling 830 thousand barrels (119 thousand tons) of Kuwaiti oil into the sea. This was the first major tanker accident. However during World War II, German U–boat attacks on tankers, between January and June of 1942, off the United States East Coast, spilled 590 thousand tons of oil. Although the *Exxon Valdez* was widely publicized as a major spill of 35 thousand tons in 1989, it is dwarfed by the deliberate dumping of oil from Sea Island into the Persian Gulf on January 25, 1991. It is estimated that the spill equaled almost 1.5 million tons of oil.

Date	Cause	Thousands tons spilled
1/42-6/42	German U-boat attacks on tankers off the East Coast of U.S. during World War II.	590
3/18/67	Tanker *Torrey Canyon* grounds off Land's End in the English Channel.	119
3/20/70	Tanker *Othello* collides with another ship in Tralhavet Bay, Sweden.	60–100
12/19/72	Tanker *Sea Star* collides with another ship in Gulf of Oman.	115
5/12/76	*Urquiola* grounds at La Coruna, Spain	100
3/16/78	Tanker *Amoco Cadiz* grounds off Northwest France.	223
6/3/79	Itox I oil well blows in Southern Gulf of Mexico.	600
7/79	Tankers *Atlantic Express* and *Aegean Captain* collide off Trinidad and Tobago.	300
2/19/83	Blowout in Norwuz oil field in the Persian Gulf.	600
8/6/83	Fire aboard *Castillo de Beliver* off Cape Town, South Africa.	250
1/25/91	Iraq begins deliberately dumping oil into Persian Gulf from Sea Island, Kuwait.	1,450

Sources: Miller, E. Willard. *Environmental Hazards: Toxic Waste and Hazardous Material*, p. 96; *The Universal Almanac 1992*, p. 537.

634 / What are flue gas "scrubbers"?

The scrubbing of flue gases refers to the removal of sulfur dioxide (SO_2) and nitric oxide (NO) which are major components of air pollution. Wet scrubbers use a chemical solvent or lime, limestone, sodium alkali, or diluted sulfuric acid to remove the SO_2 formed during combustion. Dry scrubbing uses either a lime/limestone slurry or ammonia sprayed into the flue gases.

Sources: Miller, E. Willard. *Environmental Hazards: Air Pollution*, pp. 44-45.

635 / Which pollutants lead to indoor air pollution?

Indoor air pollution, also known as "tight building syndrome," results from conditions in modern, high energy efficiency buildings, which have reduced outside air exchange, or have inadequate ventilation, chemical contamination, and microbial contamination. Indoor air pollution can produce various symptoms, such as headache, nausea, and eye, nose, and throat irritation. In addition houses are affected by indoor air pollution emanating from consumer and building products and from tobacco smoke. Below are listed some pollutants found in houses:

Pollutant	Sources	Effects
Asbestos	Old or damaged insulation, fireproofing, or acoustical tiles.	Many years later, chest and abdominal cancers and lung diseases.
Biological pollutants	Bacteria, mold and mildew, viruses, animal dander and cat saliva, mites, cockroaches, and pollen.	Eye, nose, and throat irritation; shortness of breath; dizziness; lethargy; fever; digestive problems; asthma; influenza and other infectious diseases.
Carbon monoxide	Unvented kerosene and gas heaters; leaking chimneys and furnaces; wood stoves and fireplaces; gas stoves; automobile exhaust from attached garages; tobacco smoke.	At low levels, fatigue at higher levels, impaired vision and coordination; headaches; dizziness; confusion; nausea. Fatal at very high concentrations.
Formaldehyde	Plywood, wall paneling, particle board, fiber-board; foam insulation; fire and tobacco smoke; textiles, and glues.	Eye, nose, and throat irritations; wheezing and coughing; fatigue; skin rash; severe allergic reactions; may cause cancer.
Lead	Automobile exhaust; sanding or burning of lead paint; soldering.	Impaired mental and physical development in children; decreased coordination and mental abilities; kidneys, nervous system, and red blood cells damage.
Mercury	Some latex paints.	Vapors can cause kidney damage; long–term exposure can cause brain damage.
Nitrogen dioxide	Kerosene heaters, unvented gas stoves and heaters; tobacco smoke.	Eye, nose, and throat irritation; may impair lung function and increase respiratory infections in young children.
Organic gases	Paints, paint strippers, solvents, wood preservatives; aerosel sprays; cleansers and disinfectants; moth repellants; air fresheners; stored fuels; hobby supplies; dry-cleaned clothing.	Eye, nose, and throat irritation; headaches; loss of coordination; nausea; damage to liver, kidney, and nervous system; some organics cause cancer in animals and are suspected of causing cancer in humans.
Pesticides	Products used to kill household pests and products used on lawns or gardens that drift or are tracked inside the house.	Irritation to eye, nose, and throat; damage to nervous system and kidneys; cancer.
Radon	Earth and rock beneath home; well water; building materials.	No immediate symptoms; estimated to cause about 10% of lung cancer deaths; smokers at higher risk.

Air pollution sources

Sources: Cassens, B.J. *Preventive Medicine and Public Health*, p. 214; *The Guinness Book of Answers*, pp. 536-537.

RECYCLING, CONSERVATION, AND GARBAGE *See also: ENERGY—CONSUMPTION AND CONSERVATION*

▽636▽ *What is the NIMBY syndrome?*

NIMBY is the acronym for "Not In My Back Yard," referring to major community resistance to new incinerator sitings, landfills, prisons, roads, etc. NIMFY is "Not In My Front Yard."

Sources: *Acronyms, Initialisms & Abbreviations Dictionary 1993*, vol. 1, p. 2388.

▽637▽ *What do the numbers inside the recycling symbol on plastic containers mean?*

The Society of the Plastics Industry has developed a voluntary coding system for plastic containers to assist recyclers in sorting plastic containers. The symbol is designed to be imprinted on the bottom of the plastic containers. The numerical code appears inside a three-sided triangular arrow. A guide to what the numbers mean is listed below.

Code	Material	Examples
1	Polyethylene terephthalate (PET)	Soft drink bottles
2	High-density polyethylene (HDPE)	Milk and water jugs

Code	Material	Examples
3	Vinyl	Shampoo bottles
4	Low-density polyethylene (LDPE)	Ketchup bottles
5	Polypropylene	Squeeze bottles
6	Polystyrene	Fast-food packaging
7	Other	

Sources: Branson, Gary D. *The Complete Guide to Recycling at Home*, p. 54.

Codes for plastic containers

▽ 638 ▽

How many paper mills in the United States use waste paper to produce new products?

Of the approximately 600 mills in the United States producing pulp, paper, paper-board, or building products, 200 depend almost entirely on waste paper for their raw material. Another 300 mills use 10 to 50% waste paper in their manufacturing processes.

Sources: *Pennsylvania Woodland News*, vol. 5, no. 1 (January/February 1991), p. 2.

▽ 639 ▽

Are cloth diapers or disposable diapers better for the environment?

This is a complex matter with both alternatives having an environmental impact. Disposable diapers make up 2% of the total solid waste while cloth diapers account for only 1% of the solid waste. In 1990, 16 to 17 billion disposable diapers were sold. However, cleaning cloth diapers requires the use of detergents (polluting-agent) and hot water (energy user). In addition, if a professional diaper service is used, then there is extra gasoline for delivery and the exhaust from trucks contributes to air pollution.

Sources: *Popular Science*, vol. 237 (October 1990), p. 57; *Science*, vol. 249 (September 28, 1990), p. 1481.

640
How critical is the problem of landfilling in the United States?

According to one source, landfilling is currently over-used in the United States, but will continue to be an essential component of waste management. There are more than nine thousand landfills in the United States. In 1960, 62% of all garbage was sent to landfills; in 1980, the figure was 81%; and in 1988 it decreased to 73%. After as much waste as possible has been reduced, recovered, reused, or converted to energy, there will still be some waste that cannot be disposed of in any other way except in landfills. However, landfill capacity is already dwindling severely in the most populous regions of the country as shown in the accompanying illustration.

Sources: Stilwell, E. Joseph, et al. *Packaging for the Environment*, pp. 104-105; *The Universal Almanac 1992*, p. 538.

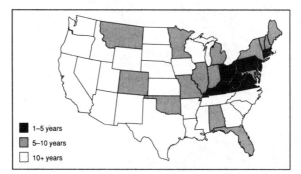

Landfill capacity left

641
How many automobile tires are scrapped each year and what can be done with them?

Some 279 million tires are scrapped each year. The aim of the Tire Recycling Incentives Act of 1990 is to motivate people to find suitable uses for the scrap tires. Some firms are turning waste tires into energy while others are turning them into powder, pellets, and other compound forms for many new uses.

Sources: *Environment*, vol. 31, no. 5 (June 1990), p. 23.

642
What is the major use for recycled telephone books?

They can be recycled to make record album and CD covers.

Sources: *The Next Step: 50 More Things You Can Do to Save the Earth*, p. 74.

643
How much waste paper does a newspaper generate?

An average yearly newspaper subscription (for example *The San Francisco Chronicle*) received every day, produces 550 pounds (250 kilograms) of waste paper (per subscription per year). The average *New York Times* Sunday edition produces 8 million pounds (3.6 million kilograms) of waste paper.

Sources: *Audubon Magazine*, vol. 92 (March 1990), p. 4.

644
How much does packaging contribute to municipal solid waste?

Packaging accounts for 30.3% of municipal solid waste in 1990. The material components of this figure are paper (47.7%), glass (24.5%), plastic (14.5%), steel (6.5%), wood (4.5%), and aluminum (2.3%).

Sources: Stilwell, E. Joseph, et al. *Packaging for the Environment*, p. 2.

645
How much newspaper must be recycled to save one tree?

One 35- to 40-foot (10.6- to 12-meter) tree produces a stack of newspapers 4 feet (1.2 meters) thick; this much newspaper must be recycled to save a tree.

Sources: *Heloise Hints for a Healthy Planet*, p. 67.

646
Is the United States a leader in the percentage of municipal solid waste that is incinerated?

Incinerating 15% of its total municipal solid waste, the United States ranks fifth. First place Switzerland incinerates 74% of its waste; Japan incinerated 66%; Sweden, 50%; and France, 15%.

Sources: Stilwell, E. Joseph, et al. *Packaging for the Environment*, p. 103.

647
What problems may be encountered when polyvinyl chloride (PVC) plastics are burned?

Chlorinated plastics, such as PVC, contribute to the formation of hydrochloric acid gases. They also may be a part of a mix of substances containing chlorine that form a precursor to dioxin in the burning process. Polystyrene, polyethylene and polyethylene terephthalate (PET) do not produce these pollutants.

Sources: Wolf, Nancy. *Plastics*, p. 54.

648

How much solid waste is generated annually in the United States?

The United States produces 4.543 million tons of solid waste annually.

Type of waste	Million tons	Percent
Agriculture	2,340	52
Mineral industries (mining and milling waste)	1,620	36
Industrial (nonhazardous)	225	5
Municipal (domestic)	180	4
Utility	90	2
Hazardous	45	2
Low-level radioactive	3	0.0007

Of the 179.6 million tons of municipal solid wastes in 1988 in the above table, paper waste ranked the highest.

Municipal Solid Waste

Type of waste	Million tons
Paper	71.8
Yard wastes	31.6
Rubber, Textile, Wood, etc.	20.8
Metals	15.3
Plastics	14.4
Food wastes	13.2
Glass	12.5

Sources: Miller, E. Willard. *Environmental Hazards: Toxic Waste and Hazardous Material*, p. 4; *The World Almanac and Book of Facts 1992*, p. 657.

Where does garbage go

649

How much space does a recycled ton of paper save in a landfill?

Each ton saves more than three cubic yards of landfill space.

Sources: *50 Simple Things You Can Do to Save the Earth*, p. 70.

▽ **650**

What is a WOBO?

A WOBO (*world* b*ottle*) is the first mass-produced container designed for secondary use as a building product. It was conceived by Albert Heineken of the Heineken beer family. The beer bottles were designed in a special shape to be used, when empty, as glass bricks for building houses. The actual building carried out with WOBOs was only a small shed and a double garage built on the Heineken estate at Noordwijk, near Amsterdam. Although not implemented, WOBO was a sophisticated and intelligent design solution to what has emerged as a major environmental issue in recent years.

Sources: Pawley, Martin. *Building For Tomorrow: Putting Waste to Work*, pp. 152-154; Pawley, Martin. *Garbage Housing*, pp. 23, 25-30.

▽ **651**

How many trees are saved by recycling paper?

One ton (907 kilograms) of recycled waste paper spares 17 trees.

Sources: Nobile, Philip. *Complete Ecology Fact Book*, p. 379.

▽ **652**

Where are the six nuclear dump sites in the United States?

In 1991, the six nuclear waste dump sites are in West Valley, New York; Sheffield, Illinois; Maxey Flats, Kentucky; Barnwell, South Carolina; Beatty, Nevada; and Richland, Washington.

Sources: *USA Today* (July 22, 1991), p. A-9.

▽ **653**

Which states have the greatest number of hazardous waste sites?

The top five are New Jersey, with 109 hazardous waste sites, Pennsylvania, with 95, New York, 83, California, 88, and Michigan, 77.

Sources: Berliner, Barbara. *The Book of Answers*, p. 16; *The World Almanac and Book of Facts 1992*, p. 657.

▽ **654**

How much garbage does the average American generate?

According to one study, Americans produce about 230 million tons of refuse a year—5.1 pounds per person a day or about 1,900 pounds per year. Another survey reported that a typical suburban family of three generated 40 pounds of garbage weekly with the components of the content given in percentages below:

Type of solid waste	Percent
Aluminum	1%
Polystyrene meat trays, cups, egg cartons, and packing	3%
Disposable diapers	3%
Wood, textiles, old clothing	5%
Metal cans and nails	5%
Plastic soda bottles and bags	5%
Glass	8%
Miscellaneous	10%
Food	11%
Paper	21%
Yard waste and grass clippings	23%

An analysis by the U.S. Environmental Protection Agency published in 1990, indicated not only an increase in the amount of garbage per person, but also a change in the content of it as well:

Garbage in pounds per person per day

Waste materials	1960	1970	1980	1988
Total nonfood product wastes	1.65	2.26	2.57	2.94
Paper and paperboard	0.91	1.19	1.32	1.60
Glass	0.20	0.34	0.36	0.28
Metals	0.32	0.38	0.35	0.34
Plastics	0.01	0.08	0.19	0.32
Rubber and leather	0.06	0.09	0.10	0.10
Textiles	0.05	0.05	0.06	0.09
Wood	0.09	0.11	0.12	0.14
Other	0.00	0.02	0.07	0.07
Other Wastes				
Food wastes	0.37	0.34	0.32	0.29
Yard wastes	0.61	0.62	0.66	0.70
Miscellaneous inorganic waste	0.04	0.05	0.05	0.06
Total waste generated	2.66	3.27	3.61	4.00

Sources: *Good Housekeeping*, vol. 209 (September 1989), p. 272; *The 1992 Information Please Environmental Almanac*, p. 108; *Rush to Burn*, pp. 21, 29.

655
How is nuclear waste stored?

Nuclear wastes consist either of fission products formed from atom splitting of uranium, cesium, strontium, or krypton, or from transuranic elements formed when uranium atoms absorb free neutrons. Wastes from transuranic elements are less radioactive than fission products; however, these elements remain radioactive far longer—hundreds of thousands of years. The types of waste are irradiated fuel (spent fuel) in the form of 12–foot–long rods, high–level radioactive waste in the form of liquid or sludge, and low–level waste (non-transuranic or legally high-level) in the form of reactor hardware, piping, toxic resins, water from fuel pool, etc.

In the United States most of the spent fuel has been left for 10 years or more in water-filled pools at the individual plant sites. They are waiting permanent disposal that is set for 1998 by a mandate in the Nuclear Waster Policy Act (1982). Most low-level radioactive waste has been stored in steel drums in shallow landfills at the six nuclear dump sites and at the Hanford Nuclear Reservation in the state of Washington. Most high-level nuclear waste has been stored in double–walled stainless-steel tanks surrounded by 3 feet (1 meter) of concrete. The current best storage method, developed by the French in 1978, is to incorporate the waste into a special molten glass mixture, then enclosing it with a steel container and burying it in special pits.

Sources: *How in the World?*, pp. 122-123; Naar, Jon. *Design for a Livable Planet*, pp. 161-163.

ZOOS *See:* THE ENVIRONMENT—ECOLOGY, RESOURCES, ETC.

▼ FOOD AND NUTRITION

BEVERAGES *See also* FOOD AND NUTRITION—DAIRY PRODUCTS

656
How does a whole pear get into a bottle of pear brandy?

Pear growers, such as those in France, tie a bottle on a pear branch over a blossom or a tiny developing fruit, which proceeds to grow and ripen inside the bottle. Bottle and pear are sent to brandy makers where distilled brandy is added. The contents are sealed and then shipped to distributors. Connoisseurs of brandy find this a conversation piece more than anything else, since the pear does not guarantee high quality.

Sources: Sutton, Caroline, and Duncan M. Anderson, *How Do They Do That?*, p. 27.

657
What is the formula for Coca-Cola®?

Only one or two senior corporate officers know the exact proportions of the ingredients. The cola has never been patented, so its composition remains a secret. Fourteen of its ingredients—including caramel, sugar, coca leaves (with cocaine content removed), cola nut extract, citric acid and sodium nitrate, caffeine, phosphoric acid, lemon, orange, lime, cassia (type of cinnamon), nutmeg oils, etc., vanilla, and glycerine—are generally well-known. The name of the fifteenth ingredient, known only as "7X," is one of the best-guarded secrets in the industry. The ingredients are mixed with carbonated water.

Sources: *How in the World?*, p. 389; Room, Adrian. *Dictionary of Trade Name Origins*, p. 56.

658
Has cocaine ever been an ingredient in Coca-Cola®?

Yes. A pharmacist in Atlanta, Georgia, Dr. John Styth Pemberton, in 1885, mixed a number of ingredients including purified cocaine to create Coca-Cola®. The name was derived from two of the constituents of the drink; extract from *coca* leaves (cocaine) and extract from the *cola* nut. Around 1899, Asa Candler, an Atlanta pharmacist who bought Pemberton's business, was convinced by Benjamin F. Thomas and Joseph Whitehead of Chattanooga, Tennessee, to join them in bottling the drink. The cocaine was removed from the soft drink shortly before the passage in 1907 of the Pure Food and Drug Act.

Sources: Burnam, Tom. *The Dictionary of Misinformation*, p. 48; Flynn, John C. *Cocaine*, p. 27; Room, Adrian. *Dictionary of Trade Name Origins*, p. 56.

659
What is the beverage known as Moxie®?

Moxie® is the tradename for a fairly tart or bitter soft drink which has been popular as far back as 1884. The word itself has come to mean "courage," "nerve," or "shrewdness."

Sources: Mariani, John F. *The Dictionary of American Food and Drink*, p. 260.

660
What are the ingredients in beer?

Beer is made from water, malt, sugar, hops, yeast (dried beer yeast, not baker's or wine yeast), a little salt, and a little citric acid. Malt gives beer its characteristic body and flavor. Hops, the ripe and carefully dried fruit of the herb *Humulus lupulus*, give beer flavor and bitterness; hops are the single most expensive ingredient in beer.

Sources: Abel, Bob. *The Book of Beer*, p. 64.

661
Why do bubbles rise in a glass of beer?

Beer bubbles are made of carbon dioxide, which having a lower density than the surrounding beer, causes these bubbles to rise. The bubbles grow as they accumulate carbon dioxide while they rise through the beer.

Sources: *Physics Today*, vol. 44 (October 1991), pp. 48-52.

662
How many calories are there in light beer?

In order for a company to call a beer "light," it must have ⅓ fewer calories than the same company's regular beer. Regular beers average about 150 calories; light beers average about 100 calories. Light beers usually contain a lower percentage of alcohol than their regular counterparts do.

Sources: Blocksma, Mary. *Reading the Numbers*, p. 112.

▽ 663 ▽
What is wassail?

This warmed drink, once popular in England on special occasions such as Christmas and Twelfth Night, is made from ale or wine flavored with spices (such as cinnamon, nutmeg, and cloves) and usually contains roasted apples. Originally it was used as a toast to someone's health or for good luck. The word itself is derived from the Old Norse word *vesheill*, which means "be in good health." Often at Christmas time in Old England, groups of carolers would go *a-wassailing*, travelling from house to house caroling their hosts in return for a cup of this delicious drink.

Sources: *Woman's Day Encyclopedia of Cookery*, vol. 12, p. 1918.

▽ 664 ▽
How many gallons are in a barrel of beer?

There are 31 gallons of beer in a U.S. barrel.

Sources: Zimmerman, *Conversion Factors and Tables*, p. 13.

▽ 665 ▽
Who drinks the most coffee in the world?

Currently the United States imports about ½ of the world's coffee. The top ten imports of coffee are as follows:

Country	Percent of world's coffee imports
United States	30.0
West Germany	13.6
France	9.1
Japan	8.4
Italy	7.2
Netherlands	4.2
United Kingdom	3.8
Spain	3.8
Sweden	2.8
USSR	2.8

Sources: *Prevention's Giant Book of Health Facts*, p. 109.

666
How is coffee decaffeinated?

There are two methods generally used. In the first, coffee beans are placed in a rotating drum and softened by steam for 30 minutes. Then the beans are rinsed in a chemical solvent such as methylene chloride for approximately 10 hours. When the chemical has absorbed the caffeine from the beans, both caffeine and chemical are drained away. The beans are steamed for another 8 to 12 hours so that any remaining solvent evaporates. A minute amount of the chemical—approximately 0.1 parts per million—remains in the beans.

In the second method—the water process—green coffee beans are soaked in hot water until the caffeine is drawn from the beans. The drained water containing the caffeine is treated with a chemical, such as methylene chloride, which absorbs the caffeine. Then the mixture is heated until both the chemical and caffeine evaporate. The remaining water is returned to the beans, wherein, according to coffee experts, the beans regain most of their flavor.

Sources: *Prevention's Giant Book of Health Facts*, p. 111.

667
How much caffeine is in coffee, tea, and cocoa?

Beverage	Caffeine in milligrams	
	Average	Range
Coffee (5 ounce cup)		
Brewed, drip method	115	60–180
Brewed, percolator	80	40–170
Instant	65	30–120
Decaffeinated, brewed	3	2–5
Decaffeinated, instant	2	1–5
Tea (5 ounce cup)		
Brewed, major U.S. brands	40	20–90
Brewed, imported brands	60	25–110
Instant	30	25–50
Cocoa (5 ounce cup)	4	2–20

Sources: *FDA Consumer*, vol. 21 (December 1987/January 1988), p. 24.

668
How is egg cream made?

This New York City soda fountain drink, popular since the 1930s, does not contain any eggs. It is a mixture of milk and chocolate syrup into which seltzer water is spritzed, causing the mixture to foam. Egg cream is so named for its frothy head that resembles beaten egg whites.

Sources: Herbst, Sharon Taylor. *Food Lover's Companion*, p. 153.

What is the difference between bottled waters such as Perrier® and Polar Spring Water®?

All bottled water is processed in some way. Many bottled waters such as Perrier® are carbonated, either through natural carbonation or from carbonation added in processing the water or from both. Polar Spring Water® is not carbonated. Perrier® comes from a spring, Polar Spring Water® is taken from a well. In addition to springs and wells, bottled water can originate from artesian wells or municipal water supplies.

Sources: Von Wiesenberger, Arthur. *The Pocket Guide to Bottled Water*, pp. 2, 5, 83, 86.

Is there any purpose to serving nuts in a bar other than to generate thirst?

The Roman naturalist Pliny and centuries of folk medicine have suggested that almonds will prevent intoxication and hangovers.

Sources: *Garden*, vol. 62, no. 2 (March/April 1982), p. 29.

What is mead?

Mead is an alcoholic drink made by fermenting a solution of water and honey with malt and yeast. Sometimes mead is flavored with wild flowers or spices. One of the oldest liquors, the name is derived from the Sanskrit *madhu*, meaning honey. In medieval Europe, spiced mead was also called *metheglin*, deriving from the Welsh word *meddyglyn*.

Sources: Mendelsohn, Oscar. *The Dictionary of Drink and Drinking*, pp. 215-216; *Woman's Day Encyclopedia of Cookery*, vol. 7, p. 1116.

What are the different sizes of wine bottles?

The standard wine bottle is 757 milliliters, which is almost equivalent to an "American fifth" (4/5 of a quart or 1 pint 9.6 fluid ounces or 25.6 fluid ounces). Other bottle sizes used to contain wine are:

Split—1/4 of a standard wine bottle or 6.4 ounces

Half-bottle—1/2 of a standard wine bottle or 12.8 fluid ounces or 4/5 pint

Magnum—1.5 liters equivalent to 2 standard bottles (usually champagne)

Jeroboam—4 bottles in one or 1.6 magnums, or 6.4 pints (usually champagne and brandy)

Gallon—5 bottles in one

Rehoboam—6 bottles in one or 3 magnums (usually champagne)

Methuselah—8 bottles in one or 4 magnums (usually champagne)

Salmanazar—12 bottles in one or 6 magnums (usually champagne)

Balthazar—16 bottles in one or 8 magnums (usually champagne)

Nebuchadnezzar—20 bottles in one or 10 magnums (usually champagne)

Sources: Herbst, Sharon Tyler. *Food Lover's Companion*, p. 502; Rovin, Jeff. *Laws of Order*, p. 50; *The World Almanac and Book of Facts 1992*, p. 305.

673
What is the difference between a vintner, a viticulturist, and an enologist?

A vintner is a winemaker; a viticulturist is a grape grower. An enologist is one who directs and coordinates all activities concerned with production of wine; a wine taster is an enologist.

Sources: *Dictionary of Occupational Titles*, pt. 1, p. 124; Seldon, Philips. *The Vintage Magazine Consumer Guide to Wine*, pp. 3, 18; Vine, Richard P. *Commercial Winemaking, Processing and Controls*, p. 460.

674
What does the term proof on a liquor bottle mean?

In the United States, the term is a measurement of strength of (ethyl) alcohol (C_2H_5OH) contained in a volume of liquor at 68°F (20°C). One degree of proof equals 0.5 of 1% of alcohol. For example, a bottle labeled "80 proof" contains 40% alcohol.

Sources: *Guinness Book of Records 1992*, p. 537; *Mr. Boston Official Bartender's Guide*, p. 212.

DAIRY PRODUCTS

675
Why is salt added to ice in an ice cream freezer?

Ice cream mix starts to freeze at 27°F (-3°C) so the container must be cooled below that temperature. Water freezes at 32°F (0°C). Even melted ice water is warmer than 0°C, and is too warm to freeze ice cream. Salt water has a lower freezing temperature than plain water. When salt is added to ice, this mixture becomes cold enough to freeze ice cream.

Sources: *Foodworks*, p. 72.

▽ **676**

What is casein and how is it used?

Casein is the main protein in milk and is not found in any food other than milk. It gives milk its white color and coagulates in milk as little white flecks when milk becomes slightly sour. Prior to the discovery of plastics, casein was used extensively in various manufacturing processes such as button-making. It is now dried and used as powdered creamer for tea and coffee and is also used in the manufacture of paint and glue.

Sources: Cleveland-Peck, Patricia. *Making Cheeses, Butters, Cream and Yogurt at Home*, pp. 12-13.

677

How does butter differ from margarine and oleo?

Butter is made from dairy cream, and United States law requires it to be at least 80% butterfat, with 18% water and 2% milk solids (casein, milk sugar, lactose, etc.). The first palatable margarine was invented by French chemist Hippolyte Mege-Mouries in 1869. It is a non-dairy product containing at least 80% fat. Generally, the fat content is derived from plant oils, such as corn, soybean, safflower, or cottonseed. The name "oleomargarine" was required by law until 1952. Since that time, the terms "margarine" and "oleomargarine" have been used interchangeably.

Sources: Bailey, Janet. *Keeping Food Fresh*, p. 142; McGee, Harold. *On Food and Cooking*, pp. 19-23.

678

Is it possible to freeze eggs?

Yes, but never freeze eggs in their shells; the contents will expand, causing the shells to break. Liquid egg white can be frozen as is. When thawed, it will act just as if it had been taken right from the shell. Whole eggs and, to a greater extent, egg yolks, become more viscous after they have been frozen. To prevent this, mix the eggs or yolks thoroughly with sugar or salt before freezing. Add either ½ teaspoon of salt or 2 to 3 tablespoons of sugar per cup of eggs. The method chosen depends on the intended use of the eggs. One cup of eggs is equal to five large eggs.

Sources: Bailey, Janet. *Keeping Food Fresh*, pp. 169-170.

679

How can the freshness of an egg be tested?

Changes that take place in an egg as time goes by can be observed without opening it up. One way is to place the egg in water. If it rests flat on its side on the bottom, it is fresh. If it tends to stand up at the broad end, it is stale. If it floats under water, it is several weeks old. If it floats on top, it is several months old and may be bad. Inspectors judge an egg's freshness by *candling* it—holding it up to an intense light to see what is inside. When the yolk is obscured by cloudy white, or when the pocket of air is small (dime-sized), the egg is fresh. Grades are assigned based on freshness determined by candling: Grade AA eggs are freshest, Grade A are slightly older, and Grade B are older and can be slightly misshapen. Any dating on the egg carton is based on the day the eggs were packed (not laid). A three-figure code for the packing date represents the day of the year—numbered consecutively from 001.

Sources: Bailey, Janet. *Keeping Food Fresh*, pp. 165-167; Stobart, Tom. *The Cook's Encyclopedia*, p. 178.

680
Is there a difference in nutritional value between brown-shelled eggs and white-shelled eggs?

There is no difference either in nutritional value or in flavor between brown-shelled and white-shelled eggs. Shell color is no indication of quality. The color of the eggshell is determined by the breed of hen. Most chicken eggs are white or brown. Brown eggs come in a variety of shades from light brown to dark brown. The Single Comb White Leghorn is the dominant breed in the United States because it lays white eggs. In New England, where brown eggs bring a premium price, Rhode Island Reds, New Hampshires, and Plymouth Rocks are popular.

Sources: *Egg Science and Technology*, pp. 13, 66; Hillman, Howard. *Kitchen Science*, p. 125.

681
How do the holes form in Swiss cheese?

Swiss cheese, also called Emmentaler cheese, is a cow's milk cheese that originated in the Emme River Valley in Switzerland. Its curd, formed from rennet, is shaped into large wheels about 36 inches (90 centimeters) in diameter with a thickness of 6 inches (15 centimeters). The wheels are salted in strong brine and wrapped to prevent drying. Complete ripening, taking three to six months, occurs in humidity- and temperature-controlled rooms, where the microbial enzymes present in the "green" cheese slowly change the cheese's composition, its texture, and its flavor. Different bacteria used especially in the ripening process give different cheeses their characteristic flavors. In Swiss cheese, the enzymes are *Streptococcus thermophilus*, *Lactobacillus bulgaricus*, and *Propionibacterium shermanii*. The latter bacterium lives on lactic acid excreted by the first two bacteria, giving off large amounts of propionic acid and carbon dioxide gas. The gas collects in large pockets to form the "eyes" or holes in the cheese.

Sources: McGee, Harold. *On Food and Cooking*, pp. 48-49; *New Encyclopedia Brittanica*, vol. 4, p. 477.

682
What is the temperature that milk must be heated to be pasteurized?

Most milk sold for direct human consumption is pasteurized—heated hot enough and long enough to destroy all disease–causing organisms, especially the heat resistant *Mycobacterium tuberculosis*. Pasteurization, a process invented by the French chemist, Louis Pasteur (1822-1895), also retards spoilage by killing most of the bacteria that affect the milk's freshness and by inactivating the inherent milk enzymes (especially the fat–splitters) that make milk unpalatable. One standard method among the different combinations of temperature and time is to heat milk in a batch process to 144°F (62°C) for 30 minutes (at that temperature). In the continuous process, the milk is heated to at least 160°F (71.7°C) for at least 15 seconds. The first method stays well below the temperature at which a "cooked" flavor develops, whereas the temperature of the second method of continuous process approaches this temperature limit.

Sources: *Encyclopedia of Chemical Technology*, vol. 15, p. 530; McGee, Harold. *On Food and Cooking*, p. 13.

DIET *See: FOOD AND NUTRITION—NUTRITION*

FOOD ADDITIVES AND SUBSTITUTES

683
How much sweeter than sugar is aspartame?

Aspartame, or Nutrasweet®, is 180 times sweeter than sugar. Created in 1965 by the pharmaceutical firm G.D. Searle and Company, it was hailed as a low-calorie alternative to the controversial saccharin.

Sources: Hackh, Ingo W.D. *Grant & Hackh's Chemical Dictionary*, p. 55; Harris, Henry. *Good Old-Fashioned Yankee Ingenuity*, p. 33.

684
From what is the fat substitute Simplesse® made?

Simplesse® is a blend of proteins from egg whites and milk, with processed sugar, into a creamy liquid. Approved by the United States Food and Drug Administration as the first low-calorie fat substitute in 1990, it was developed by the Nutrasweet Company.

Sources: *The New York Times*, (February 23, 1990), p. A1; *Yearbook of Science and the Future 1991*, p. 356.

685 What is guar gum?

Guar gum is a food additive that is used as an emulsifier, firming agent, stabilizer, or thickener. It is added to such foods as baked goods, beverages, breakfast cereals, dairy products, jellies, sauces, and processed vegetables.

Sources: Lewis, Richard J. *Food Additives Handbook*, p. 237.

686 How is sodium erythorbate used?

Sodium erythorbate, relative of ascorbic acid (vitamin C), is synthesized in various ways starting with sugar, citrus fruits, hip berries, or fresh tea leaves. Since the 1950s, it has been used in cured meats as a color fixative and to keep meat from turning color. There is no relationship between it and earthworms.

Sources: *FDA Consumer*, vol. 20 (June 1979), p. 20; *Pittsburgh Press* (March 17, 1981), p. A16.

687 What is saffron?

Saffron is the world's most expensive spice—ounce for ounce, it is more precious than gold. A pound (.4536 kilogram) of saffron threads costs well over $2,000. Saffron threads are the roasted stigmas (female organs) of a small purple crocus (*Crocus sativus*). Packed by hand it takes at least 4,000 stigmas to make 1 ounce (28.3 grams) of this spice. Aromatic when dried, with a pungent smell, it adds flavor and coloring in cooking, and is also used as a dye and a medicine. Originating in the Near East, and known to the Assyrians as early as 4000 B.C.E., the term comes from the Arabic word *za'faran* (meaning yellow). The main producers of saffron today are Spain and Kashmir.

Sources: *Smithsonian*, vol. 19, no. 5 (August 1988), pp. 105-107; *Woman's Day Encyclopedia of Cooking*, vol. 10, pp. 15-77.

688 When was the GRAS list established?

The *Generally Recognized As Safe* (GRAS) list was established by the United States Congress in 1958 to designate food additives safe for human consumption. The original list consisted of substances that had been added to food over a long time, which were recognized as safe by qualified scientists. Such items as salt, pepper, yeast, common spices, and flavorings are on the GRAS list. These items are exempt from pre-market clearance. The list has been reviewed and revised by the Food and Drug Administration, using a panel of expert pharmacologists and toxicologists.

Sources: Lewis, Richard J. *Food Additives Handbook*, pp. 5-6; *McGraw-Hill Encyclopedia of Science and Technology*, vol. 7, pp. 281-282; Winter, Ruth. *A Consumer's Dictionary of Food Additives*, pp. 16-17.

689
What is MSG?

MSG (*Monosodium glutamate*) is a white crystalline powder derived from glutamic acid, a common amino acid. It occurs naturally in seaweed, soybeans, and beets or can be produced by fermenting glucose or by processing wheat, corn, soybean, or sugar beet proteins. In 1909 Kikunae Ikeda isolated it as a flavoring agent. It is added to foods, especially high-protein or processed foods, to improve or enhance flavors. In the 1950s, a bacterium was isolated that synthesizes and excretes glutamic acid when fed the nitrogen in ammonium ions (NH_4+). Today most MSG is made using this fermentation technique, with world production at 250 thousand tons (226,800 kilograms). There is evidence that high doses of MSG might cause physical problems but it continues to be on the list of food additives generally regarded as safe by the United States Food and Drug Administration (FDA).

Sources: Freydberg, Nicholas. *The Food Additives Book*, pp. 555-556; McGee, Harold. *On Food and Cooking*, p. 570.

690
How is garlic used?

Garlic, a bulb and cousin to the onion, is used in a variety of ways. Garlic is used in various food products, such as spices, non-alcoholic beverages, candy, baked goods, relishes, condiments, meats/meat products, fats and oils, snack foods, and gravies. Garlic can also be found in health food products.

Garlic is also used in folk medicine to treat coughs, colds, bronchitis, toothaches, earaches, dandruff, high blood pressure, arteriosclerosis, hysteria, and cancer. In Chinese medicine, garlic is also used for diarrhea, dysentery, tuberculosis, bloody urine, hepatitis, trachoma, ringworm, and diphtheria.

Sources: Brody, Jane E. *Jane Brody's Good Food Book*, pp. 149-151; Leung, Albert Y. *Encyclopedia of Common Natural Ingredients*, pp. 177-178.

691
What is the difference between an herb and a spice?

Spices are pungent or aromatic seasonings obtained from the bark, buds, fruit, roots, seeds or stems of various plants and trees; whereas, herbs usually come from the leafy part of a plant.

Sources: Herbst, Sharon Tyler. *Food Lover's Companion*, p. 440.

692
Why does the smoking of meat act as a preservative?

Smoked meat (or any food) is preserved because the traces of formaldehyde present in the wood smoke arrest bacterial development.

Sources: Kendig, Frank and Richard Hutton. *Life-Spans*, p. 112.

FOOD PROCESSING AND STORAGE *See also: FOOD AND NUTRI-TION—BEVERAGES; FOOD AND NUTRITION—DAIRY PRODUCTS; FOOD AND NUTRITION—MEAT, POULTRY, AND FISH*

693

What is a hypobaric system used for food storage?

Hypobaric controlled atmosphere systems are a precisely controlled combination of low pressure, low temperature, high humidity, and ventilation that extends up to six times the length of time a perishable commodity remains fresh. Atmospheres are tailored to each perishable item. For example, iceberg lettuce uses oxygen, carbon dioxide, and nitrogen. Seafood, avocados, cherries, limes, mushrooms, tomatoes, chicken, pork, and lamb may be kept under hypobaric conditions.

Sources: *Van Nostrand Reinhold Encyclopedia of Chemistry*, p. 482.

694

How is acidulated water used?

Acidulated water is cold water to which vinegar, lemon juice, or lime juice has been added to prevent cut fruit or vegetables from discoloring. The proportions are 1 tablespoon (14.8 milliliters) of juice in one quart (946 milliliters) of water.

Sources: McCully, Helen. *Nobody Ever Tells You These Things About Food and Drink*, p. 21.

695

How long can foods be stored in the refrigerator or freezer?

Fruits

Apples—Store loose in refrigerator for 2 to 4 weeks.

Citrus fruits—Oranges, lemons, limes, grapefruits, tangerines can be refrigerated from 10 to 14 days, but not in plastic bags.

Pineapples—Can be refrigerated from 3 to 5 days.

Berries and cherries—Keep covered in the refrigerator, unwashed. Use within 2 to 4 days. Cranberries may be stored from a week to a month.

Grapes and plums—Use within 3 to 5 days, store in perforated plastic bags.

Apricots, peaches, pears, nectarines and melons—Can be refrigerated from 3 to 5 days.

Avocados and bananas—Can be refrigerated when ripe; use within a week.

Juices—Opened canned or bottled juice can be refrigerated from 7 to 10 days.

Vegetables

Potatoes and sweet potatoes (in perforated plastic bags), onions, and winter squash—do better stored in a cool basement (not less than 50°F) rather than in a refrigerator; tomatoes are best stored uncovered at room temperature rather than refrigerated.

Green peas and lima beans—Can be stores in their pods in plastic bags 2 to 3 days in the refrigerator.

Corn—Store in the husk but use as soon as possible. The sugar in corn quickly turns to starch, causing a loss of flavor.

Carrots, radishes, turnips, beets, and parsnips—Remove leafy tops and store in refrigerator in plastic about 7 to 14 days.

Peas and beans (without pods)—Store in plastic bags, 3 to 5 days.

Broccoli, Brussel sprouts, green peppers, scallions, and summer squash—Store in plastic bag in the crisper, about 3 to 5 days; green peppers, store in paper bag.

Sprouts—7 to 10 days in plastic bag.

Mushrooms—keep unwashed in open container or paper bag and cover loosely with cheesecloth 2 to 3 days.

Lettuce, salad greens, chards, spinach, and other greens—Store in the crisper (drawer at refrigerator bottom) or in sealed plastic bags away from fruit (gives off ethylene gas). Iceberg and romaine keep 7 to 10 days; butterhead, endives, spinach, kale, collards, and chard keep 2 to 4 days.

Cucumbers (in perforated plastic bag), cabbage, cauliflower, celery, and green and wax beans—Store unwashed for up to a week (celery and cabbage up to 2 weeks).

Asparagus—Keeps 4 to 6 days in plastic bags in refrigerator but stalks become tougher with each day. Keep tips dry.

Meats, Poultry and Fish

Bacon and ham—Refrigerate for up to 7 days, and 1 month in freezer except whole non-canned hams (3 months).

Ground meat—If wrapped in plastic, refrigerate from 1 to 2 days, if wrapped in paper, it must be used right away or rewrapped in plastic or waxed paper. Ground meat can be frozen in plastic wrap for 2 weeks. It will last from 2 to 3 months if frozen in a aluminum foil, freezer paper, or a plastic freezing bag.

Cold cuts and frankfurters—Bologna and frankfurters can be refrigerated 4 to 7 days and frozen for 2 months; cold cuts can be refrigerated from 3 to 4 days and frozen for 1 month.

Roasts and steaks—Can be refrigerated for 3 to 5 days. Beef steaks and roasts can be kept frozen in plastic wrap from 6 to 12 months; lamb and veal roasts for 6 to 9 months; pork roasts for 3 to 6 months.

Chops and smaller cuts—keep 3 to 4 months in freezer.

Poultry—Should be used within a day or two after refrigeration. Chicken and turkey may be stored in plastic bags in the freezer from 9 to 12 months; ducks and geese can be frozen up to 4 to 6 months.

Fish—Wrapped at 0°F (-32°C) in refrigerator (on ice) it will keep from 1 to 2 days. Frozen fatty fish keeps 3 months frozen; lean fish keeps 6 months; frozen, cooked shellfish keeps 2 to 3 months.

Dairy Products

Milk and cream—Keeps about 5 to 7 days in the refrigerator and 3 months in the freezer. Studies show that milk kept in cardboard containers retains more vitamin C than milk in plastic jugs. Buttermilk keeps 1 to 2 weeks in the refrigerator and 3 months in the freezer.

Hard cheeses (Cheddar, Parmesan and Swiss)—May be kept for several months, if tightly wrapped, in the refrigerator. Grated cheese should be kept refrigerated in a tightly covered jar.

Soft cheeses—Keep tightly covered in the refrigerator. Cottage cheese will last 1 week; cream cheese can be kept 1 to 2 weeks.

Butter and margarine—Will keep about 30 to 90 days in the refrigerator and 6 to 9 months in the freezer.

Eggs—Whole eggs should be used within 1 to 4 weeks of purchase, but freshness deteriorates after one week. Whole eggs cannot be frozen.

Sour cream—Refrigerate 1 to 3 weeks.

Yogurt—Refrigerate 1 to 2 weeks and 1 to 1½ months in freezer.

Ice Cream—Keeps 1 to 2 months in freezer.

Breads, Flours and Cereals

Soft-crusted breads—Can be refrigerated from 4 to 7 days or 3 months in freezer. Keep securely wrapped.

Yeast breads—Refrigerate from 7 to 14 days; freeze for 3 months.

White, wheat, and potato flour—Store in a tight container from 6 to 12 months in refrigerator.

Rye, oat, and soy flour—Keeps 2 to 3 months refrigerated.

Crackers (opened)—Can be refrigerated 3 to 4 months.

Wheat germ—Keeps 2 to 3 months in refrigerator in tightly sealed jar or bag.

Dough—Bread dough keeps 2 months in freezer if tightly sealed; cookie dough keeps 9 to 12 months in freezer; unbaked pie shells keep 1 to 2 days refrigerated and 4 to 6 months frozen.

Whole grain cereals—Keep 4 to 5 months in refrigerator.

Cooked cereals—Keep 2 to 3 days in refrigerator.

Miscellaneous

Coffee—Fresh ground or opened ground canned coffee can be refrigerated or frozen from 7 to 14 days; whole beans can be frozen 3 to 4 months.

Peanut butter—Commercial opened peanut butter will keep 3 to 4 months in the refrigerator.

Shelled nuts—Cashews, macadamia nuts, pecans, and roasted peanuts can be refrigerated for 6 months and frozen from 9 to 12 months. Mixed nuts, shelled and unshelled almonds, and walnuts can be refrigerated from 9 to 12 months and frozen for 12 months Pistachios can be refrigerated for 3 months.

Pies—Non-cream pies keep 4 to 5 days refrigerated and 6 months frozen; cream pies can only be refrigerated 2 to 3 days.

Sources: Bailey, Janet. *Keeping Food Fresh*, pp. 9-60, 76-115, 127-150, 171, 196-197, 207-208, 223-224, 239-241, 250–254, 263-272, 304, 311; Nugent, Nancy. *Food and Nutrition*, pp. 65–69; Rombauer, Irma S. *Joy of Cooking*, pp. 826-827.

696 ▽ What makes a good cold cellar?

Ideally, a cold cellar or root cellar for storing fruits and vegetables should have high humidity, low temperature, and some ventilation. The humidity should be around 90%, the temperature between 32°F (0°C) and 40°F (4.4°C), with enough air circulation to remove the ethylene gas that fruits give off. Most modern houses have dry warm basements that are not useful for fruit and vegetable storing.

Sources: Bailey, Janet. *Keeping Food Fresh*, p. 331; Rodale, Robert. *The Best Gardening Ideas I Know*, pp. 76-79.

697 ▽ How is food freeze-dried?

The freeze-drying food preservation process, first used in the 1950s, can be used on 600 different foods. It is a two-step process of rapid freezing and subsequent dehydration (removal of moisture from a substance). To preserve its nutrients, flavor, etc., food is fast-frozen by placing the item between two hollow plates that contain a refrigerant liquid in a tightly-sealed chamber. While the food is being frozen, a high-powered pump creates a vacuum in the chamber. Then the refrigerant liquid in the hollow plates is replaced with warm gas, which converts the ice in the food *directly* into vapor. This dehydration process takes 20 hours. The dehydrated food can be reconstituted by the addition of boiling water. The food must be packaged to seal out oxygen and moisture, which would cause the food to decay.

Sources: *How in the World*, p. 379.

698 ▽ When were frozen foods first available commercially?

Clarence Birdseye (1887-1956) discovered that quick freezing of food enabled the food to retain its flavor. The faster the freezing, the less danger of ice crystals rupturing the cell walls of the food item and releasing natural juices. He patented his quick-freezing process of placing food between two metal plates having a below 0°F(-18°C) temperature, In 1929 he sold his company to Postum Cereal Co., which founded General Foods. At the time of the sale it was agreed that Birdseye's name (as two words) would become the brand name for the frozen food items—peas, spinach, raspberries, cherries, fish, and meat—that were marketed in September, 1930.

Sources: DeBono, Edward. *Eureka!*, p. 116; Harris, Harry. *Good Old-Fashioned Yankee Ingenuity*, pp. 133-134; Robertson, Patrick. *The Book of Firsts*, pp. 75-76.

699
How long will food keep in the refrigerator or freezer after the power goes off?

Assuming the refrigerator door remains closed, food in the refrigerator will stay chilled for six hours after the power goes off. Adding bags of regular ice on the upper shelves will help keep the temperature down. After power has been restored, check the temperature in various parts of the refrigerator to make certain that milk products, meat, fish, poultry, mayonnaise, and cooked food have been kept colder than 45°F (7°C). Likewise, if the food in the freezer is allowed to thaw and warm up to above 45°F (7°C), harmful bacteria will thrive and food will spoil. A fully-loaded freezer will stay cold for up to two days, a half-loaded one will do so for one day, but only if the freezer is not opened. Dry ice can be added to extend this time period. A 10-cubic-foot freezer, fully loaded, will need 25 pounds (11.34 kilograms) of dry ice to keep food cold for up to three days. Always wear gloves when handling dry ice; cover the food first with cardboard and place the dry ice on the cardboard. Once the ice is in place, only open the freezer to add more ice.

Sources: Bailey, Janet. *Keeping Food Fresh*, pp. 376-378; *Rodale's Complete Home Products Manual*, pp. 125, 253.

700
Which foods cannot be frozen?

Bananas, cream fillings and puddings, custards, gelatin dishes, hard-cooked eggs, cooked potatoes, raw vegetables, such as celery and radishes, and soft cheeses.

Sources: McCully, Helen. *Nobody Ever Tells You These Things About Food and Drink*, p. 11.

701
Can food be left in the can once it has been opened?

Yes, but it must be securely covered and refrigerated. Modern cans are lined with a thin layer of pure tin that in no way affects food.

Sources: McCully, Helen. *Nobody Ever Tells You These Things About Food and Drink*, p. 9.

702

How many ounces are there in the common can sizes?

No. 300	14 to 16 fluid ounces
No. 303	16 to 17 fluid ounces
No. 1 tall	16 fluid ounces
No. 2	1 pound 4 ounces or 1 pint 2 fluid ounces
No. 2½	1 pound 13 ounces or 28 fluid ounces
No. 3	46 fluid ounces
No. 10	96 fluid ounces

Sources: Blocksma, Mary. *Reading the Numbers*, p. 23.

703

Who invented the tin can?

The English inventor, Peter Durand, created the "tin canister" in 1810. The cans were handmade and expensive. Machine-stamped cans were introduced in 1847. In 1959 Ermal Cleon Fraze (1913-1989) invented the flip-top can.

Sources: Harris, Harry. *Good Old-Fashioned Yankee Ingenuity*, p. 127; Sanders, Dennis. *The First of Everything*, p. 17; Stilwell, E. Joseph, et al. *Packaging for the Environment*, p. 77.

704

How many cans are produced annually in the United States?

Nearly 112 billion cans were produced in the United States during 1989, and almost 75% of those were used by the beer and soft drink industry. In terms of total units, aluminum dominated, accounting for nearly 75% of all metal containers and 97% of all beverage cans.

Sources: Stilwell, E. Joseph, et al. *Packaging for the Environment*, p. 70.

705

How can one prevent plastic wrap from sticking to itself or to the box?

Storing plastic wrap in the refrigerator will allow it to unroll more easily, and it won't stick to itself or to the box. The stickiness of the wrap is due to static electricity. The film acquires a negative electric charge through friction, displacing electrons from the surface of the second piece of wrap, causing it to have a positive electric charge. The two pieces become bonded by electrical attraction. Plastic wrap is made from either of two plastics PVC (*polyvinyl chloride* softened with a plasticizer) or already-soft polyethene.

Sources: *How in the World?*, p. 23; McCully, Helen. *Nobody Ever Tells You These Things About Food and Drink*, p. 16.

FRUITS AND VEGETABLES

706
Is the tomato a fruit or a vegetable?

Botanists class the tomato (*Lycopericon esculentum*) technically as a berry, which in a strictly scientific sense, is a simple fruit. But horticulturists and the United States Department of Agriculture in its bulletins and reports class the tomato as a vegetable. On May 10, 1893, the United States Supreme Court ruled that the tomato is a vegetable and that it was subject to import duties under the United States Tariff Act of 1883, which levied duties on imported vegetables but not on fruits.

Sources: Stimpson, George. *Information Roundup*, p. 210.

707
Why were tomatoes called "love apples"?

Tomatoes (*Lycopersicon esculentum*), of the nightshade family, were cultivated in Peru and introduced into Europe by Spanish explorers. Tomatoes were brought via Morocco into Italy where they were called *pomi de Mori* (apples of the Moors). The French called the tomato *pommes d'amore* (apples of love). This may be because tomatoes were thought by some to have aphrodisiac powers, or it may simply be a corruption of the Italian term.

Sources: Herbst, Sharon Tyler. *Food Lover's Companion*, p. 473; Simon, Andre L., and Robin Howe. *Dictionary of Gastronomy*, unpaged alphabetical entry.

708
What is an ugli?

An ugli (*Citrus reticulata*) is a tropical fruit originating in Jamaica. Possibly a hybrid between a grapefruit and a tangerine, this unattractive fruit with a loose, wrinkled, yellowish-red skin with a greenish tinge, ranges in size between that of a navel orange and a giant grapefruit. Its acid-sweet flavor suggests grapefruit with hints of orange.

Sources: Wyman, Donald. *Wyman's Gardening Encyclopedia*, p. 1135; Simon, Andre L., and Robin Howe. *Dictionary of Gastronomy*, unpaged alphabetical entry.

709
Why can't fresh pineapple be used in jello?

A pineapple (*Ananas comosus*) contains a protease, an enzyme that breaks down proteins. This must be destroyed by boiling if the fruit is to be used in a gelatin preparation—gelatin is a protein derived from meat. The protease will simply digest the gelatin away, leaving a soupy, unappealing dish.

Sources: McGee, Harold. *On Food and Cooking*, p. 188.

710 ▽
What is a legume?

As the third largest family among flowering plants and the second most important to the human diet (grasses are first), a legume is any of thousands of plant species that have seed pods that split along both sides when ripe. The seeds have twice as much protein as grains, much iron, and B vitamins. Some of the more common legumes used for human consumption are beans, lentils, peanuts, peas, and soybeans. Others, such as clover and alfalfa, are used as animal fodder.

Legume type	Date of domestication	Place of domestication
Broad bean	3000 B.C.E.	Near East
Chick pea	5000 B.C.E.	Near East
Common bean	6000 B.C.E.	Central and South America
Lentil	7000 B.C.E.	Near East
Lima bean	6000 B.C.E.	Central and South America
Mung bean	1000 B.C.E.	Asia
Pea	6000 B.C.E.	Near East
Soybean	1000 B.C.E.	Asia

Sources: Herbst, Sharon Tyler. *Food Lover's Companion*, p. 256; McGee, Harold. *On Food and Cooking*, pp. 249-258.

711 ▽
What is Indian corn?

Indian corn is flint corn, a hard starch corn that predominated in the northern United States and Canada centuries before the Europeans arrived. Early native growers produced red, blue, black, yellow, white, and multicolored corns. Later growers continued to cultivate and selectively breed this corn, producing ears of every conceivable hue. It is usually this colored flint corn, also called Indian corn, that is used in autumn decorations.

Sources: Rupp, Rebecca. *Blue Corn and Square Tomatoes*, p. 194.

712 ▽
How many kinds of artichokes are there?

Originating in Canada, the vegetable Jerusalem artichoke (*Helianthus Tuberosus*), also called a sunchoke, is a variety of the sunflower. This lumpy brown-skinned tuber, that often resembles ginger root, may be eaten raw in a salad or boiled or steamed. The Italian name for artichoke is *girasole*, which the British corrupted into *Jerusalem*. The globe or leafy artichoke, considered the true artichoke, is the bud of a large plant (*Cynara scolymus*) from the thistle family. Now grown mainly in California, it was brought to the New World by the Spanish; it originated in Africa and Europe. The meat on the petals and the heart are edible; the fuzzy choke is inedible. The Chinese (or Japanese) artichoke, a native of China and Japan, has a taste similar to the Jerusalem artichoke. It may be eaten raw, boiled, baked, or steamed.

Sources: Simon, Andre L., and Robin Howe. *Dictionary of Gastronomy*, unnumbered alphabetical entry "Artichoke"; *Woman's Day Encyclopedia of Cookery*, vol. 6, p. 984.

ARTICHOKES (globe, Japanese, Jerusalem)

713

What is the difference between a fruit and a vegetable?

In botany, a fruit is the ovary of a plant—the section of the plant that surrounds the seeds. All other edible plant parts are considered vegetables. By this definition, tomatoes, pumpkins, eggplants, peppers, peapods, cucumbers, and corn kernels are fruits. Common usage holds that a plant is a vegetable if it is usually eaten as part of a meal's main course. A fruit is eaten as a dessert or a between meal snack.

Sources: Hillman, Howard. *Kitchen Science*, p. 130; McGee, Harold. *On Food and Cooking*, pp. 124-126.

714

Which foods originated in the New World?

Corn and white potatoes have become staple crops in many countries. Sweet potatoes, kidney beans, and peanuts also originated in America. The list also includes lima beans, tomatoes, sweet peppers, most kinds of squash and pumpkins, avocados, cocoa, pineapples, and many lesser crops.

Food	Place & time of origin	Notes
Corn (Maize)	Under cultivation by 3500 BCE in Central America.	Spread rapidly to many Indian cultures in New World (Incas, Mayas, Aztec, etc.). Columbus brought it to Europe.

Food	Place & time of origin	Notes
Potato	Native to Central & South America; cultivated more than 4,000 years ago in mountainous areas (up to 15,000 ft.) where corn could not grow—staple crop of Incas.	Spanish explorers brought plant to Europe about 1570.
Pineapple	Native to South America	Introduced into Europe before 1535; brought to Hawaii in 18th century.
Sweet potato	Native to Central America	Introduced into Europe by Columbus.
Pumpkin & squashes	Native to all the Americas; 25 species cultivated for 9,000 years.	
Green bean	Originated in Central America; cultivated in Mexico for 7,000 years.,	Brought to Europe by Spanish and Portuguese in 16th century.
Lima bean	Native to tropical America; cultivated in Peru since 5,000 or 6,000 B.C.E.	The name comes from Lima, the capitol of Peru.
Sweet or bell pepper	Tropical America; cultivated long before European arrival.	
Tomato	Native to Andes Region; cultivated in Mexico	Brought to Europe in 1523.
Avocado	Cultivated in Central America for 7,000 years	Introduced into Jamaica in 1650, Florida and California in 19th century.
Common bean (Kidney, navy, pinto, & black)	Cultivated in SW Mexico 7,000 years ago and spread north and south about 2,000 years ago.	Exported to Europe by Spanish explorers.
Nuts		
Brazil nut	Native to Amazon region of Brazil.	Export began in 1835.
Cashew nut	Native to Amazon region of Brazil.	Portuguese took it to India in 16th century.
Hickory nut	Native to North America, east of Mississippi River.	Includes hickory, pecan, and walnut
Peanut	Native to South America as far back as 800 B.C.E.	Portuguese took it to East Africa.
Sunflower seeds	Cultivated in American Southwest.	Brought to Europe in 1510.
Cocoa bean	Originated in the river valleys of South America; carried to Mexico by the Mayas in 7th centuryv C.E. where cultivated.	Columbus brought back beans to Europe in 1502; but Cortez and other Spaniards finally understood its use in 1519.

Sources: *Can Elephants Swim?*, p. 57; McGee, Harold. *On Food and Cooking*, pp. 128-132, 188-193, 200-202, 240, 254, 266-271, 397.

▽ **715** ▽

Will an orange ripen after it is picked?

An orange (*citrus sinesis*), unlike many fruits, does not continue to ripen after being picked; it is either tree-ripened, or not ripened at all. But some green color on its skin does not mean that it is unripe. The skin of a Florida orange is commonly dyed "orange" to have a pleasing consumer appeal. The orange, a berry in the botanical classification of fruits, originated in southern China about 2200 B.C.E., and the Romans introduced it into Europe in the 1st century C.E. Christopher Columbus brought it to Hispaniola in 1493, and in 1539, de Soto planted seedlings in St. Augustine, Florida. Its cultivation slowly moved westward to California, where it did not thrive. In 1873 two navel orange trees from Brazil (via Washington, D.C.) were planted in California. These orange trees, called Washington navel, are the ancestors of all California navel oranges.

Sources: Brody, Jane E. *Jane Brody's Good Food Book*, p. 158; Panati, Charles. *Panati's Browser's Book of Beginnings*, pp. 103-104; Root, Waverly L. *Food*, p. 306.

HERBS *See:* FOOD AND NUTRITION—FOOD ADDITIVES AND SUBSTITUTES

KITCHEN EQUIPMENT, MEASUREMENTS, ETC.

716
Who invented Tupperware®?

In the 1940s, Earl S. Tupper (d.1983), an American molding engineer, saw the potential of using polyethylene for making bowls, food storage items, etc. In 1945, he produced his first item—a seven ounce (207 milliliter) bathroom tumbler—in the Tupperware® line, followed by polyethylene bowls that had a revolutionary new vacuum–creating seal. Tupper devised a marketing plan to sell the products through in-home sales parties in 1946. In 1958 Tupper sold his business to Rexall Drugs.

Sources: Panati, Charles. *Panati's Extraordinary Origins of Everyday Things*, pp. 129-130; Room, Adrian. *Dictionary of Trade Name Origins*, p. 176.

717
What is the origin of the chef's hat?

The tall white hat is a copy of the hat worn by Orthodox priests. To escape persecution, some famous cooks sought refuge in monasteries. The cooks sought a distinction from other monks, which was found in the innovation of a white hat instead of the conventional black. This was authorized and is now the white hat worn by chefs the world over.

Sources: Simon, André L., and Robin Howe. *Dictionary of Gastronomy*, p. 114.

If a recipe calls for a pinch, or a few grains, how much is it?

A few grains means less than ⅛ of a teaspoon. This is equivalent to pinch. Other measurements worth noting are listed below:

Volume

Measure	Equivalent
Pinch (a few grains)	less than ⅛ teaspoon
Dash	8 drops or about ⅛ teaspoon
½ teaspoon	30 drops
3 teaspoons	1 tablespoon
½ tablespoon	1½ teaspoons
2 tablespoons	1 fluid ounce
3 tablespoons	1½ fluid ounces or 1 jigger
4 tablespoons	¼ cup or 2 fluid ounces
5 tablespoons + 1 teaspoon	⅓ cup
10⅔ tablespoons	⅔ cup
16 tablespoons	1 cup
1 cup	½ pint or 8 fluid ounces
2 cups	1 pint or 16 fluid ounces
2 pints	1 quart
1 quart	2 pints or 4 cups or 32 fluid ounces
4 quarts	1 gallon
2 dry pints	1 dry quart
8 dry quarts	1 peck
2 pecks	1 pail
4 pecks	1 bushel

Weight

Measures	Equivalent
2 ounces	⅛ pound
4 ounces	¼ pound
8 ounces	½ pound
16 ounces	1 pound

Sources: Brody, Jane. *Jane Brody's Good Food Book*, p. 279; Kogelman, Stanley and Barbara R. Heller. *The Only Math Book You'll Ever Need*. p. 206; Rovin, Jeff. *Laws of Order*, pp. 44-45.

How many cups are there in a baking pan?

Pan size in inches	Pan type	Volume in cups
8 x 8 x 1½	square	6
8 x 8 x 2	square	8
9 x 9 x 1½	square	8
9 x 9 x 2	square	10
11 x 7 x 2	rectangular	6
13 x 9 x 2	rectangular	15
8 x 4 x 2½	loaf	4

Pan size in inches	Pan type	Volume in cups
8½ x 4½ x 2½	loaf	6
9 x 5 x 3	loaf	8
1¾ x ¾	mini muffin cup	⅛
2¾ x 1⅛	muffin cup	¼
2¼ x 1⅜	muffin cup	scant ½
8 x 1¾	pie plate	3
9 x 1½	pie plate	4
9 x 2	pie plate (deep dish)	6
8 x 1½	cake	4
8 x 2	cake	7
9 x 1½	cake	6
9 x 2	cake	8½
10 x 2	cake	10¾
9 x 3	bundt	9
10 x 3½	bundt	12
8 x 3	tube	9
9 x 3	tube	10
10 x 4	tube	16
9½ x 2½	springform	10
10 x 2½	springform	12

Sources: Herbst, Sharon Tyler. *Food Lover's Companion*, p. 536.

720
How many liters are there in one gallon?

Converting to Metric

Known measure	Multiply by	To get
Teaspoons	4.93	Milliliters
Tablespoons	14.79	Milliliters
Fluid ounces	29.57	Milliliters
Cups	236.59	Milliliters
Cups	0.236	Liters
Pints	473.18	Milliliters
Pints	0.473	Liters
Quarts	946.36	Milliliters
Quarts	0.946	Liters
Gallons	3.785	Liters
Ounces	28.35	Grams
Pounds	0.454	Kilograms
Inches	2.54	Centimeters
Fahrenheit	subtract 32; then multiply by 5, and divide by 9	Celsius (Centigrade)

Converting from Metric

Known measure	Divide by	To get
Milliliters	4.93	Teaspoons
Milliliters	14.79	Tablespoons
Milliliters	29.57	Fluid ounces

Converting from Metric

Known measure	Divide by	To get
Milliliters	236.59	Cups
Liters	0.236	Cups
Milliliters	473.18	Pints
Liters	0.473	Pints
Milliliters	946.36	Quarts
Liters	0.946	Quarts
Liters	3.785	Gallons
Grams	28.35	Ounces
Kilograms	0.454	Pounds
Centimeters	2.54	Inches
Celsius (Centigrade)	Multiply by 9; then divide by 5, and add 32	Fahrenheit

Sources: Herbst, Sharon Tyler. *Food Lover's Companion*, p. 532.

▽ **721** ▽

Why does water not put out an oil fire?

Since the specific gravity of oil is less than water, it will float; because of this, water will not put out oil fires. Instead, the burning oil will float on top of the water and spread. Throwing sand, table salt, or baking soda on the flames will extinguish them. Also a damp towel spread over the fire will cool and smother it but all the water must be wrung out.

Sources: Rosenfeld, Sam. *Science Experiments with Water*, pp. 53-54.

MEASUREMENTS, *See:* FOOD AND NUTRITION—KITCHEN EQUIP-MENT, MEASUREMENTS, ETC.

MEAT, POULTRY, AND FISH

▽ **722** ▽

What is a beefalo?

Beefalo is a cross between the American bison (commonly called a buffalo) and cattle. The dark red meat beefalo is very lean and has a somewhat stronger flavor than beef. It may be cooked in any manner suitable for beef.

Sources: Herbst, Sharon Tyler. *Food Lover's Companion*, p. 32.

723

How hot must the temperature of meat be to ensure that it is safe to eat?

Meat and poultry cooked throughout to these internal temperatures are generally safe to eat.

Meat	Celsius	Fahrenheit
Fresh Beef		
Rare	60°	140°
(Rare beef cooked to only		
140°F means some food		
poisoning organisms may survive)		
Medium	71°	160°
Well Done	77°	170°
Ground Beef	77°	170°
Fresh Veal	77°	170°
Fresh Lamb		
Medium	77°	170°
Well Done	82°	180°
Fresh Pork	77°	170°
Poultry		
Chicken	82–85°	180–185°
Turkey	82–85°	180–185°
Boneless		
Turkey Roasts	77–80°	170–175°
Stuffing (inside or outside bird)	74°	165°
Cured Pork		
Ham, Raw (Cook before eating)	77°	170°
Ham, Fully cooked (Heat before serving)	60°	140°
Shoulder (Cook before eating)	77°	170°

Sources: *The Safe Food Book*, p. 11.

724

What is muktuk?

Muktuk is the skin of the narwhal, an unusual whale of the Arctic. It is eaten as food by the native peoples.

Sources: *World Book Encyclopedia*, vol. 14, pp. 21–22.

725

What is a prawn?

This term can have several definitions:

1. A species that is part of the lobster family. These prawns have bodies shaped like tiny Maine lobsters, including minuscule claws. The meat has a sweet delicate flavor. They are 6 to 8 inches (15 to 20 centimeters) long and have pale–red bodies deepening to dark–red tails. The crustaceans included are the

Dublin bay prawn, Danish lobster, Italian scampi, langostino (Spanish), langoustine (French), Caribbean lobsterette, and Florida lobsterette.

2. "Freshwater" prawns (macrobrachium) which migrate from saltwater to freshwater to spawn. They look like a cross between a shrimp and a lobster. Their bodies have narrower abdomens and longer legs than shrimps.

3. The term "prawn" is also loosely used to describe any large shrimp, especially those that number 15 or fewer to 1 pound (0.4536 kilograms).

Sources: Herbst, Sharon Tyler. *Food Lover's Companion*, pp. 367–368.

726 / What are sea legs or sea claws?

Sea legs are a mixture known as "surimi," processed in Japan. It consists mainly of pollack, a cousin of cod fish, which is mixed with the meat of the sea crab and given body by the addition of starches and stiffeners, usually egg white. Dyed a pretty pink, this agglomeration is formed into the shape of a crab leg, and frozen. It is far cheaper than real crabmeat and frequently passes for the real thing.

Sources: Chalmers, Irena. *Great American Food Almanac*, p. 180.

727 / Is the puffer fish, eaten as a delicacy in Japan, toxic?

The tiger puffer (*Fugu rubripes rubripes*), also known as *fugu* or *tora fugu*, contains a powerful natural poison. It takes a special skill to prepare a dish that is definitely safe to eat. When imported into the United States, it is subject to rigid laboratory testing, and must be FDA–approved and guaranteed by the Japanese government to be toxin–free.

Sources: McClane, Albert J. *McClane's Fish Buyer's Guide*, p. 76; *Science 82*, vol. 3 & 4 (May 1982), p. 68.

728 / Which fish have the most nutrients and are the least susceptible to contamination?

The purer and most nutritious fish are cod, halibut, red snapper, haddock, salmon, sea bass, and sea trout. They all swim at the middle to upper level of the sea. The more fatty the fish, the more flavorful the fish is and the more vitamins and nutrients it has. Fat content of a fish generally decreases in relation to the depth of the fish's environment. Lobster, shrimp, mussels, and clams are scavengers that live off debris at the bottom of the ocean and are more likely to be contaminated. For instance, bottom–dwelling shellfish ingest toxic mercury wastes that settle on the ocean floor. Fish that eat these toxic shellfish likewise have higher concentrations of toxins and the fish such as tuna, sharks, and sword-fish, that ingest the toxic fish, in turn increase their mercury contamination. Because these bigger fish live longer than smaller fish, they can eat more toxic fish and have high mercury levels within their tissues. Tuna fish and swordfish are the most frequently contaminated with mercury poisoning.

Sources: D'Adamo, James. *One Man's Food—Is Someone Else's Poison*, pp. 41–42; Hillman, Howard. *Kitchen Science*, pp. 76–77, 85–86.

729
Are oysters good to eat only in months ending with an "R"?

Now oysters can be eaten any month of the year. The belief that oysters should not be eaten in May, June, July, and August stems from pre–refrigeration days when with the warm weather came more danger of contaminated oysters with harmful toxins. However since oysters tend to be more watery and less meaty dur-ing the summer months when they spawn, the dictum still carries some meaning in terms of good taste.

Sources: Cazeau, Charles J. *Science Trivia*, pp. 126–127; Hillman, Howard. *Kitchen Science*, p. 96.

MISCELLANEOUS FOODS

730
How are angels on horseback made?

Angels on horseback are oysters wrapped in bacon, skewered, broiled, and served on hot buttered toast. *Devils on horseback* are prunes stuffed with anchovy fillets or anchovy olives, wrapped in bacon, baked in an oven, and also served on toast. Both are popular in England.

Sources: Bailey, Adrian. *Mrs. Bridges' Upstairs Downstairs Cookery Book*, p. 143; McCully, Helen. *Nobody Ever Tells You These Things About Food and Drink*, p. 139.

731
Why are pistachio nuts dyed red?

In Turkey and Iran, where pistachio trees grow, the nuts are roasted in the sun in a salt brine, which turns the shells a pinkish color. American importers and California producers attempted to imitate this color and in the process came out with the now familiar shocking red. The California Pistachio Commission states that these nuts are still dyed red because many people find this form familiar and they are easier to spot in a bowl of mixed nuts.

Sources: Herbst, Sharon Tyler. *Food Lover's Companion*, p. 353; McCully, Helen. *Nobody Ever Tells You These Things About Food and Drink*, p. 202.

▽ **732** ╱
╱ *How are candied flowers made?*

Crystallized flower petals, used for cake and candy decorations, can be made from two different methods—each with a different storage time. To make crystallized petals that will last for months, dissolve ½ ounce of gum arabic in ¼ cup of warm water. After cooling, paint it on the petals with a paint brush. Next boil ½ cup of sugar with ¼ cup of water until it reaches 150°F. When cool, brush it on the petals, and sift superfine sugar over the top.

An easier method lasts for about a week. Beat an egg white until frothy but not whipped. With a small brush, carefully coat each petal on both sides with the egg white. Using a tea strainer, sift superfine sugar over each petal. Place on waxed paper and dry for two hours.

Sources: *The Little Scented Library*, vol. 4, p. 38.

▽ **733** ╱
╱ *Where did bagels originate, and why are genuine bagels boiled before baking?*

There are as many stories for the origin of the bagel as there are ethnic and immigrant groups in America. Most groups can claim a roll or roll-like bread with a hole as its distinguishing feature. In general though, the bagel is acknowledged as a Jewish contribution to America, and today it is a universal American food. Boiling or *kettling* in Yiddish is traditional and apparently serves to "lock" in flavor prior to baking and to give bagels their sheen.

The bagel's origin has been traced back to Austria in 1683. As a token of thanks for defeating the Turks, a Jewish baker made a hard-roll in the shape of a stirrup as a way of recognizing the king's bravery and love of riding. The word "stirrup" in Austrian-German is *beugel*. Eventually this word evolved from German to Yiddish into English as *bagel*.

Sources: Bagel, Marilyn, and Tom Bagel. *The Bagels' Bagel Book*, pp. 4-5, 11.

▽ **734** ╱
╱ *How long can chocolate be stored?*

Mass-market produced chocolate like Hershey's can last from 9 to 15 months in bars and 15 months in morsels, if stored at a constant room temperature. Mechanically-made commercial chocolates produced by smaller manufacturers have shelf lives of 3 to 4 months, if the temperature is held at 68° to 70°F (20 to 21°C) and the humidity at 50 to 60%. Smaller manufacturers that produce "home-made" chocolates estimate the shelf lives of their products at a maximum of 4 to 6 weeks. Baking chocolate, dry cocoa, or hot chocolate mix will last one year at room temperature if kept in a tightly closed container so as not to absorb moisture and odor. Unopened liquid chocolate can be stored up to 18 months, after opening, it should be stored in the refrigerator and can be kept for one year.

Sources: Bailey, Jane. *Keeping Food Fresh*, pp. 317-319; Kendig, Frank, and Richard Hutton. *Life-Spans*, pp. 157-158.

735 / What are the different kinds of sugar?

Sugar from sugar beets or sugar cane are sucrose sugar. Sucrose is 66% as sweet as fructose (sugar from fruits); but sucrose is twice as sweet as dextrose, which is a glucose sugar known as corn or grape sugar. Sucrose is three times as sweet as maltose (malt sugar) and about six times sweeter than lactose (milk sugar).

Granulated or white sugar is highly refined cane or beet sugar. Superfine sugar, known in Britain as castor sugar, is more finely granulated. Confectioners' or powdered sugar is granulated sugar that has been crushed into a fine powder. A small amount of cornstarch is added. Confectioners' sugar is known as icing sugar in Britain.

Decorating or coarse sugar (also called sugar crystals) has granules about four times larger than those of regular granulated sugar. Brown sugar is white sugar combined with molasses (a by-product of the cane sugar refining process).

Sources: Herbst, Sharon Tyler. *Food Lover's Companion*, pp. 452-453; Hillman, Howard. *Kitchen Science*, pp. 182-183.

736 / Where did pizza originate?

Pizza is thought to have evolved from early Egyptian flat bread. In ancient Pompeii, coarse bread dough was brushed with olive oil and baked. Naples, Italy is considered the place of origin of pizza as it is known today, although its date of origin is unknown. Pizza was brought to the United States by immigrants and made popular by soldiers returning from Italy after World War II.

Sources: Garrison, Webb. *How It Started*, pp. 40-41; Herbst, Sharon Tyler. *Food Lover's Companion*, p. 354.

737 / How does popcorn pop?

The starch grains in a popcorn kernel are embedded in a very strong protein matrix. When the kernel is heated in hot oil to 150°F (66°C) the water vaporizes within the high-moisture-content corn. This water vapor softens the starch, which in turn rapidly expands in volume, breaking open the tough protein matrix with tremendous pressure (the "popping" sound). As the water vapor escapes, the starch grains dry out becoming light and crisp. The idea of popping corn, perfected by the Indians, is at least five thousand years old and was simplified in the 1880s with the introduction of specially designed home and commercial machines. With the advent of the movies and later, television, the American demand for popcorn increased. In 1952, Orville Redenbacher (b. 1907) and Charles Bowman produced a corn whose kernels seldom failed to pop. Because the hybrid was expensive to produce, popcorn companies declined to sell Redenbacher's product, believing that Americans would not buy the higher-priced corn. Redenbacher marketed it himself, and it has become America's best-selling popcorn. Today the average American consumes almost two pounds of popcorn in one year.

Sources: Chalmers, Irena. *Great American Food Almanac*, p. 76; McGee, Harold. *On Food and Cooking*, p. 242; Panati, Charles. *Panati's Extraordinary Origins of Everyday Things*, pp. 390-393.

738 / What is manna that is mentioned in the Book of Exodus in the Bible?

Some sources say that manna is a sweet, granular substance secreted by two varieties of scale insects that live on tamarisk shrubs. Today in the Sinai region, this substance is collected in June; it is used to make candy.

Sources: Felton, Bruce. *Felton & Fowler's Best, Worst, and Most Unusual*, pp. 228-229.

739 / Who invented peanut butter?

The inventor of peanut butter could have been a St. Louis physician who in 1890 ground up peanuts to serve as a nutritious and easily digested food for his elderly patients. But antecedents of this peanut puree can be found earlier in South American, Asian, and African cooking in the United States. A note in an 1883 Savannah newspaper suggests that peanut butter was already a Georgian favorite. However, the peanut butter boom in the United States did begin around the time of the St. Louis "discovery." By 1900 grocers were selling it by the pound in big tubs, stirring the oil that rose to the top. In the 1920s hydrogenated fats, to prevent oil separation, and sweeteners were added. To meet the United States Food and Drug Administration Standards, peanut butter must contain 90% peanuts and have no artifical flavoring or sweetener, no preservative, color additive, or added vitamins.

Sources: Chalmers, Irena. *Great American Food Almanac*, p. 154; Garrison, Webb. *How It Started*, pp. 27-28; Harris, Harry. *Good Old-Fashioned Yankee Ingenuity*, pp. 219-220.

740 / What is canola oil?

Canola oil is the market name for the bland-tasting rapeseed oil used in cooking and salad dressings. It has become popular in the United States because it is lower in saturated fat (about 6%) than any other oil and contains more cholesterol-balancing monosaturated fat than any oil except olive oil. In addition, it contains omega-3 fatty acids—the polyunsaturated fat reputed to not only lower both cholesterol and triglyceride values, but to enhance brain growth and development as well.

Sources: Herbst, Sharon Tyler. *Food Lover's Companion*, p. 65.

741 ▽ *What are rose hips?*

Though too tart to eat raw, the ripe reddish-orange fruit or seed capsule of the rose—located between the stem and petals—is often used to make jellies and jams, syrup, tea, and wine. An excellent source of Vitamin C, rose hips can be ground into powder and compressed into tablets and sold in health food stores.

Another part of the rose is edible as well—rose petals. Rose petals may be included in a fruit or vegetable salad, soup, or puree. To prepare them for a sandwich, butter them and refrigerate. Then add enough fresh petals to cover. Ground or powdered, they can be incorporated into all cooked foods, baked or broiled meat, casseroles, baked goods, jellies and preserves, or omelets. Finely ground or dried petals may be added to cottage cheese, sour cream, or other spreads.

Rosewater is an essence distilled from rose petals and has a delicious scent and flavor of the flower. Used for extensive cosmetic uses, for finger washing, and for rinsing fine linens, it is also used in the kitchen, where it makes an admirable flavoring—used in place of various extracts or in combination with them, especially in Indian, Iranian and Turkish cooking as well as in Victorian and medieval cookery. The production of true rosewater is a long and complicated process. However, a passable substitute can be made at home. Start with 2 handfuls of scented rose petals, ½ cup of sugar, and 2 pints of boiling water. Put the rose petals into an earthenware pot. Dissolve the sugar in the boiling water and let cool until tepid. Pour over the rose petals. Let infuse for one hour. Stir gently but thoroughly, then let it set for another hour. Strain and bottle. Kept in the refrigerator, this rosewater will last for about a week.

Sources: Harris, Ben C. *Make Use of Garden Plants*, p. 167; Jones, Julia and Barbara Deere. *Royal Pleasures and Pastimes*, p. 14; *Woman's Day Encyclopedia of Cookery*, vol. 10, p. 1560.

742 ▽ *From what is tofu derived?*

Also known as soybean curd and bean curd, this custard-like white block is made from curdled soy milk. The process starts with soaked soybeans that are pureed, cooked, and filtered through cloth. The resulting milk is curdled with a coagulant, such as magnesium chloride. The formed curds are pressed into soft cakes, allowing the whey to drain off.

Tofu is perishable and should be refrigerated for no longer than a week. When storing, it should be covered with water that is changed daily. Tofu may be frozen up to 3 months, although freezing will cause it to have a chewier texture. High in protein, low in salt and calories, with no cholesterol, tofu is very bland and readily picks up the flavor of whatever it is cooked with.

Sources: Brody, Jane E. *Jane Brody's Good Food Book*, pp. 99-100.

▽ **743** ▽
How did Toll House® cookies originate?

Long before chocolate was sold in small chips, homemakers cut up bars of chocolate when making chocolate chip cookies. Ruth Wakefield (1905-1977), who ran the Toll House Inn in Whitman, Massachusetts, made these cookies popular. She sold the rights to the name "Toll House cookies" to the Nestle Company. In 1939, the company began packaging chocolate chips.

Sources: Beard, James. *James Beard's American Cookery*, p. 713; Panati, Charles. *Panati's Extraordinary Origins of Everyday Things*, pp. 414-415.

▽ **744** ▽
How are truffles found?

Truffles, found near the roots of oak, elm, aspen, and willow trees are located by specially trained dogs or pigs that scent out and dig up these underground fungi. Dogs are preferred for this activity, because unwatched pigs tend to eat their finds.

Prized by gourmets for centuries, truffles are used to flavor cooked foods. The French black truffles (Perigord variety) or the Italian white truffles (Piedmont variety) are considered the finest, but English black truffles are also hunted. Another main variety of truffle is a whitish pear-shaped mushroom, called *terfez* or lion's truffle (also called desert, sand, or Arab truffle), that grows below the soil in North Africa and the Middle East.

Sources: Simon, Andre L., and Robin Howe. *Dictionary of Gastronomy*, unpaged alphabetical entry.

▽ **745** ▽
What is a pizzelle?

A thin wafer, stove-top cookie baked in a pizzelle cookie iron.

Sources: McNair, James K. *Adventures in Italian Cooking*, p. 89.

NUTRITION

746

What are good sources of Vitamin C?

Contrary to popular opinion, the best sources of Vitamin C are green peppers, parsley, broccoli, brussels sprouts, turnip greens, and guava. Tomatoes, strawberries, and citrus fruits are good but not the best sources. Green, leafy vegetables and even potatoes are relatively good sources.

Sources: Holmes, Ann M. *Nutrition & Vitamins*, p. 100.

747

Does leaf lettuce or iceberg (head) lettuce have more vitamins?

Vitamin content of 3.5 ounces (99.22 grams)

Type	Vitamin A International units (IU)	Vitamin C milligrams
Leaf lettuce	1990	18.0
Butterhead lettuce	970	8.0
Iceberg lettuce	330	3.9

Sources: *Prevention's Giant Book of Health Facts*, p. 243.

748

How much junk food does the average American eat in a year?

The current estimate informs us that each American consumes the following in a year:

50 pounds of cookies and cakes
63 dozen doughnuts
100 pounds of refined sugar
55 pounds of fat and oil
300 containers of soda
200 sticks of chewing gum
20 gallons of ice cream
5 pounds of potato chips
18 pounds of candy
unknown quantity of popcorn, pretzels, and a wide variety of snack foods.

Sources: Elkort, Martin. *The Secret Life of Food*, 1991, p. 125.

Which foods are high in potassium?

The Food and Drug Administration has not set a recommended daily allowance for potassium, although it is recognized as an essential part of a diet. The National Research Council has established a level of about 5,600 milligrams of potassium daily in a healthy diet. The top foods having the most potassium per serving are listed below:

Food	Potassium (milligrams per serving)
Potato	844
Cantaloupe	825
Prune juice	706
Avocado	602
Watermelon	559
Raisins	545
Dates	541
Tomato juice	536
Apricots	482
Orange juice	473

Other good sources of potassium include coffee, milk, bananas, poultry, ocean fish, liver, winter squash, lima beans, navy beans, and parsnips.

Sources: Gebhardt, Susan E., and Ruth H. Matthews. *Nutritive Value of Foods*, pp. 8-9, 12-13, 18-32, 38-49, 52-61; Krantz, Les. *The Best and Worst of Everything*, p. 269; Long, James W. *The Essential Guide to Prescription Drugs*, p. 885.

Why does one need to eat fewer calories to maintain a given weight after losing weight?

Weight loss reduces maintenance needs. It takes 12 kilogram calories per pound to maintain body weight. When 10 pounds are lost, 120 fewer calories are needed.

Sources: Bodinski, Lois H. *The Nurse's Guide to Diet Therapy*, p. 76.

What are RDAs (Recommended Dietary Allowances)?

RDAs (Recommended Dietary Allowances) are estimates for meeting the nutritional needs of a group of people. They are not averages but are calculated for individuals with the highest requirements. These levels are intended to apply to persons whose physical activity is considered light and who live in temperate climates. They provide a safety margin for each nutrient above the minimum level that will maintain health. The United States RDAs are formulated by the United States Food and Drug Administration. These values are generally the highest values for each nutrient given in the RDA tables produced by the Food and Nutrition

Board of the National Research Council. The Food and Drug Administration RDAs were established primarily in connection with legal labeling requirements of foods and vitamin preparations.

Sources: Burton, Benjamin T., and Willis R. Foster. *Human Nutrition*, pp. 176-177; Dunne, Levon J. *Nutrition Almanac*, pp. 10, 118.

752 / What is the new food guide pyramid?

On November 5, 1990, the United States Department of Health and Human Services and the Department of Agriculture issued dietary guidelines with the recommendation that no more than 30% of dietary calories come from fat and no more than 10% of that from saturated fats high in cholesterol. Maximum alcohol intake is one drink per day for women and two drinks for men. Below are listed the recommendated servings and groups. This general guidelines of what to eat each day to get the nutrients needed by the body, replacing the familiar "Four food groups."

Sources: *USDA's Food Guide Pyramid*, pp. 2-3; *The World Almanac and Book of Facts 1992*, p. 256.

The food guide pyramid

753 / What is the difference between essential and non-essential amino acids?

There are twenty different amino acids that make up all the proteins in humans. These are needed to replenish tissue, red blood cells, and enzymes, etc. Of these, twelve can be made by the body. They are known as non-essential amino acids because they do not need to be obtained from the diet. The other eight (some authorities say nine), the essential amino acids, cannot be made by the body and must be obtained from the diet. Proteins in meat have all the essential amino acids; vegetables with the exception of soybeans are "incomplete proteins" in that they do not have all the essential amino acids. By combining certain foods, collectively, all essential amino acids can be ingested. For example, whole grain cereal combined with legumes and eaten together as "complimentary foods" form a complete source.

Sources: *American Medical Association Encyclopedia of Medicine*, p. 94; Hillman, Howard. *Kitchen Science*, pp. 256-257.

754 / How are food calories measured?

When calories are specified for a food, this translates into the potential energy (heat) that the food can generate when ingested. Each food calorie is equal to the amount of energy needed to raise the temperature of 1 kilogram (or 1 liter or 2.2 pounds) of water by 1°C at 1 atmospheric pressure. Since the calculation is based on a kilogram (1,000 grams), a food calorie is one thousand times as much as a "normal" calorie (unit commonly used in physics and chemistry). The food–calorie value was determined by a device called a bomb calorimeter (a sealed compartment in which a specific weight of food is completely burned). Overconsumption of calories is a common problem. If the body does not use them, they are stored as glycogen reserves and fat. More of the body's calorie consumption should come from complex dietary carbohydrates (starches) than simple carbohydrates (sugars) because simple carbohydrates tend to be "empty" or pure calories (fuel) that contain more than enough fuel but not enough of the other substances necessary to life. Below is listed the number of calories needed daily by the human body, based on a composite of sources (National Research Council Nutrition Board, etc.):

	Male		Female	
Age range	Weight in pounds	Calories needed	Weight in pounds	Calories needed
1	24	1,100	24	1,100
2-3	31	1,300	31	1,300
4-6	40	1,800	40	1,800
7-9	55	2,200	55	2,200
10-12	75	2,500	79	2,200
13-15	110	2,800	106	2,200
16-18	136	3,200	117	2,100
19-24	156	3,000	128	2,100
25-49	163	2,700	130	1,900
50-74	161	2,300	139	1,800
75+	152	2,000	141	1,500

Sources: Hillman, Howard. *Kitchen Science*, p. 271; Hopkins, Nigel J., John W. Mayne, and John R. Hudson. *The Numbers You Need*, p. 84; McGee, Harold. *On Food and Cooking*, pp. 528-529.

755

What is the daily recommended allowance of cholesterol and salt in the diet?

The American Heart Association recommends a restriction of dietary cholesterol to 300 milligrams per day and of sodium to 3 grams per day. Too much salt (in particular the sodium content of salt) in the diet causes extra water to be drawn into the blood vessels, increasing the pressure on the artery walls and causing high blood pressure. Generally 1.5 teaspoons of sodium a day is sufficient to get the 3 grams or 3,000 milligrams needed. However, the average American ingests two to four times that much. Cholesterol, an essential element for bodily functions, can be supplied by eating high dietary cholesterol foods of animal origin (egg yolks; meat, poultry, and fish, and milk and milk products), or foods having high saturated fat content (butter, cheese, whole milk, ice cream, cream, meat, poultry, and shellfish (in particular fatty meats)), and a few vegetable fats (coconut oil, cocoa butter (found in chocolate), palm and palm kernel oil). Of all the things in one's diet, saturated fats seem to raise blood cholesterol levels the most. Too high of a blood cholesterol could lead to coronary heart diseases and atherosclerosis, so researchers stress avoiding or limiting foods that contain high dietary cholesterol (organ meats, egg yolks, fat portions of meat, milk products) or high saturated fats (butter, ice cream, cream portions of milk products, fatty meats), and baked or fried products and snack foods containing coconut, palm, or palm kernel oil.

Sources: *Eating to Lower Your High Blood Cholesterol*, pp. 4-8; Hopkins, Nigel H., John W. Mayne, and John R. Hudson. *The Numbers You Need*, p. 84; Wyngaarden, James B., and Lloyd H. Smith. *Cecil Textbook of Medicine*, p. 1208.

756

How many glasses of water should humans drink every day?

In the course of a day, the body uses between 1½ and 3 quarts (1.4 to 2.8 liters) of water just in breathing, sweating, and releasing waste fluids. Less than a quart is supplied by food. The rest is from about six glasses of liquid a day. The experts recommend six to eight glasses as a safe amount. However active persons may need 15 glasses, while small and inactive persons only need two glasses.

Sources: *Vegetarian Times*, no. 176 (April 1992), pp. 10-12.

GENERAL SCIENCE AND TECHNOLOGY

AWARDS, PRIZES, AND FAIRS

757
When was the first world's fair?

The Great Exhibition of the Works of Industry of All Nations, commonly referred to as the Great Exhibition of 1851 or the Crystal Palace Exhibition, was the first true world's fair. It opened May 1, 1851, and closed October 15, 1851.

Sources: Findling, John E. *Historical Dictionary of World's Fairs and Expositions, 1851-1988*, pp. xv, 3.

758
What is contained in the time capsule buried at the site of the New York World's Fair in 1940?

The contents fall into five categories:

1. Small items of common use—Eyeglasses, nail file, safety pin, etc.
2. Textiles and materials—Cloth, iron, aluminum, leather, paper, coal.
3. Miscellaneous items—U.S. money, seeds, books, type.
4. An essay in microfilm—Books, pictures and speeches that give a description of the world in 1940.
5. Newsreel—Characteristic or significant scenes of 1940 in sound film.

Sources: Smithsonian Institute. *Annual Report, 1939*, pp. 533-554.

759
Who was the first American woman to win a Nobel prize?

For work on glycogen conversion, this Czechoslovakian-born biochemist, Gerty T. Cori (1896-1957), who shared the 1947 Nobel Prize for Physiology and Medicine with her husband, Carl (1896–1984), became the first American winner. The first native-born American woman to be honored was Rosalyn S. Yallow (b.1921), who shared the 1977 prize in physiology and medicine with two male associates. The award was given for their work on hormones in body chemistry.

Sources: Sanders, Dennis. *The First of Everything*, p. 167.

760 *What is the Guzman prize?*

It is a prize offered in France in 1901 to the first person to make contact with beings from another planet. Mars is excluded as being "too easy." The prize is as yet unclaimed.

Sources: *International Encyclopedia of Astronomy*, p. 176.

761 *Who was the first American-born winner of the Nobel Prize for Physiology or Medicine?*

The American geneticist Thomas Hunt Morgan (1866-1945) was awarded the Nobel Prize in 1933 for his study on how chromosomes function in heredity.

Sources: Magill, Frank K. *Nobel Prize Winners: Physiology or Medicine*, vol. 1, p. 359.

762 *What were the discoveries of John Bardeen, the only person to win two Nobel prizes in the same field?*

In 1956 John Bardeen (b.1908) won the Nobel Prize for Physics for his research on semiconductors and discovery of the transistor effect. His second prize, awarded in 1972, was for the microscopic theory of superconductivity.

Sources: *Physics Today*, vol. 45 (April 1992), p. 23.

763 *Which women have won the Nobel Prize?*

As of 1991, only nine women had received the Nobel Prize. Marie Curie received the prize twice.

Chemistry

Curie, Marie S. (1911)
Joliot-Curie, Irène (1935)
Hodgkin, Dorothy Mary Crowfoot (1964)

Physics

Curie, Marie S. (1903)
Mayer, Maria Goeppert (1963)

Physiology and Medicine

Cori, Gerty (1947)
Radnitz, Theresa (1947)
Yallow, Rosalyn S. (1977)
McClintock, Barbara (1983)
Levi-Montalcini, Rita (1986)
Elion, Gertrude B. (1988)

Sources: *Encyclopedia Americana*, vol. 20, pp. 393-395; *The World Almanac and Book of Facts 1992*, p. 31.

764 / *For what accomplishments did Marie Curie receive a Nobel Prize?*

Marie Curie (1867-1934) received two Nobel Prizes—one in physics in 1903 and one in chemistry in 1911. She and her husband Pierre (1859–1906), along with A.H. Becquerel (1852-1908), were awarded the prize in 1903 for their work on radioactivity. They separated minute amounts of two new highly radioactive chemical elements from uranium ore. The Curies named the elements "radium" and "polonium." In 1911, Marie Curie won the Nobel Prize for Chemistry for her discovery of the new elements, and for work in isolating and studying the chemical properties of radium.

Sources: *World Book Encyclopedia*, vol. 4, p. 1194.

765 / *Who was the first American to receive a Nobel Prize?*

Albert Abraham Michelson (1852-1931), a German-born American physicist, received the 1907 Nobel Prize in Physics, becoming the first American to win the award. The Prize recognized both his design of precise optical instruments and the accurate measurements he made with them. In association with Edward Morley (1838–1923), he performed the classic Michelson-Morley experiment for light waves and made very precise determinations of the velocity of light. This work entailed the development of extremely sensitive optical equipment.

Sources: *The Biographical Dictionary of Scientists: Physicists*, p. 113

766 / *What is the National Inventors Hall of Fame?*

The National Inventors Hall of Fame was founded in 1973 on the initiative of H. Hume Matthews, then chairman of the National Council of Patent Law

Associations (now called the National Council of Intellectual Property Law Associations). The United States Patent and Trademark Office became a co-sponsor the following year. A foundation was created in 1977 to administer it. Its purpose is to honor inventors who conceived the great technological advances which contributed to the nation's welfare. Usually the recognition is for a specific patent. Thomas A. Edison (1847-1931), the first inductee, was honored for his electric lamp, United States patent number 223,898, granted January 27, 1880. Since then 93 other inventors have been inducted into the hall (as of March 1991).

Sources: *Encyclopedia of Associations*, vol. 1, part 1, p. 579; *The National Inventors Hall of Fame 1990*, pp. 2, 62; *The World Almanac and Book of Facts 1992*, p. 199.

767
Who was the first woman to be named to the National Inventors Hall of Fame?

Gertrude Belle Elion (b.1918) was named to the National Inventors Hall of Fame in 1991 for her research at the Burroughs Wellcome Company, which led to the development of drugs to combat leukemia, septic shock, and tissue rejection in patients undergoing kidney transplants.

Sources: *Scientific American*, vol. 265 (October 1991), p. 44; *The World Almanac and Book of Facts 1992*, p. 199.

768
Which astronauts have won the Congressional Space Medal of Honor?

The Congressional Space Medal of Honor recognizes astronauts who have made outstanding contributions to their field and to the welfare of the United States. The President of the United States presents the award to the recipient in the name of the Congress.

Recipient	Award Date
Neil A. Armstrong	1 Oct 1978
Frank Borman	1 Oct 1978
Charles P. Conrad	1 Oct 1978
John H. Glenn, Jr.	1 Oct 1978
Virgil "Gus" Grissom	1 Oct 1978
Alan B. Shepard, Jr.	1 Oct 1978
John W. Young	19 May 1981

Sources: Bali, Mrinal. *Space Exploration: A Reference Handbook*, p. 138; Siegman, Gita. *Awards, Honors, & Prizes*, vol. 1, p. 656.

FAIRS *See:* GENERAL SCIENCE AND TECHNOLOGY—AWARDS, PRIZES, AND FAIRS

HOAXES, ENIGMAS, AND ACCIDENTS

769

What was the world's greatest scientific fraud?

Probably the most famous fraud was the Piltdown man, a phony human fossil "discovered" in a gravel formation in southern England in 1912. The unknown person combined an ape jaw with skull fragments of a modern man so skillfully that scientists argued over Piltdown man for about 40 years. The forger filed the teeth in the ape jaw to give them the appearance of a human wear pattern. Chemical investigations in the 1950s showed the cranium and jaw were of different ages.

Sources: Cazeau, Charles J. *Science Trivia*, p. 167.

770

What is the Bermuda Triangle?

The Bermuda Triangle is an area of the Atlantic Ocean with the corners of the triangle being at the southern Virginia coast, the Bermuda Islands, and the Florida Keys, covering about 14,000 square miles (36,260 square kilometers). On a number of occasions ships and airplanes have vanished in this area, often without a trace.

Explanations for these disappearances have ranged from the plausible (extreme air turbulence, powerful ocean currents, waterspouts, magnetic anomalies, electromagnetic storms) to the absurd (kidnappings by alien flying saucers, gateways to other dimensions). However, scientific studies have revealed no significant peculiarities about the area. Many of the so-called incidents within the Bermuda Triangle can be attributed to other causes including bad weather, malfunctioning instruments, and faulty navigation. It is also not the only part of the ocean where ships and planes have disappeared.

Sources: *Academic American Encyclopedia*, vol. 3, p. 219; Downs, Robert B. *Scientific Enigmas*, pp. 96-101.

771

Which industrial accidents have become major disasters?

Date	Place	Event	Result
9/21/21	Oppau, Germany	Explosion at a nitrate manufacturing plant destroyed plant and nearby village	561 deaths; more than 1,500 persons injured
4/16/47	Texas City, Texas	Explosion in freighter being loaded with ammonium nitrate	561 deaths; much of city destroyed

Date	Place	Event	Result
7/28/48	Ludwigshafen, Federal Democratic Republic of Germany	Vapor explosion from dimethyl ether	209 deaths
7/10/76	Seveso, Italy	Chemical reactor explosion releasing 2,3,7,8-TCDD	100,000 animals killed; 760 people evacuated; 4,450 acres contaminated
2/25/84	Cubatao, Sao Paulo, Brazil	Explosion from a gasoline leak in a pipeline burned a nearby shanty town	Greater than 500
11/19/84	San Juan, Ixtaheupec, Mexico City, Mexico 700,000 evacuated	5 million liters of liquified butane exploded at a storage facility	More than 400 deaths; 7,231 persons injured;
12/3/84	Bhopal, India	Release of methyl isocyanate from pesticide plant	More than 2,000 deaths; 100,000 injured

Sources: *The Public Health Consequences of Disasters, 1989*, p. 98.

772 ∕ Does dowsing have any scientific basis?

Dowsing is searching for water, metal ores, or other materials hidden beneath the earth's surface by using a divining rod or dowsing rod. These rods are usually forked or Y-shaped and can be made of wood or metal. The rods are carried by their tips and are believed to bend downward when they pass over hidden materials. Although studies have failed to explain scientifically the basis for dowsing, the practice is still popular throughout the world. Some civil engineering companies use dowsers to locate existing pipelines or cables when making site surveys, because often the companies find the results more reliable than those from standard detectors.

Sources: *How In the World?*, p. 1; Roberts, Kenneth L. *The Seventh Sense*, p. 316; *Science and Technology Illustrated*, vol. 8, p. 1012.

773 ∕ What is pyramid power?

An interesting belief about the strange power of pyramids was popular during the 1970s. The power of the pyramid is based on the belief that a pyramid shaped like that of the Great Pyramid of Giza is able to focus an unspecified energy in a mystical way to bring about a beneficial result. Some of the claims of its power includes keeping food fresh, making meat tender, improving the taste of wine and tobacco, purifying water, curing headaches, shortening healing time, making plants grow better, and keeping razor blades sharp. This was done by placing the item under study inside a hollow pyramid. A physicist, Dr. G. Pat Flanagan, describes experiments he conducted in his book *Pyramid Power*, published in 1973. Later researchers have debunked this theory.

Sources: *Human Behavior*, vol. 8 (April 1979), p. 37; *Mechanix Illustrated*, vol. 73 (April 1977), pp. 151, 153.

774 / What is a jackalope?

The imaginary animal is supposed to be a cross between a jack rabbit and a deer. It probably was created by a taxidermist.

Sources: *Pittsburgh Press*, (September 17, 1972), Roto Section, p. 18.

INVENTION AND DISCOVERY (GENERAL)

775 / Which inventions are credited to Thomas Jefferson, even though he never received a patent?

Thomas Jefferson (1743-1826), the third United States president, is recognized as having invented the swivel chair, pedometer, shooting stick, a hemp-treating machine, and an improvement in the moldboard of a plow.

Sources: Richardson, Robert O. *The Weird and Wondrous World of Patents*, p. 38.

776 / Which scientific and technological event was the most important in United States history?

The first United States moon landing, with the first man (Neil Armstrong, b. 1930) to step on the moon on July 20, 1969 is probably the most significant event.

Sources: Wallechinsky, David. *The Book of Lists*, p. 30.

777 / What are the ten outstanding engineering achievements of the last 25 years?

The National Academy of Engineering lists them as follows:
　　The Moon Landing
　　Application Satellites

Microprocessors
Computer-Aided Design and Manufacturing
CAT Scan
Advanced Composite Materials
Jumbo Jets
Lasers
Fiber-Optic Communication
Genetically Engineered Products

Sources: *Engineering and the Advancement of Human Welfare*, p. 1.

778 / What is a Rube Goldberg contraption?

A *Rube Goldberg contraption* is an overengineered solution to a simple problem. Ruben Lucius Goldberg (1883-1970) was a skilled engineering draftsman, a stand-up comic, a reporter, and one of the world's greatest cartoonists. His bizarre inventions mixed grand silliness with high seriousness. His devices used balloons, leaking hot water bottles, candle flames, pulleys, strings, levers, and various animals in elaborate schemes for accomplishing simple tasks.

Sources: *Inventive Genius*, p. 42.

Rube Goldberg juice squeezer

779 / What are the areas of critical technology?

There are 22 areas of technological development that the United States considers to be critical to the prosperity and security of the country. They are aeronautics; applied molecular biology; ceramics; composites; computer simulation; data storage; electronic and photonic materials; energy; environment; flexible computer integrated manufacturing; high-definition imaging and displays; high–performance computing/networking; high-performance metals and alloys; intelligent processing equipment; materials processing; medicine; micro- and nano-fabrication; microelectronics and optoelectronics; sensors and signal processing; software; surface transportation; and systems-management.

Sources: *The World Almanac and Book of Facts 1992*, p. 200.

780 / Which scientific discoveries were made accidently?

Some accidental discoveries include velcro, penicillin, x-rays, dynamite, vulcanization of rubber, synthetic dyes, rayon, saccharin, and the discovery of iodine and helium.

Sources: Roberts, Royston M. *Serendipity: Accidental Discoveries in Science*, pp. 31-33, 53-58, 66-74, 86-87, 94, 139, 150-159, 220–222.

MACHINES *See: GENERAL SCIENCE AND TECHNOLOGY—TOOLS, MACHINES, AND PROCESSES*

OFFICE ITEMS

781 / What is a Wooton Patent Desk?

With dozens of pigeonholes and compartments, hinged and rotating parts and elaborate exteriors, Wooton Patent Desks were manufactured in Indiana from 1874 to 1897 by William S. Wooton and Co. (changed in 1880 to Wooton Desk Manufacturing Co.). The design of the desk provided a solution to the businessman's problem of organizing the increasing volume of paperwork that accompanied the rapid expansion of business. These desks served well until the 1890s when the widespread use of manila folders and file drawers replaced the limited storage capacity of the desk.

Sources: *Fine Woodworking*, no. 87 (March/April 1991), pp. 52, 55.

782 / Why are the letters all mixed up on the typewriter keyboard?

Although Christopher Latham Scholes did not invent the typewriter, he made it work faster with his keyboard design. The standard keyboard (QUERTY) as used today was established by Scholes in 1872. He rearranged the keyboard from its alphabetical arrangement, which caused the bars of the machine to jam, to one in which the letter arrangement made the bars hit the inked ribbon from opposite directions, resulting in less jamming. Eventually the Remington Fire Arms Company bought the patent and mass-produced the machine in 1874. American writer Mark Twain (1835-1910) was one of the 400 customers who purchased Remington typewriters in 1874. The first 10 years of production had disappointing sales. One reason was the use of aniline-inked typewriter ribbons that would eventually fade. With the introduction in 1885 of permanent ink ribbons, the typewriter became an essential office item.

Sources: Beeching, W.A. *Century of the Typewriter*, p. 40; Robertson, Patrick. *The Book of Firsts*, pp. 193-194.

783 / *Who invented the Xerox® machine?*

After 20 years of experimentation, American physicist Chester F. Carlson (1906-1968) developed in 1938 a method of copying that used dry powder, electric charge, and light. Because nothing moist was used in the process, the procedure was called *xerography* (meaning dry writing in Greek). But Carlson could not interest the industry of this new method until the Haloid Company of Rochester, New York, acquired his patent. The company, which later became the Xerox Corporation, gave the first public demonstration of this copier at the annual meeting of the Optical Society of America, in Detroit, Michigan, on October 22, 1948.

Carlson's new method used static electricity as the basis of the xerographic process. An electrostatic charge is induced on an insulating photoconductive surface in darkness. Then this surface is exposed to light reflecting from the original (similar to a photographic plate). The resultant charge pattern is dusted with a charged powder (called toner) that is attracted to the pattern but rejected by the background. The powder pattern (the image of the original) is transferred to ordinary paper with an electrostatic charge and "fixed" on the paper by either heat or by chemicals. Finally the xerographic surface is cleaned and ready for reuse. Today, copy paper is used, which forms the photo conductive surface and consequently becomes the final copy.

Sources: Asimov, Isaac. *Asimov's Chronology of Science and Discovery*, p. 598; De Bono, Edward. *Eureka!*, pp. 61-62.

784 / *Why are most pencils yellow?*

Painting pencils yellow became a sign of quality during the 1890s. The practice had originally been done as early as 1854 in Keswick, England, most likely to cover the imperfect wood used in some pencils. It was the American Koh-I-Noor Company whose successful, high-quality yellow pencil became firmly established as a sign of "pencilness" in the minds of users; any other color was assumed to indicate an inferior pencil. Yellow has thus become established as the preferred color for a writing pencil, as it has for school buses and highway signs. It makes them all highly visible, whether on a busy desk or highway.

Very early pencils were covered, square sticks of graphite (a soft black lustrous form of carbon) that left as much of a mark on the user's hand as on the paper. Later, in England and Germany, marking sticks were wrapped tightly with string to fashion pencils, with the string slowly unwound to expose more stick. In the 16th century the Germans inserted the sticks into groves cut in narrow strips of wood.

As graphite supplies became scarce in the late 18th century, chemists sought ways to reduce the graphite content in pencils. In 1795 N.J. Conte (1755-1805) produced pencils made from ground graphite and clay that had been shaped into sticks and kiln-baked—a process still done to this very day. Today, although a pencil appears to be all one piece, it actually is formed from two pieces of wood glued together so well that the seams are not visible.

The development of the process of vulcanizing rubber in 1839 made feasible the attachment of the eraser. In 1858 Hyman L. Lipman of Philadelphia patented a pencil with an eraser glued to one end.

Sources: Harris, Harry. *Good Old-Fashioned Yankee Ingenuity*, p. 121; Panati, Charles. *Panati's Browser's Book of Beginnings*, p. 79; Petroski, Henry. *The Pencil: A History of Design and Circumstance*, pp. 138-140, 162-163.

785 / Who invented Liquid Paper®?

Betty Nesmith Graham in 1951 started using a white paint to correct her typographical errors. After her coworkers began requesting bottles of the paint, Mrs. Graham started a small company to sell the paint. Over the years her small company grew to a large industry.

Sources: Vare, Ethlie Ann. *Mothers of Invention*, pp. 38-42.

PATENTS AND TRADEMARKS (GENERAL)

786 / What was the name of the dog portrayed in the His Master's Voice trademark, associated with RCA Victor?

Little Nipper—originally painted by Francis Barraud, an Englishman, for exhibition at the Royal Academy. When Barraud's painting was denied space at the exhibition, he sold it to the Gramophone Company, Ltd., after repainting part of the picture. He painted over the cylinder machine portrayed in the original and painted in the gramophone.

Sources: Read, Oliver. *From Tin Foil to Stereo*, p. 134.

787

Which patent was issued to a United States president?

On May 22, 1849, 12 years before he became the 16th United States president, Abraham Lincoln (1809-1865), was granted United States patent number 6,469 for a device to help steamboats pass over shoals and sand bars. The device, never tested or manufactured, had a set of adjustable buoyancy chambers (made from metal and waterproof cloth) attached to the ship's sides below the waterline. Bellows could fill the chambers with air to float the vessel over the shoals and sandbars. It was the only patent ever held by a United States president.

Sources: *Inventive Genius*, p. 23.

Lincoln's shoal device

788

What types of patents are available?

There are three types:

1. Utility patents, granted for new, useful and nonobvious process, machine, manufactured article, composition, or an improvement in any of the above.
2. Design patents, granted for new, original and ornamental designs for an article of manufacture.
3. Plant patents, provided to anyone who has invented or discovered and asexually reproduced any distinct and new variety of plant.

Sources: U.S. Department of Commerce. Patent and Trademark Office. *General Information Concerning Patents*, pp. 3, 29.

789

For what is Percy Lavon Julian noted?

Dr. Percy Lavon Julian (1899-1975), a grandson of a former slave, is noted for his work on synthesizing hormones and medicinal substances, as well as for developing many industrial applications for plant products. He obtained 130 chemical patents in his lifetime. For instance, in 1935 he synthesized *physostigmine*, a powerful drug for the treatment of the disease glaucoma (build-up of pressure inside the eyeball that gradually destroys the retina, causing blindness). A year later, he isolated and prepared soybean protein to be used for paper coatings, textile sizing, and cold-water paints, and as a chemical fire-extinguishing foam. He also developed methods of synthesizing cortisone (used for treatment of rheumatoid arthritis and other inflammatory diseases) and the hormones progesterone and testosterone.

Sources: Haber, Louis. *Odyssey Black Pioneers of Science and Invention*, pp. 131-133; Sammons, Vivian O. *Blacks in Science and Medicine*, p. 140.

790 / Which patents were issued to Mark Twain?

The American writer Mark Twain (1835-1910), whose real name was Samuel Clemens, received three patents. Patent No. 121,992 issued in 1871 was for suspenders. His second, issued in 1873, was for his famous "Mark Twain's Self-Pasting Scrapbook." The third patent in 1885 was for an educational game that helped players remember important historical dates.

Sources: Levy, Richard C. *The Inventor's Desktop Companion*, p. 9.

791 / When was the first United States patent received?

The first patent was issued on July 31, 1790, to Samuel Hopkins for "making pot and pearl ashes"—a cleaning formula called potash (a key ingredient for soap-making then). The first really significant patent issued by the U.S. Patent Office was one for a cotton gin given to Eli Whitney (1765-1825) on March 14, 1794. Its importance lies in the fact that with the mechanical separation of cotton seeds from the lint, cotton became a staple crop of the South. This crop was a major factor in retaining the then-declining institution of slavery.

Between Hopkins's first patent in 1790, and Whitney's in 1794, the U.S. Patent Office issued 70 patents to inventors, some of whom have famous names (Oliver Evans, James Rumsey, John Fitch, and John Stevens) for their work with steam propulsion.

Sources: *Encyclopedia Americana*, vol. 8, p. 81; *Inventive Genius*, p. 15; *The New American State Papers, Science and Technology*, vol. 4: *Patents*, pp. 25-26.

792 / What is the name of the dog on the Cracker Jack® box?

In 1918, Sailor Jack and his dog, Bingo, first appeared on the red-white-and-blue Cracker Jack® box. The caramelized popcorn inside the box was first made in 1893 by the Rueckheim brothers for the World Columbian Exposition held in

Chicago. In 1896, they named their product "Cracker Jack" after a positive slang expression. It sold in bulk until 1899, when a wax-sealed, moisture-proof box was developed for it. During World War I, Sailor Jack and Bingo were first put on the patriotically-colored box. They were modeled after F.W. Rueckheim's grandson Robert and his dog. The tiny prizes inside each box are older, and probably date from 1913. Before this time, the box carried a discount coupon toward subsequent purchase.

Sources: Elkort, Martin. *The Secret Life of Food*, p. 99; Jaramillo, Alex. *Cracker Jack Prizes*, pp. 7-8, 10-12; Panati, Charles. *Panati's Extraordinary Origins of Everyday Things*, p. 395.

793
When was the first patent covering an animal issued?

On April 12, 1988, patent history was made when a patent was issued covering specifically genetically-engineered mice for cancer research.

Sources: Levy, Richard C. *The Inventor's Desktop Companion*, p. 10.

794
How long does a patent, a trademark, or a copyright protect the holder?

The term of a utility or a plant patent is 17 years from the date of issue, subject to the payment of maintenance fees. The term of a design patent is 14 years from the date of issue and is not subject to the payment of maintenance fees. The term of a trademark is potentially infinite.

The term of a copyright is more complex, but basically stated it is: a work that was created on or after January 1, 1978 is usually given a term enduring for the author's life, plus an extra 50 years after the author's death. There are special terms for works registered before 1978, works of joint authors, etc.

Sources: Levy, Richard C. *Inventing and Patenting Sourcebook*, p. 55.

795
Must all United States patent applications be accompanied by a drawing?

It is required by law in most cases that a drawing of the invention accompany the patent application. Composition of matter or some processes *may* be exempt, although in these cases the commissioner of Patents and Trademarks may require a drawing for a process when it is useful. The required drawing must show every feature of the invention specified in the claims, and be in a format specified by the Patent and Trademark Office. Drawings must be in black and white unless waived by the Deputy Assistant Commissioner for Patents. Color drawings are permitted for plant patents where color is a distinctive characteristic.

Sources: U.S. Department of Commerce. Patent and Trademark Office. *General Information Concerning Patents*, pp. 16, 30.

796 What cannot be patented?

Many things are not open to patent protection:

The laws of nature, physical phenomena and abstract ideas.

A new mineral or a new plant found in the wild.

Inventions useful solely in the utilization of special nuclear material or atomic energy for weapons.

A machine that is not *useful* (i.e. does not have a useful purpose or does not operate to perform the intended purpose).

Methods of doing business.

Printed matter.

In the case of mixtures of ingredients, such as medicines, a patent cannot be granted unless the effect of mixture is more than the effect of its components.

Human beings cannot be patented.

Mere substitution of one material for another or changes in size to a previously known useful invention without "novelty."

Sources: Levy, Richard C. *The Inventor's Desktop Companion*, p. 10.

797 How do copyrights differ from patents and trademarks?

Copyrights are very different from patents and trademarks. A patent primarily prevents inventions, discoveries, etc., of useful processes, machines, etc., from being manufactured, used, or marketed without the permission of the patent holder. It is a grant of property rights to the inventor and heirs by the government. A trademark is a word, name, or symbol to indicate origin or source of goods and to distinguish the products and services of one company from those of another. Copyrights protect the form of expression rather than the subject matter of the writings or artistic works. They protect the original work of authors and other creative people against copying and unauthorized public performance and recording.

Sources: U.S. Department of Commerce. Patent and Trademark Office. *General Information Concerning Patents*, p. 2.

798 How many United States patents have been given?

Over five million patents have been given by the United States Patent Office since its inception. Below are listed the total number given from the period 1901-1990.

Year	Total number of patents granted
1901	25,546
1921	37,798

Year	Total number of patents granted
1930	45,226
1940	42,238
1950	43,040
1960	47,170
1965	62,857
1970	64,427
1980	61,800
1985	71,700
1987	83,000
1988	77,900
1989	102,712
1990	90,366

Sources: *The Universal Almanac 1992*, p. 522; *The World Almanac and Book of Facts 1992*, p. 199.

PRIZES *See: GENERAL SCIENCE AND TECHNOLOGY—AWARDS, PRIZES, AND FAIRS*

PUBLICATIONS AND SOCIETIES

799
When was the first scientific journal published?

A proposal to publish a French weekly scientific journal was submitted in 1664 by Sir Denis de Sallo, Counselor of the Court of Parliament under Louis XIV (1638-1715). In August 1664, a privilege was signed establishing *Le Journal des Scavans* (Journal of Learned Men), and the privilege was registered in December 1664. The first weekly issue of the *Journal des Scavans* was published on January 4, 1665.

Sources: Subramanyam, Krishna. *Scientific and Technical Information Resources*, p. 31.

800
What is the Flat Earth Society?

The Flat Earth Society International, founded in 1800, is composed of individuals characterized by a seeking for truth and the denial of "imaginary" theories. They believe the "spinning ball" theory of the earth is absurd, and that the earth is a flat, infinite world.

Sources: Cazeau, Charles J. *Science Trivia*, p. 56; *Encyclopedia of Associations*, vol. 1, pt. 1, p. 727.

QUOTATIONS, HUMOR, ETC.

801

Isaac Asimov suggested a quick way to identify a chemist or a non–chemist. What is it?

Show the person the word "unionized". The non-chemist will pronounce the word "YOON-yun-ized." The chemist, however, will say without a moment's hesitation "un-EYE-on-ized."

Sources: Asimov, Isaac. *Asimov on Chemistry*, p. 90.

802

What is the verse written by Hilaire Belloc about a microbe?

Hilaire Belloc (1870-1953), a French-born British poet, essayist, and historian, wrote the following lines from the poem, "The Microbe" in his book, *Cautionary Verses*:

The Microbe is so very small
You cannot make him out at all.
But many people hope
To see him through a microscope.
His jointed tongue lies beneath
A hundred curious rows of teeth;
His seven tufted tails with lots
Of lovely pink and purple spots,
On each of which a pattern stands,
Composed of forty separate bands;
His eyebrows of a tender green;
But Scientists, who ought to know,
Assure us that they must be so...
Oh! let us never, never doubt
What nobody is sure about!

Sources: *The Macmillan Dictionary of Quotations*, p. 501.

803

Albert Einstein commented on science in many of his writings. What are some of his thoughts?

The German-born American physicist, Albert Einstein (1879-1955), said "The whole of science is nothing more than a refinement of everyday thinking."

"One thing I have learned in a long life that all our science, measured against reality, is primitive and childlike—and yet it is the most precious thing we have."

Sources: *Scientific Quotations*, p. 52.

804

What did the poet Lord Byron write about gravity?

George Gordon Byron (1788-1824), the English poet known as Lord Byron, wrote the following lines:

> When Newton saw an apple fall, he found...
> A mode of proving that the earth turn'd round
> In a most natural whirl, called "gravitation";
> And thus is the sole mortal who could grapple,
> Since Adam, with a fall or with an apple.

Sources: Asimov, Isaac. *Isaac Asimov's Book of Science and Nature Questions*, p. 113.

805

Who did Jonathan Swift call the best doctors in the world?

English churchman and writer, Jonathan Swift (1667-1745) wrote, "The best doctors in the world are Doctor Diet, Doctor Quiet, and Doctor Merryman."

Sources: *The Macmillan Dictionary of Quotations*, p. 166.

806

What did Plato have to say about mathematicians?

In the *Republic*, Book VII, the Greek philosopher Plato (ca. 428–348 B.C.E.) wrote, "I have hardly ever known a mathematician who was capable of reasoning."

Sources: Bartlett, John. *Familiar Quotations*, p. 84.

807

Who said "Wine is the most beautiful and most hygienic of beverages?"

Louis Pasteur (1822-1895), a French chemist who demonstrated in the fermentation of wines and beers that microorganisms were responsible for the substantive changes rather than spontaneous generation or chemical processes. The scientific community, exemplified by the pre-eminent organic chemists in Europe, Justus von Liebig (1803-1873) and Friedrich Wohler (1800-1882), was convinced that fermentation was a purely chemical process. Early ridicule met theorists who proposed that living cells were involved in fermentation.

Sources: Daintith, John, and Amanda Isaac. *Medical Quotes*, p. 59.

TERMS, THEORIES, SYMBOLS, ETC.

▽ 808 ▽
What is High Technology or High Tech?

This buzz term used mainly by the lay media (as opposed to scientific, medical, or technological media) appeared in the late 1970s. It was initially used to identify the newest, "hottest" application of technology to fields such as medical research, genetics, automation, communication systems, and computers. It usually implied a distinction between technology to meet the information needs of society and traditional heavy industry which met more material needs. By the mid-1980s, the term had become a catch-all applying primarily to the use of electronics (especially computers) to accomplish everyday tasks.

Sources: *Van Nostrand's Scientific Encyclopedia*, vol. 1, p. 1452; Winkler, Connie. *Careers in High Tech*, p. 1.

▽ 809 ▽
How is a therblig defined?

Frank Bunker Gilbreth (1868-1924), the founder of modern motion study technique, called the fundamental motions of the hands of a worker *therbligs* (Gilbreth spelled backwards). He concluded that any and all operations are made up of series of these 17 divisions. The 17 divisions are search, select, grasp, reach, move, hold, release, position, pre-position, inspect, assemble, disassemble, use, unavoidable delay, avoidable delay, plan, and test to overcome fatigue.

Sources: Niebel, Benjamin W. *Motion and Time Study*, pp. 13, 165-166.

▽ 810 ▽
What does the term ergonomics mean?

The study of human capability and psychology in relation to the working environment and the equipment operated by the worker is variously known as ergonomics, human engineering, human factors engineering, engineering psychology, or biotechnology. Ergonomics is based on the premise that tools humans use and the environment they work in should be matched with their capabilities and limitations, rather than forcing humans to adapt to the physical environment. Researchers in ergonomics try to determine optimum conditions in communication, cognition, reception of sensory stimuli, physiology, and psychology, and examine the effect of adverse conditions. Specific areas of study include design of work areas (including seats, desks, consoles, and cockpits) in terms of human physical size, comfort, strength, and vision; effects of physiological stresses such as work speed, work load, decision making, fatigue, and demands on memory and perception; and design of visual displays to enhance the quality and speed of interpretation.

Sources: *The McGraw-Hill Dictionary of Scientific and Technical Terms*, p. 660; *McGraw-Hill Encyclopedia of Science and Technology*, vol. 8, pp. 520-523.

811 / What is the new science of chaos?

Chaos or chaotic behavior is the behavior of a system whose final state depends very sensitively on the initial conditions. The behavior is unpredictable and cannot be distinguished from a random process, even though it is strictly determinate in a mathematical sense. Chaos studies the complex and irregular behavior of many systems in nature, such as changing weather patterns, flow of turbulent fluids, and swinging pendulums. Scientists once thought they could make exact predictions about such systems, but found that a tiny difference in starting conditions can lead to greatly different results. Chaotic systems do obey certain rules which mathematicians have described with equations, but the science of chaos demonstrates the difficulty of predicting the long-range behavior of chaotic systems.

Sources: *The McGraw-Hill Dictionary of Scientific and Technical Terms*, p. 329; *World Book Encyclopedia*, vol. 3, p. 376.

TOOLS, MACHINES, AND PROCESSES

812 / What is a donkey engine?

A donkey engine is a small auxiliary engine which is usually portable or semi–portable. It is powered by steam, compressed air, or other means. It is often used to power a windlass or lift cargo on shipboard.

Sources: *The McGraw-Hill Dictionary of Scientific and Technical Terms*, p. 569.

813 / How does a breathalyser determine the alcohol level of the breath?

Breathalysers used by police are usually electronic, using the alcohol blown in through the tube as fuel to produce electric current. The more alcohol the breath contains, the stronger the current. If it lights up a green light, the driver is below the legal limit and has passed the test. An amber light means the alcohol level is near the limit; a red light above the limit. The device has a platinum anode which causes the alcohol to oxidize into acetic acid with its molecules losing some electrons. This sets up the electric current. Earlier breathalysers detected alcohol by color change. Orange–yellow crystals of a mixture of sulfuric acid and potassium dichromate in a blowing tube turn to blue–green chromium sulfate and colorless potassium sulphate when the mixture reacts with alcohol, which changes into acetic acid (vinegar). The more crystals that change color, the higher the alcohol level in the body.

Sources: *How in the World?*, p. 251.

814 / How does a four–stroke differ from a two–stroke engine?

A four–stroke engine functions by going through four cycles: 1) the intake stroke draws a fuel–air mixture in on a down stroke; 2) the mixture is compressed on an upward stroke; 3) the mixture is ignited causing a down stroke; and 4) the mixture is exhausted on an up stroke. A two–stroke engine combines the intake and compression strokes (1 and 2) and the power and exhaust strokes (3 and 4) by covering and uncovering ports and valves in the cylinder wall. Two–stroke engines are typically used in small displacement applications, such as chain saws, some motorcycles, etc.

Sources: *Reader's Digest Fix-It-Yourself Manual*, pp. 370–371.

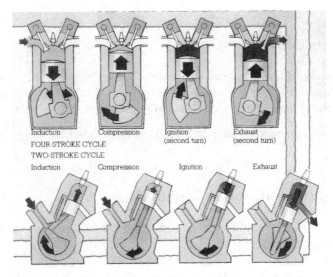

815 / What is DNA fingerprinting?

DNA, or genetic, fingerprinting is a method of determining identity, family relationships, etc. Formulated by Alec Jefferys, a British geneticist, the genetic fingerprint is based on the assumption that every person (except identical twins) has an unique sequence of DNA (Deoxyribonucleic acid)—a substance present in the nucleus of every cell that determines individual characteristics.

Within the DNA molecule, the sequence of the genetic information is repeated many times along the DNA structure, which resembles an endless twisting ladder. The length of the sequence, the number of repetitions, and its precise location within the DNA chain seems to be unique in all cases but identical twins, with the odds being 30 billion to one. A process has been developed that translates these sequences into a visual record that resembles a series of bars on x–ray film. In this process the technician isolates the DNA from blood, saliva, hair follicles, or semen, then the DNA strand is separated into thousands of shorter pieces by placing the DNA into an enzyme solution. Finally the fragments, placed in a gelatin–like material, is subjected to a strong electrical current that separates the fragments according to size and electrical behavior.

In criminal investigations, hair, blood, and skin samples left by the criminal can yield a DNA fingerprint, which can be matched against any suspect's DNA fingerprint. DNA fingerprinting is also used to determine the father in paternity cases since the DNA strand contains the genetic coding of the parents—half the chromosomes were taken from each of the parents.

Sources: Hazen, Robert M., and James Trefil. *Science Matters*, pp. 241–242; *How in the World*, pp. 94–95; Tunnell, James E. *Latest Intelligence*, p. 59.

816 / What is the lost wax process?

This process is used for making valve parts, small gears, magnets, surgical tools, and jewelry. It involves the use of a wax pattern between a two–layered mold. The wax is removed by melting and replaced with molten metal; hence the name "lost wax."

Sources: *The Making, Shaping, and Treating of Steel*, p. 1219; *The McGraw-Hill Dictionary of Scientific and Technical Terms*, p. 1109.

817 / In which industries are robots being used?

A robot is a device that can execute a wide range of maneuvers under the direction of a computer. Reacting to feedback from sensors or by reprogramming, a robot can alter its maneuvers to fit a changed task or situation. The worldwide population of robots is about 250 thousand with approximately 65% being used in Japan and 14% being used in the United States (the second largest user).

Robots are used to do welding, painting, drilling, sanding, cutting, and moving tasks in manufacturing plants. Automobile factories account for more than 50% of robot use in the United States. Robots can work in environments that are extremely threatening to humans or in difficult physical environments. They clean up radioactive areas, extinguish fires, disarm bombs, and load and unload explosives and toxic chemicals. Robots can process light–sensitive materials, such as photographic films that require near darkness—a difficult task for human workers. They can perform a variety of tasks in underwater exploration and in mines (a hazardous, low–light environment). There are only 100 robots in mines, but they produce about ⅓ of the coal coming from these mines. In the military and security fields, robots are used to sense targets or as surveillance devices. In the printing industry, robots perform miscellaneous tasks, such as sorting and tying bundles of output material, delivering paper to the presses, and applying book covers. In research laboratories, small desktop robots prepare samples and mix compounds.

Sources: Cone, Robert J. *How the New Technology Works*, pp. 67–70.

818

Famous movie robots such as R2D2 and C3PO from Star Wars are from current movies. How far back do robots go in the movies?

In 1886, the French movie *l'Eve Futur* had a Thomas Edison-like mad scientist building a robot in the likeness of a woman. A British lord falls in love with her, in a variation on the Pygmalion theme. True working robots, in contrast to these entertainment devices, are strictly functional in appearance, looking more like machines rather than human beings or animals. However, one early exception to this is the very human–looking "Scribe" built in 1773 by two French inventors, Pierre and Henri-Louis Jacquel-Droz, a father and son team, who produced "The Automaton" to dip a quill pen into an inkwell and write a text of 40 characters maximum.

Sources: Cone, Robert J. *How the New Technology Works*, p. 67; *International Encyclopedia of Robotics*, vol. 2, p. 967.

819

What does the acronym "LASER" stand for?

A "LASER" (*L*ight *A*mplification by *S*timulated *E*mission of *R*adiation) is a device that produces a narrow, uniform, high–intensity beam of light made up of one very pure color (one wave length) that can be directed over small and large distances. This light beam is so intense that it can vaporize the hardest and most heat–resistant materials. In 1960 Theodore Maiman (b. 1927) demonstrated a working model of a laser, based on the principles of the maser (a device utilizing single–wave length microwaves, developed by Charles Townes (b. 1915) in 1953) using the wave frequencies of visible light. Ordinary light consists of relatively short wave packets; laser light is made up of long stretches of waves of even amplitude (height) and frequency (width), called "coherent light" because the wave packets or photons seem to stick together. The power of the laser lies in this concentration of photons, which strike the target surface simultaneously. Maiman used a ruby crystal to produce photons by exciting the chromium atoms in the ruby from a low–energy to a high–energy state. After a few thousandths of a second, the atoms revert back to their normal state and emit an identical photon. This photon can stimulate the chromium atoms to emit a photon. The beam is built up by billions of these paired photons which are reflected back and forth by mirrors until they emerge at one end in bursts of red light.

LASER Applications

Semiconductor industry (Ultraviolet lasers)	Micromachining of components and circuit boards
Telecommunications (diode laser)	Development of optical fiber communications in the long–distance telephone networks
Retailing	Bar code scanners
Printing	Desktop printers, type–setting machines and color scanners
Military	Target designators
Medical	Diagnostic and therapeutics

Sources: Asimov, Isaac. *Asimov's New Guide to Science*, p. 454; Cone, Robert J. *How the New Technology Works*, pp. 39–42; *G.K. Hall Encyclopedia of Modern Technology*, p. 104.

820
What does "cc" mean in engine sizes?

Cubic centimeters (cc), as applied to an internal combustion engine, is a measure of the combustion space in the cylinders. The size of an engine is measured by theoretically removing the top of a cylinder, pushing the piston all the way down and then filling the cylinder with liquid. The cubic centimeters of liquid displaced (spilled out) when the piston is returned to its high position is the measure of the combustion volume of the cylinder. If a motorcycle has four cylinders, each displacing 200 ccs, it has an 800 cc engine. Automobile engines are sized essentially the same way.

Sources: Blocksma, Mary. *Reading the Numbers*, pp. 124–126.

821
Who was the father of cybernetics?

Norbert Weiner (1894–1964) is considered the creator of cybernetics. Derived from the Greek word *kubernetes*, meaning steerman or helmsman, cybernetics is concerned with the common factors of control and communication in living organisms, automatic machines, or organizations. These factors are exemplified by the skill used in steering a boat, in which the helmsman uses continual judgement to maintain control. The principles of cybernetics are used today in control theory, automation theory, and computer programs to reduce many time-consuming computations and decision-making processes formerly done by people.

Sources: Ashby, W. Ross. *An Introduction to Cybernetics*, pp. 1-5; Trask, Maurice. *The Story of Cybernetics*, p. 9.

Who invented the electron microscope?

The theoretical and practical limits to the use of the optical microscope were set by the wavelength of light. When the oscilloscope was developed, it was realized that cathode–ray beams could be used to resolve much finer detail because their wavelength was so much shorter than that of light. In 1928, Ernst Ruska (1906–1988) and Max Knoll, using magnetic fields to "focus" electrons in a cathode–ray beam, produced a crude instrument which gave a magnification of 17 and by 1932 they developed an electron microscope having a magnification of 400. By 1937 James Hillier (b. 1915) advanced this magnification to 7,000. The 1939 instrument Vladimir Zworykin (1889–1982) developed gave 50 times more detail than any optical microscope ever could with a magnification up to two million. The electron microscope revolutionized biological research, especially in the areas of cell structures, proteins, and viruses, where for the first time, scientists could "see" molecules of these minute particles.

Sources: Cipolla, Carlo M. and Derek Birdsall. *The Technology of Man*, p. 236.

What is a power take–off?

The standard power take–off is a connection that will turn a shaft inserted through the rear wall of a gear case. It is used to power accessories such as a cable control unit, a winch, or a hydraulic pump. Farmers use the mechanism to pump water, grind feed, or saw wood. A power take–off drives the moving parts of mowing machines, hay balers, combines, and potato diggers.

Sources: Nichols, Herbert L. *Moving the Earth, 1976* p. 15–4; *World Book Encyclopedia*, vol. 19, pp. 360, 457.

824
What are the six simple machines?

All machines and mechanical devices, no matter how complicated, can be reduced to some combinations of six basic, or simple, machines. The lever, the wheel and axle, the pulley, the inclined plane, the wedge, and the screw were all known to the ancient Greeks who learned that a machine works because an "effort," which is exerted over an "effort distance," is magnified through "mechanical advantage" to overcome a "resistance" over a "resistance distance." Some consider there to be only five simple machines, and regard the wedge as a moving inclined plane.

Sources: Macauley, David. *The Way Things Work*, pp. 14-73; O'Brien, Robert. *Machines*, p. 16.

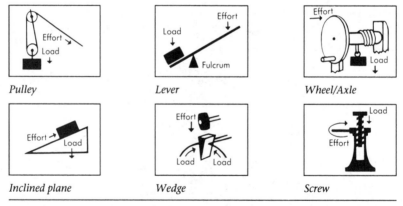

Pulley   Lever   Wheel/Axle

Inclined plane   Wedge   Screw

Illustrations from *The World Book Encyclopedia*. Copyright © by World Book, Inc. Reprinted by permission of the publisher.

825
Which tools were used by Neanderthal man?

Neanderthal tool kits are termed *Mousterian* from finds at LeMoustier in France (traditionally dated to the early Fourth Glacial Period, around 40,000 B.C.E.).

Using fine–grained glassy stone like flint and obsidian, Neanderthals improved on the already old, established *Levallois* technique for striking one or two big flakes of predetermined shape from a prepared core. They made each core yield many small, thin sharp–edged flakes, which were then trimmed to produce hide scrapers, points, backed knives, stick sharpeners, tiny saws, and borers. These could have served for killing, cutting, and skinning prey, and for making wooden tools and clothing.

Sources: Lambert, David. *Field Guide to Early Man*, p. 144.

826 / What is sintering?

Sintering is the bonding together of compacted powder particles at temperatures below the melting point. This bonding produces larger forms such as cakes and pellets. Sintering is used in powder metallurgy technology, which is defined as the production of useful artifacts from metal powder without passing through the molten state. The heating process in which the powder particles are welded together is called sintering, and the components produced are referred to as sintered parts. These parts are normally quite small, and some typical examples made today are shock absorber pistons, belt pulleys, small helical gears, drive gears for chainsaws, and automotive pump gears. Because they are molded, sintered parts can have extremely complex shapes, and do not require machining. The toughness and high strength properties of sintered parts make them especially good for today's high technology systems.

Sources: *Dictionary of Mining, Mineral and Related Terms*, p. 1018; Dowson, Gordon. *Powder Metallurgy: The Process and Its Products*, pp. 1-2, 8-9; *Encyclopedic Dictionary of Science*, p. 218.

827 / How did Frank B. Gilbreth use his cyclograph?

The American engineer and efficiency expert, Frank Bunker Gilbreth (1868–1924) used his cyclograph, or "motion recorder," to refine his study of time and motion. An ordinary camera and a small electric bulb was all he needed to show the path of movement. The light patterns reveal all hesitation or poor habits interfering with a worker's dexterity.

Sources: Giedion, Siegfried. *Mechanization Takes Command*, p. 103.

828 / What is holography?

Hungarian–born scientist, Dennis Gabor, invented the technique of holography (image in the round) in 1947, but it was not until 1961 when Emmet Leith and

Juris Upatnieks produced the modern hologram using a laser, which gave the hologram the strong, pure light it needed. Three–dimensions are seen around an object because light waves are reflected from all around it, overlapping and interfering with each other. This interaction of these collections of waves, called wave fronts, give an object its light, shade, and depth. A camera cannot capture all the information in these wave fronts, so it produces two dimensional objects. Holography captures the depth of an object by measuring the distance light has travelled from the object.

A simple hologram is made by splitting a laser light into two beams through a silvered mirror. One beam, called the object beam, lights up the subject of the hologram. These light waves are reflected onto a photographic plate. The other beam, called a reference beam, is reflected directly onto the plate itself. The two beams coincide to create, on the plate, an "interference pattern." After the plate is developed, a laser light is projected through this developed hologram at the same angle as the original reference beam, but from the opposite direction. The pattern scatters the light to create a projected, three–dimensional, ghost–like image of the original object in space.

Sources: *How in the World?*, p. 229.

▽ 829 ▽ *Who invented the compound microscope?*

The principle of the compound microscope, in which two or more lenses are arranged to form an enlarged image of an object, occurred independently, at about the same time, to more than one person. Certainly many opticians were active at the end of the 16th century, especially in Holland, in the construction of telescopes, so that it is likely that the idea of the microscope may have occurred to several of them independently. In all probability, the date may be placed within the period 1590-1609, and the credit should go to three spectacle makers in Holland. Hans Janssen, his son Zacharias (1580–1638), and Hans Lippershey (1570–1619) have all been cited at various times as deserving chief credit. An Englishman, Robert Hooke (1635-1703) was the first to make the best use of a compound microscope, and his book, *Micrographia*, published in 1665, contains some of the most beautiful drawings of microscopic observations ever made.

Sources: Bradbury, Savile. The Evolution of the Microscope, pp. 21, 39-67.

TRADEMARKS *See: GENERAL SCIENCE AND TECHNOLOGY—PATENTS AND TRADEMARKS (GENERAL)*

WEAPONS *See also:* CARS, BOATS, PLANES, AND TRAINS—MILITARY
VEHICLES

830
How did the bazooka get its name?

It was coined by American comedian Bob Burns (1893-1956). As a prop in his act
he used a unique musical instrument that was long and cylindrical and resembled
an oboe. When United States Army soldiers in World War II were first issued hol-
low-tube rocket launchers, they named them bazookas because of their similarity
to Burns' instrument.

Sources: *Science and Technology Illustrated*, p. 354.

Bazooka/Rocket Launcher

831
*Which warfare innovations were introduced during the American Civil
War?*

Barbed wire, trench warfare, hand grenades, land mines, armored trains, ironclad
ships, aerial reconnaissance, submarine vessels, machine guns, and even a primitive flamethrower were products of the American Civil War, 1861-1865.

Sources: Newton, Michael. *Armed and Dangerous*, p. 17.

832
Who invented the Bowie knife?

A popular weapon of the American West, the Bowie knife was named after Jim
Bowie (1796-1836), who was killed at the Alamo. According to most reliable
sources, his brother Rezin Bowie might have been the actual inventor. The knife's

blade measured up to 2 inches (5 centimeters) wide and its length varied from 9 inches to 15 inches (23 to 38 centimeters).

Sources: DeBono, Edward. *Eureka!*, p. 218; Wallechinsky, David. *The Book of Lists*, p. 89.

833
When was the Colt revolver patented?

The celebrated six-shooter of the American West was named after its inventor, Samuel Colt (1814-1862). Although he did not invent the revolver, he perfected the design which he first patented in 1835 in England and then the United States in the following year. Colt had hoped to mass-produce this weapon, but failed to get enough backing to acquire the necessary machinery. Consequently, the guns, made by hand, were expensive and attracted only limited orders. In 1847, the Texas Rangers ordered one thousand Colt pistols enabling Colt to finally set up assembly line manufacture in a plant in Hartford, Connecticut.

Sources: *New Illustrated Science and Invention Encyclopedia*, vol. 24, p. 3241.

834
Who was known as the Cannon King?

Alfred Krupp (1812-1887), whose father Friedrich (1787-1826) established the family's cast-steel factory in 1811, began manufacturing guns in 1856. Krupp supplied large weapons to so many nations that he became known as the "Cannon King." Prussia's victory in the Franco-German War of 1870-1871 was largely the result of Krupp's field guns. In 1933 when Hitler came to power, this family business began manufacturing a wide range of artillery. Alfried Krupp (1907-1967), the greatgrandson of Alfred, supported the Nazis in power and accrued staggering wealth for the company. The firm seized property in occupied countries and used slave labor in its factories. After the war, Alfred was imprisoned for twelve years and had to forfeit all his property. Granted amnesty in 1951, he restored the business to its former position by the early 1960s. On his death in 1967, however, the firm became a corporation and the Krupp family dynasty ended.

Sources: Hogg, Ian V. *The Illustrated Encyclopedia of Artillery*, p. 184; Manchester, William R. *The Arms of Krupp, 1587-1968*, p. 93.

835
What was the Manhattan Project?

The Manhattan Engineer District was the formal code name for the United States government project to develop an atomic bomb during World War II. It soon became known as the Manhattan Project—a name taken from the location of the office of Colonel James C. Marshall, who had been selected by the United States Army Corp of Engineers to build and run the bomb's production facilities. When the project was activated by the United States War Department in June 1942, it came under the direction of Colonel Leslie R. Groves (1896-1970).

The first major accomplishment of the project's scientists was the successful initiation of the first self-sustaining nuclear chain reaction, done at a University of Chicago laboratory on December 2, 1942. The project tested the first experimental detonation of an atomic bomb in a desert area near Alamogordo, New Mexico on July 16, 1945. The test site was called Trinity, and the bomb generated an explosive power equivalent to between 15 thousand and 20 thousand tons (15,240 and 20,320 tonnes) of TNT. Two of the project's bombs were dropped on Japan the following month (Hiroshima on August 6 and Nagasaki on August 9, 1945) resulting in the Japanese surrender to end World War II.

Sources: Beyer, Don E. *The Manhattan Project*, p. 36; Rhodes, Richard. *The Making of the Atomic Bomb*, pp. 436-442.

836 / How did Big Bertha get its name?

This was the popular name first applied to the 42–centimeter (about 16.5 inches) howitzers used by the Germans and Austrians in 1914, the year World War I started. Subsequently the term included other types of huge artillery pieces used during World War I and II. The large gun was built by the German arms manufacturer Friedrich A. Krupp (1854-1902), and was named after his only child, Bertha Krupp (1886-1957).

These large guns were used to destroy the concrete and steel forts defending Belgium in 1914, and their shells weighed 205 pounds (930 kilograms) and were nearly as tall as a man. Since it was slow and difficult to move such guns, the use of them was practical only in static warfare. In World War II, bomber aircraft took away this role of long-range bombardment from these cumbersome guns.

Sources: Hogg, Ian V. *The Illustrated Encyclopedia of Artillery*, p. 90; *Weapons: An International Encyclopedia from 5000 B.C. to 2000 A.D.*, p. 188.

837 / What is a nuclear winter?

The term *nuclear winter* was coined by American physicist Richard P. Turco in a 1983 article in the journal *Science*, in which he describes a hypothetical post-nuclear war scenario having severe world-wide climatic changes: prolonged periods of darkness, below-freezing temperatures, violent windstorms, and persistent radioactive fallout. This would be caused by billions of tons of dust, soot, and ash being tossed into the atmosphere, accompanied by smoke and poisonous fumes from firestorms. In the case of a severe nuclear war, within a few days, the entire northern hemisphere would be under a blanket so thick that as little as 1/10 of 1%

of available sunlight would reach the earth. Without sunlight, temperatures would drop well below freezing for a year or longer, causing dire consequences for all plant and animal life on earth.

Reaction to this doomsday prediction lead critics to coin the term *nuclear autumn*, which downplayed such climatic effects and casualties. In January 1990, the release of *Climate and Smoke: An Appraisal of Nuclear Winter*, based on five years of laboratory studies and field experiments, reinforced the original 1983 conclusions.

Sources: Ali, Sheikh R. *The Peace and Nuclear War Dictionary*, pp. 202-203; Brennan, Richard P. *Dictionary of Scientific Literacy*, pp. 220-221; Williams, Gene B. *Nuclear War, Nuclear Winter*, pp. 11-14.

838 / Who invented the machine gun?

The first successful machine gun, invented by Richard J. Gatling (1818-1903), was patented in 1862 during the Civil War. Its six barrels were revolved by gears operated by a hand crank, and it fired 1,200 rounds per minute. Although there had been several partially successful attempts at building a multi-firing weapon, none were able to overcome the many engineering difficulties until Gatling. In his gear-driven machine, cocking and firing were performed by cam action. The United States Army officially adopted the gun on August 24, 1866.

The first automatic machine gun was a highly original design by Hiram S. Maxim (1840-1915). In 1884, the clever Maxim designed a portable, single-barreled automatic weapon that made use of the recoil energy of a fired bullet to eject the spent cartridge and load the next.

The original "tommy gun" was the Thompson Model 1928 SMG. This 45 caliber machine gun, designed in 1918 by General John Taliaferro Thompson (1860-1940), was to be used in close-quarter combat. The war ended before it went into production however, and Thompson's Auto Ordnance Corporation did not do well, until the gun was adopted by American gangsters during Prohibition. The image of a reckless criminal spraying his enemies with bullets from his hand-held "tommy gun" became a symbol of the depression years. The gun was modified several times and was much used during World War II.

Sources: Chinn, George M. *The Machine Gun*, vol. 1, pp. 24, 48-63, 123-149; Ellis, John. *The Social History of the Machine Gun*, pp. 149-165; *Weapons: An International Encyclopedia from 5000 B.C. to 2000 A.D.*, p. 149.

839 / How far can a Pershing missile travel?

The surface-to-surface nuclear missile, which is 34.5 feet long (10.5 meters) and weighs 10 thousand pounds (4,536 kilograms), has a range of about 1,120 miles (1,800 kilometers). It was developed by the United States Army in 1972. Other surface-to-surface missiles are the Polaris, with a range of 2,860 miles (4,600 kilometers); the Minuteman, with a range of 1,120 miles (1,800 kilometers); the Tomahawk, 2,300 miles (3,700 kilometers); the Trident, 4,600 miles (7,400 kilometers); and the Peacemaker, 6,200 miles (10 thousand kilometers).

Sources: Berliner, Barbara. *The Book of Answers*, p. 20; *Oxford Illustrated Encyclopedia of Invention and Technology*, p. 230.

840 ▽ Who invented mine barrage?

In 1777, David Bushnell (1742?-1824) conceived the idea of floating kegs containing explosives which would ignite upon contact with ships.

Sources: Gies, Joseph, and Frances Gies. *The Ingenious Yankee*, pp. 22-23.

HEALTH AND MEDICINE

ACCIDENTS *See: HEALTH AND MEDICINE—HEALTH HAZARDS, RISKS, ETC.*

AGING *See: THE HUMAN BODY—FUNCTIONS, PROCESSES, AND CHARACTERISTICS*

AWARDS *See: GENERAL SCIENCE AND TECHNOLOGY—AWARDS, PRIZES, AND FAIRS*

DEATH, CAUSES OF *See: HEALTH AND MEDICINE—HEALTH HAZARDS, RISKS, ETC.*

DENTAL CARE *See: HEALTH AND MEDICINE—HEALTH CARE*

DIAGNOSTIC EQUIPMENT, TESTS, ETC.

841

What is nuclear magnetic resonance imaging?

Magnetic resonance imaging (MRI), sometimes called nuclear magnetic resonance imaging (NMR), is a non-invasive, non-ionizing diagnostic technique. It is useful in detecting small tumors, blocked blood vessels, or damaged vertebral disks. Because it does not involve the use of radiation, it can often be used where x-rays are dangerous. Large magnets beam energy through the body causing hydrogen atoms in the body to resonate. This produces energy in the form of tiny electrical signals. A computer detects these signals, which vary in different parts of the body and according to whether an organ is healthy or not. The variation enables a picture to be produced on a screen and interpreted by a medical specialist.

The concept of using MRI to detect tumors in patients was proposed by Raymond Damadian in a 1927 patent application. The fundamental MRI imaging concept used in all present day MRI instruments was proposed by Paul Lauterbar in an article in *Nature* in 1973. The main advantages of MRI are that it not only gives superior images of soft tissues (like organs), but can measure dynamic physiological changes in a non-invasive manner (without penetrating the body in any way).

Sources: *Medicine*, p. 31; Webster, John G. *Encyclopedia of Medical Devices and Instrumentation*, pp. 1833, 1839.

842 ⟋
Is there a name for the heart-monitoring machine that people sometimes wear for a day or two while carrying on their normal activities?

A portable version of the electrocardiograph (ECG) designed by J.J. Holter is called a Holter monitor. Electrodes attached to the chest are linked to a small box containing a recording device. The device records the activity of the heart.

Sources: *Mayo Clinic Family Health Book*, p. 818.

843 ⟋
What is the meaning of the numbers in a blood pressure reading?

When blood is forced into the aorta, it exerts a pressure against the walls; this is referred to as blood pressure. In the reading the upper number, the systolic, measures the pressure during the period of ventricular contraction. The lower number, the diastolic, measures the pressure when blood is entering the relaxed chambers of the heart. While these numbers can vary due to age, sex, weight, and other factors, the normal blood pressure is around 110/60 to 140/90 millimeters of mercury.

Sources: Melonakos, K. *Saunders Pocket Reference for Nurses*, p. 14; *Mosby's Medical, Nursing and Allied Health Dictionary*, pp. 367, 1146.

844 ⟋
How is blood pressure measured?

A sphygmomanometer is the device used to measure blood pressure. It was invented in 1881 by an Austrian named Von Bash. It consists of a cuff with an inflatable bladder that is wrapped around the upper arm, a rubber bulb to inflate the bladder, and a device that indicates the pressure of blood. Measuring arterial tension (blood pressure) of a person's circulation is achieved when the cuff is applied to the arm over the artery and pumped to a pressure which occludes or blocks it. This gives the systolic measure or the maximum pressure of the blood which occurs during contraction of the ventricles of the heart. Air is then released from the cuff until the blood is first heard passing through the opening artery (called Korotkoff sounds). This gives diastolic pressure, or the minimum value of blood pressure that occurs during the relaxation of the arterial-filling phase of the heart muscle.

Sources: *The American Medical Association Encyclopedia of Medicine*, p. 927; Graf, Rudolf F., and George J. Whalen. *The Reston Encyclopedia of Biomedical Engineering Terms*, pp. 107, 361, 373.

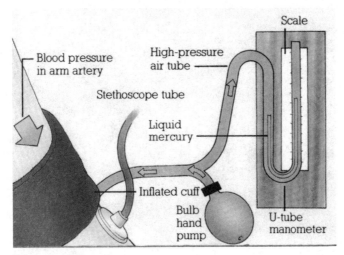

Sphygmomanometer in use

845
What is the meaning of the medical abbreviation NYD?

Not yet diagnosed.

Sources: Logan, Carolynn. *Logan's Medical and Scientific Abbreviations*, p. 385.

846
Is MRI (magnetic resonance imaging) the same as an x-ray?

Magnetic resonance imaging (MRI) uses magnetic waves instead of radiation to produce pictures of the body's internal organs and functions. What distinguishes MRI from computerized x-ray scanners is that most x-ray studies cannot distinguish between a living body and a cadaver, while MRI "sees" the difference between life and death in great detail. More specifically, it can discriminate between healthy and diseased tissues with more sensitivity than conventional radiographic instruments like x-rays or CAT scans. CAT (*computerized axial tomography*) scanners have been around since 1973 and are actually glorified x-ray machines. They offer 3-dimensional viewing but are limited because the object imaged must remain still.

Ultrasound is another type of 3-D computerized imaging. Using brief pulses of ultrahigh frequency acoustic waves (lasting 0.01 second), it can produce a sonar map of the imaged object.

Sources: McAleer, Neil. *The Body Almanac*, pp. 323–328.

6 **CAT SCANNER**
Computerized Axial Tomography involves passing very weak X-rays through the body. The patient lies inside a large rotating drum containing the X-ray source and recording camera. These move around and along the body, taking several pictures from different angles. The pictures are then combined by computer to show a "slice" through the body. Like an ordinary X-ray, dense parts of the body show up lighter. The CAT scanner can distinguish between different soft tissues, such as muscles and nerves. The procedure is painless and takes only a few minutes.

Supercooled superconducting magnet in low-loss cryostat
Electrical power
Shim and gradient coils control beam
Computer
Machine scans successive "slices"

▽ **847** ▽
What are the normal test ranges for total cholesterol, triglycerides, low density lipoproteins (LDL), and high density lipoproteins (HDL)?

Normal blood cholesterol levels can range from 150 to 289 milligrams per deciliter. In order to lower the risk of coronary heart disease, the recommended total cholesterol level is less than 200 milligrams per deciliter. A range of 200 to 239 is considered to be "borderline" and anything above 240 milligrams per deciliter is considered to be high. LDL–cholesterol in the blood seems to encourage the formation of artery-clogging fatty deposits, so the levels of it should be below 130 milligrams per deciliter (130 to 159 is considered borderline and greater than 160 is high risk). However the average LDL–cholesterol level is around 150 milligrams per deciliter, with the normal range occurring between 100 and 200. The average HDL–cholesterol level is 55 milligrams per deciliter with the norm range being between 30 and 80. Anything less than 35 is considered a risk factor for a heart attack. However the ratio of total cholesterol to HDL–cholesterol (which washes out excess cholesterol from the blood) should be 3.5 to 4.5. Triglycerides (fatty substances consisting of fatty acids and glycerol that are usually stored as fat as an energy reserve) range from 30 to 145 milligrams per deciliter.

Sources: *Eating to Lower Your High Blood Cholesterol*, pp. 2–4; *Mayo Clinic Family Health Book*, p. 789; Pinkney, Cathey. *The Patient's Guide to Medical Tests*, pp. 214, 216.

848 / What is the instrument a doctor uses to check reflexes?

A plessor or plexor or percussor is a small hammer, usually with a soft rubber head, used to tap the part directly. Also called a reflex hammer or a percussion hammer, it is used by a doctor to elicit reflexes by tapping on tendons. In the most common test, the patient sits on a surface high enough to allow his legs to dangle freely, and the physician lightly taps the patellar tendon, just below the kneecap. This stimulus briefly stretches the quadriceps muscle on top of the thigh. The stretch causes the muscle to contract, which makes the leg kick forward. The time interval between the tendon tap and the start of the leg extension is about 50 microseconds. That interval is too short for the involvement of the brain and is totally reflexive. This test indicates the status of an individual's reflex control of movement.

Sources: Carlson, Neil R. *Foundations of Physiological Psychology*, pp. 208–209; *Stedman's Medical Dictionary*, p. 1215.

DISEASES, DISORDERS, AND OTHER HEALTH PROBLEMS

849 / Was the disease pellagra ever a common disease in the United States?

Pellagra, characterized by diarrhea, skin problems, and dementia, occurred frequently years ago in rural cotton mill towns, mining and sawmill camps, and in orphanages, prisons, and poorhouses. In these places, poverty and tradition reduced the diet to corn meal, fat salt pork, and carbohydrates with very few green vegetables especially during the winter months. Pellagra is brought on by an inadequate diet with not enough vitamins, especially vitamin B.

Sources: Dublin, Louis I. *Factbook on Man from Birth to Death*, pp. 348–349.

850 / What is carpal tunnel syndrome?

Carpal tunnel syndrome occurs when a branch of the median nerve in the forearm is compressed at the wrist as it passes through the tunnel formed by the wrist bones (or carpals), and a ligament that lies just under the skin. The syndrome occurs most often in middle age and more so in women than men. The symptoms are intermittent at first, then become constant. Numbness and tingling begin in the thumb and first two fingers; then the hand and sometimes the whole arm becomes painful. Treatment involves wrist splinting, weight loss, control of edema; treatments for arthritis may help also. If not, a surgical procedure in which the ligament at the wrist is cut, can relieve pressure on the nerve.

Sources: Columbia University College of Physicians and Surgeons. *Complete Home Medical Guide*, p. 599.

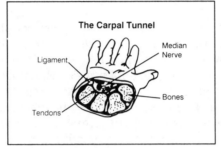

851 / What is lactose intolerance?

Lactose, the principal sugar in cow's milk, found only in milk and dairy products, requires the enzyme lactase for human digestion. Lactose intolerance occurs when the lining of the walls of a person's small intestine does not produce normal amounts of this enzyme. Lactose intolerance causes abdominal cramps, bloating, diarrhea, and excessive gas when more than a certain amount of milk is ingested. Most people are less able to tolerate lactose as they grow older.

A person having lactose intolerance need not eliminate dairy products totally from the diet. Decreasing the consumption of milk products, drinking milk only during meals, and getting calcium from cheese, yogurt, and other dairy products having lower lactose values are options. Another alternative is to buy a commercial lactose preparation that can be mixed into milk. These preparations convert lactose into simple sugars that can be easily digested.

Sources: *Mayo Clinic Family Health Book*, pp. 470, 623.

852 / What is narcolepsy?

Although most people think of a narcoleptic as a person who falls asleep at inappropriate times, victims of narcolepsy also share other symptoms, including excessive daytime sleepiness, hallucinations, and cataplexy (a sudden loss of muscle strength following an emotional event). Persons with narcolepsy experience an uncontrollable desire to sleep, sometimes many times in one day. Episodes may last from a few minutes to several hours.

Sources: *The Almanac of Science and Technology,* p. 136; *Mosby's Medical, Nursing, and Allied Health Dictionary,* p. 792.

853 / Who was Typhoid Mary?

Mary Mallon (1855–1938), a cook who lived in New York City around 1900, was identified as a chronic carrier of the typhoid bacilli. She was the cause of at least 53 outbreaks of typhoid fever. The name "Typhoid Mary" became known around the world as a symbol for the chronic carrier of disease.

Sources: Dowling, Harry F. *Fighting Infection,* pp. 17–18; *Guinness Book of Records 1992,* pp. 182–183.

854 / How does jet lag affect one's body?

The physiological and mental stress encountered by airplane travellers when crossing four or more time zones is commonly called *jet lag.* Patterns of hunger, sleep, elimination, along with alertness, memory, and normal judgement, may all be affected. There are more than 100 biological functions that fluctuate during the 24-hour cycle (circadian rhythm) that can become desynchronized. Most people's bodies adjust at a rate of about one hour per day. Thus after four time zone changes, the body will require about four days to return to its usual rhythms. Flying eastward is often more difficult than flying westward, which adds hours to the day.

Sources: *Mayo Clinic Family Health Book,* pp. 341–342; Bair, Frank E. *The Weather Almanac,* p. 229.

855 / What is the medical name for a sinking feeling in the pit of the stomach?

Epigastric sensation.

Sources: *Taber's Cyclopedia Medical Dictionary,* p. 1658.

856 / What is pica?

Pica refers to the craving for unnatural or non-nutritious substances. It is named after the magpie (*Pica pica*) which has a reputation for sticking its beak into all kinds of things to satisfy its hunger or curiosity. It can happen in both sexes, all races, and in all parts of the world, but is especially noted in pregnant women.

Sources: *FDA Consumer*, vol. 19 (December 1985/January 1986), p. 29.

857
Why do deep-sea divers get the bends?

Bends is a painful condition in limbs and abdomen. It is caused by the formation and enlargement of bubbles of nitrogen in blood and tissues as a result of rapid reduction of air pressure. This condition can develop when a diver ascends too rapidly after being exposed to increased pressure. Severe pain will develop in the muscles and joints of the arms and legs. More severe symptoms include vertigo, nausea, vomiting, choking and shock, and sometimes death. Bends is also known as caisson disease, tunnel disease, and diver's paralysis.

Sources: *Taber's Cyclopedic Medical Dictionary*, p. 203.

858
What is progeria?

Progeria is premature old age. There are two distinct forms of the condition, both of which are extremely rare. In Hutchinson-Gilford syndrome, aging starts around the age 4, and by 10 or 12, the affected child has all the external features of old age, including gray hair, baldness, and loss of fat, resulting in thin limbs and sagging skin on the trunk and face. There are also internal degenerative changes, such as atherosclerosis (fatty deposits lining the artery walls). Death usually occurs at puberty. Werner's syndrome, or adult progeria, starts in early adult life and follows the same rapid progression as the juvenile form. The cause of progeria is unknown.

Sources: *The American Medical Association Encyclopedia of Medicine*, pp. 822–823.

859
How many people in America are estimated to have Alzheimer's disease?

Researchers have concluded that 10% of people over 65 have Alzheimer's disease, including nearly 50% of people over 85. The National Institute on Aging estimated that four million Americans have Alzheimer's disease. Some experts, however, do not agree with these figures.

Alzheimer's disease is a progressive condition in which nerve cells in the brain degenerate and the brain substance shrinks. Although the cause is unknown, some theorize that it is toxic poisoning by a metal such as aluminum. Others believe it to be of genetic origin. There are three stages; in the first, the person becomes forgetful; in the second, the patient experiences severe memory loss and disorientation, lack of concentration, loss of ability to calculate and find the right word to use (dysphasia), anxiety and sudden personality changes. In the third stage the patient is severely disoriented and confused, suffers from hallucinations and delusions, has severe memory loss and the nervous system declines, with regression into infantile behavior, violence, etc., requiring hospital care.

Sources: *The American Medical Association Encyclopedia of Medicine*, p. 91; *Yearbook of Science and the Future 1991*, pp. 380–381.

860
What is monkey-paw?

A contracture of the hand resulting from median nerve palsy; the thumb cannot be opposed to the tips of the fingers.

Sources: *Stedman's Medical Dictionary*, p. 977.

861
Which medical condition is associated with Abraham Lincoln's awkward appearance?

Abraham Lincoln (1809-1865) probably had Marfan's syndrome (*Arachnodactyly*) which abnormally lengthens the bones. It is a rare inherited degenerative disease of the connective tissue. Besides the excessively long bones, there are chest deformities, scoliosis (spine curvature), an arm span that can exceed height, eye problems (especially myopia or nearsightedness), abnormal heart sounds, and sparse subcutameous fat. In 1991, researchers identified the gene behind the disease.

Sources: *Diseases and Disorders Handbook*, pp. 467–468; Loomer, Alice. *Famous Flaws*, p. 145.

862
How is an iatrogenic illness defined?

It is an adverse mental or physical condition caused by the effects of treatment by a physician or surgeon. The term implies that it could have been avoided by judicious care on the part of the physician.

Sources: *Taber's Cyclopedia Medical Dictionary*, p. 885.

863
What is pelvic inflammatory disease?

Pelvic inflammatory disease (PID), a term used for a group of infections in the female organs, for example, inflammations of the Fallopian tubes, cervix, uterus, and ovaries. It is the most common cause of female infertility today. PID is most often found in sexually active women under the age of 25 and almost always results from gonorrhea or chlamydia, but women who have IUDs are also at risk. Aerobic or anaerobic organisms such as *Neisseria gonorrhoeae* that causes acute cervicitis, (a foul-smelling vaginal discharge accompanied by itching, burning, and pelvic discomfort, often resulting in infertility and spontaneous abortion) or common bacteria such as staphylocci, chlamydiae, and coliforms to produce acute endometritis (oozing vaginal discharge with ulteration, lower abdominal pain and spasms, fever, etc.) Chronic endometritis is becoming increasingly common because of the widespread use of IUDs. Untreated PID could be fatal.

Sources: *Diseases and Disorders Handbook*, pp. 555–557; Galperin, Anne. *Gynecological Disorders*, p. 75.

864 ▽ *What is meant by* vectors *in medicine?*

A vector is an animal that transmits a particular infectious disease. A vector picks up disease organisms from a source of infection, carries them within or on its body and later deposits them where they infect a new host. Mosquitoes, fleas, lice, ticks, and flies are the most important vectors of disease to humans.

Sources: *The American Medical Association Encyclopedia of Medicine*, p. 1041.

865 ▽ *How are warts caused?*

A wart is a lump on the skin produced when one of the 30 types of papillomavirus invades skin cells and causes them to multiply rapidly. There are several different types of warts: common warts, usually on injury sites; flat warts on hands, accompanied by itching; digitate warts having fingerlike projections; filiform warts on eyelids, armpits, and necks; plantar warts on the soles of the feet; and genital warts, pink cauliflowerlike areas that if occurring in a woman's cervix, could predispose her to cervical cancer. Each is produced by a specific virus and most usually symptomless. Wart viruses are spread by touch or by contact with the skin shed from a wart.

Sources: *The American Medical Association Encyclopedia of Medicine*, pp. 1069–1070; *The American Medical Association Family Medical Guide*, p. 252.

866 ▽ *What is dyslexia and what causes it?*

Dyslexia covers a wide range of language difficulties. In general, a person with dyslexia cannot group the meaning of sequences of letters, words, or symbols or the concept of direction. The condition can affect people of otherwise normal intelligence. Dyslexic children may reverse letter and word order, make bizarre spelling errors, and may not be able to name colors or write from dictation. It may be caused by minor visual defects, emotional disturbance, or failure to train the brain. New evidence shows that a neurological disorder may be the underlying cause. Approximately 90% of dyslexics are male.

The term *dyslexia* (of Greek origin) was first suggested by Professor Rudolph of Stuttgart, Germany in 1887. The earliest references to the condition date as far back as 30 C.E. when Valerius Maximus and Pliny described a man who lost his ability to read after being struck on the head by a stone.

Sources: *The American Medical Association Encyclopedia of Medicine*, pp. 382–383; Millichap, J. Gordon. *Dyslexia as the Neurologist and Educator Read It*, p. 3; *The New Good Housekeeping Family Health and Medical Guide*, p. 141.

867
What was the contribution of Dr. Gorgas to the building of the Panama Canal?

Dr. William C. Gorgas (1854–1920) brought the endemic diseases of Panama under control by destroying mosquito breeding grounds, virtually eliminating yellow fever and malaria. His work was probably more essential to the completion of the canal than any engineering technique.

Sources: Asimov, Isaac. *Asimov's Biographical Encyclopedia of Science and Technology*, p. 453; Hawkes, Nigel. *Structures*, p. 139; Williams, Trevor. *A Biographical Dictionary of Scientists*, p. 220.

868
How is the term zoonosis defined?

A zoonosis is any infectious or parasitic disease of animals that can be transmitted to humans. Examples include cat-scratch fever, psittacosis from parrots, and trichinosis from pigs.

Sources: *The American Medical Association Encyclopedia of Medicine*, p. 1088.

869
Who were the "Blue People" in Appalachia?

The "Blue People" were descendants of Martin Fugate, a French immigrant to Kentucky. He had a recessive gene that limited or stopped the body's production of the enzyme diaphorase. Diaphorase breaks down methemoglobin into hemoglobin in red blood cells. When the enzyme is not present, a disproportionate amount of methemoglobin remains in the blood, giving the cells a bluish tint, rather than the normal pink associated with Caucasians. The condition is strictly one of pigment and does not deprive the person of oxygen.

Despite bluish color, there are no known health risks associated with the deficiency. Fugate's family suffered from the condition because of excessive inbreeding. When both spouses had the recessive gene, their children would be blue. As the family became mobile following World War II and moved out of their Kentucky valley, the inbreeding ceased. As of 1982 there were only two to three members of the family with the condition.

Sources: *Science 82*, vol. 3, no. 9 (November 1982), pp. 34–39.

870
What is "tango foot"?

Officially known as *tibialis anticus*, tango foot is a strain of the leg muscles resulting from dancing the tango and maxixe.

Sources: Plunkett, Edward R. *Folk Name & Trade Diseases*, p. 282.

871
Why is neurofibromatosis called "Elephant Man's disease"?

Neurofibromatosis or Von Recklinghausen's disease (for the German doctor who first described the disease in detail) severely deformed the Englishman Joseph Merrick (1862–1889), who was referred to as "the Elephant Man." Neurofibromatosis is an uncommon inherited developmental disorder of the nervous system, bones, muscles, and skin. Occurring in one in every three thousand births, its symptoms general appear during childhood or adolescence. It is characterized by numerous soft, fibrous swellings that grow on nerve trunks of extremities, head, neck, and body; *cafe au lait* spots (pale, coffee-colored patches) on the skin of the trunk and pelvis; scoliosis; and short stature. In most cases, neurofibromatosis affects only the skin. But if the soft fibrous swellings occur in the central nervous system they can cause epilepsy, affect vision and hearing, and can also shorten life. Surgical removal of neurofibromas is necessary only if they cause complications, or are likely to do so.

Sources: *The American Medical Association Encyclopedia of Medicine*, pp. 722–723; *Disease and Disorders Handbook*, pp. 506–507; Drimmer, Frederick. *The Elephant Man*, p. 138.

872
What is Lou Gehrig's disease?

Sometimes called Lou Gehrig's disease, *a*myotrophic *l*ateral *s*clerosis (ALS) is a motor neuron disease of middle or late life. It results from a progressive degeneration of nerve cells controlling voluntary motor functions that ends in death 3 to 10 years after onset. There is no cure for it. At the beginning of the disease, the patient notices weakness in the hands and arms, with involuntary muscle quivering and possible muscle cramping or stiffness. Eventually all four extremities become involved. As nerve degeneration progresses, disability occurs and physical independence declines until the patient, while mentally and intellectually aware, can no longer swallow or move.

Sources: *American Medical Association Encyclopedia of Medicine*, p. 696, Columbia University College of Physicians and Surgeons. *Complete Home Medical Guide*, p. 599.

873 ▽ Why is Legionnaire's disease known by that name?

Legionnaire's disease was first identified in 1976 when a sudden, virulent outbreak of pneumonia took place at a hotel in Philadelphia, Pennsylvania, where delegates to an American Legion Convention were staying. The cause was eventually identified as a previously unknown bacterium that was given the name *Legionnella pneumophilia*. The bacterium probably was transmitted by an airborne route. It can spread through cooling tower or evaporation condensers in air-conditioning systems, and has been known to flourish in soil and excavation sites. Usually the disease occurs in late summer or early fall and its severity ranges from mild to life-threatening with a mortality rate as high as 15%. Symptoms include diarrhea, anorexia, malaise, headache, generalized weakness, recurrent chills and fever accompanied by cough, nausea, and chest pain. Antibiotics such as Erythroycin® are administered along with other therapies (fluid replacement, oxygen, etc.) that treat the symptoms.

Sources: *Diseases and Disorders Handbook*, pp. 439–441; *Mayo Clinic Family Health Book*, p. 921.

874 ▽ What is "Christmas factor"?

In the clotting of blood, factor IX, or the Christmas factor, is a coagulation factor present in normal plasma, but deficient in the blood of persons with hemophilia B or Christmas disease. It was named after a man named Christmas who, in 1952, was the first patient in whom this genetic disease was shown to be distinct from hemophilia (another genetic blood-clotting disease in which the blood does not have factor VIII).

Sources: *The American Medical Association Encyclopedia of Medicine*, p. 277; *Mosby's Medical, Nursing, and Allied Health Dictionary*, p. 459.

875 ▽ Which name is now used as a synonym for leprosy?

Hansen's disease is the name of this chronic, systemic infection characterized by progressive lesions. Caused by a bacterium, *Mycabacterium leprae*, that is transmitted through airborne respiratory droplets, the disease is not highly contagious.

Continuous close contact is needed for transmittal. Antimicrobial agents, such as sulfones (dapsone in particular), are used to treat the disease.

Sources: *Diseases and Disorders Handbook*, pp. 440–441; *Merck Manual of Diagnosis and Therapy*, p. 127.

876
What are dust mites?

Dust mites are microscopic arachnids (members of the spider family), commonly found in house dust. Dust mite allergen is probably one of the most important causes of asthma (breathlessness and wheezing caused by the narrowing of small airways of the lungs) in North America, as well as the major cause of common allergies (exaggerated reactions of the immune system to exposure of offending agents).

Sources: Godish, Thad. *Indoor Air Pollution Control*, pp. 55–56.

877
How many types of herpes virus are there?

There are five human herpes viruses all characterized by an eruption of small, usually painful, skin blisters:

Herpes simplex type 1—causes recurrent cold sores and infections of the lips, mouth, and face. The virus is contagious and spreads by direct contact with the lesions or fluid from the lesions. Cold sores are usually recurrent at the same sites and reoccur where there is an elevated temperature at the affected site; such as a fever or prolonged sun exposure. Occasionally this virus may occur on the fingers with a rash of blisters. If the virus gets into the eye, it could cause conjunctivitis, or even a corneal ulcer. On rare occasions, it can spread to the brain to cause encephalitis.

Herpes simplex type 2—causes genital herpes and infections acquired by babies at birth. The virus is contagious and can be transmitted by sexual intercourse. The virus produces small blisters in the genital area that burst to leave small painful ulcers, which heal within 10 days to three weeks. Headache, fever, enlarged lymph nodes, and painful urination are the other symptoms.

Varicella-zoster (herpes zoster)—causes chicken pox and shingles. Shingles can be caused by the dormant virus in certain sensory nerves that re-emerge with the decline of the immune system (because of age, certain diseases, and the use of immunosuppressants), excessive stress, or use of corticosteroid drugs. The painful rash of small blisters dry and crust over, eventually leaving small pitted scars. The rash tends to occur over the rib area or a strip on one side of the neck or lower body. Sometimes it involves the lower half of the face and can affect the eyes. Pain that can be severe and longlasting affects about half of the sufferers and is caused by nerve damage.

Epstein-Barr—causes infectious mononucleosis (acute infection having high fever, sore throat and swollen lymph glands, especially in the neck, which occurs mainly during adolescence) and is associated with Burkitt's lymphoma (malignant tumors of the jaw or abdomen that occur mainly in African children and in tropical areas).

Cytomegalovirus—usually no symptoms but enlarges the cells it infects; it can cause birth defects when a pregnant mother infects her unborn child.

Herpes gestationis is a rare skin-blister disorder occurring only in pregnancy and is not related to the herpes simplex virus.

Sources: *The American Medical Association Encyclopedia of Medicine,* pp. 331, 415, 536–537, 693; Balfour, Henry H. *Herpes Diseases and Your Health,* p. 3.

878 ▽ What is Factor VIII?

Factor VIII is one of the enzymes involved in the clotting of blood. Hemophiliacs lack this enzyme and are at high risk of bleeding to death unless they receive supplemental doses of Factor VIII. The lack of Factor VIII is due to a defective gene, which shows a sex-linked inherited pattern, affecting males (one in 10 thousand). Females can carry the gene. Hemorrhage into joints and muscles usually make up the majority of bleeding episodes.

Sources: *The Almanac of Science and Technology,* p. 104; *The American Medical Association Encyclopedia of Medicine,* p. 530.

879 ▽ Which disease is the deadliest?

The most deadly infectious disease was the pneumonic form of the plague, the so-called Black Death of 1347–1351, with a mortality rate of 100%. Today, the disease with the highest mortality (almost 100%) is rabies in humans when it prevents the victim from swallowing water. This disease is not to be confused with being bitten by a rabid animal. With immediate attention, the rabies virus can be prevented from invading the nervous system and the survival rate in this circumstance is 95%. The virus AIDS (*acquired immunodeficiency syndrome*) first reported in 1981 is caused by the HIV (*human immunodeficiency virus*). The World Health Organization (WHO) reported 345,553 HIV positive cases worldwide as of April 1, 1991, and estimated that there are 8 to 10 million people world-wide infected with HIV, with 1 million of these being in the United States. The United States has 171,876 cases of HIV positive with the number of deaths being 108,731. There has not been a case of recovery from AIDS yet.

Sources: *Guinness Book of Records 1992,* pp. 181–182; *The World Almanac and Book of Facts 1992,* p. 268.

880
What is the medical term for a heart attack?

Myocardial infarction is the term used for a heart attack in which part of the heart muscle's cells die as a result of reduced blood flow through one of the main arteries (many times due to atherosclerosis); the outlook for the patient is dependant on the size and location of the blockage and extent of damage, but 33% of patients die within 20 days after the attack; it is a leading cause of death in the United States.

Sources: *Diseases and Disorders Handbook*, pp. 491–496.

881
How are the forms of cancer classified?

The many forms of cancer (malignant tumors) are classified into four groups: carcinomas, which involve the skin and skin-like membranes of the internal organs; sarcomas, which involve the bones, muscles, cartilage, and fat; leukemias, which involve the white blood cells; and lymphomas, which involve the lymphatic system.

Sources: Schneider, Herman. *The Harper Dictionary of Science in Everyday Language*, p. 49.

882
Which species of mosquito causes malaria and yellow fever in humans?

The bite of the female mosquito of the genus *Anopheles* can contain the parasite of the genus *Plasmodium*, which causes malaria, a serious tropical infectious disease affecting 200 to 300 million people worldwide. More than one million African babies and children die from the disease annually. The *Aedes aegypti* transmits yellow fever, a serious infectious disease characterized by jaundice, giving the patient yellowish skin; 10% of the patients die.

Sources: *Guinness Book of Records 1992*, p. 52; Passarin d'Entreves, P. *The Secret Life of Insects*, pp. 29–30.

883
What is Chinese restaurant syndrome?

Monosodium glutamate (MSG), a commonly used flavor-enhancer, is thought to cause flushing, headache, and numbness about the mouth in susceptible people. Because many Chinese restaurants use MSG in food preparation, these symptoms may appear after a susceptible person eats Chinese food.

Sources: *Mayo Clinic Family Health Book*, p. 470.

What is the phobia of number 13 called?

Triskaidekaphobia is a superstitious fear of the number 13. A phobia (a tense and irrational fear of an object, situation, or organism) can develop for a wide range. The list below demonstrates the variety:

Phobia subject	Phobia term
Animals	Zoophobia
Beards	Pogonophobia
Books	Bibliophobia
Churches	Ecclesiaphobia
Dreams	Oneirophobia
Flowers	Anthophobia
Food	Sitophobia
Graves	Taphophobia
Infection	Nosemaphobia
Lakes	Limnophobia
Leaves	Phyllophobia
Lightning	Astraphobia
Men	Androphobia
Money	Chrometophobia
Music	Musicophobia
Sex	Genophobia
Shadows	Sciophobia
Spiders	Arachnophobia
Sun	Heliophobia
Touch	Haptophobia
Trees	Dendrophobia
Walking	Basiphobia
Water	Hydrophobia
Women	Gynophobia
Work	Ergophobia
Writing	Graphophobia

Sources: *The Guinness Book of Answers*, pp. 160–162; *The McGraw-Hill Dictionary of Scientific and Technical Terms*, p. 1971.

Was Napoleon poisoned?

The most common opinion today is that Napoleon Bonaparte (1769–1821), Emperor of France from 1804–1815, died of a cancerous, perforated stomach. A significant minority of doctors and historians have made other claims ranging from various diseases to malign neglect to outright homicide. A Swedish toxicologist, Sten Forshufvud, advanced the theory that Napoleon died of arsenic poisoning, administered by an agent of the French Royalists who was planted in Napoleon's household during his final exile on the island of St. Helena.

Sources: Karlen, Arno. *Napoleon's Glands*, p. 15; Weider, Ben, and David Hapgood. *The Murder of Napoleon*.

886

Which disease is the most common?

The most common noncontagious disease is periodontal disease, such as gingivitis or inflammation of the gums. Few people in their lifetime can avoid the effects of tooth decay. The most common contagious disease in the world is coryza or the common cold.

Sources: *Guinness Book of Records 1992*, p. 181.

887

What is the chemical composition of kidney stones?

About 80% are calcium, mainly calcium oxalate and/or phosphate; 5% are uric acid; 2% are amino acid cystine; the remainder are magnesium ammonium phosphate. About 20% of these stones are infective stones, linked to chronic urinary infections, and contain a combination of calcium, magnesium, and ammonium phosphate produced from the alkalinity of the urine and bacteria action on urea (a substance in urine).

Sources: *The American Medical Association Encyclopedia of Medicine*, p. 225; *Merck Manual of Diagnosis and Therapy*, p. 1641.

888

What are HeLa cells?

HeLa cells, used in many biomedical experiments, were obtained from a cervical carcinoma in a black woman named *He*nriette *La*cks. Epithelial tissue obtained by biopsy became the first continuously cultured human malignant cells.

Sources: *Stedman's Medical Dictionary*, p. 268.

889

How does a stroke occur?

A stroke, or *c*erebro*v*ascular *a*ccident (CVA), is a medical emergency produced by a blood clot that lodges in an artery and blocks the flow of blood to a portion of the brain. It produces symptoms ranging from paralysis of limbs and loss of speech to unconsciousness and death. Less commonly, a stroke may be the result of bleeding into the substance of the brain. Strokes are fatal in about 33% to 50% of the cases and 500 thousand cases a year make it the third most common cause of death. Certain factors increase the risk of stroke such as high blood pressure (hypertension), which weakens artery walls, and atherosclerosis (thickening of the lining of artery walls), which narrows arteries.

Sources: *Diseases and Disorders Handbook*, pp. 128–132; *Mayo Clinic Family Healthbook*, p. 1328.

How are burns classified?

Type	Effects
First-degree	Sunburn; steam. Reddening and peeling. Affects epidermis (top layer of skin). Heals within a week.
Second-degree	Scalding; holding hot metal. Deeper burns causing blisters. Affects dermis (deep skin layer). Heals in two to three weeks.
Third-degree	Fire. A full layer of skin is destroyed. Requires a doctor's care and grafting.
Circumferential	Any burns (often electrical) that completely encircle a limb or body region (such as the chest), which can impair circulation or respiration; requires a doctor's care; fasciotomy (repair of connective tissues) is sometimes required.
Chemical	Acid, alkali. Can be neutralized with water (for up to a half hour). Doctor's evaluation recommended.
Electrical	Destruction of muscles, nerves, circulatory system, etc., below the skin. Doctor's evaluation and ECG monitoring required.

If more than 10% of body surface is affected in second- and third-degree burns, shock can develop when large quantities of fluid (and its protein) are lost. When skin is burned, it cannot protect the body from airborne bacteria.

Sources: Rovin, Jeff. *Laws of Order*, p. 102.

891
What are the symptoms and signs of AIDS?

The early symptoms (AIDS-related complex, or ARC, symptoms) include night sweats, prolonged fevers, severe weight loss, persistent diarrhea, skin rash, persistent cough, and shortness of breath. The diagnosis changes to AIDS (acquired immunodeficiency syndrome) when the immune system is affected and the patient becomes susceptible to opportunistic infections and unusual cancers, such as herpes viruses (herpes simplex, herpes zoster, cytomegalovirus infection), *candida albicans* (fungus) infection, *cryptosporidium enterocolitis* (protozoan intestinal infection), *pneumocystis carinii* pneumonia (PCP, a common AIDS lung infection), toxoplasmosis (protozoan brain infection), *progressive multifocal leukoencephalopathy* (PML, a central nervous system disease causing gradual brain degeneration), *mycobacterium avium intracellulare* infection (MAI, a common generalized bacterial infection), and Kaposi's sarcoma (a malignant skin cancer characterized by blue-red nodules on limbs and body, and internally in the gastrointestinal and respiratory tracts, where the tumors cause severe internal bleeding). More than 75% of AIDS victims die within 2 years of diagnosis.

The signs of AIDS are generalized swollen glands, emaciation, blue or purple-brown spots on the body, especially on the legs and arms, prolonged pneumonia, and oral thrush.

Sources: *Diseases and Disorders Handbook*, pp. 11–17; Gong, Victor. *AIDS: Facts and Issues*, p. 7.

892
What is anorexia?

Anorexia simply means a loss of appetite. Anorexia nervosa is a psychological disturbance that is characterized by an intense fear of being fat. It usually affects teenage or young adult females. This persistent "fat image," however untrue in reality, leads the patient to self-imposed starvation and emaciation (extreme thinness) to the point where one-third of the body weight is lost. There are many theories on the causes of this disease, which is difficult to treat and can be fatal. Between 5% and 10% of patients hospitalized for anorexia nervosa later die from starvation or suicide. Symptoms include a 25% or greater weight loss (for no organic reason) coupled with a morbid dread of being fat, an obsession with food, an avoidance of eating, compulsive exercising and restlessness, binge eating followed by induced vomiting, and/or use of laxatives or diuretics.

Sources: *The American Medical Association Encyclopedia of Medicine*, pp. 112–113; *Diseases and Disorders Handbook*, pp. 47–49.

893
How is Lyme disease carried?

The cause of the disease is the spirochete *Borrelia burgdorferi* that is transmitted to humans by the small tick *Ixodes dammini* or other ticks in the Ixodidae family. The tick injects spirochete–laden saliva into the bloodstream or deposits fecal matter on the skin. This multisystemic disease usually begins in the summer with a skin lesion called *erythema chronicum migrans* (ECM), followed by more lesions, a molar rash, conjuctivitis, and urticaria. The lesions are eventually replaced by small red blotches. Other common symptoms in the first stage include fatigue, intermittent headache, fever, chills, and achiness.

In stage two, which can be weeks or months later, cardiac or neurologic abnormalities sometimes develop. In the last stage (weeks or years later) arthritis develops with marked swelling, especially in the large joints. If tetracycline, penicillin, or erythromycin is given in the early stages, the later complications can be minimized. High dosage of intravenously-given penicillin can also be effective on the late stages.

Sources: *Diseases and Disorders Handbook*, pp. 457–458.

894

If the sap of the poison ivy plant touches the skin, will a rash develop?

Studies show that 85% of the population will develop an allergic reaction if exposed to poison ivy, but this sensitivity varies with each individual according to circumstance, age, genetics, and previous exposure. The poison comes mainly from the leaves when the allergens touch the skin, a red rash with itching, and burning, will develop; skin blisters will develop usually within 6 hours to several days after exposure. Thorough washing of the affected area with mild soap within five minutes of exposure can be effective; sponging with alcohol and applying a soothing and drying lotion, such as calamine lotion, is the prescribed treatment for light cases. If the affected area is large, fever, headache, and generalized body weakness may develop. For severe reactions, a physician should be consulted, who may prescribe a corticosteroid drug. Clothing that touched the plants should also be washed.

Sources: American Academy of Dermatology. *Poison Ivy*; *First Aid Book*, pp. 220–221.

DRUGS, MEDICINES, ETC.

895

What is China white?

China white is a powerful synthetic heroin and even small amounts can be fatal. The chemical name is 3-methylfentanyl.

Sources: Tunnell, James E. *Latest Intelligence*, p. 39.

896

From what plant is taxol extracted?

Taxol is produced from the bark of the western or Pacific yew (*Taxus brevifolia*). It has been shown to inhibit the growth of HeLa cells (human tumor cells) and is a promising new treatment for several kinds of cancer. The bad news is that it is in short supply and may not be able to be synthesized on a commercial scale due to its molecular configuration.

Sources: *Audubon Magazine*, vol. 94, no. 2 (March/April 1992), pp. 79, 81.

897

Who was the first to use chemotherapy?

Chemotherapy is the use of chemical substances to treat diseases, specifically malignant diseases. The drug must interfere with the growth of bacterial, parasitic, or tumor cells, without significantly affecting host cells. Especially effective in types of cancer as leukemia and lymphoma, chemotherapy was introduced in medicine by the German physician Paul Ehrlich (1854–1915).

Sources: *Encyclopedic Dictionary of Science*, pp. 61, 84.

898 / *What is the meaning of the abbreviations often used by a doctor when writing a prescription?*

Latin phrase	Shortened form	Meaning
quaque hora	qh	every hour
quaque die (de-uh)	qd	every day
bis in die	bid	twice a day
ter in die	tid	three times a day
quarter in die	qid	four times a day
pro re nata	prn	as needed
ante cibum	a.c.	before meals
post cibum	p.c.	after meals
per os	p.o.	by mouth
nihil per os	n.p.o.	nothing by mouth
signetur	sig	let it be labeled
statim	stat	immediately
ad libitum	ad lib	at pleasure
hora somni	h.s.	at bedtime
cum		with
sine		without
guttae	gtt	drops
semis		a half
et	et	and

Sources: Cartwright, Rudolph. *Doctors Are Easy to Understand If You Speak Their Language*, pp. 7–8.

899 / *Why are anabolic steroids harmful?*

Anabolic (protein-building) steroids are drugs that mimic the effects of testosterone and other male sex hormones. They can build muscle tissue, strengthen bone, and speed muscle recovery following exercise or injury. They are sometimes prescribed to treat osteoporosis in postmenopausal women and some types of anemia. Some athletes use anabolic steroids to build muscle strength and bulk, and to allow a more rigorous training schedule. Weight lifters, field event athletes, and body builders are most likely to use anabolic steroids. The drugs are banned from most organized competitions because of the dangers they pose to health and to prevent an unfair advantage.

Adverse effects include hypertension, acne, edema, and damage to liver, heart, and adrenal glands. Psychiatric symptoms can include hallucinations, paranoid delusions, and manic episodes. In men, anabolic steroids can cause infertility, impotence, and premature balding. Women can develop masculine characteristics, such as excessive hair growth, male-pattern balding, disruption of menstruation, and deepening of the voice. Children and adolescents can develop problems in growing bones, leading to short stature.

Sources: *The American Medical Association Encyclopedia of Medicine*, pp. 935, 940–941, 1049–1050; Edelson, Edward. *Sports Medicine*, p. 78; *Prevention's Giant Book of Health Facts*, p. 214.

900 / What are monoclonal antibodies?

Monoclonal antibodies are artificially produced antibodies, designed to neutralize a specific foreign protein (antigen). Cloned cells (genetically identical) are stimulated to produce antibodies to the target antigen. Most monoclonal antibody work so far has used cloned cells from mice infected with cancer. In some cases they are used to destroy cancer cells directly; in others they carry other drugs to combat the cancer cells.

Sources: *The American Medical Association Encyclopedia of Medicine*, p. 115; Galperin, Anne. *Gynecological Disorders*, pp. 85–86.

901 / What is meant by the term "orphan drugs"?

Orphan drugs are intended to treat diseases that affect fewer than 200 thousand Americans. With little chance of making money, a drug company is not likely to undertake the necessary research and expense of finding drugs which might treat these diseases. The Orphan Drug Act of 1983 offers a number of incentives to drug companies to encourage development of these drugs. The act has provided hope for millions of people with rare and otherwise untreatable conditions.

Sources: *Current Health*, vol. 16, no. 8 (April 1990), p. 14.

902 / How is patient-controlled analgesia (PCA) administered?

This a drug delivery system that dispenses a preset intravenous (IV) dose of a narcotic analgesic for reduction to pain, when the patient pushes a switch on an electric cord. The device consists of a computerized pump with a chamber containing a syringe holding up to 60 milliliters of a drug. The patient administers a dose of narcotic when the need for pain relief arises. A lockout interval device automatically inactivates the system if the patient tries to increase the amount of narcotic within a preset time period.

Sources: *Mosby's Medical, Nursing, and Allied Health Dictionary*, p. 888.

903
How is the drug Clomid® used to treat infertility?

In female infertility, Clomid® acts by stimulating gonadotropin release, which then stimulates ovulation. In men, the drug is used to stimulate sperm production. Multiple births may occur as a result of the treatment.

Sources: *The American Medical Association Encyclopedia of Medicine*, p. 284; *Current Medical Diagnosis and Treatment 1991*, p. 524.

904
What is pharmacognosy?

It is the science of natural drugs and their physical, botanical, and chemical properties.

Sources: *Taber's Cyclopedic Medical Dictionary*, p. 1380.

905
Which tests detect cocaine use?

Urine tests detect cocaine used by a person within the past 24 to 36 hours. Hair analysis can detect cocaine used more than a year ago because hair grows only about 0.5 inch (1.27 centimeters) per month.

Sources: *U.S.A. Today*, (December 15, 1989), p. A-1.

906
Where is marijuana grown legally in the United States?

The federal marijuana farm at the University of Mississippi was established in 1968. This farm supplies most of the marijuana (*Cannabis sativa*) used for official medical research in America.

Sources: Wallace, Irving. *The Book of Lists #2*, p. 88.

907
How long do chemicals from marijuana stay in the body?

When marijuana is smoked, *tetra*hydro*c*annabinol (THC), its active ingredient, is absorbed primarily in the fat tissues. The body transforms the THC into metabolites which can be detected by urine tests for up to a week. Tests involving radioactively labeled THC have traced the metabolites for up to a month. The retention of labeled THC in humans is about 40% at three days and 30% at one week.

Sources: U.S. Department of Health and Human Services. *Marijuana*, unpaged.

908
Which drugs are prescribed the most?

Brand name	Therapy
Zantac®	Treats ulcers
Cardizem®	Treats cardiovascular problems
Tagamet®	Treats ulcers
Vasotec®	Treats cardiovascular problems
Ceclor®	Antibiotic
Mevacor®	Lowers serum cholesterol
Naprosyn®	Anti-arthritic drug
Capoten®	Treats cardiovascular problems
Tenormin®	Treats cardiovascular problems
Procardia®	Treats cardiovascular problems

Sources: *American Health*, vol. 10 (January-February 1991), p. 16.

909
Who discovered the antibiotic streptomycin?

The Russian–born microbiologist Selman A. Waksman (1888–1973) coined the term antibiotic and subsequently discovered streptomycin in 1943. In 1944 Merck and Company agreed to produce it to be used against tuberculosis and tuberculosis meningitis. Because the drug can adversely affect hearing and balance, careful control of dosage is needed.

Sources: *Dictionary of American Medical Biography*, vol. 2, pp. 769–770; Harris, Harry. *Good Old–Fashioned Yankee Ingenuity*, pp. 279–280.

910
How does RU-486, the abortion pill, cause a woman to abort a fetus?

A pill containing RU-486 (mifepristone) causes the abortion of a fetus by providing the fertilized embryo with an incompatible uterine environment to terminate the pregnancy within 45 days of fertilization. It is currently used in China and France.

Sources: Graedon, Joe, and Dr. Teresa Graedon. *Graedon's Best Medicine*, p. 355; *Science*, vol. 246 (November 24, 1989), p. 994.

911
Who discovered penicillin?

British bacteriologist Sir Alexander Fleming (1881–1955) discovered the bacteria–killing property of penicillin in 1928. Fleming noticed that no bacteria grew around bits of the *penicillium notatum* fungus that accidentally fell into a bacterial culture in his laboratory. However, although penicillin was clinically used, it was not until 1941 that Dr. Howard Florey (1898–1968) purified and tested it. The first large-scale plant to produce penicillin was constructed under the direction of Dr. Ernest Chain (1906–1979). By 1945, penicillin was commercially available. In that year, Chain, Florey, and Fleming received a joint Nobel Prize for their work on penicillin.

Sources: Robertson, Patrick. *The Book of Firsts*, pp. 124-125; Sanders, Dennis. *The First of Everything*, p. 381.

912 *Who developed the poliomyelitis vaccine in America?*

Immunologist Jonas E. Salk (b. 1914) developed the first vaccine (made from a killed-virus) against poliomyelitis. In 1952 he prepared and tested the vaccine and in 1954 massive field tests were successfully undertaken. Two years later immunologist Dr. Albert Sabin (b. 1906) developed an oral vaccine made from inactivated live viruses of three polio strains.

Sources: *Encyclopedia Americana*, vol. 24, p. 154; Harris, Harry. *Good Old-Fashioned Yankee Ingenuity*, p. 227.

913 *How long can a prescription drug be kept?*

Generally a prescription drug should not be more than one year old. Some over-the-counter medications have an expiration date on their box or container. A cream should not be used if it has separated into its components. Although some general guidelines are given below, "When in doubt, throw it out."

Remedy	Maximum shelf life (years)
Cold tablets	1 to 2
Laxatives	2 to 3
Minerals	6 or more
Nonprescription painkiller tablets	1 to 4
Prescription antibiotics	2 to 3
Prescription antihypertension tablets	2 to 4
Travel sickness tablets	2
Vitamins (protected from heat, light, and moisture)	6 or more

Sources: *New York Public Library Desk Reference*, p. 616.

914 *How many of the medications used today are derived from plants?*

Of the more than 250 thousand known plant species, less than 1% have been thoroughly tested for medical applications. Yet out of this tiny portion have come 25% of our prescription medicines. The United States National Cancer Institute has identified three thousand plants from which anti-cancer drugs are or can be made. This includes ginseng (*Panax quinquefolius*), Asian mayapple (*Podophyllum hexandrum*), western yew (*Taxus brevifolia*), and rosy periwinkle. 70% of these three thousand come from rain forests, which also are a source of countless other drugs for diseases and infections. Rain forest plants are rich in so-called secondary metabolites, particularly alkaloids, which biochemists believe the plants produce to protect them from disease and insect attack.

Sources: *McGraw-Hill Yearbook of Science and Technology 1991*, pp. 202–203; *National Wildlife*, vol. 30, No. 3 (April/May 1992), p. 44; *New Scientist*, vol. 131, no. 1782 (August 1991), pp. 34–39.

915
What are the common medication measures?

Approximate equivalents of apothecary measures are given below.

Apothecary volume

Volume	Equivalent
1 minim	0.06 milliliters or .02 fluid drams or 0.002 fluid ounces
1½ minims	0.1 milliliters
15 minims	1 milliliter
480 minims	1 fluid ounce
1 dram	3.7 milliliters or 60 minims
1 t (teaspoon)	60 drops
3 t (teaspoons)	½ ounce
1 T (tablespoon)	½ ounce
2 T (tablespoons)	1 ounce
1 C (cup)	8 ounces or 30 milliliters

Apothecary weights

Weight	Equivalent
1 grain	60 milligrams or ½ dram
60 grains	1 dram or 3.75 grams
8 drams	1 ounce or 30 grams

Sources: Hegstad, Lorrie N. *Essential Drug Dosage Calculations*, p. 23; Solomon, Eldra Pearl, and Gloria A. Phillips. *Understanding Human Anatomy and Physiology*, p. 369.

FIRST AID, POISONS, ETC.

916
Which first aid measures can be used for a bite by a black widow spider?

The black widow spider (*Latrodectus mactans*) is common throughout the United States. Its bite is severely poisonous, but no first aid measures are of value. Age, body size, and degree of sensitivity determine the severity of symptoms, which include an initial pinprick with a dull numbing pain, followed by swelling. An ice cube may be placed over the bite to relieve pain. Between 10 and 40 minutes after the bite, severe abdominal pain and rigiditiy of stomach muscles develops. Muscle spasms in the extremities, ascending paralysis, and difficulty in swallowing and breathing follow. The mortality rate is less than 1%; but anyone who has been bitten should see a doctor; the elderly, infants, and those with allergies are most at risk, and should be hospitalized.

Sources: *Diseases and Disorders Handbook*, pp. 412-416; *Merck Manual of Diagnosis and Therapy*, p. 2572.

Black widow spider

▽ **917** ▽ How can a tick be removed?

A tick, common in woods and forests througout the United States, fastens itself to a host with its teeth, then secretes a cementlike material to reinforce its hold. This flat brown speckled 0.25-inch insect can transmit diseases such as Rocky Mountain spotted fever and Lyme disease. To remove a tick cover it with mineral, salad, or machine oil to block its breathing pores. If it does not disengage after about 30 minutes, use tweezers and grasp the tick as close to the skin as possible. Pull it away with a steady pressure, or lift the tick slightly upward and pull it parallel to the skin until the tick detaches. Do not twist or jerk the tick. Wash the bite site and hands well with soap and water and apply alcohol. If necessary apply a cold pack to reduce pain.

Sources: *Diseases and Disorders Handbook*, pp. 412-413; Thygerson, Alton L. *First Aid Essentials*, pp. 115-116.

▽ **918** ▽ Have childhood deaths from poisoning decreased since the introduction of childproof containers?

After childproof packaging was required on all drugs and medications beginning in 1973, the childhood poisoning death rate declined dramatically. A 50% decrease was noted in 1973–1976 and the decline has continued. Other factors in this decline have been the development of poison control centers, changes in products to reduce poisonous agents, and the introduction of single dose packages.

Sources: Baker, Susan P., et al. *The Injury Fact Book*, pp. 207–208.

919 / Which part of mistletoe is poisonous?

The white berries contain toxic amines, which cause acute stomach and intestinal irritation with diarrhea and a slow pulse. Mistletoe should be considered a potentially dangerous Christmas decoration, especially if children are around.

Sources: Blackwell, Will H. *Poisonous and Medicinal Plants*, pp. 241-242.

920 / How deadly is strychnine?

The fatal dose of strychnine or deadly nightshade (the plant from which it is obtained) is 15 to 30 milligrams. It causes severe convulsions and respiratory failure. If the patient lives for 24 hours, recovery is probable.

Sources: Dreisbach, Robert H. *Handbook of Poisoning*, pp. 418-419.

921 / What is the poison on arrows used by South American Indians to kill prey and enemies?

The botanical poison used by the Aucas and similar tribes in the South American jungles is curare. It is a sticky, black mixture with the appearance of liquorice and is processed from either of two different vines. One is a liana (*Chondodendron tomentosum*); the other is a massive tree-like vine (*Strychonos quianensis*).

Sources: Ricciuti, Edward R. *The Devil's Garden: Facts and Folklore of Perilous Plants*, pp. 119-120.

922 / How deadly is Amanita phalloides?

The poisonous mushroom *Amanita phalloides* has a fatality rate of about 50%. Ingestion of part of one mushroom may be sufficient to cause death. Over 100 fatalities occur each year from eating poisonous mushrooms, with more than 90% caused by the *Amanita phalloides* group.

Sources: Dreisbach, Robert H. *Handbook of Poisoning*, pp. 504-505; Parker, Sybil P. *Synopsis & Classification of Living Organisms*, vol. 1, p. 254.

923
What is the Heimlich maneuver?

This effective first-aid technique to resuscitate choking and drowning victims was introduced by Dr. Henry J. Heimlich (b. 1920) of Xavier University, in Cincinnati, Ohio. It is a technique for removing a foreign body from the trachea or pharynx where it is preventing flow of air to the lungs. When the victim is in the vertical position, the maneuver consists of applying subdiaphragmatic pressure by wrapping one's arms around the victim's waist from behind, making a fist with one hand and placing it against the victim's abdomen between the navel and the rib cage, clasping one's fist with the other hand and pressing in with a quick, forceful thrust. Repeat several times if necessary. When the victim is in the horizontal position (which some experts recommend), the rescuer straddles the victim's thighs.

Sources: Harris, Harry. *Good Old-Fashioned Yankee Ingenuity*, pp. 143-144; *Taber's Cyclopedic Medical Dictionary*, pp. 793-795.

DOING THE HEIMLICH HUG WITH VICTIM STANDING OR SITTING

DOING THE HEIMLICH HUG WITH VICTIM LYING ON HIS BACK

924
How is activated charcoal used medically?

Activated charcoal is an organic substance, such as burned wood or coal, that has been heated to approximately 1,000°F (537°C) in a controlled atmosphere. The result is a fine powder containing thousands of pores that have great absorbent qualities to rapidly absorb toxins and poisons. Activated charcoal is used medically in the treatment of drug overdoses and poisonings.

Sources: Tichy, William. *Poisons: Antidotes and Anecdotes*, p. 69.

925
How long does it take to bleed to death?

Serious bleeding requires immediate attention and care. If a large blood vessel is severed or lacerated, a person can bleed to death in one minute or less. Rapid loss of one quart or more of the total blood volume often leads to irreversible shock and death.

Sources: Thomas, Lowell J. *First Aid for Backpackers and Campers*, p. 7; Thygerson, Alton L. *First Aid Essentials*, p. 53.

926 / What is the deadliest natural toxin?

Botulinal toxin, produced by the bacterium *Clostridium botulinum*, is the most potent poison of humans. It has an estimated lethal dose in the bloodstream of 10^{-9} milligrams per kilogram. It causes botulism, a severe neuroparalytic disease that travels to the junctions of skeletal muscles and nerves, where it blocks the release of the neurotransmitter acetylcholine, causing muscle weakness and paralysis, and impairing vision, speech, and swallowing. Death occurs when the respiratory muscles are paralyzed; this usually occurs during the first week of illness. Mortality from botulism is about 25%.

Because the bacterium can form the toxin only in the absence of oxygen, canned goods and meat products wrapped in airtight casings are potential sources of botulism. The toxin is more likely to grow in low-acid foods, such as mushrooms, peas, corn, or beans rather than high-acid foods like tomatoes. However, some new tomato hybrids are not acidic enough to prevent the bacteria from forming the toxin. Foods being canned must be heated to a temperature high enough and for a long enough time to kill the bacteria present. Suspect food includes any canned or jarred food product with a swollen lid or can. Ironically, this dreaded toxin in tiny doses is being used to treat disorders that bring on involuntary muscle contractions, twisting, etc. The United States Food and Drug Administration (FDA) has approved the toxin for: treatment of strabismus (misalignment of the eyes), blepharospasm (forcible closure of eyelids), and hemifacial spasm (muscular contraction on one side of the face).

Sources: *Cecil Textbook of Medicine*, p. 1633; Hillman, Harold. *Kitchen Science*, pp. 267-268; *New Scientist*, vol. 128 (December 8, 1990), p. 24.

927 / How can the amount of lead in tap water be reduced in an older house having lead-containing pipes?

The easiest way is to let the tap run until the water becomes very cold before using it for human consumption. By letting the tap run, water that has been in the lead–containing pipes for awhile is flushed out. Also, cold water, being less corrosive than warm, contains less lead from the pipes. Lead (Pb) accumulates in the blood, bones, and soft tissues of the body as well as the kidneys, nervous system, and blood–forming organs. Excessive exposure to lead can cause seizures, mental retardation, and behavior disorders. Infants and children are particularly susceptible to low doses of lead and suffer from nervous system damage.

Another source of lead poisoning is old flaking lead paint. Lead oxide and other lead compounds were added to paints before 1950 to make the paint shinier and more durable. 14% of the lead ingested by humans comes from the seam soldering of food cans, according to the United States *Food* and *Drug* Administration (FDA). The FDA has proposed a reduction in this lead to 50% over the next five years. Improperly glazed pottery can be a source of poisoning. Acidic liquids such as tea, coffee, wine, and juice can break down the glazes so that the lead can leak out of the pottery. The lead is ingested little by little over a period of time. People can be exposed to lead in the air. Lead gasoline additives, nonferrous smelters, and battery plants are the most significant contributors of atmospheric lead emissions.

Sources: *Current Health 2*, vol. 18, no. 6 (February 1992), p. 29; Hunter, Linda Mason. *The Healthy Home*, pp. 108–110.

928 / How did lead contribute to the fall of the Roman Empire?

Some believe Romans from the period around 150 B.C.E. may have been victims of lead poisoning. Symptoms of lead poisoning include sterility, general weakness, apathy, mental retardation, and early death. The lead could have been ingested in water taken from lead-lined water pipes or from food cooked in their lead-lined cooking pots. It is thought that this inadvertent toxic food additive may have contributed to the decline and fall of the Roman Empire. However, this could be only speculation since only a few Roman emperors displayed derangement. If lead pipes contributed to lead poisoning, then many Romans should have exhibited the symptoms, and all the emperors that lived in the same dwelling should be equally susceptible.

Sources: Cazeau, Charles. *Science Trivia*, p. 122; Elkort, Martin. *The Secret Life of Food*, p. 115.

HEALTH CARE

929 / How many people visit a dentist regularly?

Only about half the population visit a dentist as often as once a year.

Sources: *Reader's Digest Consumer Adviser*, p. 301.

930 / What was the first medical college in the United States?

The College of Philadelphia Department of Medicine, now the University of Pennsylvania School of Medicine, was established on May 3, 1765. The first commencement was held June 21, 1768, when the first medical diplomas were presented to the ten members of the graduating class.

Sources: Kane, Joseph N. *Famous First Facts*, p. 383.

---

931 / Who was the first woman physician in the United States?

Elizabeth Blackwell (1821-1910), received her degree in 1849 from Geneva Medical College in New York. After overcoming many obstacles, she set up a small practice which expanded into the New York Infirmary for Women and Children, which featured an all-female staff.

Sources: Sanders, Dennis. *The First of Everything*, p. 156.

---

932 / What is the difference between the degrees doctor of dental surgery (DDS) and doctor of medical dentistry (DMD)?

The title depends entirely on the school's preference in terminology. The degrees are equivalent.

Sources: *Reader's Digest Consumer Adviser*, p. 302.

---

933 / What is the number of active physicians per 100 thousand population in the United States?

There were 223.4 active physicians per 100 thousand population in the United States in 1990. This figure is projected to rise to 247.8 per 100 thousand population by the year 2000.

Sources: *Universal Healthcare Almanac*, Table 7.2.1

---

934 / Which symbol is used to represent medicine?

The staff of Aesculapius has represented medicine since 800 B.C.E. It is a single serpent wound around a staff. The Caduceus, the twin-serpent magic wand of the god Hermes or Mercury, came into use after 1800 and is commonly used today. The serpent has traditionally been a symbol of healing, and it is an old belief that eating part of a serpent would bring the power of healing to the ingester. Early Greeks saw in the serpent regenerative powers expressed by the serpent's periodic sloughing of its skin and venerated the serpent. Later the Greek god of medicine, Asklepius, called Aesculapius by the Romans, performed his functions in the form of a serpent. Sometimes this god is represented in art as an old man with a staff, around which is coiled a serpent.

Sources: *The Journal of the American Medical Association*, vol. 253 (April 26, 1985), p. 2369; Whittick, Arnold. *Symbols, Signs and Their Meaning*, p. 258.

Symbol of Asclepius

Caduceus

935 / *Who is generally regarded as the father of medicine?*

Hippocrates (ca.460–ca.377 B.C.E.), a Greek physician, holds this honor. Greek medicine, previous to Hippocrates, was a mixture of religion, mysticism, and necromancy. Hippocrates established the rational system of medicine as a science, separating it from religion and philosophy. Diseases had natural causes and natural laws: they were not the "wrath of the gods." Hippocrates believed that the four elements (earth, air, fire, and water) were represented in the body by four body fluids (blood, phlegm, black bile, and yellow bile) or "humors." When they existed in harmony within the body, the body was in good health. The duty of the physician was to help nature to restore the body's harmony. Diet, exercise, and moderation in all things kept the body well, and psychological healing (good attitude toward recovery), bed rest, and quiet were part of his therapy. Hippocrates was the first to recognize that different diseases had different symptoms; and he described them in such detail that the descriptions generally would hold today. His descriptions not only included diagnosis but prognosis.

Sources: Downs, Robert B. *Landmarks in Science*, pp. 21-24.

936 / *What is the difference between homeopathy, osteopathic medicine, naturopathy, and chiropractic?*

Developed by a German physician, Christian F.S. Hahneman (1755-1843), the therapy of homeopathy treats patients with small doses of two thousand substances. Based on the principle that "like cures like," the medicine used is one that produces the same symptoms in a healthy person that the disease is producing in the sick person.

Chiropractic is based on the belief that disease results from the lack of normal nerve function. Relying on physical manipulation and adjustment of the spine for therapy, rather than on drugs or surgery, this therapy, used by the ancient Egyptians, Chinese, and Hindus, was rediscovered in 1895 by Daniel David Palmer (1845-1913).

Developed in the United States by Dr. Andrew Skill, osteopathic medicine recognizes the role of the musculoskeletal system in healthy function of the human body. The physician is fully licensed and uses manipulation techniques as well as traditional diagnostic and therapeutic procedures. Osteopathy is practiced as part of standard Western medicine.

Naturopathy is based on the principle that disease is due to the accumulation of waste products and toxins in the body. Practitioners believe that health is maintained by avoiding anything artificial or unnatural in the diet or in the environment.

Sources: *The American Medical Association Encyclopedia of Medicine,* pp. 270, 544, 716, 755; *The Guinness Book of Answers,* pp. 167-168.

▽ **937** ▽
What is the Hippocratic Oath?

An oath demanded of physicians who are entering practice, which can be traced back to the Greek physician and teacher, Hippocrates (ca.460–ca.377 B.C.E.). The oath reads as follows:

> I swear by Apollo the physician, by Aesculapius, Hygeia, and Panacea, and I take to witness all the gods, all the goddesses, to keep according to my ability and my judgement the following Oath:
>
> To consider dear to me as my parents him who taught me this art; to live in common with him and if necessary to share my goods with him; to look upon his children as my own brothers, to teach them this art if they so desire without fee or written promise; to impart to my sons and the sons of the master who taught me and the disciples who have enrolled themselves and have agreed to rules of the profession, but to these alone, the precepts and the instruction. I will prescribe regiment for the good of my patients according to my ability and my judgement and never to harm anyone. To please no one will I prescribe a deadly drug, nor give advice which may cause his death. Nor will I give a woman a pessary to procure abortion. But I will preserve the purity of my life and my art. I will not cut for stone, even for patients in whom the disease is manifest; I will leave this operation to be performed by practitioners (specialists in this art). In every house where I come

I will enter only for the good of my patients, keeping myself far from all intentional ill-doing and all seduction, and especially from the pleasures of love with women or with men, be they free or slaves. All that may come to my knowledge in the exercise of my profession or outside of my profession or in daily commerce with men, which ought not to be spread abroad, I will keep secret and will never reveal. If I keep this oath faithfully, may I enjoy life and practice my art, respected by all men and in all times; but if I swerve from it or violate it, may the reverse be my lot.

The oath varies slightly in wording among different sources.

Sources: *Medical Dictionary*, pp. 716-717.

938
What is the difference between an ophthalmologist, optometrist, and optician?

An ophthalmologist is a physician who specializes in care of the eyes. Ophthalmologists conduct examinations to determine the quality of vision and the need for corrective glasses or contact lenses. They also check for the presence of any disorders, such as glaucoma or cataracts. Ophthalmologists may prescribe glasses and contact lenses, medication, or perform surgery, as necessary.

An optometrist is a specialist trained to examine the eyes and to prescribe, supply, and adjust glasses or contact lenses. Because they are not physicians, optometrists may not prescribe drugs or perform surgery. An optometrist refers patients requiring these types of treatment to an ophthalmologist.

An optician is a person who fits, supplies, and adjusts glasses or contact lenses. Because their training is limited, opticians may not examine or test eyes or prescribe glasses or drugs.

Sources: *The American Medical Association Encyclopedia of Medicine*, pp. 745-746.

939
When was the first blood bank opened?

Several sites claim the distinction. Some sources list the first blood bank as opening in 1940 in New York City under the supervision of Dr. Richard C. Drew (1904–1950). Others list an earlier date of 1938 in Moscow at the Sklifosovsky Institute (Moscow's central emergency service hospital) founded by Professor Sergei Yudin. The term *blood bank* was coined by Bernard Fantus who set up a centralized storage depot for blood in 1937 at the Cook County Hospital in Chicago, Illinois.

Sources: Berliner, Barbara. *The Book of Answers*, p. 77; Robertson, Patrick. *The Book of Firsts*, p. 26.

940
Who was the father of modern medicine?

Thomas Sydenham (1624–1689), who was also called the English Hippocrates, reintroduced the Hippocratic method of accurate observation at the bedside, and recording of observations, to build up a general clinical description of individual diseases. He is also considered one of the founders of epidemiology.

Sources: *The Great Scientists*, vol. 11, p. 108.

941
How many nursing homes are there in the United States?

In 1990, there were 15,607 nursing homes in the United States; 11,486 of them were profit–making facilities and 4,121 were non–profit facilities.

Sources: *Universal Healthcare Almanac*, Table 9.3.1.

942
How many health maintenance organizations (HMOs) are there in the United States and has this number changed over the past decade?

In 1990 the total number of health maintenance organizations (HMOs) of all types was 556. This represented a 135% increase from 236 in 1980.

Sources: *Universal Healthcare Almanac*, Table 3.1.1.

HEALTH HAZARDS, RISKS, ETC. *See also: HEALTH AND MEDICINE— FIRST AID, POISONS, ETC.*

943
What are the leading causes of death for young persons?

For persons under the age of 21 accidents are the leading cause of death. Motor vehicle accidents account for almost half of all accidents involving this age group. However, the home appears to be the most dangerous place for infants (0 to 1 years). From birth to age one, the most prominent cause of accidental death is choking on food or objects placed in the mouth. Motor vehicle accidents rank second, and suffocation (from smothering by bedclothes, plastic bags, etc.) is the third leading cause. Fourth are fires, burns, etc.; the fifth is drowning. For children aged 1 to 4 years, the ranking is (1) motor vehicle accidents; (2) drowning; (3) fires and burns; (4) choking; and (5) falls. For ages 5 through 14, the order is (1) motor vehicle accidents; (2) drowning; (3) firearms; (4) poisoning; and (5) fires, burns, etc. From ages 15 to 24, motor vehicle accidents ranked first, drownings and poisoning tied for second and third place, and falls and fires, burns, etc,. tied for fourth and fifth positions.

Sources: *The American Medical Association Encyclopedia of Medicine*, p. 60; National Safety Council. *Accident Facts 1989*, p. 23; *The World Almanac and Book of Facts 1992*, p. 941.

944 / What are the leading causes of death in the United States?

Of the 2,162 thousand deaths in 1990, the leading cause of death was heart disease. Below are listed the four major causes of death in the United States:

Cause	Number	Percent of total deaths
Heart disease	725,010	33.5
Cancer	506,000	23.4
Cerebrovascular diseases	145,340	6.7
Accidents	93,550	4.3

Sources: *The World Almanac and Book of Facts 1992*, p. 951.

945 / Does cigarette smoking impair bone healing?

In addition to heart disease, a variety of cancers, and the harm it does the body's ability to heal skin wounds, cigarette smoking also slows down the healing time for bone fractures. A 5–centimeter bone gap, which would ordinarily heal in 10 months in a non–smoker, will take 15 months in a smoker.

Sources: *Science News*, vol. 141 (February 29, 1992), p. 133.

946 / What is the composition of cigarette smoke?

Cigarette smoke contains about four thousand chemicals. Carbon dioxide, carbon monoxide, methane, and nicotine are some of the major components, with lesser amounts of acetone, acetylene, formaldehyde, propane, hydrogen cyanide, toluene, and many others.

Sources: *Involuntary Smoking*, p. 6; *Mayo Clinic Family Health Book*, p. 409; *Prevention's Giant Book of Health Facts*, p. 198.

947 / Why does the risk of cancer diminish rapidly after one quits the cigarette smoking habit?

Exposure of a premalignant cell to a promoter converts the cell to an irreversibly malignant state. Promotion is a slow process, and exposure to the promoter must be sustained for a certain period of time. This requirement explains why the risk of cancer diminishes rapidly after one quits smoking; both cancer initiators and promoters appear to be contained in tobacco smoke.

Sources: Zakrzewski, Sigmund F. *Principles of Environmental Toxicology*, p. 69.

948 / What percentage of the American public are smokers?

Smokers constitute a minority and a shrinking one—apparently around 30% of the adult American population (1985). Smoking has only recently been categorized as harmful to health. Studies have linked cigarette smoking to lung cancer, pipe and cigar smoking to oral cavity cancers and to respiratory diseases such as chronic bronchitis and emphysema and coronary heart disease.

Sources: *Clearing the Air: Perspectives on Environmental Tobacco Smoke*, p. 89.

949 / Which "risk factors" affects one's health?

Such characteristics as age, gender, work, family history, behavior, and body chemistry are some of the risk factors to consider when deciding whether one is at risk for various conditions. Some risk factors are statistical, describing trends among large groups of people but not giving information about what will happen to individuals. Other risk factors might be described as causative—exposure to them has a direct effect on whether or not the person will become sick.

Sources: Ringler, Carol Ann. *Are You At Risk?*, p. XIII.

950 / What are the odds against being struck by lightning?

606,944 to 1 against.

Sources: *The Odds on Virtually Everything*, p. 181.

951 / Which sport has the highest rate of injuries and what kind of injury is most common?

Football players suffer more injuries than other athletes, collectively. They have 12 times as many injuries as do basketball players, who have the next highest rate of injury. Knee problems are the most common type of injury. Two-thirds of basketball players' injuries and one-third of football players' injuries are knee-related.

Sources: *Prevention's Giant Book of Health Facts*, p. 215.

952

What are the odds on being killed on a motorcycle?

1,250 to 1 against.

Sources: *The Odds on Virtually Everything*, p. 162.

953

Which direction of impact results in the greatest number of fatalities in automobile crashes?

Frontal crashes are responsible for the largest percentage of fatalities in passenger cars.

Sources: Baker, Susan P., et al. *The Injury Fact Book*, p. 235.

Car impact in fatal accidents

954

Did raising the speed limit on rural interstate highways from 55 to 65 miles per hour have an effect on the accident and death rate?

There was an estimated 20% to 30% increase in deaths on those roads and a 40% increase in serious injuries when the speed limit was raised from 55 miles (88.5 kilometers) per hour to 65 miles (104.5 kilometers) per hour on rural interstate highways.

Sources: Baker, Susan P., et al. *The Injury Fact Book*, p. 251.

Are men or women more accident-prone?

Women drive and even cross the street more safely than men. Men account for 70% of pedestrian fatalities since 1980. Between the ages of 18 and 45, males outnumber females as fatal crash victims by almost three to one according to the National Highway Traffic Safety Administration. Accidental deaths of all types—from falls, firearms, drownings, fires, even food and other poisonings—also are more common among men than women.

Sources: *McCall's*, vol. 119, no. 3 (December 1991), p. 20.

What are the leading causes of stress?

In 1967, when they conducted a study of the correlation between significant life events and the onset of illness, Dr. Thomas H. Holmes and Dr. Richard H. Rahe from the University of Washington compiled a chart of the major causes of stress, with assigned point values. They published their findings on stress effects as "The Social Readjustment Scale," printed in *The Journal of Psychosomatic Research*. The researchers calculated that a score of 150 points indicated a 50–50 chance of the respondent developing an illness or a "health change." A score of 300 would increase the risk to 90%. Of course, many factors enter into an individual's response to a particular event, so this scale, partially represented below, can only be used as a guide.

Event	Point value
Death of spouse	100
Divorce	73
Marital separation	65
Jail term or death of close family member	63
Personal injury or illness	53
Marriage	50
Fired at work	47
Marital reconciliation or retirement	45
Pregnancy	40
Change in financial state	38
Death of close friend	37
Mortgage over $10,000	31
Foreclosure of mortgage or loan	30
Outstanding personal achievement	28
Trouble with boss	23
Change in work hours or conditions or change in residence or schools	20
Vacation	13
Christmas	12
Minor violations of the law	11

Sources: Barnard, Christiaan. *The Body Machine*, p. 166; *The World Almanac and Book of Facts 1992*, p. 253.

957
Can ozone be harmful to humans?

Ozone (O_3) in the lower atmosphere contributes to air pollution. It is formed by chemical reactions between sunlight and oxygen in the air in the presence of impurities. Such ozone can damage rubber, plastic, and plant and animal tissue. Exposure to certain concentrations can cause headaches, burning eyes, and irritation of the respiratory tract in many individuals. Asthmatics and others with impaired respiratory systems are particularly susceptible. Exposure to low concentrations for only a few hours can significantly affect normal persons while exercising. Symptoms include chest pain, coughing, sneezing, and pulmonary congestion.

Sources: *McGraw–Hill Encyclopedia of Science and Technology*, vol. 12, pp. 637-638; *World Book Encyclopedia*, vol. 14, p. 902.

958
How does the United States Environmental Protection Agency (EPA) classify carcinogens?

A carcinogen is an agent that can produce cancer (a malignant growth or tumor that spreads throughout the body, destroying tissue). The EPA classifies chemical and physical substances by their toxicity to humans.

EPA classification system for carcinogens

Group A. Human carcinogen

This classification indicates that there is sufficient evidence from epidemiological studies to support a cause–effect relationship between the substance and cancer.

Group B. Probable human carcinogen

B_1: Substances are classified as B_1 carcinogens on the basis of sufficient evidence from animal studies, and limited evidence from epidemiological studies.

B_2: Substances are classified as B_2 carcinogens on the basis of sufficient evidence from animal studies, with inadequate or nonexistent epidemiological data.

Group C. Possible human carcinogen

For this classification, there is limited evidence of carcinogenicity from animal studies and no epidemiological data.

Group D. Not classifiable as to human carcinogenicity

The data from human epidemiological and animal studies are inadequate or completely lacking, so no assessment as to the substance's cancer–causing hazard is possible.

Group E. Evidence of noncarcinogenicity for humans

Substances in this category have tested negative in at least two adequate (as defined by the EPA) animal cancer tests in different species and in adequate epidemiological and animal studies. Classification in group E is based on available evidence; substances may prove to be carcinogenic under certain conditions.

Sources: Harte, John. *Toxics A to Z*, p. 32.

▽ **959** ╱
Is there more violence on the streets and in mental institutions when there is a full moon?

A review of 37 studies that attempted to correlate the phases of the moon with violent crime, suicide, crisis center hotline calls, psychiatric disorders, and mental hospital admissions found that there is absolutely no relation between the moon and the mind.

Sources: *Prevention's Giant Book of Health Facts*, p. 87.

▽ **960** ╱
Can owning a pet be beneficial to your health?

As a result of several studies, researchers now believe that regular contact with pets can reduce heart rate, blood pressure, and levels of stress. In a study of 93 heart attack patients, only 1 of 18 pet owners died compared to 1 of 3 patients who did not have pets. Pets offer constancy, stability, comfort, security, affection, and intimacy.

Sources: Hunter, Linda Mason. *The Healthy Home*, pp. 222–227; *Prevention's Giant Book of Health Facts*, p. 433.

▽ **961** ╱
Do electric and magnetic fields produced by power transmission lines present a health hazard?

No study has produced evidence to allow a firm conclusion on this question. Scientists cannot agree on the significance of the inconsistent and puzzling findings. On electromagnetic fields, in general, a 1990 study by the United States Environmental Protection Agency (EPA) noted a *possible* statistically significant link between cancer and exposure to *extremely low frequency* (ELF) electromagnetic fields. ELF waves are nonionizing electromagnetic radiation that closely resemble the body's micropulsations. Biological studies have yet to prove such a connection. So far the ELF risks are controversial. Effects (from those studies supporting ELF hazards) run the gamut from headaches, to miscarriages, to cancer. In addition *electromagnetic radiation* (EMR) from devices such as electric blankets, video display terminals, microwaves, toasters, and hair dryers, powered by *alter-*

nating current (AC) power lines have all been targets of study to better understand the health hazards that might be posed.

Sources: Hunter, Linda Mason. *The Healthy Home*, pp. 100–101; *Science*, vol. 245 (July 21, 1989), p. 241; *Scientific American*, vol. 264 (March 1991), pp. 30–33.

▽ **962** /
Does using aluminum cooking utensils for food preparation have any effect on health?

This question has not yet been resolved. Studies indicate that patients with Alzheimer's disease show an increased level of aluminum in their brains. However, research scientists do not consider the evidence presently available to be strong enough to definitely state that aluminum plays a causal role in this disease.

Aluminum can be ingested from a variety of other sources—antacids, some baking powders, food additives such as potassium alum (the flour whitener) and aluminum calcium silicate (a table salt additive to keep salt free–running), and aluminum foil and anti–perspirants. Most of the aluminum taken into the body is excreted and the remainder is stored in the lungs, brain, liver, and thyroid gland. There are some aluminum–induced diseases. The aluminum content in the water in dialysis machines has induced dementia (a form of mental disorder) in some patients undergoing this treatment. Aluminum hydroxide in antacids can cause tiredness, weakness, and loss of appetite; aluminum chloride in anti–perspirants can cause allergic dermatitis (itching, inflamed skin).

Sources: *The American Medical Association Encyclopedia of Medicine*, pp. 89–90; *Harvard Health Letter*, vol. 15, no. 11 (October 1990), pp. 1–2.

▽ **963** /
Why is exposure to asbestos a health hazard?

Exposure to asbestos has long been known to cause asbestosis. This is a chronic, restrictive lung disease caused by the inhalation of tiny mineral asbestos fibers which scar lung tissues. Asbestos has also been linked with cancers of the larynx, pharynx, oral cavity, pancreas, kidneys, ovaries, and gastrointestinal tract. The American Lung Association reports that prolonged exposure doubles the likelihood that a smoker will develop lung cancer. It takes cancer 15 to 30 years to develop from asbestos. These fibers were used in building materials between 1900 and the early 1970s as insulation for walls and pipes, as fireproofing for walls and fireplaces, in soundproofing and acoustic ceiling tiles, as a strengthener for vinyl flooring and joint compounds, and as a paint texturizer. Asbestos poses a health hazard only if the tiny fibers are released into the air, but this can happen with any normal fraying or cracking. Asbestos removal aggravates this normal process and multiplies the danger level—it should only be handled by a contractor trained in handling asbestos. Once released, the particles can hang suspended in the air for more than 20 hours.

Sources: Hunter, Linda Mason. *The Healthy House*, pp. 80–85; Miller, E. Willard. *Environmental Hazards: Toxic Waste and Hazardous Material*, pp. 81–83.

964 / How does the blood alcohol level affect the body and behavior?

The effects of drinking alcoholic beverages depend on the amount of actual ethyl alcohol consumed and body weight. The level of alcohol in the blood is calculated in milligrams (1 milligram equals 1/30,000 of an ounce) of pure (ethyl) alcohol per deciliter (3.5 fluid ounces), commonly expressed in percentages.

Number of drinks	Blood alcohol level	Effect
1	0.02–.03%	Changes in behavior, coordination, and ability to think clearly
2	0.05%	Sedation or tranquilized feeling
3	0.08–0.10%	Legal intoxication in many states
5	0.15–0.20%	Person is obviously intoxicated and may show signs of delirium
12	0.30–0.40%	Loss of consciousness
24	0.50%	Heart and respiration become so depressed that they cease to function and death follows

Sources: *New York Public Library Desk Reference*, pp. 649–650; *Prevention's Giant Book of Health Facts*, p. 41.

965 / How is human exposure to radiation measured?

The *r*adiation *a*bsorbed *d*ose (rad) and the *r*oentgen *e*quivalent *m*an (rem) were used for many years to measure the amount and effect of ionizing radiation absorbed by humans. While officially replaced by the gray and the sievert, both are still used in many reference sources. The rad equals the energy absorption of 100 ergs per gram of irradiated material (an erg is a unit or work or energy). The rem is the absorbed dose of ionizing radiation that produces the same biological effect as 1 rad of x-rays or gamma rays (which are equal). The rem of x-rays and gamma rays is therefore equal to the rad; for each type of radiation, the number of rads is multiplied by a specific factor to find the number of rems. The millirem, 0.001 rems, is also frequently used; the average radiation dose received by a person in the United States is about 180 millirems per year.

In the SI system (*Système International d'Unités*, or International System of Units), the gray and the sievert are used to measure radiation absorbed; these units have largely superseded the older rad and rem. The gray (Gy), equal to 100 rads, is now the base unit. It is also expressed as the energy absorption of 1 joule per kilogram of irradiated material. The sievert (Sv) is the absorbed dose of radiation that produces the same biological effect as 1 gray of x-rays or gamma rays. The sievert is equal to 100 rems, and has superseded the rem. The becquerel (Bq) measures the radioactive strength of a source, but does not consider effects on tissue. One becquerel is defined as one disintegration (or other nuclear transformation) per second.

Sources: *The American Medical Association Encyclopedia of Medicine*, pp. 844–845.

966 / Why is radon a health hazard?

Radon is a colorless, odorless, tasteless radioactive gaseous element produced by the radioactive decay of radium. It has three naturally-occurring isotopes found in many natural materials, such as soil, rocks, well water, and building materials. Because the gas is continually released into the air, it makes up the largest source of radiation that humans receive. Some believe that radon may be a significant cancer cause, especially lung cancer. It has been estimated that it may cause as much as 10%, or 5 thousand to 20 thousand cases, of lung cancer deaths annually. Smokers seem to be at a higher risk than non–smokers. The *Environmental Protection Agency* (EPA) recommends that in radon testing, the level should not be more than 4 picocuries per liter. The estimated national average is 1.5 picocuries per liter. Because EPA's "safe level" is equivalent to 200 chest x–rays per year, some others believe that lower levels are appropriate. The *American Society of Heating, Refrigeration, and Air–Conditioning Engineers* (ASHRAE) recommends 2 picocuries/liter. The EPA estimates that nationally 8% to 12% of all houses are above the 4 picocuries/liter; whereas in another survey in 1987, it was estimated that 21% of homes were above this level.

Sources: Godish, Thad. *Indoor Air Pollution Control*, p. 34; Hunter, Linda Mason. *The Healthy Home*, p. 69.

967 / How much radiation does the average dental x–ray emit?

Dental examinations are estimated to contribute 0.15 millirems per year to the average genetically–significant dose, a small amount when compared to other medical x–rays.

Sources: Bomberger, Audrey S., and Betty A. Dannenfelser. *Radiation and Health*, p. 58.

968 / What is the estimated radiation dose to which the average person in the United States is exposed each year?

Of the approximately 160 millirems of radiation exposure per year, 50% comes from natural sources and 50% from human activities. Medical diagnosis and therapy account for more than 90% of the man–made dose. Cosmic rays, solar radiation, upper atmospheric radiation, and natural radioactivity are sources of ionizing radiation (which creates ions through ejection of electrons from atoms of the body's molecules, which may cause biological damage). Medical x–rays and radioactive isotopes are human sources of ionizing radiation.

Sources: *The American Medical Association Encyclopedia of Medicine*, pp. 844–845; *The Nuclear Waste Primer*, p. 17.

969 ▽ What is "good" and "bad" cholesterol?

Chemically a lipid, cholesterol is an important constituent of body cells. This fatty substance, produced mostly in the liver, is involved in bile salt and hormone formation, and in the transport of fats in the bloodstream to the tissues throughout the body. Both cholesterol and fats are transported around as lipoproteins (units having a core of cholesterol and fats in varying proportions with an outer wrapping of carrier protein (phospholoids and apoproteins)). An overabundance of cholesterol in the bloodstream can be an inherited trait, can be triggered dietary intake, or can be the result of a metabolic disease, such as diabetes mellitus. Fats (from meat, oil, and dairy products) strongly affect the cholesterol level. High cholesterol levels in the blood may lead to a narrowing of the inner lining of the coronary arteries from the build–up of a fatty tissue called atheroma. This increases the risk of coronary heart disease or stroke. However, if most cholesterol in the blood is in the form of *high density lipoproteins* (HDL), then it seems to protect against arterial disease. HDL picks up cholesterol in the arteries and brings it back to the liver for excretion or reprocessing. HDL is referred to as "good cholesterol." Conversely, if most cholesterol is in the form of *low density lipoproteins* (LDL), or *very low density lipoproteins* (VLDL), then arteries can become clogged. "Bad cholesterol" is the term used to refer to LDL and VLDL.

Sources: *The American Medical Association Encyclopedia of Medicine*, p. 275; Asimov, Isaac. *The Human Body*, pp. 179–183; Brennan, Richard B. *Dictionary of Scientific Literacy*, p. 137.

Normal Artery Wall

Artery Wall Thickened By Fatty Deposits

Normal Open Channel For Arterial Blood

Abnormal Narrowed Channel For Arterial Blood

Atheroma buildup

LIFE EXPECTANCY *See: THE HUMAN BODY—FUNCTIONS, PROCESSES, AND CHARACTERISTICS*

MEDICAL DIAGNOSIS *See: HEALTH AND MEDICINE—DIAGNOSTIC EQUIPMENT, TESTS, ETC.*

MEDICAL TESTS *See: HEALTH AND MEDICINE—DIAGNOSTIC EQUIPMENT, TESTS, ETC.*

NURSING HOMES *See: HEALTH AND MEDICINE—HEALTH CARE*

POISONS *See: HEALTH AND MEDICINE—FIRST AID, POISONS, ETC.*

SURGERY AND OTHER NON-DRUG TREATMENTS

▽ 970

What are the most frequent surgical procedures for which second opinions are most commonly advised?

Back surgery	Hernia repair
Breast surgery	Hip reconstruction
Bunion removal	Hysterectomy
Cataract removal	Knee surgery
Dilatation and curettage	Nasal septum repair
Gall bladder removal	Prostate removal
Heart surgery	Tonsillectomy
Hemorrhoid removal	Varicose vein removal

Sources: McCarthy, Eugene J., et al. *The Second Opinion Handbook*, p. 33.

▽ 971

Does catgut really come from cats?

Catgut, an absorbable sterile strand, is obtained from collagen derived from healthy mammals. It was originally prepared from the submucous layer of the intestines of sheep. It is used as a surgical ligature.

Sources: *Dorland's Illustrated Medical Dictionary*, p. 284.

972 / What is lithotripsy?

Lithotripsy is the use of ultrasonic or shock waves to pulverize kidney stones (calculi), allowing the small particles to be excreted or removed from the body. There are two different methods: *extracorporeal shock wave lithotripsy* (ESWL) and percutaneous lithotripsy. The ESWL method, used on smaller stones, breaks up the stones with external shock waves from a machine called a lithotripter. This technique has eliminated the need for more invasive stone surgery in many cases. For larger stones, a type of endoscope, called a nephroscope, is inserted into the kidney through a small incision. The ultrasonic waves from the nephroscope shatter the stones, and the fragments are removed through the nephroscope.

Sources: *The American Medical Association Encyclopedia of Medicine*, pp. 643–644.

973 / Why are eye transplants not available?

It is because the eye's retina is part of the brain, and the retina's cells are derived from brain tissue. Retinal cells and the cells that connect them to the brain are the least amenable to being manipulated outside the body.

Sources: *The New York Times* (December 17, 1991), p. B1.

974 / What is the longest operation on record?

The longest surgical operation was performed on Mrs. Gertrude Levandowski of Burnips, Michigan, for removal of an ovarian cyst. She lost 308 pounds during the operation; her weight was 616 pounds before it started. The operation lasted 96 hours and took place from Feb. 4 to 8, 1951 in Chicago, Illinois.

Sources: *Guinness Book of Records 1992*, p. 187.

975 / Who developed psychodrama?

Psychodrama was developed by Jacob L. Moreno, a psychiatrist born in Romania, who lived and practiced in Vienna until he came to the United States in 1925. He soon began to conduct psychodrama sessions and founded a psychodrama institute in Beacon, New York, in 1934. Largely because of Moreno's efforts it is practiced worldwide.

Psychodrama is defined as a method of group psychotherapy in which personality makeup, interpersonal relationships, conflicts, and emotional problems are explored by means of special dramatic methods.

Sources: *Comprehensive Textbook of Psychiatry*, vol. 2, p. 1531.

976
Who was the first Black surgeon to do heart surgery?

Dr. Daniel Hale Williams (1858–1931) was a pioneer in open heart surgery. In 1893, he was able to save a knifing victim by opening up the patient's chest with the help of a surgical team, exposing the beating heart. He sewed the knife wound a fraction of an inch from the heart without the aid of x-rays, blood transfusions, or anesthetics.

Sources: Carwell, Hattie. *Blacks in Science: Astrophysicist to Zoologist*, pp. 57–58.

977
How is a mustard plaster made?

Mix 1 tablespoon of powdered mustard and 4 tablespoons of flour. Add enough warm water to make a runny paste. Put the mixture in a folded cloth and apply to the chest. Add olive oil first to the skin if the patient has a delicate skin.

Sources: Buchman, Dian Dincin. *Dian Dincin Buchman's Herbal Medicine*, pp. 99–100.

978
Who received the first heart transplant?

On December 3, 1967, in Capetown, South Africa, Dr. Christiaan Barnard (b. 1922) and a team of 30 associates performed the first heart transplant. In a five–hour operation the heart of Denise Ann Darval, age 25, an auto accident victim, was transplanted into the body of Louis Washansky, a 55-year-old wholesale grocer. Washansky lived for 18 days before dying from pneumonia.

The first heart transplant performed in the United States was on a 2½-week-old baby boy at Maimonides Hospital, Brooklyn, New York, on December 6, 1967, by Dr. Adrian Kantrowitz (b. 1918). The baby boy lived 6½ hours. The first adult to receive a heart transplant in the United States was Mike Kasperak, age 54, at the Stanford Medical Center in Palo Alto, California, on January 6, 1968. Dr. Norman Shumway performed the operation. Mr. Kasperak lived 14 days. Since then until February 1990 there have been 7,831 heart transplants performed. In 1989, 1,621 transplants were done.

Sources: *Guinness Book of Records 1992*, p. 188; Sanders, Dennis. *The First of Everything*, p. 384.

979
What is the longest coma on record?

Elaine Esposito (1934–1978) of Tarpon Springs, Florida, went into a coma after appendix surgery at age six. She died after having been in a coma for 37 years, 111 days.

Sources: *Guinness Book of Records 1992*, p. 183.

980 When was the first artificial heart used?

On December 2, 1982, Dr. Barney B. Clark (1921–1983), a 61–year–old retired dentist, became the first human to receive a permanently implanted artificial heart. It was known as the Jarvik-7 after its inventor, Dr. Robert Jarvik (b. 1946). The 7½ hour operation was performed by Dr. William DeVries (b. 1943), a surgeon at the University of Utah Medical Center. Dr. Clark died on March 23, 1983, 112 days later. In Louisville, Kentucky, William Schroeder (1923–1986) survived 620 days with an artificial heart (November 25, 1984 to August 7, 1986). On January 11, 1990, the United States Food and Drug Administration (FDA) recalled the Jarvik-7, which had been the only artificial heart approved by the FDA for use.

Sources: *Guinness Book of Records 1992*, p. 189; Tortora, G.J. *Principles of Anatomy and Physiology*, p. 462.

981 What is reflexology?

Reflexology—an ancient Chinese and Egyptian therapy—is the application of specific pressures to reflex points in the hands and feet. The reflex points relate to every organ and every part of the body. Massaging of the reflex points is done to prevent or cure diseases. Dr. William Fitzgerald and Eunice D. Ingham introduced reflexology into the United States earlier this century.

Sources: Dougans, Inge. *Reflexology*, p. 3; *The Guinness Book of Answers*, p. 168.

982 What are the most common surgical procedures performed in the United States?

In 1990, these are the most-performed operations:

1. *Caesarean section* (with 909 thousand performed)
2. *Gall bladder removal* (with 498 thousand performed)
3. *Heart catheterization* (with 481 thousand performed)
4. *Hysterectomy* (with 437 thousand performed)

Sources: *USA Today* (October 19, 1991), p. D-1.

983 Who was the first test-tube baby?

Louise Brown, born on July 25, 1978, is the first baby produced from fertilization done in vitro—outside the mother's body. *In vitro fertilization* (IVF) occurs in a glass dish, not a test tube, where eggs from the mother's ovary are combined with the father's sperm (in a salt solution). Fertilization should occur within 24 hours and when cell division begins, these fertilized eggs are placed in the mother's womb (or possibly another woman's womb).

Sources: DuVall, Nell. *Domestic Technology*, p. 325; *How in the World*, pp. 276–277.

984

What is the most popular type of plastic surgery?

The 1988 statistics from the American Academy of Facial Plastic and Reconstructive Surgery indicate that the most popular plastic surgical procedure is rhinoplasty or a "nose job." Sometimes in conjunction with this surgery mentoplasty (chin implantation) is done as well.

Procedure	Number performed
Nose surgery (rhinoplasty)	99,358
Head/neck tumors	70,247
Head/neck trauma	43,012
Injectable fillers (zyderm)	42,529
Head/neck reconstructive	41,929
Eyelid surgery (blepharoplasty)	38,732
Scar revision	30,187
Facial fractures	26,711
Facelift	21,025
Face sanding (dermabrasion)	15,393

Sources: Krantz, Les. *The Best and Worst of Everything*, pp. 309–310.

985

What is a negative ion generator?

A negative ion generator is an electrostatic air cleaner which sprays a continuous fountain of negatively charged ions into the air. Some researchers say that the presence of these ions causes a feeling of well-being, increased mental and physical energy, and relief from some of the symptoms of allergies, asthma, and chronic headaches. It may also aid in healing of burns and peptic ulcers.

Sources: Greenfield, Ellen J. *House Dangerous*, pp. 104–105; Pearce, B.G. *Health Hazards of VDT's?*, p. 178.

986

How much does the cardiac pacemaker weigh?

The modern pacemaker generator is a hermetically-sealed titanium metal can weighing from 1 to 4.5 ounces (30 to 130 grams) and powered by a lithium battery that can last 2 to 15 years. Used to correct an insufficient or irregular heartbeat, the cardiac pacemaker corrects the low cardiac rhythm through electrical stimulation, which increases the contractions of the heart muscle. The contraction and expansion of the heart muscle produces a heartbeat, one of three billion in an average lifetime, which pumps blood throughout the body. A pacemaker has gold or platinum electrodes, conducting wires, and a pacing box (miniature generator). In both kinds of pacemakers—those implanted in the chest or external—the electrode is attached to the heart's right ventricle (chamber), either directly through the chest or threaded through a vein.

Sources: Sutton, Carol, and Duncan M. Anderson. *How Do They Do That?*, pp. 28–30; *Van Nostrand's Scientific Encyclopedia*, vol. 1, p. 220.

987
Prior to the condom, what was the main contraceptive practice?

Contraceptive devices have been used throughout recorded history. The most traditional of such devices was a sponge soaked in vinegar. The condom was named for its English inventor, the personal physician to Charles II (1630–1685), who used a sheath of stretched, oiled sheep intestine to protect the king from syphilis. Previously penile sheaths were used, such as the linen one made by Italian anatomist, Gabriel Fallopius (1523–1562), but they were too heavy to be successful.

Sources: Panati, Charles. *Panati's Extraordinary Origins of Everyday Things*, pp. 331–334; *Technology and Culture*, vol. 22, January 1981, p. 104.

988
How much do organ transplant operations generally cost in the United States and how successful are they?

Two American doctors, Joseph E. Murray and E. Donnall Thomas shared the 1990 Nobel Prize for Medicine for their breakthroughs in organ transplantation. Murray performed the first successful organ transplant in 1954 when he transferred a healthy kidney from one identical twin to the other, and Thomas pioneered bone marrow transplants as a cure for leukemia. Murray recognized that close genetic makeup helped to overcome the body's natural rejection process of transplanted organs. In 1967 cardiac patients with failing hearts were encouraged when Dr. Christian Barnard (b. 1922) performed the first successful human heart transplant. By 1970, however, the mortality rates of heart transplant patients were so high that leading medical centers were considering dropping such operations.

The primary difficulty was the refusal of the body's immune system to accept the organ and not treat it as a "foreign invader." The promise of transplants could not be realized until an effective immunosuppressant could be discovered. In 1979 Dr. Jean Borel of the Sandoz, Ltd., pharmaceutical company won a United States patent and clinical testing rights for his immunosuppressant cyclosporine (derived from soil fungal compound). Although there was some concern that the drug could induce T-lymphocytes in the immune system to produce a form of cancer called lymphoma, the drug was successful in increasing transplant survival rates. With the drug and improved techniques, kidney transplant survival rates went from 53%-83% to 73%-90% and one-year heart transplant survival rates went from 63% to 81%. One-year liver transplant survival percentages went from 32% to 54%-71%. By 1990 many once-difficult operations became relatively routine. For example, three hundred thousand cornea transplants have been done with a success rate of 90%.

Dr. Thomas Starzl developed FK-506, a new immunosuppressant, that is 50 to 100 times more powerful than cyclosporine. In addition organs now can be preserved in ViaSpan®, a DuPont organ preservative that keeps organs awaiting transplant viable for a day or longer; now time and distance are less serious obstacles to matching organs with potential recipients. On April 15, 1991, according to the United Network for Organ Sharing, there were 22,909 Americans waiting for donor organs: 18,366 for kidneys; 2,007 for hearts; 1,386 for livers; 570 for pancreas; 411 for lungs; and 169 for heart-lung. Below are listed some types of transplants, their costs, and success rates (as of 1988).

Organ	Number of transplants	Cost range	Mean cost	Success rate (%)
Heart	1,689	$50,000–287,000	$148,000	82.8
Heart-lung	67	135,000–250,000	210,000	56.9
Kidney	8,942	25,000–130,000	51,000	92–97
Liver	2,191	66,000–367,000	235,000	76.0
Lung	91	N.A.	225,000	48.4
Pancreas	418	51,000–135,000	70,000	89.1

Sources: Asimov, Isaac. *The Human Body*, pp. 317–322; *The Universal Almanac 1992*, p. 265.

989

What is synthetic skin?

The material consists of a very porous collagen fiber bonded to a sugar polymer (*glycoaminoglycan*) obtained from shark cartilage. It is covered by a sheet of silicon rubber. It was developed by Ioannis V. Yannas and his colleagues at Massachusetts Institute of Technology around 1895.

Sources: *The Almanac of Science and Technology*, pp. 192–193.

WEIGHT *See:* THE HUMAN BODY—FUNCTIONS, PROCESSES, AND CHARACTERISTICS

THE HUMAN BODY

BODY FLUIDS

990

What is the normal pH of blood, urine, and saliva?

Normal pH of arterial blood is 7.4; pH of venous blood is about 7.35. Normal urine pH averages about 6.0. Saliva has a pH between 6.0 and 7.4.

Sources: Guyton, Arthur C. *Textbook of Medical Physiology*, pp. 331, 340, 711.

991

How many miles of blood vessels are contained in the body?

If they could be laid end to end, the blood vessels would total about 60,000 miles (97,000 kilometers).

Sources: *World Book Encyclopedia*, vol. 2, p. 424.

992

Which of the major blood types are the most common in the United States?

Blood type	Frequency in U.S.
O+	37.4%
O-	6.6%
A+	35.7%
A-	6.3%
B+	8.5%
B-	1.5%
AB+	3.4%
AB-	0.6%

In the world, the preponderance of one blood group varies greatly by locality. Group O is generally the most common (46%), but in some areas Group A predominates.

Sources: *Current Health*, vol. 16 (October 1989), p. 12; *Guinness Book of Records 1992*, p. 179.

993 — How much blood is in the average human body?

A man weighing 154 pounds (69.8 kilograms) would have about 5.5 quarts of blood. A woman weighing 110 pounds (49.8 kilograms) would have about 3.5 quarts.

Sources: McAleer, Neil. *The Body Almanac*, p. 143.

994 — How does the body introduce oxygen to the blood and where does this happen?

Blood entering the right side of the heart (right auricle or atrium) contains carbon dioxide, a waste product of the body. The blood travels to the right ventricle, which pushes it through the pulmonary artery to the lungs. In the lungs, the carbon dioxide is removed and oxygen is added to the blood. Then the blood travels through the pulmonary vein carrying the fresh oxygen to the left side of the heart, first to the left auricle where it goes through a one-way valve into the left ventricle, which must push the oxygenated blood to all portions of the body (except the lungs) through a network of arteries and capillaries. The left ventricle must contract with six times the force of the right ventricle, so its muscle wall is twice as thick as the right.

Sources: Asimov, Isaac. *The Human Body*, pp. 156-166.

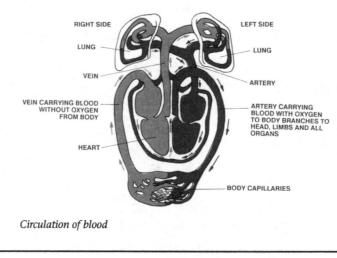

Circulation of blood

995
What are the blood group combinations that can normally be used to prove that a man is not the father of a particular child?

If the mother is	and the child is	the father can be	but not
O	O	O, A, or B	AB
O	A	A or AB	O or B
O	B	B or AB	O or A
A	O	O, A, or B	AB
A	A	any group	
A	B	B or AB	O or A
A	AB	B or AB	O or A
B	O	O, A, or B	AB
B	B	any group	
B	A	A or AB	O or B
B	AB	A or AB	O or B
AB	AB	A, B, or AB	O

Sources: Smith, Anthony. *The Body*, p. 249.

996
Which blood type is the rarest?

The rarest blood type is Bombay blood (subtype h-h), found only in a Czechoslovakian nurse in 1961 and in a brother and sister named Jalbert living in Massachusetts in 1968.

Sources: *Prevention's Giant Book of Health Facts*, p. 397.

997
What is the amount of carbon dioxide found in normal blood?

Carbon dioxide normally ranges from 19 to 25 mm per liter in arterial blood and 22 to 30 mm per liter in venous blood.

Sources: Pinckney, Cathey. *The Patient's Guide to Medical Tests*, p. 72.

998
What are the four humors of the body?

The four constituent humors of the body were identified as blood, phlegm, yellow bile, and black bile, originating in the heart, brain, liver, and spleen, respectively. Empedocles of Agrigentum (504-433 B.C.E.) probably originated the theory in which he equated the body fluids to the four elements of nature: earth, fire, air, and water. These humors could determine the health of the body and the personality of the person as well. To be in good health the humors should be in harmony within the body. Ill health could be remedied by treatments to realign the humors and reestablish the harmony.

Sources: Ackerknecht, Erwin H. *A Short Story of Medicine*, p. 47.

999
How similar are seawater and blood?

Component	Seawater (grams/liter)	Blood (grams/liter)
Na (sodium)	10.7	3.2–3.4
K (potassium)	0.39	0.15–0.21
Ca (calcium)	0.42	0.09–0.11
Mg (magnesium)	1.34	0.012–0.036
Cl (chloride)	19.3	3.5–3.8
SO_4 (sulfate)	2.69	0.16–0.34
CO_3 (carbonate)	0.073	1.5–1.9
Protein		70.0

Sources: Sherman, Irwin W. *Biology: A Human Approach*, p. 276.

BONES *See: THE HUMAN BODY—MUSCLES, BONES, AND NERVES*

DEATH, CAUSES OF *See: HEALTH AND MEDICINE—HEALTH HAZARDS, RISKS, ETC.*

EARS AND EYES *See: THE HUMAN BODY—SENSES AND SENSE ORGANS*

FUNCTIONS, PROCESSES, AND CHARACTERISTICS

1000
What is cryonic suspension?

Cryonic suspension, the controversial process of freezing and storing bodies for later revival, has been practiced since the late 1960s. Ordinarily, people are placed in cryonic suspension only after they are pronounced legally dead. Suspended animation, as the long-term storage of humans is generally known, has been a topic of scientific speculation since the beginning of the modern era.

Sources: *Search for Immortality*, pp. 96-97.

1001
How much heat is lost through the head when a person is not wearing a hat?

Between 7% and 55% of total body heat can be lost through the head. The amount of blood going to the head is controlled by cardiac output and the harder the body works, the more blood is circulated to the head, which is quickly radiated away.

Sources: *Parents*, vol. 64 (December 1989), p. 242.

1002
Who were the original Siamese twins?

Chang and Eng (1811–1874) were the original Siamese twins. Although born in Siam (now Thailand), they were actually three-quarters Chinese.

Sources: Burnam, Tom. *The Dictionary of Misinformation*, p. 229.

1003
Why do people snore?

Snoring is produced by vibrations of the soft palate, usually caused by any condition that hinders breathing through the nose. It is more common while sleeping on the back.

Sources: *The American Medical Association Encyclopedia of Medicine*, p. 922.

1004
How loud can a snore sound?

Research has indicated that a snore can reach 69 decibels as compared to 70 to 90 decibels for a pneumatic drill.

Sources: McAleer, Neil. *The Body Almanac*, p. 178.

1005
What is REM sleep?

REM sleep is *r*apid *e*ye *m*ovement sleep. It is characterized by faster breathing and heart rates than NREM (*n*on*r*apid *e*ye *m*ovement) sleep. The eyes move rapidly, and dreaming, often with elaborate story lines, occurs. The only people who do not have REM sleep are those who have been blind from birth. REM sleep occurs in about 4 to 5 periods, varying from 5 minutes to about an hour, growing progressively longer as sleep continues.

Scientists do not understand why dreaming is important, but they think the brain is either cataloging the information it picked up during the day and throwing out the data it does not want, or is creating scenarios to work through situations causing emotional distress. Regardless of the reasons, most people who are deprived of sleep or dreams become disoriented, unable to concentrate, and may even have hallucinations.

Sources: *The American Medical Association Encyclopedia of Medicine*, p. 915; Rovin, Jeff. *Laws of Order*, p. 112.

1006 / Is it true that people need less sleep as they get older?

As a person ages, the time spent in sleeping changes. The following table shows how long a night's sleep generally lasts.

Age	Sleep time (in hours)
1-15 days	16-22
6-23 months	13
3-9 years	11
10-13 years	10
14-18 years	9
19-30 years	8
31-45 years	7.5
45-50 years	6
50+ years	5.5

Sources: Kendig, Frank, and Richard Hutton. *Life Spans*, p. 25.

1007 / What are endorphins?

They are chemical substances called polypeptides, produced in the brain. They act as opiates, in that they produce an absence to the sensation of pain by binding to opiate receptor sites involved in pain perception. This action increases the threshold for pain.

Sources: *Taber's Cyclopedic Medical Dictionary*, p. 593.

1008 / What is the antigen-antibody reaction?

The immune system has two main components: white blood cells and antibodies circulating in the blood. The antigen-antibody reaction forms the basis for this immunity. When an antigen (*antibody gen*erator), such as a harmful bacterium, virus, fungus, parasite, or other foreign substance invades the body, a specific antibody is generated to attack the antigen. The antibody is produced by beta-lymphocytes (B-cells) in the spleen or lymph nodes. An antibody may either destroy the antigen directly or it may "label" it so that a white blood cell (called a microphage, or scavenger cell) can engulf the foreign intruder. After a human has been exposed to an antigen, a later exposure to the same antigen will produce a faster immune system reaction. The necessary antibodies will be produced more rapidly and in larger amounts. Artificial immunization uses this antigen-antibody reaction to protect the human body from certain diseases, by exposing the body to a safe dose of antigen to produce effective antibodies as well as a "readiness" for any future attacks of the harmful antigen.

Sources: *Mayo Clinic Family Healthbook*, pp. 448-449; *Oxford Illustrated Encyclopedia of Invention and Technology*, p. 177; *Taber's Cyclopedic Medical Dictionary*, p. 113.

1009 / Who coined the term homeostasis?

Walter Bradford Cannon (1871–1945) who elaborated on Claude Bernard's (1813–1878) concept of the *milieu intérieur* (interior environment) used the term *homeostasis* to describe the body's ability to maintain a relative constancy in its internal environment.

Sources: *Dictionary of Scientific Biography*, vol. 15, suppl. 1, p. 73.

1010 / How do T cells differ from B lymphocytes?

T cells, responsible for dealing with most viruses, some bacteria and fungi, and for cancer surveillance, are one of the two main classes of lymphocytes. As one variety of white blood cells, lymphocytes are part of the body's immune system; the immune system fights invading organisms that have penetrated the body's general defenses. T lymphocytes, or T cells, compose about 80% of the lymphocytes circulating in the blood. They have been "educated" in the thymus to perform particular functions. Killer T cells are sensitized to multiply when they come into contact with antigens (foreign proteins) on abnormal body cells (cells that have been invaded by viruses, cells in transplanted tissue, or tumor cells). These killer T cells attach themselves to the abnormal cells and release chemicals (lymphokines) to destroy them. Helper T cells assist killer cells in their activities and control other aspects of the immune response. When B lymphocytes, which compose about 10% of total lymphocytes, contact the antigens on abnormal cells, the lymphocytes enlarge and divide to become plasma cells. Then the plasma cells secrete vast numbers of immunoglobulins or antibodies into the blood, which attach themselves to the surfaces of the abnormal cells, to begin a process that will lead to the destruction of the invaders.

Sources: *The American Medical Association Encyclopedia of Medicine*, pp. 655-656; *Current Medical Diagnosis and Treatment*, pp. 558-559.

1011

When a person swallows solid or liquid food, what prevents it from going down the windpipe?

Once food is chewed, voluntary muscles move it to the throat. In the pharynx (throat) automatic involuntary reflexes take over. The epiglottis closes over the larynx (voice box) which leads to the windpipe. A sphincter at the top of the esophagus relaxes, allowing the food to enter the digestive tract.

Sources: *The American Medical Association Encyclopedia of Medicine*, p. 958.

EPIGLOTTIS (OPEN)
TRACHEA (WINDPIPE)
ESOPHAGUS (FOODPIPE)

EPIGLOTTIS (CLOSED)
TRACHEA (WINDPIPE)

Epiglottis opened and closed

1012

Who were the doctor and patient involved in the first studies on digestion performed by direct observation of the patient's stomach?

Alexis St. Martin, a French Canadian, was accidentally wounded by a shotgun blast in 1822. Fortunately, Dr. William Beaumont (1785-1853), an Army surgeon, was nearby and began treatment of the wound immediately. St. Martin's recuperation lasted nearly three years, and the enormous wound healed except for a small opening leading into his stomach. A fold of flesh covered this opening; when this was pushed aside the interior of the stomach was exposed to view. Dr. Beaumont began a series of experiments and observations which formed the basis of our modern knowledge of digestion. Today, the use of x-rays and other medical instruments provides the same diagnostic function.

Sources: Haggard, Howard W. *The Lame, the Halt, and the Blind*, pp. 365-371.

1013

How long does it take food to digest?

The stomach holds a little under two quarts (1.9 liters) of semi-digested food that stays in the stomach three to five hours. The stomach releases food slowly to the rest of the digestive tract. Fifteen hours or more after the first bite started down the alimentary canal the final residue of the food is passed along to the rectum and is excreted through the anus as feces.

Sources: McAleer, Neil. *The Body Almanac*, p. 186.

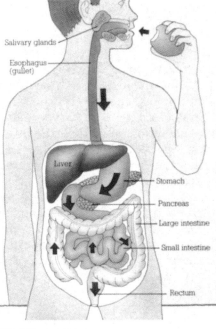

Digestion process

▽ **1014** ▽
Considering that 64% of all Americans are overweight, who is the heaviest person that ever lived?

John Brower Minnoch (1941–1983) of Bainbridge Island, Washington, weighed 976 pounds (443 kilograms) in 1976 and was estimated to have weighed more than 1,387 pounds (630 kilograms) when he was rushed to the hospital in 1978 with heart and respiratory failure. Much of his weight was due to fluid retention. After two years on a hospital diet, he was discharged at 476 pounds (216 kilograms). He had to be readmitted, however, after reportedly gaining 197 pounds (87 kilograms) in seven days. In 1983 when he died, he weighed 798 pounds (362 kilograms).

The heaviest woman ever recorded was Rosie Carnemolla (b. 1944) of Poughkeepsie, New York, who weighed 850 pounds (386 kilograms). When Mrs. Percy Pearl Washington, who suffered from polydipsia (excessive thirst), died in a Milwaukee hospital in 1972, the scales only registered 800 pounds (363 kilograms) maximum, but she was credited with weighing 880 pounds (400 kilograms).

Sources: *Guinness Book of Records 1992*, p. 156; *Prevention's Giant Book of Health Facts*, pp. 546–547.

1015
What are the types of human body shapes?

The best known example of body typing (classifying body shape in terms of physiological functioning, behavior, and disease resistance), was devised by American psychologist William Herbert Sheldon (1898–1977). Sheldon's system, known as somatotyping, distinguishes three types of body shapes, ignoring overall size: endomorph, mesomorph, and ectomorph. The extreme endomorph tends to be spherical: a round head, a large fat abdomen, weak penguinlike arms and legs, with heavy upper arms and thighs but slender wrists and ankles. The extreme mesomorph is characterized by a massive cubical head, broad shoulders and chest, and heavy muscular arms and legs. The extreme ectomorph has a thin face, receding chin, high forehead, a thin narrow chest and abdomen, and spindly arms and legs. In Sheldon's system there are mixed body types, determined by component ratings. Sheldon assumed a close relationship between body build and behavior and temperament. This system of body typing has many critics.

Sources: *Encyclopedia Americana*, vol. 4, p. 133.

SHELDON'S CLASSIFICATION OF BODY TYPES

ENDOMORPH   MESOMORPH   ECTOMORPH

1016
In the United States, what is the average height and weight for a man and a woman?

The average female is 5 feet 3¾ inches (1.62 meters) tall and weighs 135 pounds (61.24 kilograms). The average male is 5 feet 9 inches (1.75 meters) tall and weighs 162 pounds (73.48 kilograms). Between 1960 and 1990 the average

American male became two inches (5.08 centimeters) taller and 27 pounds (12.25 kilograms) heavier, while the average American woman grew two inches (5.08 centimeters) taller and gained one pound (0.45 kilograms).

Sources: Asimov, Isaac. *The Human Body*, p. 314; Diagram Group. *Comparisons*, pp. 72–73.

1017 / Who is called the father of physiology?

As an experimenter, Claude Bernard (1813–1878) enriched physiology by his introduction of numerous new concepts into the field. The most famous of these concepts is that of the *milieu intérieur* or internal environment. The complex functions of the various organs are closely interrelated and are all directed to maintaining the constancy of internal conditions despite external changes. All cells exist in this aqueous (blood and lymph) internal environment, which bathes the cells and provides a medium for the elementary exchange of nutrients and waste material.

Sources: Gascoigne, Robert M. *A Chronology of the History of Science 1450–1900*, p. 253; Guyton, Arthur C. *Textbook of Medical Physiology*, p. 3.

1018 / How many chromosomes are there in a human body cell?

A human being normally has 46 chromosomes (23 pairs) in all but the sex cells. One of each pair is inherited from the mother's egg; the other, from the father's sperm. When the sperm and egg unite in fertilization, they create a single cell, or zygote, with 46 chromosomes. When cell division occurs, the 46 chromosomes are duplicated and the process is repeated billions of times over, with each of the cells containing the identical set of chromosomes. Only the gametes, or sex cells, are different. In their cell division, the members of each pair of chromosomes are separated and distributed to different cells. Each gamete has only 23 chromosomes.

Chromosomes contain thousands of genes, which have information for a specific trait. That information is in the form of a chemical code, and the chemical compound that codes this genetic information is *deoxyribonucleic acid* or DNA. A gene can be seen as a sequence of DNA that is coded for a specific protein. These proteins determine specific physical traits (such as height, body shape, color of hair, eyes, skin, etc.), body chemistry (blood type, metabolic functions, etc.), and some aspects of behavior and intelligence. More than 150 human disorders are inherited, and genes are thought to determine susceptibility to many diseases.

Sources: Soloman, Eldra Pearl, and Gloria A. Phillips. *Understanding Human Anatomy and Physiology*, pp. 358–359; *World Book Encyclopedia*, vol. 3, p. 538.

1019

What are desirable weights for men and women?

Below is listed the desirable weight ranges for adult men (21 years and older) and women (25 years and older). Height is measured with shoes on: women with two–inch heels and men with one–inch heels. Weight is for clothed figures. For nude weight, subtract two to four pounds for women and five to seven pounds for men. Women 18 to 25 years old can use the table by subtracting one pound for each year under the age of 25.

Men

Height Feet Inches	Small frame	Medium frame	Large frame
5 2	112–120	118–129	126–141
5 3	115–123	121–133	129–144
5 4	118–126	124–136	132–148
5 5	121–129	127–139	135–152
5 6	124–133	130–143	138–156
5 7	128–137	134–147	142–161
5 8	132–141	138–152	147–166
5 9	136–145	142–156	151–170
5 10	140–150	146–160	155–174
5 11	144–154	150–165	159–179
6 0	148–158	154–170	164–184
6 1	152–162	158–175	168–189
6 2	156–167	162–180	173–194
6 3	160–171	167–185	178–199
6 4	164–175	172–190	182–204

Women

Height Feet Inches	Small frame	Medium frame	Large frame
4 10	92–98	96–107	104–119
4 11	94–101	98–110	106–122
5 0	96–104	101–113	109–125
5 1	99–107	104–116	112–128
5 2	102–110	107–119	115–131
5 3	105–113	110–122	118–134
5 4	108–116	113–126	121–138
5 5	111–119	116–130	125–142
5 6	114–123	120–135	129–146
5 7	118–127	124–139	133–150
5 8	122–131	128–143	137–154
5 9	126–135	132–147	141–158
5 10	130–140	136–151	145–163
5 11	134–144	140–155	149–168
6 0	138–148	144–159	153–173

Sources: Sherman, Irwin W. *Biology: A Human Approach*, p. 243.

▽ 1020 ▽

What is the average lifespan of cells in the human body?

Cell type	Length of time
Blood cells	
red blood cells	120 days
lymphocytes	Over 1 year
other white cells	10 hours
platelets	10 days
Bone cells	25–30 years
Brain cells	Lifetime
Colon cells	3–4 days
Skin cells	19–34 days
Spermatozoa	2–3 days
Stomach cells	2 days

Sources: Diagram Group. *Comparisons*, p. 175; *Guinness Book of Records 1992*, p. 180.

▽ 1021 ▽

How many cells are in the human body?

Sources give figures that vary from 50 to 75 trillion cells.

Sources: Asimov, Isaac. *The Human Body*, p. 79; *New Encyclopedia Britannica*, vol. 6, p. 134; Van Amerogen, C. *The Way Things Work Book of the Body*, p. 13.

▽ 1022 ▽

What chemicals constitute the human body?

About 24 elements are used by the body in its functions and processes

Major elements		
Element	Percentage	Function
Oxygen	65.0	Part of all major nutrients of tissues; vital to energy production
Carbon	18.5	Essential life element of proteins, carbohydrates, and fats; building blocks of cells
Hydrogen	9.5	Part of major nutrients; building blocks of cells
Nitrogen	3.3	Essential part of proteins, DNA, RNA; essential to most body functions
Calcium	1.5	Form nonliving bone parts; a messenger between cells
Phosphorous	1.0	Important to bone building; essential to cell energy

Potassium, sulfur, sodium, chlorine, and magnesium each occur at 0.35% or less. There are also traces of iron, cobalt, copper, manganese, iodine, zinc, fluorine, boron, aluminum, molybdenum, silicon, chromium, and selenium.

Sources: Brandreth, Gyles. *Your Vital Statistics*, p. 14.

1023
In addition to left or right–handedness, what other preferences do people have?

Most people have a preferred eye, ear, and foot. In one study, for example, 46% were strongly right–footed, while 3.9% were strongly left–footed; furthermore 72% were strongly right–handed and 5.3% strongly left–handed. However estimates vary about the proportion of left–handers to right–handers, but it may be as high as one in ten. Some 90% of healthy adults use the right hand for writing; two–thirds favor the right hand for most activities requiring coordination and skill. There are no male–female difference in these proportions.

Sources: *The American Medical Association Encyclopedia of Medicine*, p. 506; Brandreth, Gyles D. *Your Vital Statistics*, p. 84; Springer, Sally P. *Left Brain, Right Brain*, pp. 137, 140.

1024
What causes protruding ears?

If a fold paralleling a part of the outer roll of the ear is sparse, but not absent, the ear will protrude. This trait usually runs in certain families.

Sources: Simon, Gilbert. *The Parent's Pediatric Companion*, p. 72.

1025
How many calories does a person burn while sleeping?

A 150 pound (68 kilograms) person burns 1.0 calories per minute during bed rest. Approximate caloric values of other activities for a person weighing 150 pounds are given below. Values may vary, depending on the vigor of the exercise, air temperature, clothing, etc.

Activity	Calories used per hour
Basketball	500
Bicycling (5½ mph)	210
(13 mph)	660
Bowling	220–270
Calisthenics	300
Digging	360–420
Gardening	200
Golfing (using power cart)	150–220
(pulling cart)	240–300
(carrying clubs)	300–360
Football	500
Handball (social)	600–660
(competitive)	>660
Hoeing	300–360
Housework	180
Jogging (5–10 mph)	500–800

Activity	Calories used per hour
Lawn mowing (power)	250
(hand)	420–480
Raking leaves	300–360
Sitting	100
Skiing (cross–country)	600–660
(downhill)	570
Snow shoveling	420–480
Square dancing	350
Standing	140
Swimming moderately	500–700
Tennis (doubles)	300–360
(singles)	420–480
Vacuuming	240–300
Volleyball	350
Walking (2 mph)	150-240
(3.5 mph)	240-300
(4 mph)	300-400
(5 mph)	420-480

Sources: Agress, Clarence M. *Energetics*, p. 35; Brody, Jane E. *Jane Brody's Good Food Book*, p. 226; *New York Public Library Desk Reference*, p. 648.

1026 ▽ Who is the world's oldest person?

The greatest authenticated age to which any human has lived is 120 years 237 days in the case of Shigechiyo Izumi (1871-1986) of Japan.

Sources: *Guinness Book of Records 1992*, p. 166.

1027 ▽ How many calories does the average college student expend in a day?

The average college student weighing 115 pounds (52 kilograms) expends 2,093 calories a day. The total is derived from the total number of calories used per pound (listed below) multiplied by the weight of the student.

Activity	Hours spent in activity	Calories per pound per hour	Total calories used per pound
Asleep	8	0.4	3.2
Lying still, awake	1	0.5	0.5
Dressing and undressing	1	0.9	0.9
Studying in class, eating, studying, talking	8	0.7	5.6
Walking	1	1.5	1.5
Standing	1	0.8	0.8
Driving a car	1	1.0	1.0
Running	.05	4.0	2.0

Activity	Hours spent in activity	Calories per pound per hour	Total calories used per pound
Playing ping–pong	0.05	2.7	1.3
Writing	2	0.7	1.4
TOTAL	24	——	18.2

Sources: Sherman, Irwin W. *Biology: A Human Approach*, p. 80.

1028 / How much of the body remains after cremation?

The solids left after cremation are about 3% of the former body weight, so a corpse weighing 150 pounds (68 kilograms) produces about five pounds (2 kilograms), or about 200 cubic inches, of remains.

Sources: Phipps, William E. *Cremation Concerns*, p. 33.

HAIR *See:* THE HUMAN BODY—SKIN, HAIR, AND NAILS

IMMUNE SYSTEM *See:* THE HUMAN BODY—FUNCTIONS, PROCESSES, AND CHARACTERISTICS

MUSCLES, BONES, AND NERVES

1029 / How many muscles are in the human body?

There are about 656 muscles in the body, although some authorities make this figure even as high as 850 muscles. No exact figure is available because authorities disagree about which are separate muscles and which ones slip off larger ones. Also, there is a wide variability from one person to another, though the general plan remains the same.

There are three body systems in which muscles are used: the skeletal muscles move various parts of the body, are striped or striated fibers and are called voluntary muscles because the person controls their use. The second are smooth muscles found in the stomach and intestinal walls, vein and artery walls, and in various internal organs. Called involuntary muscles, they are not generally controlled by the person. The last are the cardiac muscles, or the heart muscles, containing striped and involuntary muscles.

Sources: *The American Medical Association Encyclopedia of Medicine*, pp. 703-706; Brandreth, Gyles D. *Your Vital Statistics*, p. 17; Cody, John. *Visualizing Muscles*, p. 5.

▽1030▽ What is an ecorche?

An ecorche is a flayed figure, a three-dimensional representation of the human body, usually made of plaster, with the envelope of skin and fat removed. Its intent is to depict the surface muscles with precise anatomical correctness.

Sources: Cody, John. *Visualizing Muscles*, p. 115.

Muscles of the body (ecorche)

▽1031▽ How many muscles does it take to produce a smile and a frown?

There are 17 smiling muscles. The average frown needs 43.

Sources: Brandreth, Gyles. *Your Vital Statistics*, p. 17; McAleer, Neil. *The Body Almanac*, p. 111.

How many bones are in the human body?

Babies are born with about 300 to 350 bones, but many of these fuse together between birth and maturity to produce an average adult total of 206. Bone counts vary according to the method used to count them, because some systems treat as multiple bones a structure that other systems treat as a single bone with multiple parts.

Location	Number
Skull	22
Ears (pair)	6
Vertebrae	26
Sternum	3
Throat	1
Pectoral girdle	4
Arms (pair)	60
Hip bones	2
Legs (pair)	58
Total	206

Sources: Brandreth, Gyles. *Your Vital Statistics*, p. 14; Solomon, Eldra Pearl, and Gloria A. Phillips. *Understanding Human Anatomy and Physiology*, pp. 77-80.

THE SKELETON

skull

7 cervical vertebrae
clavicle
scapula
sternum
ribs
12 thoracic vertebrae
humerus
5 lumbar vertebrae
radius
pelvis
sacrum
coccyx
ulna
femur
patella
tibia
fibula

1033

What is the most commonly broken bone?

The clavicle (collar bone) is one of the most frequently fractured bones in the body. Fractured clavicles are caused by either a direct blow or a transmitted force resulting from a fall on the outstretched arm.

Sources: Thygerson, Alton L. *First Aid Essentials*, p. 162.

1034

What are the hamstring muscles?

There are three hamstring muscles, located at the back of the thigh. They flex the leg on the thigh such as in the process of kneeling.

Sources: Memmler, Ruth L. *Structure and Function of the Human Body*, p. 73; Rothenberg, Mikel A. *Dictionary of Medical Terms for the Non-Medical Person*, p. 200.

1035

Why does excessive exercise cause muscles to become stiff and sore?

If muscles are worked too hard, the cells run out of oxygen. This starts a fermentation process which produces lactic acid. The build-up of lactic acid in the muscles causes soreness and stiffness.

Sources: Trefil, James. *1001 Things Everyone Should Know About Science*, p. 95.

1036

Why do humans have wisdom teeth?

Although they are useless to humans now, they must have had some purpose in the past. Nature rarely produces extraneous features. Perhaps when primitive man ate tough meats, the extra molars in the mouth, now known as wisdom teeth, helped to chew up the meat fibers. As humans evolved, the brain became larger and the face position moved further downward and inward. Also the protruding jawbones of early humans gradually moved backward, leaving no room for these third molars or wisdom teeth.

Sources: Feldman, David. *Why Do Clocks Run Clockwise? and Other Imponderables*, p. 137.

1037

What is the largest nerve in the body?

The sciatic nerve is the largest in the human body—about as thick as a lead pencil—0.78 inches (1.98 centimeters). It is a broad, flat nerve composed of nerve fibers, and it runs from the spinal cord down the back of the each leg.

Sources: McAleer, Neil. *The Body Almanac*, p. 79.

1038

What is the name of the small fleshy mass hanging from the back of the mouth?

The uvula is a small, soft structure, hanging from the free edge of the soft palate. It is composed of muscle, connective tissue, and mucous membrane.

Sources: *Taber's Cyclopedic Medical Dictionary*, pp. 1961-1962.

1039

How much force does a human bite generate?

All the jaw muscles working together can close the teeth with a force as great as 55 pounds (25 kilograms) on the incisors or 200 pounds (90.7 kilograms) on the molars. A force as great as 268 pounds (122 kilograms) for molars has been reported.

Sources: Best, Charles H. *Best and Taylor's Physiological Basis of Medical Practice*, p. 642; Guyton, Arthur C. *Basic Human Physiology*, p. 659.

NAILS *See: THE HUMAN BODY—SKIN, HAIR, AND NAILS*

NERVES *See: THE HUMAN BODY—MUSCLES, BONES, AND NERVES*

NUTRITION *See: FOOD AND NUTRITION—NUTRITION*

ORGANS AND GLANDS *See also: THE HUMAN BODY—SENSES AND SENSE ORGANS; THE HUMAN BODY—SKIN, HAIR, AND NAILS*

1040

What was the likely purpose of the human appendix?

Experts can only theorize on its use. It may have had the same purpose it does in present-day herbivores, where it harbors colonies of bacteria that help in the digestion of cellulose in plant material. Another theory suggests that tonsils and the appendix might manufacture the antibody-producing white blood cells called B-lymphocytes. However B-lymphocytes could also be produced by the bone marrow. The third theory is that the appendix may "attract" body infections to localize the infection in one spot that is not critical to body functioning. The earliest surgical removal of the appendix was by Claudries Amyand (1680-1740) in England in 1736.

Sources: Barnard, Christiaan. *The Body Machine*, p. 127; Feldman, David. *Do Penguins Have Knees?*, pp. 152-153; *Guinness Book of Records 1992*, p. 187.

1041 / Which gland is the largest?

The liver is the largest gland and the second largest organ after the skin. At 3 to 4 pounds (1.35 to 1.8 kilograms) the liver is seven times larger than it needs to be to perform its estimated 500 functions. It is the main chemical factory of the body. A ducted gland that produces bile to break down fats and reduce acidity in the digestive process, the liver is also a part of the circulatory system. It cleans poisons from the blood and regulates blood composition.

Sources: McAleer, Neil. *The Body Almanac*, p. 100.

1042 / What regulates body temperature in humans?

The hypothalamus gland controls internal body temperature. The gland responds to sensory impulses from temperature receptors in the skin and in the deep body regions. The hypothalamus establishes a "set point" for the internal body temperature, then constantly compares this with its own actual temperature. If the two do not match, the hypothalamus activates either temperature-decreasing or temperature-increasing procedures to bring them into alignment.

Sources: Guyton, Arthur C. *Textbook of Medical Physiology*, p. 802; Selkurt, Ewald E. *Physiology*, p. 169.

1043 / Are the lungs identical?

No, the right lung is shorter than the left by 1 inch (2.5 centimeters); however, its total capacity is greater. The right lung has three lobes, the left lung has two. The maximum capacity of the lung averages about 6,500 cubic centimeters or about 1.7 gallons (6.4 liters).

Sources: Asimov, Isaac. *The Human Body*, p. 145; Gray, Henry. *Anatomy of the Human Body*, p. 1148.

1044 / What is the basic unit of the brain?

Neurons are the nerve cells that are the major constituent of the brain. At birth the brain has the maximum number of neurons—20 billion to 200 billion neurons. Thousands are lost daily, never to be replaced and apparently not missed, until the cumulative loss builds up in very old age.

Sources: *The Human Brain*, pp. 16-17.

1045

Whose brain is larger: Neanderthal man or modern man?

The capacity of the skull of "classic" Neanderthal man was often larger than that of modern man. The capacity was between 1,350 to 1,700 cubic centimeters with the average being 1,400 to 1,450 cubic centimeters. The mean cranial capacity of modern man is 1,370 cubic centimeters, with a range of 950 to 2,200. However, brain size alone is not an index of intelligence.

Sources: Jelinek, Jan. *The Pictorial Encyclopedia of the Evolution of Man*, pp. 92-93; Kennedy, Kenneth A.R. *Neanderthal Man*, p. 40.

1046

What is the largest organ in the human body?

The largest and heaviest human organ is the skin, with a total surface area of about 20 square feet (1.9 square meters) for an average person or 25 square feet (2.3 square meters) for a large person and a weight of 5.6 pounds (2.7 kilograms) on the average. Although generally it is not thought of as an organ, medically it is. An organ is a collection of various tissues integrated into a distinct structural unit and performing specific functions. After the skin, the second largest and heaviest organ is the liver, weighing 2.5 to 3.3 pounds (1.1 to 1.5 kilograms). The liver is the body's chemical factory and regulates the levels of most of the body's chemicals.

Sources: *The American Medical Association Encyclopedia of Medicine*, p. 749; *Guinness Book of Records 1992*, pp. 183-184; McAleer, Neil. *The Body Almanac*, p. 100.

1047

What are the seven endocrine glands?

The major endocrine glands include the pituitary, thyroid, parathyroids, adrenals, pancreas, testes, and ovaries. These glands secrete hormones into the blood system, which generally stimulate some change in metabolic activity:

Pituitary—secretes ACTH to stimulate the adrenal cortex; produces aldosterone to control sodium and potassium reabsorption by the kidneys; FSH to stimulate gonad function and prolactin to stimulate milk secretion of breasts; TSH to stimulate thyroid gland to produce thyroxin; LH to stimulate ovulation in females and testerone production in males; GH to stimulate general growth. Stores oxytocin for uterine contraction.

Thyroid gland—secretes thyroxine (T_3) and triiodothyronine (T_4) to stimulate metabolic rate, especially in growth and development, and secretes calcitonin to lower blood-calcium levels.

Parathyroids—secrete hormone PTH to increase blood-calcium levels; activates vitamin D and stimulates calcium reabsorption in kidneys.

Adrenals—secrete epinephrine and norepinephrine to help the body cope with stress, raise blood pressure, heart rate, metabolic rate, raise blood sugar levels, etc. Aldosterone secreted by the adrenal cortex maintains sodium-phosphate balance in kidneys and cortisol helps the body adapt to stress, mobilizes fat, and raises blood sugar level.

Pancreas—secretes insulin to control blood sugar levels, stimulates glycogen production, fat storage, and protein synthesis. Glucagon secretion raises blood sugar level and mobilizes fat.

Ovaries and testes—secrete estrogens, progesterone, or testosterone to stimulate growth and reproductive processes.

Sources: *Science and Technology Illustrated*, vol. 9, p. 1136; Solomon, Eldra Pearl, and Gloria A. Phillips. *Understanding Human Anatomy and Physiology*, pp. 178-179.

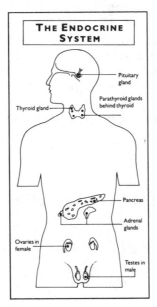

1048 / *How hard does the heart work?*

The heart squeezes out about 2½ ounces (70.8 grams) of blood at every beat. It daily pumps at least 2,500 gallons (9,450 liters) of blood, which weighs 20 tons (18,144 kilograms). On the average the adult heart beats 70 to 75 times a minute.

The rate of the heartbeat is determined in part by the size of the organism. Generally the smaller the size, the faster the heartbeat. Thus women's hearts beat 6 to 8 times a minute more than men's hearts do. At birth the heart of a baby can beat as fast as 130 times per minute.

Sources: Asimov, Isaac. *The Human Body*, p. 167; Brandreth, Gyles D. *Your Vital Statistics*, p. 53; Fisher, Arthur. *The Healthy Heart*, p. 11.

PHYSICAL CHARACTERISTICS See: THE HUMAN BODY—FUNCTIONS, PROCESSES, AND CHARACTERISTICS

SENSES AND SENSE ORGANS

1049 / How do colors affect one's moods?

According to the American Institute for Biosocial Research, "colors are electromagnetic wave bands of energy." Each color has its own wavelength. The wave bands stimulate chemicals in your eye, sending messages to the pituitary and pineal glands. These master endocrine glands regulate hormones and other physiological systems in the body. Stimulated by response to colors, glandular activities can alter moods, speed up heart rates, and increase brain activity.

Sources: *Health*, vol. 14 (July 1982), pp. 36-39.

1050 / What is the difference in the functions of the rods and cones found in the eyes?

Rods are concerned with light and dark distinction and cones with color perception.

Sources: *Encyclopedic Dictionary of Science*, p. 66.

1051 / Why do wheels appear to turn backwards in motion pictures?

This is due to the persistence of vision of the human eye. Motion pictures consist of a series of still pictures projected in rapid succession, usually at a speed of 24 frames per second. In row A of this drawing, each successive film frame shows the spokes of the wheel only slightly turned from its previous position. The eye fuses these steps into a forward motion. But if the car travels at a speed in which every successive frame shows the spokes turned almost full circle, as in row B, the eye fuses the different positions into backward motion.

Sources: *Encyclopedia Americana*, vol. 19, pp. 540-541; Strasser, Alex. *The Work of the Science Film Maker*, p. 35.

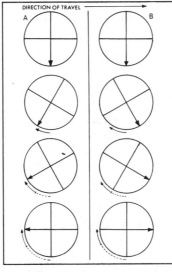

Persistence of vision

1052 / Why do all newborn babies have blue eyes?

The color of the iris gives the human eye its color. The amount of dark pigment, melanin, in the iris is what determines its color. In newborns the pigment is concentrated in the folds of the iris. When a baby is a few months old, the melanin moves to the surface of the iris and gives the baby his or her permanent eye color.

Sources: McAleer, Neil. *The Body Almanac*, p. 41.

1053 / Are more people nearsighted or farsighted?

About 30% of Americans are nearsighted to some degree. About 60% are farsighted. If the light rays entering the pupils of the eye converge exactly on the retina, then a sharply-focused picture is relayed to the brain. But if the eyeball is shaped differently, the focal point of the light rays is too short or too long, and vision is blurred. Convex lenses for farsightedness correct a too–long focal point. Concave lenses correct nearsightedness when the focal point of light rays, being too short, converge in front of the retina.

Sources: *Prevention's Giant Book of Health Facts*, p. 518.

1054

What are phosphenes?

If the eyes are shut tightly, the lights seen are phosphenes. Technically, the luminous impressions are due to the excitation of the retina caused by pressure on the eyeball.

Sources: Wallechinsky, David, et al. *The Book of Lists*, p. 163.

1055

How often does the human eye blink?

The rate of blinking varies, but on the average the eye blinks once every five seconds or 17 thousand times each day or 6¼ million times a year.

Sources: Brandreth, Gyles. *Your Vital Statistics*, p. 26.

1056

What are the floaters that move around on the eye?

Floaters are semi–transparent specks perceived to be floating in the field of vision. Some originate with red blood cells that have leaked out of the retina. The blood cells swell into spheres, some forming strings, and float around the areas of the retina. Others are shadows cast by the microscopic structures in the *vitreous humor*, a jellylike substructure located behind the retina. A sudden appearance of a cloud of dark floaters, if accompanied by bright light flashes, could indicate retinal detachment.

Sources: *The American Medical Association Encyclopedia of Medicine*, pp. 458-459; *Scientific American*, vol. 246 (April 1981), p. 150.

Parts of the eye

1057
What does it mean to have 20/20 vision?

Many people think that with 20/20 vision the eyesight is perfect, but it actually means that the eye can see clearly at 20 feet what a normal eye can see clearly at that distance. Some people can see even better—20/15, for example. With their eagle eyes, they can view objects from 20 feet away with the same sharpness that a normal-sighted person would have to move in to 15 feet to achieve.

Sources: *Prevention's Giant Book of Health Facts*, p. 519.

1058
Who invented bifocal lenses?

The original bifocal lens was invented in 1784 by Benjamin Franklin (1706–1790). At that time, the two lenses were joined in a metallic frame. In 1899, J.L. Borsch welded the two lenses together. One-part bifocal lenses were developed by Bentron and Emerson in 1910 for the Carl Zeiss Company.

Sources: *Dictionary of Visual Science*, p. 388; Giscard-d'Estaing, Valerie-Anne. *The World Almanac Book of Inventions*, p. 221.

1059
What are the three bones in the ear called?

The three bones are the malleus, which means hammer; incus, which means anvil; and stapes, which means stirrup. The bones look somewhat like the objects for which they are named. The stapes is the smallest bone in the body, measuring 1.02 to 1.34 inches (2.6 to 3.4 centimeters) and weighing 0.00071 to 0.0015 ounces (0.002 to 0.004 grams). These three tiny bones in the middle ear conduct sound vibrations from the outer to inner ear.

Sources: *Guinness Book of Records 1992*, pp. 144; *World Book Encyclopedia*, vol. 6, pp. 7-8.

Diagram of ear (a) auricle (b) cartilaginous part of auditory meatus (c) bony part of auditory meatus (d) tympanic membrane or eardrum, (e), (f), (g) chain of ossicles (hammer, anvil and Sylvius bones) in (h) tympanic cavity, (i) stirrup bone, (j) semicircular duct, (k) oval window, (l) Eustachian tube, (m) utricle and saccule (n) perilymph, (o) endo-lymphatic duct, (p) auditory nerve (q) cohlear duct

1060
How broad is the range of sound frequency that most people can hear?

Sounds with frequencies from about 20 to 20,000 hertz can be heard by most people. A hertz is a measure of sound frequencies. Environmental sound is measured in decibels (db) to calculate its loudness. Hearing starts with 0 db; every increase of 10 units is equivalent to a tenfold increase. In comparison, leaves rustling = 10 db, a typical office is 50 db, pneumatic drills = 80 db, riveting machines = 110 db, and a jet takeoff at 200 feet (61 meters) measures 120 db. Noise above 70 db is harmful to hearing; noise at 140 db is physically painful.

Sources: Hunter, Linda Mason. *The Healthy Home*, pp. 205-209; *World Book Encyclopedia*, vol. 18, p. 600.

1061
What are the primary sensations of taste?

The four primary categories are sweet, sour, salty and bitter. The intensity and location of these areas on the tongue varies from person to person. Some of the nine thousand taste buds are located in the other areas of the mouth as well. The lips (usually very salt-sensitive), the inner cheeks, the underside of the tongue, the back of the throat, and the roof of the mouth are some examples. The sense of taste is intimately associated with the sense of smell, so that foods taste bland to someone suffering from a cold. Also important are the appearance, texture, and temperature of food.

Sources: *The American Medical Association Encyclopedia of Medicine*, p. 965; Guyton, Arthur C. *Textbook of Medical Physiology*, p. 581; Hillman, Howard. *Kitchen Science*, pp. 283-284.

SKIN, HAIR, AND NAILS

1062
Can a person get a suntan through a window?

No. The ultraviolet radiation which causes suntan does not pass through glass. A tan is the skin's natural defense against these harmful sun rays. Pigment cells in the lower skin level makes more pigment and melanin to absorb the ultraviolet rays. As the pigment spreads into the top layer of the skin, it becomes darker.

Sources: Flatow, Ira. *Rainbows, Curve Balls and Other Wonders of the Natural World Explained*, pp. 36-39; *Science Digest*, vol. 93 (January 1985), p. 76.

1063
How much skin does an average person have?

There are about 20 square feet or 2 square meters of skin covering the average human body. Weighing 6 pounds (2.7 kilograms), the skin is composed of two main layers: the epidermis (outer layer) and the dermis (inner layer). The epidermis layer is replaced continually as new cells, produced in the stratum basale, mature and are pushed to the surface by the newer cells beneath; the entire epidermis is replaced in about 27 days. The dermis, the lower layer, contains nerve endings, sweat glands, hair follicles, and blood vessels. The upper portion of the dermis has small fingerlike projections called "papillae" which extend into the upper layer. The capillaries in these papillae deliver oxygen and nutrients to the epidermis cells and also function in temperature regulation. The patterns of ridges and grooves visible on the skin of the soles, palms, and fingertips are formed from the tops of the dermal papillae.

Sources: McAleer, Neil. *The Body Almanac*, pp. 100, 103; Solomon, Eldra Pearl, and Gloria A. Phillips. *Understanding Human Anatomy and Physiology*, pp. 60-61.

Parts of the skin

1064 / How can tattoos be removed?

Tattoos can be removed by skin grafting, by infrared coagulation treatment, or by salabrasion in which table salt is scrubbed into the anaesthetized skin. The hypertonic irritant solution disperses the tattoo pigment particles. This procedure will produce scars.

Sources: Burton, J.L. *Essentials of Dermatology*, p. 275.

1065 / What is the purpose of goose-bumps?

The puckering of the skin that takes place when goose-flesh is formed is the result of contraction of the muscle fibers in the skin. This muscular activity will produce more heat, and raise the temperature of the body.

Sources: *Marshall Cavendish Illustrated Encyclopedia of Family Health*, vol. 19, p. 2052.

1066

Do the nails and hair of a dead person continue to grow?

Between 12 and 18 hours after death, the body begins to dry out. That causes the tips of the fingers and the skin of the face to shrink, creating the illusion that the nails and hair have grown.

Sources: *Prevention's Giant Book of Health Facts*, p. 156.

1067

How fast do finger nails grow?

Healthy nails grow about 0.8 inches (2 centimeters) each year. The middle finger-nail grows the fastest, because the longer the finger, the faster its nail growth. Fingernails grow four times as fast as toenails.

Sources: *Harvard Medical School Health Letter*, vol. 9, no. 7 (May 1984), p. 3; Kendig, Frank and Richard Hutton. *Life Spans*, p. 24; Van Amerongen, C. *The Way Things Work Book of the Body*, p. 416.

1068

What is dermatoglyphics?

It is the study of the skin ridge patterns on fingers, toes, palms of hands, and soles of feet. The patterns are used as a basis of identification and also have diagnostic value because of association between certain patterns and chromosomal abnormalities.

Sources: *Mosby's Medical, Nursing, and Allied Health Dictionary*, p. 355.

1069

What information can a forensic scientist determine from a human hair?

A single strand of human hair can identify the age and sex of the owner, drugs and narcotics the individual has taken, and through DNA evaluation and sample comparisons, from whose head the hair came.

Sources: Tunnell, James E. *Latest Intelligence*, p. 103.

1070

How many hairs does the average person have on their head?

The amount of hair covering varies from one individual to another. An average person has about 100 thousand hairs on their scalp. Most redheads have about 90 thousand hairs, blonds have about 140 thousand, and brunettes fall in between these two figures. Most people shed between 50 to 100 hairs daily.

Sources: Brandreth, Gyles D. *Your Vital Statistics*, p. 22; Margo. *Growing New Hair!*, p. 40.

1071 ⟋
How much does human hair grow in a year?

Each hair grows about 5 inches (12.7 centimeters) every year.

Sources: McAleer, Neil. *The Body Almanac*, p. 106.

1072 ⟋
Who first classified fingerprints as a means of identification?

It is generally acknowledged that Francis Galton (1822–1911) was the first to classify fingerprints. However, his basic ideas were further developed by Sir Edward Henry (1850–1931), who devised a system based on the pattern of the thumb print. In 1901 in England, Henry established the first fingerprint bureau with Scotland Yard called the Fingerprint Branch.

Sources: Block, E.B. *Fingerprinting: Magic Weapon Against Crime*, p. 7.

MATHEMATICS, NUMBERS, AND COMPUTERS

COMPUTERS

1073
What is meant by fifth generation computers? *What are the other four generations?*

The evolution of computers has advanced so much in the past few decades that "generations" are used to describe these important advances:

First generation computer—a mammoth computer using vacuum tubes, drum memories, and programming in machine code as its basic technology. Univax 1, used in 1951, was one of the earliest of these vacuum-tube based electronic computers. The generation starts at the end of World War II and ends about 1957.

Second generation computer—a computer using discrete transistors as its basic technology. Solid-state components replaced the vacuum tubes during this period from 1958 to 1963. Magnetic core memories store information. This era includes the development of high-level computer languages.

Third generation computer—a computer having integrated circuits, semiconductor memories, and magnetic disk storage. New operating systems, minicomputer systems, virtual memory, and timesharing are the advancements of this period from 1963 to 1971.

Fourth generation computer—a computer using microprocessors and large scale integrated chips as its basic technology, which made computers accessible to a large segment of the population. Networking, improved memory, database management systems, and advanced languages mark the period from 1971 to the end of the 1980s.

Fifth generation computer—a computer which uses inference to draw reasoned conclusions from a knowledge base, and interacts with its users via an intelligent user interface to perform such functions as speech recognition, machine translation of natural languages, and robotic operations. These computers using artificial intelligence have been under development since the early 1980s, especially in Japan, as well as in the United States and Europe. However, in

1991, Japan began a new ten-year initiative to investigate neural networks, which will probably divert resources from development of the fifth generation as traditionally defined.

Sources: Bishop, Peter. *Fifth Generation Computers*, pp. 155, 159, 161; Edmunds, Robert A. *The Prentice-Hall Encyclopedia of Information Technology*, p. 219; Longley, Dennis and Michael Shain. *Van Nostrand Reinhold Dictionary of Information Technology*, pp. 215, 224, 449, 505.

1074 / What is a pixel?

A pixel (*pix* [for picture] *el*ement) is the smallest element on a video display screen. A screen contains thousands of pixels, each of which can be made up of one or more dots or a cluster of dots. On a simple monochrome screen, a pixel is one dot; the two colors of image and background are created when the pixel is switched either on or off. Some monochrome screen pixels can be energized to create different light intensities, to allow a range of shades from light to dark. On color screens, three dot colors are included in each pixel—red, green, and blue. The simplest screens have just one dot of each color, but more elaborate screens have pixels with clusters of each color. These more elaborate displays can show a large number of colors and intensities. On color screens, black is created by leaving all three colors off; white by all three colors on; and a range of grays by equal intensities of all the colors.

The most economical displays are monochrome, with one bit per pixel, with settings limited to on and off. High-resolution color screens, which can use a million pixels, with each color dot using four bytes of memory, would need to reserve many megabytes just to display the image.

Sources: Freedman, Alan. *The Computer Glossary*, p. 461.

1075 / Where did the term *bug* originate?

The slang term *bug* is used to describe problems and errors occurring in computer programs. The term may have originated during the early 1940s at Harvard University when a computer malfunctioned. A dead moth found in the system was thought to be the cause of the system's failure. This famous carcass, taped to a page of notes, is preserved with the trouble log notebook at the Virginia Naval Museum.

Sources: Cortada, James W. *Historical Dictionary of Data Processing: Technology*, p. 49.

1076 / A lot of people have heard of ENIAC, the first large electronic computer. What was MANIAC?

MANIAC (*m*athematical *a*nalyzer, *n*umerator, *i*ntegrator, *a*nd *c*omputer) was built at the Los Alamos Scientific Laboratory under the direction of Nicholas C.

Metropolis between 1948 and 1952. It was one of several different copies of the high-speed computer built by John von Neumann (1903-1957) for the Institute for Advanced Studies (IAS). It was constructed primarily for use in the development of atomic energy applications, specifically the hydrogen bomb.

It was originated with the work on ENIAC (*e*lectronic *n*umerical *i*ntegrator *a*nd *c*omputer), the first fully operational large-scale electronic digital computer. ENIAC was built at the Moore School of Electrical Engineering at the University of Pennsylvania between 1943 and 1946. Its builders, John Prosper Eckert, Jr., and John William Mauchly (1907–1980), virtually launched the modern era of the computer with ENIAC.

Sources: Cortada, James W. *Historical Dictionary of Data Processing: Technology*, pp. 156 and 258.

1077 / What is a computer "virus" and how is it spread?

Taken from the obvious analogy with biological viruses, a computer "virus" is a program that searches out other programs and "infects" them by replicating itself in them. When the programs are executed, the embedded virus is executed too, thus propagating the "infection." This normally happens invisibly to the user. A virus cannot infect other computers, however, without assistance. It is spread when users communicate by computer, often when they trade programs. The virus may do nothing but propagate itself and then allow the program to run normally. Usually, however, after propagating silently for a while, it starts doing things from inserting "cute" messages to destroying all of the user's files. Computer "worms" and "logic bombs" are similar to viruses, but they do not replicate themselves within programs as viruses do. A logic bomb does its damage immediately—destroying data, inserting garbage into data files, or reformatting the hard disk; a worm can alter the program and database either immediately or over a period of time.

In the 1990s, viruses, worms, and logic bombs have become such a serious problem, especially among IBM PC and Macintosh users, that the production of special detection and "inoculation" software has become an industry.

Sources: Freedman, Alan. *The Computer Glossary*, pp. 346, 636-637, 657; Raymond, Eric S. *The New Hacker's Dictionary*, p. 371.

1078 / Is an assembly language the same thing as a machine language?

While the two terms are often used interchangeably, an assembly language is a more "user friendly" translation of a machine language. A machine language is the collection of patterns of bits recognized by a *c*entral *p*rocessing *u*nit (CPU) as instructions. Each particular CPU design has its own machine language. The machine language of the CPU of a microcomputer generally includes about 75 instructions; the machine language of the CPU of a large mainframe computer may include hundreds of instructions. Each of these instructions is a pattern of 1's and 0's that tells the CPU to perform a specific operation.

An assembly language is a collection of symbolic, mnemonic names for each instruction in the machine language of its CPU. Like the machine language, the assembly language is tied to a particular CPU design. Programming in assembly language requires intimate familiarity with the CPU's architecture, and assembly language programs are difficult to maintain and require extensive documentation.

The computer language C, developed in the late 1980s, is now frequently used instead of assembly language. It is a high-level programming language that can be compiled into machine languages for almost all computers, from micro-computers to mainframes, because of its functional structure.

Sources: Freedman, Alan. *The Computer Glossary*, pp. 33, 79, 354; Hofstadter, Douglas R. *Gödel, Escher, Bach: An Eternal Golden Braid*, pp. 289-291; Shore, John. *The Sachertorte Algorithm*, pp. 145–148.

1079 / Who invented the COBOL computer language?

COBOL (*c*ommon *b*usiness *o*riented *l*anguage) is a prominent computer language designed specifically for commercial uses, created in 1960 by a team drawn from several computer makers and the Pentagon. The best-known individual associated with COBOL was then-Lieutenant Grace M. Hopper (1906-1992) who made fundamental contributions to the United States Navy standardization of COBOL. COBOL excels at the most common kinds of data processing for business—simple arithmetic operations performed on huge files of data. The language endures because its syntax is very much like English and because a program written in COBOL for one kind of computer can run on many others without alteration.

Sources: *Understanding Computers: Illustrated Chronology and Index*, p. 47; Williams, Michael R. *A History of Computing Technology*, p. 247.

1080 / What does DOS stand for?

DOS stands for *d*isk *o*perating *s*ystem, a program that controls the computer's transfer of data to and from a hard or floppy disk. Frequently it is combined with the main operating system. The operating system was originally developed at Seattle Computer Products as SCP-DOS. When IBM decided to build a personal computer and needed an operating system, it chose the SCP-DOS after reaching an agreement with the Microsoft Corporation to produce the actual operating system. Under Microsoft, SCP-DOS became MS-DOS which IBM referred to as PC-DOS (*p*ersonal *c*omputer), and which everyone eventually simply called DOS.

Sources: Freiberger, Paul, and Michale Swaine. *Fire in the Valley: The Making of the Personal Computer*, pp. 269, 272, 276; Rosenberg, Jerry M. *Dictionary of Computers, Data Processing, and Telecommunications*, p. 157.

1081

Who invented the computer?

Computers developed from calculating machines. One of the earliest mechanical devices for calculating, still widely used today, is the abacus—a frame carrying parallel rods on which beads or counters are strung. Herodotus, the Greek historian who lived around 400 B.C.E., mentions the use of the abacus in Egypt. In 1617, John Napier (1550–1617) invented "Napier's Rods"—marked pieces of ivory for multiples of numbers. In the middle of the same century, Blaise Pascal (1623–1662) produced a simple mechanism for adding and subtracting. Multiplication by repeated addition was a feature of a stepped drum or wheel machine of 1694 invented by Gottfried Wilhelm Leibniz (1646–1716). In 1823, the English visionary Charles Babbage (1792–1871) persuaded the British government to finance an "analytical engine." This would have been a machine that could undertake any kind of calculation. It would have been driven by steam, but the most important innovation was that the entire program of operations was stored on a punched tape. Babbage's machine was not completed and would not have worked if it had been. The standards required were far beyond the capabilities of the engineers of the time, and in any case, rods, levers, and cogs move too slowly for really quick calculations. Only electrons, which travel at near the speed of light, are rapid enough. Although he never built a working computer, Babbage thought out many of the basic principles that guide modern computers.

Based on the concepts of British mathematician Alan M. Turing (1912–1954), the earliest programmable electronic computer was the 1,500-valve "Colossus," formulated by Max Newman (1897–1985), built by T.H. Flowers, and used by the British government in 1943 to crack the German codes generated by the coding machine "Enigma."

Sources: De Bono, Eduard. *Eureka!*, p. 214; *Guinness Book of Records 1992*, p. 287; *Purnell's Encyclopedia of Inventions*, p. 96.

1082

What is the origin of the expression "Do not fold, spindle, or mutilate"?

This is the inscription on an IBM punched card. Frequently, office workers organize papers and forms by stapling or folding them together, or by impaling them on a spindle. Because Hollerith (punched) card readers scan uniform rectangular holes in a precise arrangement, any damage to the physical card makes it unusable. In the 1950s and 1960s, when punched cards became widespread, manufacturers printed a warning on each card; IBM's "Do not fold, spindle, or mutilate" was the best-known. In 1964, the student revolution at the University of California, Berkeley used the phrase as a symbol of authority and regimentation.

Sources: *Scientific Quotations: The Harvest of a Quiet Eye*, p. 80.

1083
Who was the first programmer?

According to historical accounts, Lord Byron's daughter, Augusta Ada Byron, the Countess of Lovelace, was the first person to write a computer program for Charles Babbage's (1792–1871) "analytical engine." This machine, never built, was to work by means of punched cards that could store partial answers that could later be retrieved for additional operations, and that would print results. Her work with Babbage and the essays she wrote about the possibilities of the "engine" established her as a "patron saint," if not a founding parent, of the art and science of programming. The programming language called "Ada" was named in her honor by the United States Department of Defense. In modern times the honor goes to Commodore Grace Murray Hopper (1906–1992) of the United States Navy. She wrote the first program for the Mark I computer.

Sources: Jespersen, James. *RAMS, ROMS, and Robots*, p. 59; Rheingold, Howard. *Tools for Thought*, p. 31; Rochester, Jack B. *The Naked Computer*, pp. 43, 45.

1084
How is a byte defined?

A byte, a common unit of computer storage, holds the equivalent of a single character, such as a letter ("A"), a number ("2"), a symbol ("$"), a decimal point, or a space. It is usually equivalent to eight "data bits" and one "parity bit." A bit (a *binary* digi*t*), the smallest unit of information in a digital computer, is equivalent to a single "0" or "1". The parity bit is used to check for errors in the bits making up the byte. Although eight data bits per byte is the most common size, computer manufacturers are free to define a differing number of bits as a byte. Six data bits per byte is another common size.

Sources: Freedman, Alan. *The Computer Glossary*, pp. 66, 77; *Understanding Computers: Illustrated Chronology and Index*, p. 16; Wyatt, Allen L. *Computer Professional's Dictionary*, p. 60.

1085
What was the first major use for punched cards?

Punched cards were a way of programming, or giving instructions to, a machine. In 1801, Joseph Marie Jacquard (1752-1834) built a device that could do automated pattern weaving. Cards with holes were used to direct threads in the loom, creating predefined patterns in the cloth. The pattern was determined by the arrangement of holes in the cards, with wire hooks passing through the holes to grab and pull through specific threads to be woven into the cloth.

By the 1880s, Herman Hollerith (1860–1929) was using the idea of punched cards to give machines instructions. He built a punched card tabulator that processed the data gathered for the 1890 United States Census in six weeks (three times the speed of previous compilations). Metal pins in the machine's reader passed through holes punched in cards the size of dollar bills, momentarily closing electric circuits. The resulting pulses advanced counters assigned to details such as income and family size. A sorter could also be programmed to pigeonhole cards according to pattern of holes, an important aid in analyzing census statistics. Later, Hollerith founded Tabulating Machines Co., which in 1924 became IBM. When IBM adopted the 80-column punched card (measuring 7⅜ by 3¼ inches, or 18.7 by 8.25 centimeters, and .007 inches, or .018 centimeters, thick), the de facto industry standard was set, which has endured for decades.

Sources: Cortada, James W. *Historical Dictionary of Data Processing: Technology*, p. 308; Graham, John. *Facts On File Dictionary of Telecommunications*, p. 127; *Understanding Computers: Illustrated Chronology and Index*, pp. 9, 11.

1086 / What is Hopper's rule?

Electricity travels one foot in a nanosecond (a billionth of a second). This is one of a number of rules compiled for the convenience of computer programmers. This is also considered to be a fundamental limitation on the possible speed of a computer—signals in an electrical circuit cannot move any faster.

Sources: Fisher, David J. *Rules of Thumb for Engineers and Scientists*, p. 87; Trefil, James. *1001 Things Everyone Should Know About Science*, p. 129.

1087 / What does it mean to "boot" a computer?

Booting a computer is starting it, in the sense of turning control over to the operating system. The term comes from bootstrap, because bootstraps allow an individual to pull on boots without help from anyone else. Some people prefer to think of the process in terms of using bootstraps to lift oneself off the ground, impossible in the physical sense, but a reasonable image for representing the process of searching for the operating system, loading it, and passing control to it. The commands to do this are embedded in a *read only memory* (ROM) chip that is automatically executed when a microcomputer is turned on or reset. In mainframe or minicomputers, the process usually involves a great deal of operator input. A cold boot powers on the computer and passes control to the operating system; a warm boot resets the operating system without powering off the computer.

Sources: Freedman, Alan. *The Computer Glossary*, pp. 68, 113, 647; Hofstadter, Douglas R. *Gödel, Escher, Bach: An Eternal Golden Braid*, pp. 24, 293–294.

1088
Who invented the computer mouse?

A computer "mouse" is a hand-held input device that, when rolled across a flat surface, causes a cursor to move in a corresponding way on a display screen. A prototype mouse was part of an input console demonstrated by Douglas C. Englehart in 1968 at the Fall Joint Computer Conference in San Francisco. Popularized in 1984 by the Macintosh from Apple Computer, the mouse was the result of 15 years devoted to exploring ways to make communicating with computers simpler and more flexible.

The physical appearance of the small four-sided box with the dangling tail-like wire suggested the informal name of "mouse," which quickly superseded the formal name.

Sources: *Compute*, vol. 13 (August 1991), p. 52; *Understanding Computers: Illustrated Chronology and Index*, pp. 62, 73.

1089
Who coined the term technobabble?

John A. Barry used the term *technobabble* to mean the pervasive and indiscriminate use of computer terminology, especially as it is applied to situations that have nothing at all to do with technology. He first used it in the early 1980s.

Sources: Barry, John A. *Technobabble*, pp. xiii, 3–5.

1090
What is an algorithm?

An algorithm is a set of clearly defined rules and instructions for the solutions of a problem. It is not necessarily applied only in computers, but can be a step-by-step procedure for solving any particular kind of problem. A nearly 4,000-year-old Babylonian banking calculation inscribed on a tablet is an algorithm, as is a computer program which consists of the step-by-step procedures for solving a problem.

The term is derived from the name of Muhammad ibn Musa al Kharizmi (ca. 780–ca.850), a Baghdad mathematician who introduced Hindu numerals (including 0) and decimal calculation to the west. When his treatise was translated into

Latin in the 12th century, the art of computation with Arabic (Hindu) numerals became known as *algorism*.

Sources: Asimov, Isaac. *Asimov's Biographical Encyclopedia of Science and Technology*, p. 48; *Understanding Computers: Computer Languages*, pp. 17-19; *Understanding Computers: Illustrated Chronology and Index*, p. 8.

1091 / What is a hacker?

A hacker is a skilled computer user. The term originally denoted a skilled programmer, particularly one skilled in machine code and with a good knowledge of the machine and its operating system. The name arose from the fact that a good programmer could always hack an unsatisfactory system around until it worked.

The term later came to denote a user whose main interest is in defeating password systems. The term has thus acquired a pejorative sense, with the meaning of one who deliberately and sometimes criminally interferes with data available through telephone lines. The activities of such hackers have led to considerable efforts to tighten security of transmitted data.

Sources: Sinclair, Ian R. *The HarperCollins Dictionary of Computer Terms*, p. 105.

1092 / For what purpose was MADAM designed?

MADAM (Manchester *a*utomatic *d*igit*a*l *m*achine) is a chess–playing machine designed by Alan M. Turing (1912–1954) in 1950. Turing was one of the first individuals to program a computer to play chess. His machine was a very poor chess player and made foolish moves. After several moves the machine would be forced to give up. Today it is possible to play a fairly advanced game of chess with a computer. However, no machine has been designed that analyzes every possible strategy corresponding to any move. It would take 10^{108} years to play all the possible games if we had a machine that could play a million games a second.

Sources: *The Biographical Dictionary of Scientists: Mathematicians*, p. 126; Newmark, Joseph. *Mathematics As a Second Language*, p. 344.

1093 / What is E mail?

Electronic mail, also known as E mail or E–mail, uses communication facilities to transmit messages. Many systems use computers as transmitting and receiving interfaces, but fax communication is also a form of E mail. A user can send a message to a single recipient, or to many. Different systems offer different options for sending, receiving, manipulating text, and addressing. For example a message can be "registered," so that the sender is notified when the recipient looks at the message (though there is no way to tell if the recipient has actually read the message). Many systems allow messages to be forwarded. Usually messages are stored in a simulated "mailbox" in the network server or host computer; some systems announce incoming mail if the recipient is logged onto the system. An organization (such as a corporation, university, or professional organization) can provide electronic mail facilities; there are also national and international networks. In order to use E mail, both sender and receiver must have accounts on the same system or on systems connected by a network.

Sources: Cassel, Don. *Understanding Computers*, p. 147; Freedman, Alan. *The Computer Glossary*, p. 217; Longley, Dennis, and Michael Shain. *Van Nostrand Reinhold Dictionary of Information Technology*, p. 188.

1094 / *What is meant by* default?

As used in computer, communication, and data processing systems, a default is the response that a system "assumes" in the absence of specific instructions. It can be a value (6), a condition (true), or an action (close file).

Sources: Longley, Dennis, and Michael Shain. *Van Nostrand Reinhold Dictionary of Information Technology*, p. 148; Weik, Martin H. *Communications Standard Dictionary*, p. 253.

1095 / *How is a hard disk different from a floppy disk?*

Both types of disk use a magnetic recording surface to record, access, and erase data, in much the same way as magnetic tape records, plays, and erases sound or images. A read/write head, suspended over a spinning disk, is directed by the *central processing unit* (CPU) to the sector where the requested data is stored, or where the data is to be recorded. A hard disk uses rigid aluminum disks coated with iron oxide for data storage. It has much greater storage capacity than several floppy disks (from 10 to hundreds of megabytes). While most hard disks used in microcomputers are "fixed" (built into the computer), some are removable. Mini-computer and mainframe hard disks include both fixed and removable (in modules called disk packs or disk cartridges). A floppy disk, also called a diskette, is made of plastic film covered with a magnetic coating, which is enclosed in a non-removable plastic protective envelope. Floppy disks vary in storage capacity from 100 thousand bytes to more than 2 megabytes. Floppy disks are generally used in minicomputers and microcomputers.

　　In addition to storing more data, a hard disk can provide much faster access to storage than a floppy disk. A hard disk rotates from 2,400 to 3,600 *revolutions per minute* (rpm) and is constantly spinning (except in laptops, which conserve battery life by spinning the hard disk only when in use). An ultra-fast hard disk has a separate read/write head over each track on the disk, so that no time is lost in positioning the head over the desired track; accessing the desired sector takes only milliseconds, the time it takes for the disk to spin to the sector. A floppy disk does not spin until a data transfer is requested, and the rotation speed is only about 300 rpm.

Sources: Downing, Douglas, and Michael Covington. *Dictionary of Computer Terms*, pp. 131, 148; Freedman, Alan. *The Computer Glossary*, pp. 245–246, 275, 359–361; Hopkins, Nigel J., John W. Mayne, and John R. Hudson. *The Numbers You Need*, pp. 286–287.

1096 / Why does a computer floppy disk have to be "formatted"?

A disk must first be organized to that data can be written to it and retrieved from it. The data on a floppy disk or a hard disk is arranged in concentric tracks. Sectors, which can hold blocks of data, occupy arc-shaped segments of the tracks. Most floppy disks are soft-sectored, and formatting is necessary to record sector identification so that data blocks can be labeled for retrieval. Hard-sectored floppy disks use physical marks to identify sectors; these marks cannot be changed, so the disks cannot be reformatted. The way that sectors are organized and labeled dictates system compatibility: disks formatted for DOS computers can only be used in other DOS machines; those formatted for Macintoshes can only be used in other Macintoshes. Formatting erases any pre-existing data on the disk. Hard disk drives are also formatted before being initialized, and should be protected so that they are not reformatted unintentionally.

Sources: DeVoney, Chris. *MS-DOS User's Guide*, pp. 395–402; Freedman, Alan, *The Computer Glossary*, pp. 191, 246, 275, 359–360; Longley, Dennis, and Michael Shain. *Van Nostrand Reinhold Dictionary of Information Technology*, pp. 223, 450, 512.

1097 / How much data can a floppy disk hold?

The three common floppy disk (diskette) sizes vary widely in storage capacity.

Envelope size (inches)	Storage capacity
8	100,000–500,000 bytes
5.25	100 kilobytes–1.2 megabytes
3.5	400 kilobytes–more than 2 megabytes

An 8-inch or 5-inch diskette is enclosed in a plastic protective envelope, which does not protect the disk from bending or folding; parts of the disk surface are also exposed, and can be contaminated by fingerprints or dust. The casing on a 3.5-inch floppy disk is rigid plastic, and includes a sliding disk guard that protects the disk surface, but allows it to be exposed when the disk is inserted in the disk drive. This protection, along with the increased data storage capacity, makes the 3.5-inch disk currently the most popular.

Sources: Freedman, Alan. *The Computer Glossary*, pp. 245–246, 380–381.

1098 / What is a silicon chip?

A silicon chip is an almost pure piece of silicon, usually less than one centimeter square and about half a millimeter thick. It contains hundreds of thousands of microminiature electronic circuit components, mainly transistors, packed and interconnected in layers beneath the surface. These components can perform control, logic, and/or memory functions. There is a grid of thin metallic strips on the surface of the chip; these wires are used for electrical connections to other devices. The silicon chip was developed independently by two researchers: Jack Kilby of Texas Instruments in 1958, and Robert Noyce (b. 1927) of Fairchild Semiconductor in 1959.

While silicon chips are essential to most computer operation today, a myriad of other devices depend on them as well, including calculators, microwave ovens, automobile diagnostic equipment, and VCRs.

Sources: Cone, Robert J. *How the New Technology Works*, pp. 8–11; Renmore, C.D. *Silicon Chips and You*, p. 19.

1099 / What are the sizes of silicon chips?

SSI—small scale integration
MSI—medium scale integration
LSI—large scale integration
VLSI—very large scale integration
ULSI—ultra large scale integration
GSI—gigascale integration

The number of components packed into a single chip is very loosely defined. ULSI units can pack millions of components on a chip. GSI, a long-term target for the industry, will potentially hold billions of components on a chip.

Sources: Cone, Robert J. *How the New Technology Works*, p. 11.

1100 / Are any devices being developed to replace silicon chips?

When transistors were introduced in 1948, they demanded less power than fragile, high-temperature vacuum tubes; allowed electronic equipment to become smaller, faster, and more dependable; and generated less heat. These developments made computers much more economical and accessible; they also made portable radios practical. However, the smaller components were harder to wire together, and hand wiring was both expensive and error-prone.

In the early 1960s, circuits on silicon chips allowed manufacturers to build increased power, speed, and memory storage into smaller packages, which required less electricity to operate and generated even less heat. While through most of the 1970s manufacturers could count on doubling the components on a chip every year without increasing the size of the chip, the size limitations of silicon chips are becoming more restrictive. Though components continue to grow smaller, the same rate of shrinking cannot be maintained.

Researchers are investigating different materials to use in making circuit chips. Gallium arsenide is harder to handle in manufacturing, but it has potential for greatly increased switching speed. Organic polymers are potentially cheaper to manufacture, and could be used for liquid-crystal and other flat screen displays, which need to have their electronic circuits spread over a wide area. Unfortunately, organic polymers do not allow electricity to pass through as well as the silicons do. Several researchers are working on hybrid chips, which could combine the benefits of organic polymers with those of silicon. Researchers are also in the initial stages of developing integrated optical chips, which would use light rather than electric current. Optical chips would generate little or no heat, would allow faster switching, and would be immune to electrical noise.

Sources: Cone, Robert J. *How the New Technology Works*, pp. 8–11; *New Scientist*, vol. 128, no. 1747 (December 15, 1990), p. 24.

MATHEMATICS *See also: MATHEMATICS, NUMBERS, AND COMPUTERS—NUMBERS*

1101
What is the probability of a successful triple play in a single baseball game?

The odds against a triple play in a game of baseball are 1,400 to 1.

Sources: *The Odds on Virtually Everything*, p. 50.

1102
How many different bridge games are possible?

Roughly fifty-four octillion different bridge games are possible.

Sources: Asimov, Isaac. *Asimov on Numbers*, p. 39.

1103
What is the ancient Greek problem of squaring the circle?

This problem was to construct, with a straightedge and compass, a square having the same area as a given circle. The Greeks were unable to solve the problem because the task is impossible, as was shown by the German mathematician Ferdinand von Lindemann (1852–1939) in 1882.

Sources: *Yearbook of Science and the Future 1991*, p. 377.

▽1104▽
If 30 people are chosen at random, what is the probability that at least two of them have their birthday on the same day?

The probability that at least two people in a group of 30 share the same birthday is about 70%.

Sources: Jacobs, Harold R. *Mathematics: A Human Endeavor*, p. 380.

▽1105▽
What are the common mathematical formulas for volume?

Volume of a sphere:
    Volume = ⅔ times pi times the cube of the radius
        $V = 4\pi r^3/3$
Volume of a pyramid:
    Volume = ⅓ times the area of the base times the height
        $V = \frac{1}{3}bh$
Volume of a cylinder:
    Volume = area of the base times the height
        $V = Ah$
Volume of a circular cylinder (with circular base):
    Volume = pi times the square of the radius of the base times the height
        $V = \pi r^2 h$
Volume of a cube:
    Volume = the length of one side cubed
        $V = S^3$

Sources: Gibson, Carol. *The Facts On File Dictionary of Mathematics*, pp. 46, 49, 150, 172; Stein, Edwin I. *Arithmetic for College Students*, pp. 265–269.

▽1106▽
When does 0 x 0 = 1?

Factorials are the product of a given number and all the factors less than that number. The notation n! is used to express this idea. For example, 5! (five factorial) is 5 x 4 x 3 x 2 x 1 = 120. For completeness, 0! is assigned the value 1, so 0 x 0 = 1.

Sources: Gibson, Carol. *The Facts On File Dictionary of Mathematics*, p. 75; Karush, William. *Webster's New World Dictionary of Mathematics*, p. 104.

1107

What are the common mathematical formulas for area?

Area of a rectangle:
> Area = length times width
>> A = lw
> Area = altitude times base
>> A = ab

Area of a circle:
> Area = pi times the radius squared
>> $A = \pi r^2$ or $A = \frac{1}{4}\pi d^2$

Area of a triangle:
> Area = one half the altitude times the base
>> $A = \frac{1}{2}ab$

Area of the surface of a sphere:
> Area = four times pi times the radius squared
>> $A = 4\pi r^2$ or $A = \pi d^2$

Sources: Stein, Edwin I. *Arithmetic for College Students*, pp. 251–260.

1108

Who discovered the formula for the area of a triangle?

Heron (or Hero) of Alexandria (1st century B.C.E.) is best known in the history of mathematics for the formula which bears his name. This formula calculates the area of a triangle with sides *a*, *b*, and *c*, with *s* = half the perimeter: $A = \sqrt{[s\,(2\text{-}a)(s\text{-}b)(s\text{-}c)]}$. The Arab mathematicians who preserved and transmitted the mathematics of the Greeks report that this formula was known earlier to Archimedes (ca. 287–212 B.C.E.), but the earliest proof now known is that appearing in Heron's *Metrica*.

Sources: Boyer, Carl B. *A History of Mathematics*, p. 193; Gibson, Carol. *The Facts On File Dictionary of Mathematics*, p. 88

1109

How is "casting out nines" used to check the results of addition or multiplication?

The method of "casting out nines" is based on the excess of nines in digits of whole numbers (the remainder when a sum of digits is divided by 9). Illustrating this process in the multiplication example below, the method begins by adding the digits in both the multiplicand (one of the terms that is being multiplied) and the multiplier (the other term being multiplied). In the example below, this operation leads to the results of "13" and "12," respectively. If these results are greater than 9 (>9), then the operation is repeated until the resulting figures are less than 9 (<9). In the example below, the repeated calculation gives the results as "4" and

"3," respectively. Multiply the resulting "excess" from the multiplicand by the excess from the multiplier (4 x 3 below). Add the digits of the result to eventually yield a number equal to or less than 9 (≤9). Repeat the process of casting out nines in the multiplication product (the result of the multiplication process). The result must equal the result of the previous set of transactions, in this case "3." If the two figures disagree, then the original multiplication procedure was done incorrectly. "Casting out nines" can also be applied to check the accuracy of the results of addition.

$$
\begin{array}{r}
328 \rightarrow 13 \rightarrow 4 \\
624 \rightarrow 12 \rightarrow \underline{3} \\
\hline
1324 \qquad 12 \rightarrow 3 \\
656 \\
\underline{1868} \\
204672 \rightarrow 21 \rightarrow 3
\end{array}
$$

Sources: Karush, William. *Webster's New World Dictionary of Mathematics*, p. 28; Lennes, N.J. *New Practical Mathematics*, p. 29.

1110 / How is the "rule of 70" used?

This rule is a quick way of estimating the period of time in which a quantity can double given the percentage of increase. Divide the percentage of increase into 70. For example, if a sum of money is invested at 6% interest, the money will double in value in 70 ÷ 6 = 11.7 years.

Sources: Hershey, Robert L. *How to Think with Numbers*, p. 22.

1111 / How is percent of increase calculated?

To find the percent of increase, divide the amount of increase by the base amount. Multiply the result by 100%. For example, a raise in salary from $10,000 to $12,000 would have percent of increase = (2,000 ÷ 10,000) x 100% = 20%.

Sources: Kogelman, Stanley, and Barbara R. Heller. *The Only Math Book You'll Ever Need*, pp. 3-4.

1112 / What is Zeno's paradox?

Zeno of Elea (ca.490–ca.425 B.C.E.), a Greek philosopher and mathematician, is famous for his paradoxes which deal with the continuity of motion. One form of the paradox is: if an object moves with constant speed along a straight line from point 0 to point 1, the object must first cover half the distance (½), then half the remaining distance (¼), then half the remaining distance (⅛), and so on without end. The conclusion is that the object never reaches point 1. Because there is always some distance to be covered, motion is impossible. In another approach to this paradox, Zeno used an allegory telling of a race between a tortoise and Achilles (who could run 100 times as fast), where the tortoise started running 10 rods in front of Achilles. Because the tortoise always advanced ⅟₁₀₀ of the distance that Achilles advanced in the same time period, it was theoretically impossible for Achilles to pass him. The English mathematician and writer Charles Dodgson (1832–1898), better known as Lewis Carroll, used the characters of Achilles and the tortoise to illustrate his paradox of infinity.

Sources: *Dictionary of Scientific Biography*, vol. 14, pp. 607-608; Hofstadter, Douglas R. *Gödel, Escher, Bach: An Eternal Golden Braid*, pp. 28–32; Motz, Lloyd, and Jefferson Hane Weaver. *Conquering Mathematics*, pp. 47-48.

1113 / What are "Napier's bones"?

In the 16th century, the Scottish mathematician John Napier (1550–1617), Baron of Merchiston, developed a method of simplifying the processes of multiplication and division, using exponents of 10, which Napier called *logarithms* (commonly abbreviated as *logs*). Using this system, multiplication is reduced to addition and division to subtraction. For example, the log of 100 (10^2) is 2; the log of 1000 (10^3) is 3; the multiplication of 100 by 1000, 100 x1000 = 100,000, can be accomplished by adding their logs (log[(100)(1000)] = log(100) + log(1000) = 2 + 3 = 5 = log (100,000). Napier published his methodology in *A Description of the Admirable Table of Logarithms* in 1614. In 1617, he published a method of using a device, made up of a series of rods in a frame, marked with the digits 1 through 9, to multiply and divide using the principles of logarithms. This device was commonly called "Napier's bones" or "Napier's rods."

Sources: Bergamini, David. *Life Science Library: Mathematics*, p. 22; Motz, Lloyd, and Jefferson Hane Weaver. *Conquering Mathematics*, pp. 29-31; Shapiro, Max S. *Mathematics Encyclopedia*, p. 131.

1114 / Who invented the slide rule?

The slide rule is based on the use of logarithmic scales, which were invented by John Napier (1550–1617), Baron of Merchiston, and published in 1614. In 1620, Edmund Gunter (1581–1626) of Gresham College, London, England, described an immediate forerunner of the slide rule, his "logarithmic line of numbers." William Oughtred (1574–1660), Rector of Aldbury, England, made the first rectilinear slide rule in 1621. This slide rule consisted of two logarithmic scales that could be manipulated together for calculation. His former pupil, Richard Delamain published a description of a circular slide rule in 1630, three years before Oughtred published a description of his invention (at least one source says that Delamain published in 1620). Oughtred accused Delamain of stealing his idea, but evidence indicates that the inventions were probably arrived at independently. The earliest existing straight slide rule using the modern design of a slider moving in a fixed stock dates from 1654. A wide variety of specialized slide rules were developed by the end of the 17th century, for trades such as masonry, carpentry, and excise tax collecting. Peter Mark Roget (1779-1869), best known for his *Thesaurus of English Words and Phrases*, invented a log-log slide rule for calculating the roots and powers of numbers in 1814. While the slide rule was popular as a calculating tool for several centuries, it has largely been superseded by the electronic calculator.

Sources: De Bono, Edward. *Eureka!*, p. 210; Mount, Ellis, and Barbara A. List. *Milestones in Science and Technology*, p. 76; *New Encyclopedia Britannica*, vol. 10, p. 140.

1115 / What is a Möbius strip?

A Möbius strip is a surface with only one side, usually made by connecting the two ends of a rectangular strip of paper after putting a half-twist (180 degrees relative to the opposite side) in the strip. Cutting a Möbius strip in half down the center of the length of the strip results in a single band with four half-twists. Devised by the German mathematician August Ferdinand Möbius (1790–1868) to illustrate the properties of one-sided surfaces, it was presented in a paper that was not discovered or published until after his death. Another 19th-century German mathematician, Johann Benedict Listing, developed the idea independently at the same time.

Sources: *The Biographical Dictionary of Scientists: Mathematicians*, pp. 92–93; Karush, William. *Webster's New World Dictionary of Mathematics*, pp. 172–173; *Mathematics Encyclopedia*, p. 127.

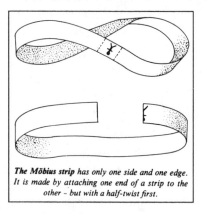

*The **Möbius strip** has only one side and one edge. It is made by attaching one end of a strip to the other – but with a half-twist first.*

1116 / How is Pascal's triangle used?

Pascal's triangle is an array of numbers, arranged so that every number is equal to the sum of the two numbers above it on either side. It can be represented in several slightly different triangles, but this is the most common form:

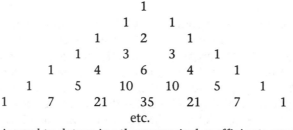

```
                    1
                1       1
            1       2       1
        1       3       3       1
    1       4       6       4       1
1       5      10      10       5       1
1       7      21      35      21       7       1
                   etc.
```

The triangle is used to determine the numerical coefficients resulting from the computation of higher powers of a binomial (two numbers added together). When a binomial is raised to a higher power, the result is expanded, using the numbers in that row of the triangle. For example, $(a + b)^1 = a^1 + b^1$, using the coefficients in the second line of the triangle. $(a + b)^2 = a^2 + 2ab + b^2$, using the coefficients in the next line of the triangle. (The first line of the triangle correlates to $(a + b)^0$.) While the calculation of coefficients is fairly straightforward, the triangle is useful in calculating them for the higher powers without needing to multiply them out. Binomial coefficients are useful in calculating probabilities; Pascal was one of the pioneers in developing laws of probability.

As with many other mathematical developments, there is some evidence of a previous appearance of the triangle in China. Around 1100 C.E., the Chinese mathematician Chia Hsien wrote on "the tabulation system for unlocking binomial coefficients"; the first publication of the triangle was probably in a book called *Piling-Up Powers and Unlocking Coefficients*, by Liu Ju-Hsieh.

Sources: Barnhart, Robert K. *The American Heritage Dictionary of Science*, p. 474; Motz, Lloyd, and Jefferson Hane Weaver. *Conquering Mathematics*, pp. 53-56; Temple, Robert. *The Genius of China*, pp. 146-147.

▽1117▽ What is the Pythagorean theorem?

In a right triangle (one where two of the sides meet in a 90 degree angle), the hypoteneuse is the side opposite the right angle. The Pythagorean theorem, also known as the rule of Pythagoras, states that the square of the length of the hypoteneuse is equal to the sum of the squares of the other two sides ($h^2 = a^2 + b^2$). If the lengths of the sides are: h = 5 inches, a = 4 inches, and b = 3 inches, then

$$h = \sqrt{a^2 + b^2} = \sqrt{4^2 + 3^2} = \sqrt{16 + 9} = \sqrt{25} = 5$$

The theorem is named for the Greek philosopher and mathematician Pythagoras (ca. 580–ca. 500 B.C.E.). Pythagoras is credited with the theory of the functional significance of numbers in the objective world and numerical theories of musical pitch. As he left no writings, the Pythagorean theorem may actually have been formulated by one of his disciples.

Sources: Asimov, Isaac. *Asimov's Biographical Encyclopedia of Science and Technology*, pp. 4–6; *The Biographical Dictionary of Scientists: Mathematicians*, pp. 109–110; Stein, Edwin I. *Arithmetic for College Students*, pp. 238–239.

1118

How long has the abacus been used? Is it still used?

The abacus grew out of early counting boards, with hollows in a board holding pebbles or beads used to calculate. It has been documented in Mesopotamia back to around 3500 B.C.E. The current form, with beads sliding on rods, dates back at least to 15th-century China. Before the use of decimal number systems, which allowed the familiar paper-and-pencil methods of calculation, the abacus was essential for almost all multiplication and division. The abacus is still used in many countries where modern calculators are not available. It is also still used in countries, such as Japan and China, with long traditions of abacus use. As recently as the mid-1970s, most Japanese shopkeepers used abaci for totalling customers' bills. While the calculator is now more widely used, many people still prefer to check the results on an abacus. At least one manufacturer offers calculators with small, built-in abaci.

Sources: *Discover*, vol. 6, (May 1985), pp. 36-43; *Oxford Illustrated Encyclopedia of Invention and Technology*, p. 1; *Science and Technology Illustrated*, pp. 20-21.

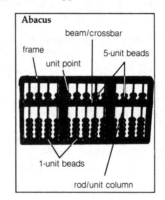

Abacus

beam/crossbar

frame

unit point

5-unit beads

1-unit beads

rod/unit column

1119

What is the difference between a median and a mean?

If a string of numbers is arranged in numerical order, the median is the middle value of the string. If there are an even number of values in the string, the median is found by adding the two middle values and dividing by two.

The arithmetic mean, also known as the simple average, is found by taking the sum of the numbers in the string and dividing by the number of items in the string. While easy to calculate for relatively short strings, the arithmetic mean can be misleading, as very large or very small values in the string can distort it. For example, the mean of the salaries of a professional football team would be skewed if one of the players was a high-earning superstar; it could be well above the salaries of any of the other players. The mode is the number in a string that appears most often.

For the string 111222234455667: The median is the middle number of the series, 3. The arithmetic mean is the sum of numbers + the number of numbers in the series, 51 + 15 = 3.4. The mode is the number that occurs most often, 2.

Sources: Brennan, Richard P. *Dictionary of Scientific Literacy*, pp. 199-200; Bush, Grace A. *Foundations of Mathematics*, pp. 157-161; Hopkins, Nigel J., John W. Mayne, and John R. Hudson. *The Numbers You Need*, pp. 275-277.

1120 / Can a person using an abacus calculate more rapidly than someone using a calculator?

In 1946, the Tokyo staff of *Stars and Stripes* sponsored a contest between a Japanese abacus expert and an American accountant using the best electric adding machine then available. The abacus operator proved faster in all calculations except the multiplication of very large numbers. While today's electronic calculators are much faster and easier to use than the adding machines used in 1946, undocumented tests still show that an expert can add and subtract faster on an abacus than someone using an electronic calculator. It also allows long division and multiplication problems with more digits than a hand calculator can accommodate.

Sources: *Discover*, vol. 6, (May 1985), pp. 37-43; *Science and Technology Illustrated*, p. 20; Shapiro, Max S. *Mathematics Encyclopedia*, p. 1.

1121 / What are Venn diagrams?

Venn diagrams are graphical representations of set theory, which use circles to show the logical relationships of the elements of different sets, using the logical operators AND, OR, and NOT. John Venn (1834-1923) first used them in his 1881 *Symbolic Logic*, in which he interpreted and corrected the work of George Boole (1815–1864) and Augustus de Morgan (1806–1871). While his attempts to clarify perceived inconsistencies and ambiguities in Boole's work are not widely accepted, the new method of diagramming is considered to be an improvement. Venn used shading to better illustrate inclusion and exclusion. Charles Dodgson (1832-1898), better known as Lewis Carroll, refined Venn's system, in particular by enclosing the diagram to represent the universal set.

Sources: *The Biographical Dictionary of Scientists: Mathematicians*, pp. 128-129.

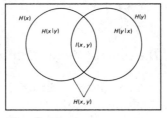

Venn diagrams

NUMBERS *See also:* MATHEMATICS, NUMBERS, AND COMPUTERS —MATHEMATICS; WEIGHTS, MEASURES, AND TIME—WEIGHTS AND MEASURES

1122

What is the value of pi out to 30 digits past the decimal point?

Pi (π) represents the ratio of the circumference of a circle to its diameter, used in calculating the area of a circle (πr^2) and the volume of a cylinder ($\pi r^2 h$) or cone. It is a "transcendental number," an irrational number with an exact value that can be measured to any degree of accuracy, but that can't be expressed as the ratio of two integers. In theory, the decimal extends into infinity, though it is generally rounded to 3.1416. Rounded to 30 digits past the decimal point, it equals 3.141592653589793238462643383279. In 1989, Gregory and David Chudnovsky, at Columbia University, New York City, calculated the value of pi to 1,011,961,691 decimal places. They performed the calculation twice on an IBM 3090 mainframe and on a CRAY–2 super computer with matching results.

Sources: Beyer, William H. *CRC Standard Mathematical Tables*, p. 5; Gibson, Carol. *The Facts On File Dictionary of Mathematics*, p. 139; *Guinness Book of Records 1992*, p. 270.

1123

Why is seven a magical number?

In magical lore and mysticism, all numbers are ascribed certain properties and energies. Seven is a number of great power, a magical number, a lucky number, a number of psychic and mystical powers, of secrecy and the search for inner truth. The origin of seven's power lies in the lunar cycle. Each of the moon's four phases lasts about seven days. The Sumerians, who based their calendar on the moon, gave the week seven days and declared the seventh and last day of each week to be uncanny. Life cycles on earth also have phases demarcated by seven, and there are seven years to each stage of human growth, seven colors to the rainbow, seven notes in the musical scale, seven petitions in the Lord's Prayer, and seven deadly sins. The seventh son of a seventh son is born with formidable magical and psychic powers. The number seven is widely held to be a lucky number, especially in matters of love and money.

Sources: Guiley, Rosemary E. *Moonscapes*, p. 123.

1124

When was a symbol for the concept "zero" first used?

Hindu mathematicians are usually given credit for developing a symbol for the concept "zero"; it appears in an inscription at Gwalior dated 870 C.E. It is certainly older than that; it is found in inscriptions dating from the 7th century in

Cambodia, Sumatra, and Bangka Island (off Sumatra). While there is no documentary evidence for the zero in China before 1247, some historians believe that it originated there, and arrived in India via Indochina.

Sources: Shapiro, Max S. *Mathematics Encyclopedia*, p. 214; Temple Robert. *The Genius of China*, p. 140.

1125
What is a perfect number?

A perfect number is a number equal to the sum of all its proper divisors (divisors smaller than the number), including 1. The first four perfect numbers are 6, 28, 496, and 8,126. The 31st discovered, which is the highest yet known, is $(2^{216,091} - 1)$ x $2^{216,090}$.

Sources: *Chambers Science and Technology Dictionary*, p. 659; *Guinness Book of Records 1992*, p. 269; *McGraw-Hill Encyclopedia of Science and Technology*, vol. 12, p. 241.

1126
What are Fibonacci numbers?

Fibonacci numbers are a series of numbers where each, after the second term, is the sum of the two preceding numbers—for example, 1, 1, 2, 3, 5, 8, 13, 21,... They were first described by Leonard Fibonacci (ca.1180–ca.1250), also known as Leonard of Pisa, as part of a thesis on series in his most famous book *Liber abaci* (The Book of the Calculator), published in 1202 and later revised by him. In addition to their function as recreational mathematics (providing entertainment to mathematicians who enjoy calculating series), Fibonacci numbers have proved useful in describing the positioning of leaves around plant stems, the spiral patterns in the heads of sunflowers, and the genealogy of honeybees.

Sources: *The Biographical Dictionary of Scientists: Mathematicians*, pp. 51–52; James, Glenn. *Mathematics Dictionary*, p. 151; Thomas, David A. *Math Projects for Young Scientists*, pp. 41–42.

1127
How large is a googol?

A googol is 10^{100} (the number 1 followed by 100 zeros). Unlike most other names for numbers, it does not relate to any other numbering scale. The American mathematician Edward Kasner first used the term in 1938; when searching for a term for this large number, Kasner asked his nephew, Milton Sirotta, then about nine years old, to suggest a name. The googolplex is 10 followed by a googol of 0's, represented as 10^{googol}.

Other very large numbers

Name	Value in powers of 10	Number of 0's	Number of groups of three 0's after 1,000
Billion	10^9	9	2
Trillion	10^{12}	12	3
Quadrillion	10^{15}	15	4
Quintillion	10^{18}	18	5
Sextillion	10^{21}	21	6
Septillion	10^{24}	24	7
Octillion	10^{27}	27	8
Nonillion	10^{30}	30	9
Decillion	10^{33}	33	10
Undecillion	10^{36}	36	11
Duodecillion	10^{39}	39	12
Tredecillion	10^{42}	42	13
Quattuor-decillion	10^{45}	45	14
Quindecillion	10^{48}	48	15
Sexdecillion	10^{51}	51	16
Septen-decillion	10^{54}	54	17
Octodecillion	10^{57}	57	18
Novemdecillion	10^{60}	60	19
Vigintillion	10^{63}	63	20
Centillion	10^{303}	303	100

The British, French, and Germans use a different system for naming denominations above one million. The googol and googolplex are rarely used outside the United States.

Sources: Asimov, Isaac. *Asimov on Numbers*, p. 44; Motz, Lloyd, and Jefferson Hane Weaver. *Conquering Mathematics*, pp. 17, 18–19; *Webster's Ninth New Collegiate Dictionary*, pp. 528, 810.

1128
What is the largest number mentioned in the Bible?

The largest number specifically named in the Bible is a thousand thousand; i.e., a million. It is found in 2 Chronicles 14:9.

Sources: Asimov, Isaac. *Asimov on Numbers*, p. 60.

1129
Is it possible to count to infinity?

No. Very large finite numbers are not the same as infinite numbers. Infinite numbers are defined as being unbounded, or without limit. Any number that can be reached by counting or by representation of a number followed by billions of 0's is a finite number

Sources: Motz, Lloyd, and Jefferson Hane Weaver. *Conquering Mathematics*, p. 17.

1130

How are names for large and small quantities constructed in the metric system?

Each prefix listed below can be used in the metric system and with some customary units. For example, centi + meter = centimeter, meaning one-hundredth of a meter.

Prefix	Power	Numerals
Exa-	10^{18}	1,000,000,000,000,000,000
Peta-	10^{15}	1,000,000,000,000,000
Tera-	10^{12}	1,000,000,000,000
Giga-	10^{9}	1,000,000,000
Mega-	10^{6}	1,000,000
Myria-	10^{5}	100,000
Kilo-	10^{3}	1,000
Hecto-	10^{2}	100
Deca-	10^{1}	10
Deci-	10^{-1}	0.1
Centi-	10^{-2}	0.01
Milli-	10^{-3}	0.001
Micro-	10^{-6}	0.000001
Nano-	10^{-9}	0.000000001
Pico-	10^{-12}	0.000000000001
Femto-	10^{-15}	0.000000000000001
Atto-	10^{-18}	0.000000000000000001

Sources: Johnstone, William D. *For Good Measure*, p. 208; *New York Public Library Desk Reference*, p. 28.

1131

What are the Roman numerals?

Roman numerals are symbols that stand for numbers. They are written using seven basic symbols: I (1), V (5), X (10), L (50), C (100), D (500), and M (1,000). Sometimes a bar is place over a numeral to multiply it by 1,000. For example, 5,000 may be written $\overline{V}$. A smaller numeral appearing before a larger numeral indicates that the smaller numeral is subtracted from the larger one. This notation is general used for 4's and 9's; for example, 4 is written IV, 9 is IX, 40 is XL, and 90 is XC.

Sources: *World Book Encyclopedia*, vol. 16, pp. 413–414.

1132

What is the largest prime number presently known?

A prime number is one that is evenly divisible only by itself and 1. The integers 1, 2, 3, 5, 7, 11, 13, 17, and 19 are prime numbers. Euclid (ca. 300 B.C.E.) proved that there is no "largest prime number," because any attempt to define the largest results in a paradox. If there is a largest prime number (P), adding 1 to the product of all primes up to and including P, $1 + (1 \times 2 \times 3 \times 5 \times \dots \times P)$, yields a number that is itself a prime number, because it cannot be divided evenly by any of the known primes.

In 1989, a team of mathematicians and computer scientists at Amdahl Corporation discovered a new largest number known to be prime. The number, $391,581 \times 2^{216193}-1$, has 65,087 decimal digits.

Source: Motz, Lloyd, and Jefferson Hane Weaver. *Conquering Mathematics*, pp. 22–25; *Yearbook of Science and the Future 1991*, p. 377.

METALS AND OTHER MATERIALS

COAL *See ENERGY—NON-NUCLEAR FUELS*

MAN-MADE PRODUCTS *See also: METALS AND OTHER MATERIALS—METALS*

1133
When and where was gunpowder probably invented?

The explosive mixture of saltpeter (potassium nitrate), sulfur, and charcoal called *gunpowder* was known in China at least by 850 C.E., and probably was discovered by Chinese alchemists searching for components to make artificial gold. Early mixtures had too little saltpeter (50%) to be truly explosive; 75% minimum is needed to get a detonation. So the first use of the mixture was in making fireworks. Later, the Chinese used it in incendiary-like weapons, for warfare. Eventually it is thought that the Chinese found the correct proportions to utilize its explosive effects in rockets and "bamboo bullets." However, some authorities still maintain that the "Chinese gunpowder" really had only pyrotechnic qualities, and "true" gunpowder was an European invention. Roger Bacon (1214–1292) had a formula for it and so might have the German monk Berthold Schwartz (1353). But its first European use depended on the development of firearms in the 14th century. Not until the 17th century was gunpowder used in peacetime in mining and civil engineering applications.

Sources: Panati, Charles. *Panati's Browser's Book of Beginnings*, pp. 346-347; Sanders, Dennis. *The First of Everything*, p. 64; Temple, Robert. *The Genius of China*, pp. 224-229.

1134
How are colored fireworks made?

Fireworks existed in ancient China in the 9th century C.E. where saltpeter (potassium nitrate), sulfur, and charcoal were mixed to produce the dazzling effects. Magnesium burns with a brilliant white light and is widely-used in making flares and fireworks. Various other colors can be produced by adding certain substances to the flame. Strontium compounds color the flame scarlet and barium compounds produce a yellowish-green color; borax produces a green color, and lithium a purple color.

Sources: Hiscox, Gardner D. *Henley's Twentieth Century Book of Formulas, Processes, and Trade Secrets*, p. 609; Vergara, William C. *Science in Everyday Life*, p. 267.

1135
Is glass a solid or a liquid?

Even at room temperature, glass appears to be a solid in the ordinary sense of the word. However, it actually is a fluid with an extremely high viscosity. It has been documented that century-old windows show signs of flow. The internal friction of fluids is called viscosity. It is a property of fluids by which the flow motion is gradually damped (slowed) and dissipated by heat. Viscosity is a familiar phenomenon in daily life. An opened bottle of wine can be poured: the wine flows easily under the influence of gravity. Maple syrup, on the other hand, cannot be poured so easily; under the action of gravity, it flows sluggishly. The syrup has a higher viscosity than the wine.

Glass is usually composed of mixed oxides based around the silicon dioxide (SiO_2) unit. A very good electrical insulator, and generally inert to chemicals, commercial glass is manufactured by the fusing of sand (silica, SiO_2), limestone ($CaCO_2$), and soda (sodium carbonate, Na_2CO_3) at temperatures around 2552°F to 2732°F (1400 to 1500°C). On cooling the melt becomes very viscous and at about 932°F (500°C, known as glass transition temperature), the melt "solidifies" to form soda glass. Small amounts of metal oxides are used to color glass, and its physical properties can be changed by the addition of substances like lead oxide (to increase softness, density, and refractive ability for cutglass and lead crystal), and borax (to lower significantly thermal expansion for cookware and laboratory equipment). Other materials can be used to form glasses if cooled sufficiently rapid from the liquid or gaseous phase to prevent an ordered crystalline structure from forming. Glasses such as obsidian occur naturally. Glass objects might have been made as early as 2500 B.C.E. in Egypt and Mesopotamia, and glass blowing developed about 100 B.C.E. in Phoenicia.

Sources: *Encyclopedia of Physical Science and Technology*, vol. 6, p. 266; Magill, Frank N. *Magill's Survey of Science. Physical Science Series*, vol. 6, p. 2685; *Oxford Illustrated Encyclopedia of Invention and Technology*, p. 156.

1136
What is crown glass?

In the early 1800s, window glass was called crown glass. It was made by blowing a bubble, then spinning it until flat. This left a sheet of glass with a bump, or crown, in the center. This blowing method of window-pane making required great skill and was very costly. Still, the finished crown glass produced a distortion through which everything looked curiously wavy, and the glass itself was also faulty and uneven. By the end of the 19th century, flat glass was mass-produced and was a common material. The cylinder method replaced the old method, and used compressed air to produce glass which could be slit lengthwise, reheated, and allowed to flatten on an iron table under its own weight. New furnaces and better polishing machines made the production of plate-glass a real industry. Today, glass is produced by a float-glass process which reheats the newly-formed ribbon and glass and allows it to cool without touching a solid surface. This produces inexpensive glass that is flat and free from distortion.

Sources: Diamond, Freda. *The Story of Glass*, pp. 111-112; Douglas, R.W., and Susan Frank. *A History of Glassmaking*, pp. 133-161; *World Book Encyclopedia*, vol. 8, p. 219.

1137 ╲╱ How is bulletproof glass made?

Bulletproof glass is composed of two sheets of plate glass with a sheet of transparent resin in between, molded together under heat and pressure. When subjected to a severe blow, it will crack without shattering. Today's bulletproof glass is a development of laminated or safety glass, invented by the French chemist Edouard Benedictus. It is basically a multiple lamination of glass and plastic layers.

Sources: Brady, George S. *Materials Handbook*, p. 550; Diamond, Freda. *The Story of Glass*, pp. 128-130.

1138 ╲╱ When were glass blocks invented?

Glass building bricks were introduced in 1931. They were invented in Europe in the early 1900s as thin blocks of glass supported by a grid. They have been in and out of favor since they were introduced. They are small, the largest being 12 inches by 12 inches (30 centimeters by 30 centimeters). They offer the following advantages: they allow natural light to filter through, but their patterns can afford privacy; they have the insulating value of a 12 inch-thick (30 centimeters) concrete wall; they absorb outside noise; and they are much more secure than ordinary flat glass.

Sources: DuVall, Nell. *Domestic Technology*, p. 254; Pennycook, Bob. *Building With Glass Blocks*, pp. 1, 3-4.

1139 ╲╱ Who invented thermopane glass?

Thermopane insulated window glass was invented by C.D. Haven in the United States in 1930. It is two sheets of glass that are bonded together in such a manner that they enclose a captive air space between the two pieces of glass. Often this space is filled with an inert gas which increases the insulating quality of the window. The insulation value of a window is determined by the infiltrated heat loss and by the conducted heat loss. Glass is also one of the best transparent materials because it allows the short wavelengths of solar radiation to pass through it, but prohibits nearly all of the long waves of reflected radiation from passing back through it.

Sources: Drake, George R. *Weatherizing Your Home*, pp. 153-154; DuVall, Nell. *Domestic Technology*, p. 254.

1140 / Who developed fiberglass?

Coarse glass fibers were used for decoration by the Egyptians before the time of Christ. Other developments were made in Roman times. Parisian craftsman Dubus-Bonnel was granted a patent for the spinning and weaving of drawn glass strands in 1836. In 1893, the Libbey Glass Company exhibited lampshades, at the World's Columbian Exposition in Chicago, that were made of coarse glass thread woven together with silk. However, this was not a true woven glass. Between 1931 and 1939, the Owens Illinois Glass Company and the Corning Glass Works developed practical methods of making fiberglass commercially. Once the technical problem of drawing out the glass threads to a fraction of their original thinness was solved—basically an endless strand of continuous glass filament as thin as 1/5000 of an inch—the industry began to produce glass fiber for thermal insulation and air filters, among other uses. When glass fibers were combined with plastics during World War II, a new material was formed. Glass fibers did for plastics what steel did for concrete—gave strength and flexibility. Glass-fiber-reinforced plastics (GFRP) became very important in modern engineering. Fiberglass combined with epoxy resins and thermosetting polyesters are now used extensively in boat and ship construction, sporting goods, automobile bodies, and circuit boards in electronics.

Sources: Diamond, Freda. *The Story of Glass*, pp. 197-224; Giscard d'Estaing, Valerie-Anne. *The World Almanac Book of Inventions*, p. 274; *Oxford Illustrated Encyclopedia of Invention and Technology.*, p. 129.

1141 / How is the glass used in movie stunts made?

The "glass" might be made of candy (sugar boiled down to a translucent pane) or plastic. This looks like glass and will shatter like glass, but will not cut a performer.

Sources: *Today's Health*, vol. 40 (May 1962), p. 33.

1142 / What is buckminsterfullerene?

It is a large molecule in the shape of a soccer ball, containing 60 carbon atoms, whose structure is the shape of a truncated icosahedron (a hollow, spherical object with 32 faces, 12 of them pentagons and the rest hexagons). This molecule was named buckminsterfullerene because of the structure's resemblance to the geodesic domes designed by American architect R. Buckminster Fuller (1895–1983). The molecule was formed by vaporizing material from a graphite surface with a laser. Large molecules containing only carbon atoms have been known to exist around certain types of cabon-rich stars. Similar molecules are also thought to be present in soot formed during the incomplete combustion of organic materials. Chemist Richard Smalley identified buckminsterfullerene in 1985 and speculated that it may be fairly common throughout the universe. Since that time, other stable, large, even-numbered carbon clusters have been produced. This new class of molecules has been called "fullerenes" since they all seem to have the structure of a geodesic dome. Buckministerfullerene (C_{60}) seems to function as an insulator, conductor, semi-conductor and superconductor in various compounds.

Sources: *The Almanac of Science and Technology*, pp. 209-210.

Buckminsterfullerene

1143
What is the float glass process?

Manufacture of high-quality flat glass, needed for large areas and industrial uses, depends on the float glass process, invented by Alistair Pilkington in 1952. The float process departs from all other glass processes where the molten glass flows from the melting chamber into the float chamber which is a molten tin pool approximately 160 feet (49 meters) long and 12 feet (3.5 meters) wide. During its passage over this molten tin, the hot glass assumes the perfect flatness of the tin surface and develops excellent thickness uniformity. The finished product is as flat and smooth as plate glass without having been ground and polished.

Sources: De Bono, Edward. *Eureka!*, p. 94; *Handbook of Glass Manufacture*, vol. 2, p. 711.

1144
How is dry ice made?

Dry ice is composed of carbon dioxide, which at normal temperatures is a gas. The carbon dioxide is stored and shipped as a liquid in tanks that are pressurized at 1073 pounds per square inch. To make dry ice, the carbon dioxide liquid is withdrawn from the tank and allowed to evaporate at a normal pressure in a porous bag. This rapid evaporation consumes so much heat that part of the liquid CO_2 freezes to a temperature of -109°F (-78°C). The frozen liquid is then compressed by machines into blocks of "dry ice" which will melt into a gas again when set out at room temperature.

It was first made commercially in 1925 by the Prest-Air Devices Company of Long Island City, New York through the efforts of Thomas Benton Slate. It was used by Schrafft's of New York in July 1925 to keep ice cream from melting. The first large sale of dry ice was made later in that year to Breyer Ice Cream Company of New York.

Sources: Kane, Joseph N. *Famous First Facts*, p. 316; Sutton, Caroline, and Duncan M. Anderson. *How Do They Do That?*, p. 154.

1145
Why is sulfuric acid so important?

Sometimes called "oil of vitriol," sulfuric acid (H_2SO_4) has become one of the most important of all chemicals. It was little used until it became essential for the manufacture of soda in the 18th century. It is prepared industrially by the reaction of water with sulfur trioxide, which in turn is made by chemical combination of sulfur dioxide and oxygen by one of two processes (the contact process or the chamber process). Many manufactured articles in common use depend in some way on sulfuric acid for their production. Its greatest use is in the production of fertilizers, but it is also used in the refining of petroleum, and production of automobile batteries, explosives, pigments, iron and other metals, and paper pulp.

Sources: McNeill, Ian. *An Encyclopedia of the History of Technology*, pp. 221–223; *Van Nostrand Reinhold Encyclopedia of Chemistry*, pp. 909–910; *World Book Encyclopedia*, vol. 18, p. 967.

1146
What is solder?

Solder is an alloy of two or more metals used for joining other metals together. The most common solder is called half and half, or "plumber's" solder, and is composed of equal parts of lead and tin. Other metals used in solder are aluminum, cadmium, zinc, nickel, gold, silver, palladium, bismuth, copper, and antimony. Various melting points to suit the work are obtained by varying the proportions of the metals.

Solder is an ancient joining method, mentioned in the Bible (*Isaiah* 41:7).

There is evidence of its use in Mesopotamia some 5 thousand years ago, and later in Egypt, Greece, and Rome. For the near future, it appears that as long as a combination of conductors, semiconductors, and insulators is used to build circuitry, based on electrical and magnetic impulses, solder will remain indispensable.

Sources: Brady, George S. *Materials Handbook*, pp. 750-752; Manko, Howard H. *Solders & Soldering*, p. xv.

1147 / When was cement first used?

Cements are finely ground powders that, when mixed with water, set to a hard mass. The cement used by the Egyptians was calcined gypsum, and both the Greeks and Romans used a cement of calcined limestone. Roman concrete (a mixture of cement, sand, and some other fine aggregate) was made of broken brick embedded in a pozzolanic lime mortar. This mortar consisted of lime putty mixed with brick dust or volcanic ash. Hardening was produced by a prolonged chemical reaction between these components in the presence of moisture. With the decline of the Roman empire, concrete fell into disuse. The first step toward its reintroduction was in about 1790, when the English engineer, John Smeaton (1724–1792), found that when lime containing a certain amount of clay was burnt, it would set under water. This cement resembled what had been made by the Romans. Further investigations by James Parker in the same decade led to the commercial production of natural hydraulic cement. In 1824, Englishman Joseph Aspdin (1799–1855) took out a patent for what he called "portland cement," a material produced from a synthetic mixture of limestone and clay. He called it "portland" because it resembled a building stone that was quarried on the Isle of Portland off the coast of Dorset. The manufacture of this cement spread rapidly to Europe and the United States by 1870. Today, concrete is often reinforced or prestressed, increasing its load-bearing capabilities.

Sources: Taylor, Walter H. *Concrete Technology and Practice*, pp. 3-4.

1148 / What is "aqua regia"?

"Aqua regia," also known as nitrohydrochloric acid, is a mixture of 1 part concentrated nitric acid and 3 parts of concentrated hydrochloric acid. The chemical reaction between the acids makes it possible to dissolve all metals except silver. The reaction of metals with nitrohydrochloric acid typically involves oxidation of the metals to a metallic ion and the reduction of the nitric acide to nitric oxide. The term comes from Latin and means royal water. It was named by the alchemists for its ability to dissolve gold, sometimes called the "royal metal."

Sources: *Van Nostrand Reinhold Encyclopedia of Chemistry*, p. 94; *McGraw-Hill Encyclopedia of Science and Technology*, vol. 2, p. 30; *World Book Encyclopedia*, vol. 1, p. 578.

1149
How was early macadam different from modern paved roads?

Macadam roads developed originally in England and France and are named after the Scottish road builder and engineer, John Louden MacAdam (1756-1836). The term "macadam" originally designated road surface or base in which clean, broken, or crushed ledge stone was mechanically locked together by rolling with a heavy weight and bonded together by stone dust screenings which were worked into the spaces and then "set" with water. With the beginning of the use of bituminous material (tar or asphalt), the term "plain macadam," "ordinary macadam," or "waterbound macadam" was used to distinquish the original type from the newer "bituminous macadam. Waterbound macadam surfaces are almost never built now in the United States, mainly because they are expensive and the vacuum effect of vehicles loosens them. Many miles of bituminous macadam roads are still in service, but their principal disadvantages are their high crowns and narrowness. Today's roads that carry very heavy traffic are usually surfaced with very durable portland cement.

Sources: Oglesby, Clarkson H., and R. Gary Hicks. *Highway Engineering*, p. 652; Scott, John S. *Dictionary of Civil Engineering*, pp. 169, 298; *Standard Handbook for Civil Engineers*, pp.16-40 – 16-41.

1150
What is Belgian block?

Belgian block is a road-building material, first used in Brussels, Belgium, and introduced into New York about 1850. Its shape is a truncated pyramid with a base of about 5 or 6 inches square (13–15 centimeters) and a depth of 7 to 8 inches (18–20.5 centimeters). The bottom of the block is not more than 1 inch (2.5 centimeters) different from the top. The original blocks were cut from trap-rock from the Palisades of New Jersey.

Belgian blocks replaced cobblestones mainly because their regular, cubical shape allowed them to remain in place better than cobblestones. They were not universally adopted, however, because they would wear round and create joints or openings which would then form ruts and hollows. Although they provided a smooth surface compared to the uneven cobblestones, they still made for a rough and noisy ride.

Sources: Byrne, Austin T. *A Treatise on Highway Construction*, pp. 81-82; Frost, Harwood. *The Art of Roadmaking*, pp. 314-315.

1151
Who developed the process for making ammonia?

Known since ancient times, ammonia (NH_3) has been commercially important for more than 100 years and has become the second largest chemical in terms of tonnage production and the first chemical in value of production. The first breakthrough in the large-scale synthesis of ammonia resulted from the work of Fritz Haber (1863–1934). In 1913, Haber found that ammonia could be produced by combining nitrogen and hydrogen ($N_2 + 3H_2 \rightleftharpoons 2NH_3$) with a catalyst (iron oxide with small quantities of cerium and chromium) at 131°F (55°C) under a pressure of about 200 atmospheres. The process was adapted for industrial-quality production by Karl Bosch (1874–1940). Thereafter, many improved ammonia-synthesis systems, based on the Haber-Bosch process, were commercialized using various operating conditions and synthesis loop-designs. One of the five top inorganic chemicals produced in the United States, it is used in refrigerants, detergents and other cleaning preparations, explosives, fabrics, and fertilizers. A little over 75% of ammonia production in the United States is used for fertilizers. It has been shown to produce cancer of the skin in humans, in doses of 1,000 milligrams (or 1 gram) per kilogram (2.2 pounds) of body weight.

Sources: *Van Nostrand Reinhold Encyclopedia of Chemistry*, pp. 58-59; Winter, Ruth. *A Consumer's Dictionary of Household, Yard and Office Chemicals*, p. 50.

▽1152▽ What is creosote?

Creosote is a yellowish poisonous oily liquid obtained from the distillation of coal tar. Coal tar constitutes the major part of the liquid condensate obtained from the "dry" distillation or carbonization of coal to coke. Crude creosote oil, also called dead oil or pitch oil, is used as a wood preservative. Railroad ties, poles, fence posts, marine pilings, and lumber for outdoor use are impregnated with creosote in large cylindrical vessels. This treatment can greatly extend the useful life of wood that is exposed to the weather.

Sources: Brady, George S. *Materials Handbook*, p. 244; *Van Nostrand Reinhold Encyclopedia of Chemistry*, pp. 264-265.

▽1153▽ How is sandpaper made?

Sandpaper is a coated abrasive which consists of a flexible-type backing (paper) upon which a film of adhesive holds and supports a coating of abrasive grains. Various types of resins and hide glues are used as adhesives. The first record of a coated abrasive is in 13th century China, when crushed seashells were bound to parchment using natural gums. The first known article on coated abrasives was published in 1808 and described how calcined, ground pumice was mixed with varnish and spread on paper with a brush. Most abrasive papers are now made with aluminum oxide or silicon carbide, although the term sandpapering is still used. Quartz grains are also used for wood polishing. The paper used is heavy, tough, and flexible, and the grains are bonded with a strong glue.

Sources: Brady, George S. *Materials Handbook*, p. 698; *Coated Abrasives: Modern Tool of Industry*, pp. 7-8; *World Book Encyclopedia*, vol. 21, p. 401.

1154
Why is titanium dioxide the most widely used white pigment?

Titanium dioxide has become the predominant white pigment in the world because of its high refractive index, lack of absorption of visible light, ability to be produced in the right size range, and its stability and nontoxicity. It is the whitest known pigment, unrivalled in respect of color, opacity, stain resistance, and durability; it is also non-toxic. The main consuming industries are paint, printing inks, plastics, and ceramics, which together account for 60% to 70% of the total demand.

Sources: *Encyclopedia of Chemical Technology*, vol. 23, p. 143; Morgans, W.M. *Outlines of Paint Technology*, p. 52; *The Timetable of Technology*, p. 28.

1155
What is the chemical formula for TNT?

TNT is the abbreviation for (2,4,6-*tri*nitrotoluene ($C_7H_5N_3O_6$). TNT is a powerful, highly explosive compound widely used in conventional bombs. Discovered by J. Wilbrand in 1863, it is made by treating toluene with nitric acid and sulfuric acids. This yellow crystalline solid with a low melting point, has low shock sensitivity and even burns without exploding. This makes it safe to handle and cast; but once detonated, it explodes violently.

Sources: *Hawley's Condensed Chemical Dictionary*, p. 1191; Kerrod, Robin. *The Concise Dictionary of Science*, p. 231; *Oxford Illustrated Encyclopedia of Invention and Technology*, p. 357.

1156
What does the symbol H_2O_2 stand for?

Hydrogen peroxide, a syrupy liquid used as a strong bleaching, oxidizing, and disinfecting agent, is the compound. It is made from barium peroxide and diluted phosphoric acid. A 3% solution of hydrogen peroxide is used medicinally as an antiseptic and germicide. Undiluted it can cause burns to human skin and mucous membranes, is a fire and explosion risk, and can be highly toxic.

Sources: *Concise Chemical and Technical Dictionary*, p. 630; Winter, Ruth. *A Consumer's Dictionary of Household, Yard and Office Chemicals*, pp. 159-160.

1157
Who invented dynamite?

Dynamite was no chance discovery but the result of a methodical search by the Swedish technologist, Alfred Nobel (1833-1896). Nitroglycerine had been discovered in 1849, by the Italian organic chemist, Ascanio Sobriero (1812–1888), but it was so sensitive and difficult to control that it was useless. Nobel sought to turn nitroglycerine into a manageable solid by absorbing it into a porous substance. In

1866–1867, he tried an unusual though by no means rare mineral, kieselguhr, and created a dough-like explosive that was controllable. He also invented a detonating cap incorporating mercury fulminate with which nitroglycerine could be detonated at will. Nobel made a great fortune and bequeathed it to a foundation for awarding prizes for contributions to science, literature, and the promotion of peace.

Sources: De Bono, Edward. *Eureka!*, pp. 76-77; *Dictionary of Scientific Biography*, vol. 10, p. 132.

1158
What is carbon black?

Carbon black is finely divided carbon produced by incomplete combustion of methane or other hydrocarbon gases (by letting the flame impinge on a cool surface). This forms a very fine pigment containing up to 95% carbon, which gives a very intense black color that is widely used in paints, inks, and protective coatings and as colorants for paper and plastics.

Sources: *Chambers Science and Technology Dictionary*, p. 133; *Van Nostrand Reinhold Encyclopedia of Chemistry*, p. 170.

1159
Who invented teflon?

In 1938, the American engineer Roy J. Plunkett (b. 1910), at DuPont de Nemours, discovered the *po*lymer of *tetra*fluor*et*hylene (PTFE) by accident. This fluorocarbon is marketed under the name of Fluon in Great Britain and Teflon® in the United States. Patented in 1939 and first exploited commercially in 1954, PTFE is resistant to all acids and has exceptional stability and excellent electrical insulating properties. It is used in making piping for corrosive materials, in insulating devices for radio transmitters, in pump gaskets, and in computer microchips. In addition, its non-stick properties make PTFE an ideal material for surface coatings. In 1956, French engineer Marc Gregoire discovered a process whereby he could fix a thin layer of teflon on an aluminum surface. He then patented the process of applying it to cookware, and the no–stick frying pan was created.

Sources Giscard d'Estaing, Valerie-Anne. *The Second World Almanac Book of Inventions*, pp. 12, 272; Panati, Charles. *Panati's Extraordinary Origins of Everyday Things*, pp. 105-107.

1160
What is the lightest solid material?

The lightest substance is silica aerogels, made of tiny spheres of bonded silicon and oxygen atoms linked together into long strands separated with air pockets. They appear almost like frozen wisps of smoke. In February, 1990, the lightest of these aerosols, having a density of only 5 ounces per cubic foot was produced at the

Lawrence Livermore Laboratory in California. It will be used in window insulation, as traps for sampling cosmic dust in space, and in liquid rocket fuel storage.

Sources: Britannica Book of the Year 1991, p. 141; *Guinness Book of Records 1992*, p. 258.

1161 / *When was plastic first invented?*

About 1850 Alexander Parkes (1813-1890) experimented with nitrocellulose (or guncotton). Mixed with camphor, it gave a hard but flexible transparent material, which he called "Parkesine." He teamed up with a manufacturer to produce it, but there was no call for it, and the firm went bankrupt. An American, John Wesley Hyatt (1837-1920), acquired the patent in 1868 with the idea of producing artificial ivory for billiard-balls. Improving the formula and with an efficient manufacturing process, he marketed the material, to make a few household articles, under the name "celluloid." Celluloid became the medium for cinematography: celluloid strips coated with a light-sensitive "film" were ideal for shooting and showing movie pictures. Celluloid was the only plastic material until 1904 when a Belgian scientist, Leo Hendrik Baekeland (1863-1944), succeeded in producing a synthetic shellac from formaldehyde and phenol. Called *bakelite*, it was the first of the "thermosetting" plastics (i.e., synthetic materials which, having once been subjected to heat and pressure, became extremely hard and resistant to high temperatures).

Sources: De Bono, Edward. Eureka!, p. 89.

1162 / *How can plastics be made biodegradable?*

Plastic does not rust nor rot. This is an advantage in its usage, but when it comes to disposal of plastic, the advantage turns into a liability. Degradable plastic has starch in it so that it can be attacked by starch-eating bacteria to eventually disintegrate the plastic into bits. Chemically degradable plastic can be broken up with a chemical solution that dissolves it. Biodegradable plastic is also used in surgery where stitching material is the type of plastic that slowly dissolves in the body fluids. Photodegradable plastic contains chemicals that slowly disintegrate when exposed to light over a period of one to three years. 25% of the plastic yokes used to package beverages are made from a plastic called Ecolyte® which is photodegradable.

Sources: How in the World?, p. 130.

1163 / *Who discovered deuterium?*

American chemist Harold C. Urey (1893-1981), 1934 winner of the Nobel Prize for Chemistry, discovered deuterium (heavy hydrogen, symbol D) in 1931 with F.G.

Brickwedde and G.M. Murphy. This isotope (a form of an element that differs in the number of neutrons and atomic weight) has twice the weight of hydrogen, while all the other isotopes differed slightly in their atomic weights. Deuterium and its oxide makes heavy water (D_2O), which is used to slow down the neutrons in atomic piles of nuclear reactors.

Sources: Dickerson, Richard Earl. *Chemical Principles*, p. 4; *The Great Scientists*, vol. 11, p. 203; Farber, Edward. *Nobel Prize Winners in Chemistry, 1901-1961*, p. 137.

1164
Who made the first successful synthetic gemstone?

In 1902, Auguste Victor Louis Verneuil (1856-1913), synthesized the first man-made gemstone, a ruby. Verneuil perfected a "flame-fusion" method of producing crystals of ruby and other corundums within a short time period.

Sources: De Bono, Edward. *Eureka!*, p. 93; Nassau, Kurt. *Gems Made by Man*, p. 27.

METALS

1165
What are the precious metals?

This is a general term for expensive metals that are used for making coins, jewelry, and ornaments. The name is limited to gold, silver, and platinum. Expense or rarity does not make a metal precious, but rather, it is a value set by law which states that the object made of these metals has a certain intrinsic value. The term is not synonymous with "noble metals," although a metal (such as platinum) may be both noble and precious.

Sources: Brady, George S. *Materials Handbook*, p. 644; Savitskii, E.M. *Handbook of Precious Metals*, pp. xv-xix.

1166
Which metal is the main component of pewter?

Tin. Roman pewter has about 70% tin. The best pewter used for expensive articles today contains 100 parts tin, 8 parts antimony, 2 parts bismuth, and 2 parts copper. This alloy is easy to work with, does not become brittle when repeatedly beaten and can be worked cold. However, it is too soft to use for heavy tools or weapons, so it is confined to domestic utensils.

Sources: De Bono, Edward. *Eureka!*, p. 87.

1167 / What are the chief gold-producing countries?

South Africa is by far the leading producer of gold; other principal areas of production are in the western United States (Utah, Nevada, California, South Dakota, Alaska), Australia, Mexico, Canada, China, India, and the Soviet Union.

1988 Top World Gold Producers

Country	Number of tons	Number of Tonnes (Metric tons)	% of world total gold production
South Africa	684.5	621.0	32.6%
USSR	308.6	280.0	14.7%
United States	226.3	205.3	10.8%
Australia	167.5	152.0	8.0%
Canada	141.1	128.5	6.8%

In the United States, Nevada is the leading gold producer, with California a distant second, followed by South Dakota in third place.

Sources: Crowson, Philip. *Minerals Handbook 1990–1991*, p. 116; *McGraw-Hill Encyclopedia of Science & Technology*, vol. 8, p. 161; *Minerals Yearbook 1989*, vol. 1, p. 457.

1168 / How thick is gold leaf?

Gold leaf is pure gold which is hammered so thin that it takes 300,000 units to make a stack one inch high. The thickness of a single gold leaf is .0000035 inches (3½ millionths of an inch), although this may vary widely according to which manufacturer makes it. Also called gold foil, it is used for architectural coverings and for hot-embossed printing on leather.

Sources: Calasibetta, Charlotte M. *Fairchild's Dictionary of Fashion*, p. 254; Le Blanc, Raymond. *Gold-Leaf Techniques*, p. 2.

1169 / What is 24 karat gold?

The term "karat" refers to the percentage of gold versus the percentage of an alloy in an object. Gold is too soft to be usable in its purest form; it has to be mixed with other metals.

Karatage	Percentage of fine gold
24	100
22	91.75
18	75
14	58.5
12	50.25
10	42

Karatage	Percentage of fine gold
9	37.8
8	33.75

Sources: Rovin, Jeff. *Laws of Order*, p. 32.

1170 / Is white gold really gold?

White gold is the name of a class of jeweler's white alloys used as substitutes for platinum. Different grades vary widely in composition, but usual alloys consist of from 20% to 50% nickel, with the balance gold. A superior class of white gold is made of 90% gold and 10% palladium. Other elements used may be copper and zinc. The main use of these alloys is to give the gold a white color.

Sources: Brady, George S. *Materials Handbook*, p. 882; Thrush, Paul W. *A Dictionary of Mining, Mineral, and Related Terms*, p. 1235.

1171 / What is German silver?

Nickel silver, sometimes known as German silver, is a silver-white alloy composed of 52% to 80% copper, 10% to 35% zinc, and 5% to 35% nickel. It may also contain a small percent of lead and tin. There are other forms of nickel silver, but the term "German silver" is the name used in the silverware trade.

Sources: Brady, George S. *Materials Handbook*, pp. 539-540; *The McGraw-Hill Dictionary of Scientific and Technical Terms*, p. 1273; Thrush, Paul W. *A Dictionary of Mining, Mineral, and Related Terms*, p. 490.

1172 / What is the Q-BOP process for steelmaking?

A variation of the *basic oxygen process* (BOP) involves blowing high-purity oxygen through a bath of molten pig iron. This is called the Q-BOP process, used in the production of low-alloy steel, which is defined as steel that has no more than 5% total combined alloying elements. It has mainly a surface hardness. The depth of hardness depends on alloy content. There are five types of processes currently employed in the production of low-alloy steels: oxygen top-blowing, known as LD (Linz–Donawitz) process, BOF (*basic oxygen furnace*), or BOP; oxygen and lime bottom blowing, known as OBM or Q- BOP; top and bottom mixed blowing; open-hearth furnace; and electric-arc furnace.

Sources: *The Making, Shaping and Treating of Steel*, pp. 24, 463.

1173
What is high speed steel?

High speed steel is a general name for high alloy steels which retain their hardness at very high temperatures and are used for metal-cutting tools. All high speed steels are based on either tungsten or molybdenum (or both) as the primary heat-resisting alloying element. These steels require a special heat so that their unique properties can be fully realized. This procedure consists of heating to a high temperature of 2150 to 2400°F (1175 to 1315°C) to obtain solution of a substantial percentage of the alloy carbides, quenching to room temperature, tempering at 1000 to 1150°F (535 to 620°C), and again cooling to room temperature.

Sources: Brady, George S. *Materials Handbook*, p. 394; *The Making, Shaping and Treating of Steel*, p. 1309.

1174
Why are alchemical symbols for metals and astrological symbols for planets identical?

The ancient Greeks and Romans knew seven metals and also knew seven "planets" (the five nearer planets plus the sun and the moon). They related each planet to a specific metal. Alchemy, originating in about the 3rd century B.C.E., focused on changing base metals, such as lead, into gold. Although at times alchemy degenerated into mysticism, it contained centuries of chemical experience, which provided the foundation for the development of modern chemistry.

English name	Chemical symbol	Latin name	Alchemical symbol
Gold	Au	*aurum*	☉ (Sun)
Silver	Ag	*argentum*	☽ (Moon)
Copper	Cu	*cuprum*	♀ (Venus)
Iron	Fe	*ferrum*	♂ (Mars)
Mercury	Hg	*hydrargyrum*	☿ (Mercury)
Tin	Sn	*stannum*	♃ (Jupiter)
Lead	Pb	*plumbum*	♄ (Saturn)

Sources: *Oxford Illustrated Encyclopedia of Invention and Technology*, p. 10; Smith, Richard Furnald. *Chemistry for the Million*, p. 15.

1175
What are the noble metals?

The noble metals are gold, silver, mercury, and the platinum group (including palladium, iridium, rhodium, ruthenium, and osmium). The term refers to those metals highly resistant to chemical reaction or oxidation (resistant to corrosion) and is contrasted to "base" metals which are not so resistant. The term has its origins in ancient alchemy whose goals of transformation and perfection were pursued through the different properties of metals and chemicals. The term is not synonomous the "precious metals," although a metal, like platinium, may be both.

Sources: Brady, George S. *Materials Handbook*, p. 644; Thrush, Paul W. *A Dictionary of Mining, Mineral, and Related Terms*, p. 753.

1176 / Which metallic element is the most abundant?

Aluminum is the most abundant metallic element on the surface of the earth and moon; it composes more than 8% of the earth's crust. It is never free in nature, combining with oxygen, sand, iron, titanium, etc.; its ores are mainly bauxites (aluminum hydroxide). Nearly all rocks, particularly igneous rocks, contain aluminum as aluminosilicate minerals. Napoleon III (1808–1883) recognized that the physical characteristic of its lightness could revolutionize the arms industry, so he granted a large subsidy to French chemist Sainte-Claire Deville (1818–1881) to develop a method to make its commercial use feasible. In 1854, Deville obtained the first pure aluminum metal through the process of reduction of aluminum chloride. In 1886, the American Charles Martin Hall (1863–1914) and the Frenchman Paul Heroult (1863–1914) independently discovered an electrolytic process to produce aluminum from bauxite. Because of aluminum's resistance to corrosion, low density, and excellent heat-conducting property it is used in cookware manufacturing and can-making industries. It is a good conductor of electricity, and is widely used in overhead cables. Aluminum alloys, such as duralumin, have high tensile strengths and are of considerable industrial importance, especially in the aerospace industry.

Sources: *Encyclopedia of Chemical Technology*, vol. 2, p. 129; Giscard d'Estaing, Valerie-Anne. *The World Almanac Book of Inventions*, p. 256; *Oxford Illustrated Encyclopedia of Invention and Technology*, p. 12.

1177 / What is technetium?

Technetium (Tc, element 43) is a radioactive metallic element that does not occur naturally either in its pure form or as compounds; it is produced during nuclear fission (nuclear reaction during which the nucleus splits almost equally). It was the first element to be made artificially in 1937 when it was isolated and extracted by C. Perrier and Emilio Segre (1905-1989). A fission product of molybdenum (Mo, element 42), Tc can also occur as a fission product of uranium (U, element 92).

Tc has found some application in diagnostic medicine. Ingested soluble technetium compounds tend to concentrate in the liver and are valuable in labeling and in radiological examination of that organ. Also, by technetium labeling of blood serum, components, diseases involving the circulatory system can be explored.

Sources: *American Medical Association Encyclopedia of Medicine*, p. 966; Asimov, Isaac. *Asimov's Biographical Encyclopedia of Science and Technology*, pp. 803-804; *Van Nostrand Reinhold Encyclopedia of Chemistry*, 4th ed., p. 926.

1178 ▽ Where were the first successful ironworks in America?

Although iron ore in this country was first discovered in North Carolina in 1585, and the manufacture of iron was first undertaken (but never accomplished) in Virginia in 1619, the first successful ironworks in America was established by Thomas Dexter and Robert Bridges near the Saugus River in Lynn, Massachusetts. As the original promoters of the enterprise, they hired John Winthrop, Jr. from England to begin production. By 1645, a blast furnace had begun operations and in 1648, a forge was working there.

Sources: Swank, James M. *History of the Manufacture of Iron in All Ages*, pp. 108-119.

1179 ▽ Which countries have uranium deposits?

Uranium, a radioactive metallic element is the only natural material capable of sustaining nuclear fission. But only one isotope, uranium-235, which occurs in one molecule out of 40 of natural uranium, can undergo fission under neutron bombardment. Mined in various parts of the world, it must then be converted during purification to uranium dioxide (UO_2). Uranium deposits occur throughout the world. The United States (especially Arizona, Colorado, New Mexico, North Carolina, and Utah), Canada, France, South Africa, Zaire, Australia, and the former Soviet Union have significant uranium resources. Canada and Zaire provide the best sources.

Sources: Fejer, Eva, and Cecilia Fitzsimons. *An Instant Guide to Rocks and Minerals*, p. 54; *Hawley's Condensed Chemical Dictionary*, p. 1207; *Oxford Illustrated Encyclopedia of Invention and Technology*, p. 369.

NATURAL SUBSTANCES *See also: ENERGY—NON-ENERGY FUELS; METALS AND OTHER MATERIALS—ROCKS AND MINERALS*

1180 ▽ What is ambergris?

Ambergris, a highly odorous waxy substance found floating in tropical seas, is a secretion from the sperm whale (*Physeter catodon*). The whale secretes ambergris to protect its stomach from the sharp bone of the cuttlefish, a squid-like sea mollusk, which it ingests. Ambergris is used in perfumery as a fixature to extend the life of a perfume and as a flavoring for food and beverages. Today ambergris is synthesized and used by the perfume trade, which has voluntarily refused to purchase ambergris to protect the sperm whales from exploitation.

Sources: Panati, Charles. *Panati's Browser's Book of Beginnings*, p. 24; Thrush, Paul W. *Dictionary of Mining, Mineral and Related Terms*, p. 33; Winter, Ruth. *A Consumer's Dictionary of Cosmetic Ingredients*, p. 32.

1181
Is lodestone a magnet?

Lodestone is a magnetic variety of natural iron oxide. It can be made to swing like a magnet, and was used by early mariners to find the magnetic north. Lodestone is frequently called a natural magnet.

Sources: Thrush, Paul W. *Dictionary of Mining, Mineral, and Related Terms*, p. 655.

1182
What is red dog?

Red dog is the residue from burned coal dumps. The dumps are composed of waste products incidental to coal mining. Under pressure in these waste dumps, the waste frequently ignites from spontaneous combustion, producing a red-colored ash, which is used for driveways, parking lots, and roads.

Sources: Brown, Victor J. *Engineering Terminology*, P. 172.

1183
Which products are made from horsehair?

Over 40 products are listed as being made from horsehair, some of which are baskets, belts, bird nests, hair and industrial brushes, buttons, carpet, curlers, fishing lines, furniture padding, hats, lariats, fishing nets, plumes for military hats or horse bridles, surgical sutures (during the Civil War), upholstery cloth, bows (for violin, cello, and viola) whips, and wigs.

Sources: Congram, Marjorie. *Horsehair: A Textile Resource*, pp. 81–82.

1184
Does any wood sink in water?

Ironwood is a name applied to many hard, heavy woods. Some ironwoods are so dense that their specific gravities exceed 1.0 and they are therefore unable to float in water. North American ironwoods include the American hornbeam, the mesquite, the desert ironwood, and leadwood (*Krugiodendron ferreum*) with a specific gravity of 1.34–1.42 to make it the heaviest in the United States.

The heaviest wood is black ironwood (*Olea laurifolia*), also called South Afri ironwood. Found in the West Indies, it has a specific gravity of 1.49 and weighs up to 93 pounds (42.18 kilograms) per foot. The lightest wood is *Aeschynomene hispida*, found in Cuba, with a specific gravity of 0.044 and a weight of 2½ pounds (1.13 kilograms) per foot. Balsa tree (*Ochroma pyramidale*) varies between 2½ (1.13 kilograms) and 24 pounds (10.88 kilograms) per foot.

Sources: *Encyclopedia Americana*, vol. 15, p. 467; *Guinness Book of Records 1992*, p. 140.

1185 / How much wood is used to make a ton of paper?

In the United States, the wood used for the manufacture of paper is mainly from small diameter bolts and pulpwood. It is usually measured by the cord or by weight. Although the fiber used in making paper is overwhelmingly wood fiber, a large percentage of other ingredients is needed. One ton of a typical paper requires 2 cords of wood, but also requires 55 thousand gallons of water, 102 pounds of sulfur, 350 pounds of lime, 289 pounds of clay, 1.2 tons of coal, 112 kilowatt hours of power, 20 pounds of dye and pigments, and 108 pounds of starch, as well as other ingredients.

Sources: Haygreen, John G. *Forest Products and Wood Science*, pp. 410–412.

1186 / Which woods are used for telephone poles?

The principal woods used for telephone poles are southern pine, Douglas fir, western red cedar, and lodgepole pine. Ponderosa pine, red pine, jack pine, northern white cedar, other cedars, and western larch are also used.

Sources: U.S. Forest Products Laboratory. *Encyclopedia of Wood*, p. 294.

1187 / Which woods are used for railroad ties?

Many species of wood are used for ties. The more common are oaks, gums, Douglas fir, mixed hardwoods, hemlock, southern pine, and mixed softwoods.

Sources: U.S. Forest Products Laboratory. *Encyclopedia of Woods*, p. 298.

1188 / What is fuller's earth?

It is a naturally-occurring white or brown clay containing aluminum magnesium silicate. Once used to clean wool and cloth (fulling), it is currently used for decolorizing oils and fats, as a pigment extender, a filter, and an absorbent, and in floor sweeping compounds.

Sources: Winter, Ruth. *A Consumer's Dictionary of Household, Yard and Office Chemicals*, p. 144.

1189 / From where do frankincense and myrrh originate?

Frankincense is an aromatic resin obtained by tapping the trunks of trees belonging to the genus *Boswellia*. The milky resin hardens when exposed to the air and forms irregular lumps, the form in which it is usually marketed.

Myrrh comes from a tree of the genus *Commiphora*, a native of Arabia and Northeast Africa. The resin is also obtained from the tree trunk.

Sources: *Popular Encyclopedia of Plants*, pp. 97, 144, 231.

1190 / Why are essential oils called "essential"?

Called essential oils because of their ease of solubility in alcohol to form essences, essential oils are used in flavorings, perfumes, disinfectants, medicine, and other products. They are naturally occurring volatile aromatic oils found in uncombined forms within various parts of plants (leaves, pods, etc.). These oils contain as one of its main ingredients a substance belonging to the terpene group. Examples of essential oils include bergamot, eucalyptus, ginger, pine, spearmint, and wintergreen oils. Extracted by distillation or enfleurage (extraction using fat) and mechanical pressing, these oils can now be made synthetically.

Sources: *Oxford Illustrated Encyclopedia of Invention and Technology*, p. 123.

1191 / Where does a luffa sponge come from?

Luffas are nonwoody vines of the cucumber family. The interior fibrous skeletons of the fruit are used as sponges. The common name is sometimes spelled loffah. Such names as dishcloth gourd, rag gourd, and vegetable sponge are other popular names.

Sources: Everett, Thomas H. *The New York Botanical Garden Illustrated Encyclopedia of Horticulture*, vol. 6, p. 2068; *The Wise Garden Encyclopedia*, p. 602.

1192 / What is diatomite?

Diatomite (also called diatomaceous earth) is a white or cream–colored, friable, porous rock composed of the fossil remains of diatoms (small water plants with silica cell walls). These fossils build up on the ocean bottoms to form diatomite, and in some places, these areas have become dry land or diatomaceous earth. Chemically inert and having a rough texture and other unusual physical properties, it is suitable for many scientific and industrial purposes, such as filtering agent; building material; heat; cold; and sound insulator; catalyst carrier; filler absorbent; abrasive; and ingredient in pharmaceutical preparations. Dynamite is made from it by soaking it in the liquid explosive nitroglycerin.

Sources: Barnhart, Robert K. *American Heritage Dictionary of Science*, p. 161; Parker, Sybil P. *McGraw-Hill Concise Encyclopedia of Science & Technology*, p. 574.

1193 / In coal mining what is meant by damp?

A poisonous or explosive gas in a mine. Carbon monoxide is known as white damp, and methane is known as firedamp. Blackdamp is formed by mine fires and explosion of firedamp in mines. It extinguishes light and suffocates its victims. The average blackdamp contains 10% to 15% carbon dioxide and 85% to 90% nitrogen.

Sources: Stevenson, L. Harold. *The Facts On File Dictionary of Environmental Science*, p. 68; Thrush, Paul W. *A Dictionary of Mining, Mineral, and Related Terms*, p. 108.

1194 / Which products come from the tropical forests?

Products from Tropical Forests

Woods	Houseplants	Spices	Foods
Balsa	*Anthurium*	Allspice	Avocado
Mahogany	Croton	Black pepper	Banana
Rosewood	*Dieffenbachia*	Cardamom	Coconut
Sandalwood	*Dracaena*	Cayenne	Grapefruit
Teak	Fiddle-leaf fig	Chili	Lemon
	Mother-in-law's tongue	Cinnamon	Lime
Fibers	Parlor ivy	Cloves	Mango
Bamboo	*Philodendron*	Ginger	Orange
Jute/Kenaf	Rubber tree plant	Mace	Papaya
Kapok	*Schefflera*	Nutmeg	Passion fruit
Raffia	Silver vase bromeliad	Paprika	Pineapple
Ramie	*Spathiphyllum*	Sesame seeds	Plantain
Rattan	Swiss cheese plant	Turmeric	Tangerine
	Zebra plant	Vanilla bean	Brazil nuts
Gums,resins			Cane sugar
Chicle latex	**Oils, etc.**		Cashew nuts
Copaiba	Camphor oil		Chocolate
Copal	Cascarilla oil		Coffee
Gutta percha	Coconut oil		Cucumber
Rubber latex	Eucalyptus oil		Hearts of palm
Tung oil	Oil of star anise		Macadamia nuts
	Palm oil		Manioc/tapioca
	Patchouli oil		Okra
	Rosewood oil		Peanuts
	Tolu balsam oil		Peppers
	Annatto		Cola beans
	Curare		Tea
	Diosgenin		
	Quinine		
	Reserpine		
	Strophanthus		
	Strychnine		
	Yang-Yang		

Sources: *The 1992 Information Please Environmental Almanac*, p. 284.

1195
How is petrified wood formed?

Water containing dissolved minerals such as calcium carbonate ($CaCO_3$) and silicate infiltrates the wood, and the minerals gradually replace the organic matter. The process takes thousands of years. After a time, the wood seems to have turned to stone because the original outer form and structure are retained during this process of petrification. The wood itself does not turn into stone.

Sources: Stimpson, George. *Information Roundup*, p. 457.

1196
Where does isinglass come from?

Isinglass is the purest form of animal gelatin. It is manufactured from the swimming bladder of the sturgeon. It is used in the clarification of wine and beer as well as in the making of some cements, jams, jellies, and soups.

Sources: *Encyclopedia of Chemical Technology*, vol. 11, p. 911; *World Book Encyclopedia*, vol. 10, p. 463.

1197
What is rosin?

Rosin is the resin produced after the distillation of turpentine, obtained from several varieties of pine trees, especially the longleaf pine (*Pinus palustris*) and the slash pine (*Pinus caribaea*). Rosin is used in varnishes, paint driers, soluble oils, paper-sizing, belt dressings, and for producing many chemicals.

Sources: Brady, George S. *Materials Handbook*, p. 685; Winburne, John N. *A Dictionary of Agricultural and Allied Terminology*, p. 651.

ROCKS AND MINERALS See also: ENERGY—NON–NUCLEAR FUELS; METALS AND OTHER MATERIALS—METALS

1198
Which diamond is the world's largest?

The Cullinan Diamond, weighing 3,106 carats, was discovered on January 25, 1905, at the Premier Diamond Mine, Transvaal, South Africa. Named for Sir Thomas M. Cullinan, chairman of the Premier Diamond Company, it was cut into 9 major stones and 96 smaller brilliants. The total weight of the cut stones was 1,063 carats, only 35% of the original weight.

Cullinan I, also known as the "Greater Star of Africa" or the "First Star of Africa," is a pear-shaped diamond weighing 530.2 carats. It is 2⅛ inches (5.4 centimeters) long, 1¾ inches (4.4 centimeters) wide, and 1 inch (2.54 centimeters) thick at its deepest point. It was presented to Britain's King Edward VII in 1907, and was set in the British monarch's sceptre with the cross. It is still the largest cut diamond in the world.

Cullinan II, also know as the "Second Star of Africa," is an oblong stone which weighs 317.4 carats. It is set in the British Imperial State Crown.

Sources: Argenzio, Victor. *Diamonds Eternal*, pp. 40-43; Field, Leslie. *The Queen's Jewels*, pp. 72-75; *How In the World?*, pp. 144-145.

1199
Besides the Cullinan Diamond, what are the largest precious stones?

The largest ruby is a 8,500 carat stone that is 5½ inches (14 centimeters) tall, carved to resemble the Liberty Bell. The largest star ruby is the 2,475 carat "Rajarathna" from India which has a six line star. The largest cut emerald was found in Carnaiba, Brazil, in August 1974. It is 86,136 carats. A 2,302 carat sapphire from at Anakie, Queensland, Australia, was carved into a 1,318 carat head of Abraham Lincoln, making it the largest carved sapphire. "The Lone Star," at 9,719.5 carats, is the largest star sapphire. The largest natural pearl is the "Pearl of Lao-tze," also called the "Pearl of Allah." Found in May 1934 in the shell of a giant clam at Palawan, Philippines, the pearl weighs 14 pounds, 1 ounce (6.38 kilograms).

Sources: *Guinness Book of Records 1992*, pp. 48-50.

1200
How are diamonds weighed?

The basic unit is a carat, which is 200 milligrams or ¹⁄₁₄₂ of an avoirdupois ounce. A well-cut round diamond of 1 carat measures almost exactly ¼ inch in diameter. Another unit commonly used is the point, which is one hundredth of a carat. A stone of 1 carat weighs 100 points.

Sources: Argenzio, Victor. *Diamonds Eternal*, pp. 9, 230.

1201
Are there any diamond mines in the United States?

The only significant diamond deposit in North America is at Murfreesboro, Arkansas. It is on government-owned land and has never been systematically developed. For a small fee, tourists can dig there and try to find diamonds. The largest crystal found there weighed 40.23 carats and was named the "Uncle Sam" diamond.

Diamonds crystallize directly from rock melts rich in magnesium and saturated with carbon dioxide gas that has been subjected to high pressures and temperatures exceeding 2559°F (1400°C). These rock melts originally came from deep in the earth's mantle at depths of 93 miles (150 kilometers).

Diamonds are minerals composed entirely of the element carbon, with an isometric crystalline structure. The hardest natural substance, gem diamonds have a density of 3.53, though black diamonds (black carbon cokelike aggregates of microscopic crystals) may have a density as low as 3.15. Diamonds have the highest thermal conductivity of any known substance. This property enables diamonds to be used in cutting tools, because they do not become hot.

Sources: Arem, Joel E. *Color Encyclopedia of Gemstones*, p. 81; Parker, Sybil P. *McGraw-Hill Concise Encyclopedia of Science and Technology*, p. 572.

1202 / What are Cape May diamonds?

They are quartz pebbles, found in the vicinity of the Coast Guard station in Cape May, New Jersey. The pebbles are polished, faceted, and sold to tourists as "Cape May diamonds."

Sources: Fletcher, Edward. *Pebble Collecting and Polishing*, p. 21.

1203 / How can a genuine diamond be identified?

There are several tests that can be performed without the aid of tools. A knowledgeable person can recognize the surface lustre, straightness and flatness of facets, and high light reflectivity. Diamonds become warm in a warm room and cool if the surroundings are cool. A simple test that can be done is exposing the stones to warmth and cold and then touching them to one's lips to determine their appropriate temperature. This is especially effective when the results of this test are compared to the results of the test done on a diamond known to be genuine. Another test is to pick up the stone with a moistened fingertip. If this can be done, then the stone is likely to be a diamond. The majority of other stones cannot be picked up in this way.

The water test is another simple test. A drop of water is placed on a table. A perfectly clean diamond has the ability to almost "magnetize" water and will keep the water from spreading. An instrument called a diamond probe can detect even the most sophisticated fakes. Gemologists always use this as part of their inspection.

Sources: Hudgeons, Marc. *The Official Investors Guide*, pp. 166-167.

1204 / How is the value of a diamond determined?

Demand, beauty, durability, rarity, freedom from defects, and perfection of cutting determine the value of a gemstone. But the major factor in establishing the price of gem diamonds is the control over output and price as exercised by the Central Selling Organization's (CSO) Diamond Trading Company Ltd. The CSO is a subsidiary of DeBeers Consolidated Mines Ltd.

Sources: *Minerals Yearbook 1989*, vol. 1, p. 443.

1205 / What are Indian Dollars?

They are six-sided disk-shaped twin crystals of aragonite ($CaCO_3$), which have altered to calcite but retained their outer form. They occur in large numbers in northern Colorado where they are known as "Indian Dollars." In New Mexico they are called "Aztec Money" and in western Kansas they are called "Pioneer Dollars."

Sources: Pearl, Richard M. *1001 Questions Answered About the Mineral Kingdom*, p. 14; Shipley, Richard M. *Dictionary of Gems and Gemology*, p. 12.

1206 / How does the emerald get its color?

Emerald is a variety of green beryl ($Be_3Al_2Si_6O_{18}$) that is colored by a trace of chromium (Cr), which replaces the aluminum (Al) in the beryl structure. Other green beryls exist; but if no chromium is present, they are, technically speaking, not emeralds.

Sources: Arem, Joel E. *Color Encyclopedia of Gemstones*, pp. 51-52.

1207 / What is the Mohs scale?

The scale is a standard of 10 minerals by which the hardness of a mineral is rated. It was introduced in 1812 by the German mineralogist Friedrich Mohs (1773-1839). The minerals are arranged from softest to hardest.

Hardness	Mineral	Comment
1	Talc	Hardness 1–2 can be scratched by a fingernail
2	Gypsum	Hardness 2–3 can be scratched by a copper coin
3	Calcite	Hardness 3–6 can be scratched by a steel pocket knife
4	Fluorite	
5	Apatite	
6	Orthoclase	Hardness 6–7 will not scratch glass
7	Quartz	
8	Topaz	Hardness 8–10 will scratch glass
9	Corundum	
10	Diamond	

Sources: Bates, Robert L. and Julia A. Jackson. *Glossary of Geology*, p. 428.

1208
How is the star in star sapphires produced?

Sapphires are composed of blue gem-quality corundum (Al_2O_3). When cut in the unfaceted cabochon (dome or convex) form, a sapphire with a trace of titanium dioxide will display a rayed (star) figure usually with six rays. The same effect produces the star in star rubies as well.

Sources: *Encyclopedia Americana*, vol. 26, p. 786; Shipley, Robert M. *Dictionary of Gems and Gemology*, pp. 13, 31, 69, 194.

1209
What is a geode?

A geode is a hollow, stone-like formation lined inside with inward projecting, small crystals. Geodes are frequently found in limestone beds and occur in many parts of the world. They average 2 to 6 inches (5 to 15 centimeters) in diameter.

Sources: American Geological Institute. *Dictionary of Geological Terms*, p. 181; *The McGraw-Hill Dictionary of Scientific and Technical Terms*, p. 800.

1210
How old are fossils?

The oldest known fossils are of single-celled organisms, blue-green algae, found in 3.2 billion-year-old cherts, shales, and sandstone from the Transvaal of South Africa. Multicellular fossils dating from about 700 million years ago are also known. The largest number of fossils come from the Cambrian era of 590 million years ago, when living organisms began to develop skeletons and hard parts. Since these parts tended to last longer than ordinary tissue, they were more likely to be preserved in the clay to become fossilized.

Sources: Fenton, Carroll L. *The Fossil Book*, pp. 91, 93; Trefil, James. *1001 Things Everyone Should Know About Science*, pp. 59-61.

1211
How are fossils formed?

Fossils are the remains of animals or plants that were preserved in rock before the beginning of recorded history. It is unusual for complete organisms to be preserved and fossils usually represent the hard parts such as bones or shells of animals and leaves, seeds, or woody parts of plants.

Some fossils are simply the bones, teeth, or shells themselves, which can be preserved for a relatively short period of time. Another type of fossil is the imprint of a buried plant or animal which decomposes leaving a film of carbon which retains the form of the organism.

Some buried material is replaced by silica and other materials which permeate the organism and replace the original material in a process called petrification. Some woods are replaced by agate or opal so completely that even the cellular structure is duplicated. The best examples of this can be found in the Petrified Forest National Park in Arizona.

Molds and casts are other very common fossils. A mold is made from an imprint, such as a dinosaur footprint, in soft mud or silt. This impression may harden, then be covered with other materials. The original footprint will have formed a mold and the sediments filling it will be a cast of the footprint.

Sources: Hamilton, William R. *The Henry Holt Guide to Minerals, Rocks, and Fossils*, p. 210; Zim, Herbert S., and Paul Shaffer. *Rocks and Minerals*, pp. 130-132.

Fossil scallops of Cretaceous age from north-central Texas.

1212
How do rocks differ?

Rocks can be conveniently placed into three groups—igneous, sedimentary, and metamorphic.

Igneous rocks, such as granite, pegmatite, rhyolite, obsidian, gabbro, and basalt are formed by the solidification of molten magma that emerges through the earth's crust via volcanic activity. The nature and properties of the crystals vary greatly, depending in part on the composition of the original magma and partly on the conditions under which the magma solidified. There are thousands of different igneous rock types. For example, granite is formed by slow cooling of molten material (within the earth). It has large crystals of quartz, feldspars, and mica.

Sedimentary rocks such as brecchia, sandstone, shale, limestone, chert, and coals are produced by the accumulation of "sediments." These are fine rock particles or fragments, skeletons of microscopic organisms, or minerals leached from rocks that have accumulated from weathering. These "sediments" are then redeposited under water and were later compressed in layers over time. The most common sedimentary rock is sandstone, which is predominantly quartz crystals.

Metamorphic rocks, such as marble, slate, schist, gneiss, quartzite, and hornsfel, are formed by the alteration of igneous and sedimentary rocks through heat and/or pressure. These physical and/or chemical changes to rocks can be exemplified by the formation of marble from thermal changes that have occurred to limestone.

Sources: Fejer, Eva, and Cecilia Fitzsimons. *An Instant Guide to Rocks and Minerals*, pp. 16-43; Hamilton, W.R. *The Henry Holt Guide to Minerals, Rocks and Fossils*, p. 146.

▽1213▽ *What is petrology and what does a petrologist do?*

Petrology is the science of rocks. A petrologist is a person who is interested in the mineralogy of rocks and who also tries to unravel the record of the geological past contained within rocks. From these rocks, a petrologist can learn about past climates and geography, past and present composition, and the conditions which prevail within the interior of the earth.

Sources: Hamilton, W.R. *The Henry Holt Guide to Minerals, Rocks, and Fossils*, p. 146.

▽1214▽ *Who was the first person to attempt a color standardization scheme?*

The German mineralogist Abraham Gottlob Werner (1750-1817) devised a method of describing minerals by their external characteristics, including color. He worked out an arrangement of colors and color names, illustrated by an actual set of minerals.

Sources: Maerz, A. *A Dictionary of Color*, p.137.

▽1215▽ *What is fool's gold?*

Iron pyrite (FeS_2) is a mineral popularly known as "fool's gold." Because of its metallic luster and pale brass yellow color it is often mistaken for gold. Real gold is much heavier, softer, not brittle, and not grooved.

Sources: Bates, Robert L. and Julia A. Jackson. *Glossary of Geology*, p. 540; Fejer, Eva, and Cecilia Fitzsimons. *An Instant Guide to Rocks and Minerals*, p. 104.

1216 / How does a rock differ from a mineral?

Mineralogists use the term "mineral" for a substance that has all four of the following features: it must be found in nature; it must be made up of substances that were never alive (organic); it has the same chemical makeup wherever it is found; and its atoms are arranged in a regular pattern and form solid crystals.

While "rocks" are sometimes described as an aggregate or combination of one or more minerals, geologists extend the definition to include clay and loose sand and certain limestones.

Sources: The American Geological Institute. *Dictionary of Geological Terms*, pp. 282, 369; Barnhart, Robert K. *The American Heritage Dictionary of Science*, p. 566.

1217 / What is galena?

Galena is a lead sulphide (PbS) and the most common ore of lead, containing 86.6% lead. Lead-gray in color, with a brilliant metallic luster, galena has a specific gravity of 7.5 and a hardness of 2.5 on the Mohs scale, and usually occurs as cubes or a modification of an octahedral form. Mined in Australia, it is also found in Missouri, Kansas, and Oklahoma, as well as Colorado, Montana, and Idaho.

Sources: The American Geological Institute. *Dictionary of Geological Terms*, p. 177; Parker, Sybil P. *McGraw-Hill Concise Encyclopedia of Science and Technology*, p. 826; Thrush, Paul W. *Dictionary of Mining, Mineral, and Related Terms*, p. 475.

1218 / What is pitchblende?

Pitchblende is a massive variety of uraninite or uranium oxide found in metallic veins. It is radioactive material and the most important ore of uranium. It is the original source of radium.

Sources: The American Geological Institute. *Dictionary of Geological Terms*, p. 450; *Hawley's Condensed Chemical Dictionary*, p. 923.

1219 / Who is regarded as the father of modern mineralogy?

Georg Bauer (1495-1555) was a German mineralogist, physician, and scholar. Also known by the Latin name Georgius Agricola, he studied mining methods and mining ores, and wrote on the classification of minerals, mineral extraction, and geological phenomena. *De re metallica*, his major work, covered the entire field of mining and metallurgy in a systematic and comprehensive manner. Bauer was one of the first to base his writings on observation and scientific inquiry.

Sources: Downs, Robert B. *Landmarks in Science*, pp. 75-77; *The Great Scientists*, vol. 1, p. 29.

1220
What is stibnite?

Stibnite is a lead-gray mineral (Sb_2S_3) with a metallic luster. It is the most important ore of antimony, and is also known as antimony glance. One of the few minerals that fuses easily in the match flame (977°F or 525°C), stibnitz has the hardness of 2 on the Mohs scale and a specific gravity of 4.5-4.6. It is commonly found in hydrothermal veins or hot springs deposits. Stibnitz is mined in Germany, Rumania, France, Bolivia, Peru, and Mexico. The Yellow Pine mine at Stibnite, Idaho is the largest producer in the United States; but California and Nevada also have deposits.

Sources: Parker, Sybil P. *McGraw-Hill Concise Encyclopedia of Science and Technology*, p. 1788. Stiegeler, Stella E. *A Dictionary of Earth Sciences*, p. 263; Thrush, Paul W. *Dictionary of Mining, Mineral, and Related Terms*, p. 1077.

1221
What is meant by the term strategic minerals?

Strategic minerals are minerals essential to the national defense—the supply of which a country uses but cannot produce itself. 33% to 50% of the 80 minerals used by industry could be classed as strategic minerals. Wealthy countries, such as the United States, stockpile these minerals to avoid any crippling effect on their economy or military strength if political circumstances were to cut off their supplies. The United States, for instance, stockpiles bauxite (14½ million tons), manganese (2.2 million tons), chromium (1.8 million tons), tin (185 thousand tons), cobalt (19 thousand tons), tantalum (635 tons), palladium (1.25 million troy ounces), and platinum (453 thousand troy ounces).

Sources: Cunningham, William, and Barbara Woodhouse Saigo. *Environmental Science: A Global Approach*, pp. 152–153; Thrush, Paul W. *Dictionary of Mining, Mineral, and Related Terms*, p. 1086.

▼ THE PLANT WORLD

FARMING *See: THE PLANT WORLD—GARDENING, FARMING, ETC.*

FLOWERS AND PLANTS *See also: THE ENVIRONMENT—EXTINCT AND ENDANGERED PLANTS AND ANIMALS; THE PLANT WORLD—GARDENING, FARMING, ETC.; THE PLANT WORLD—PHYSICAL CHARACTERISTICS, FUNCTIONS, ETC.*

1222 / *What special significance does the passionflower have?*

Spanish friars of the 16th century first gave the name to this flower. They saw in the form of the passionflower (*Passiflora*) a representation of the passion in Christ. The flowers have five petals and five sepals, which symbolize the 10 faithful apostles present at the crucifixion. The corona of five filaments was thought to resemble Christ's crown of thorns. The five stamens represented the five wounds in Christ's body, and the three stigmas stood for the nails driven into his hands and feet. Most species of passionflower are native to the tropical areas of the Western Hemisphere.

Sources: *Encyclopedia Americana*, vol. 21, p. 512; Whittick, Arnold. *Symbols, Signs and their Meaning*, pp. 234–235.

1223 / *What are the symbolic meanings of herbs and plants?*

Aloe—Healing, protection, affection
Angelica—Inspiration
Arbor vitae—Unchanging friendship
Bachelor's buttons—Single blessedness
Basil—Good wishes, love

Bay—Glory
Carnation—Alas for my poor heart
Chamomile—Patience
Chives—usefulness
Clover, White—Think of me
Coriander—Hidden worth
Cumin—fidelity
Fennel—flattery
Fern—Sincerity
Geranium, Oak-leaved—True friendship
Goldenrod—Encouragement
Heliotrope—Eternal love
Holly—Hope
Hollyhock—Ambition
Honeysuckle—Bonds of love
Horehound—Health
Hyssop—Sacrifice, cleanliness
Ivy—Friendship, continuity
Lady's mantle—Comforting
Lavender—Devotion, virtue
Lemon balm—Sympathy
Marjoram—Joy, happiness
Mints—Eternal refreshment
Morning glory—Affectation
Nasturtium—Patriotism
Oak—Strength
Oregano—Substance
Pansy—Thoughts
Parsley—Festivity
Pine—Humility
Poppy, Red—Consolation
Rose—Love
Rosemary—Remembrance
Rudbeckia—Justice
Rue—Grace, clear vision
Sage—Wisdom, immortality
Salvia, Blue—I think of you
Salvia, Red—Forever mine
Savory—Spice, interest
Sorrel—Affection
Southernwood—Constancy, jest
Sweet-pea—Pleasures
Sweet woodruff—Humility
Tansy—Hostile thoughts
Tarragon—Lasting interest

Thyme—Courage, strength
Valerian—Readiness
Violet—Loyalty, devotion
Violet, Blue—Faithfulness
Violet, Yellow—Rural happiness
Willow—Sadness
Zinnia—Thoughts of absent friends

Sources: Jayne, Kate Lindley, and Claudette Suzanne Mautor. *Living with Potpourri*, pp. 47-48; Lloyd, Elizabeth J. *Enchanted Circles*, p. 409.

◤1224◢ Is there a national flower for the United States?

The national flower of the United States is the rose, adopted on October 7, 1986.

Sources: *World Book Encyclopedia*, vol. 20, p. 91.

◤1225◢ Where are living stones found?

It is a name given to various succulent plants from the stony deserts of South Africa, which mimic their surroundings. Each shoot or plant is made up of two grossly swollen leaves virtually fused together and colored to resemble a pebble. Large daisy-like flowers are borne from between the leaf pair.

Sources: *Illustrated Dictionary of Botany*, p. 144.

◤1226◢ What is the first wildflower to bloom in the spring in the Northern United States?

The first flower of the Northern spring is curious and interesting, but little known and rarely seen, for its chosen home is the swamp. Commonly called skunk cabbage (*Spathyema foetidus*), its time of first bloom is February. The first spring flower that is generally known in New England and the Middle West is the *Hepatica*, or liverleaf, which blooms in March or early April.

Sources: Keeler, Harriet L. *Our Early Wild Flowers*, p. xi.

◤1227◢ How are carnivorous plants categorized?

Carnivorous plants, numbering between 450–500 species and 12 genera, have been classified according to the nature of their trapping mechanisms. Active traps display rapid motion in their capture of prey. The Venus fly trap (*Dionaea muscipula*) and the bladderwort (*Utricularia vulgaris*) are active traps. Semi-active traps employ a two-stage trap in which the prey is caught in the trap's adhesive fluid. As it struggles, the plant is triggered to slowly tighten its grip. The sundew (species *Drosera*) and butterwort (*Pinguicula vulgaris*) are semi-active traps. Passive traps entice insects by nectar. The insects fall into a reservoir of water and drown. An example of the passive trap is the pitcher plant (5 genera).

Sources: Howes, F.N. *A Dictionary of Useful and Everyday Plants and Their Common Names*, p. 47; Schwartz, Randall. *Carnivorous Plants*, p. 9.

1228
What is wormwood?

Artemisia absinthium, known as wormwood, is a hardy, spreading, fragrant perennial, 2 to 4 feet (61–122 centimeters) tall. It is native to Europe but widely naturalized in North America. The liqueur absinthe is flavored from this plant. Absinthe was formerly used as a bitter tonic.

Sources: Everett, Thomas H. *New York Botanical Garden Illustrated Encyclopedia of Horticulture*, vol. 1, p. 255; *Stedman's Medical Dictionary*, p. 6.

1229
What is unique about the water-lily Victoria amazonica?

It is very big! Found only on the Amazon River, this water-lily has leaves that are up to six feet in diameter. The 12 inch flowers open at dusk on two successive nights.

Sources: Wyman, Donald. *Wyman's Gardening Encyclopedia*, pp. 1162–1163.

1230
Who is known as the father of botany?

The ancient Greek thinker, Theophrastus (ca.372–ca.287 B.C.E.), is known as the father of botany. His two large botanical works, *On the History of Plants* and *On the Causes of Plants* were so comprehensive that 1,800 years went by before any new discovery in botany was made. He integrated the practice of agriculture into botany and established a theory of plant growth and the analysis of plant structure. He related plants to their natural environment and identified, classified, and described 550 different plants.

Sources: *The Biographical Dictionary of Scientists: Biologists*, p. 127; Downs, Robert B. *Landmarks in Science*, pp. 30-33.

1231

Why do cats like catnip?

Catnip (*Nepeta cataria*) is a hardy perennial herb attractive to cats. Also known as catmint or catnep, it belongs to the mint family. The whole cat family (*Felidae*) reacts to catnip. Mountain lions, lynx, tigers, and lions roll over, rub their face, extend their claws, and do a body twist when they smell catnip's pungent odor. The scent of catnip probably excites cats because it contains a chemical called trans-neptalactone, which closely resembles an excretion in the queen's urine.

Sources: *Horticulture*, vol. 59 (January 1981), p. 14; Taylor, David. *You & Your Cat*, p. 33; *The Wise Garden Encyclopedia*, pp. 208, 264.

1232

What are the parts of a flower?

Sepal—found on the outside of the bud or on the underside of the open flower. It serves to protect the flower bud from drying out. Some sepals ward off predators by their spines or chemicals.

Petals—serve to attract pollinators and are usually dropped shortly after pollination occurs.

Nectar—contains varying amounts of sugar and proteins that can be secreted by any of the floral organs. It usually collects inside the flower cup near the base of the cup formed by the flower parts.

Stamen—is the male part of a flower and consists of a filament and anther where pollen is produced.

Pistil—is the female part which consists of the stigma, style, and ovary containing ovules. After fertilization the ovules mature into seeds.

Sources: *Practical Botany*, p. 6.

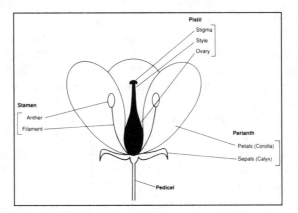

Parts of a flower

1233

What is meant by an "imperfect" flower?

An imperfect flower is one which is unisexual, having either stamens (female parts) or pistils (male parts) but not both.

Sources: Swartz, Delbert. *Collegiate Dictionary of Botany*, p. 247.

1234

How does a bulb differ from a corm, a tuber, and a rhizome?

Many times the term "bulb" is applied to any underground storage organ in which a plant stores energy for its dormant period. Dormancy is one of nature's devices a plant utilizes to get through difficult weather conditions (winter cold or summer drought).

A *true bulb*—consists of fleshy scales containing a small basal plate (a modified stem from which the roots emerge) and a shoot. The scales that surround the embryo are modified leaves that contain the nutrients for the bulb during dormancy and early growth. Some bulbs have a tunic (a paper-thin covering) around the scales. The basal plate can also hold the scales together. New bulbs form from the lateral buds on the basal plate. Tulips, daffodils, lilies, and hyacinths are examples of bulb flowers.

A *corm*—is actually a stem that has been modified into a mass of storage tissue. The eye(s) at the top of the corm is a growing point. The corm is covered by dry leaf bases similar to the tunic–covering of the bulb. Roots grow from the basal plate on the underside of the corm. New corms form on top of or beside the old one. Corm-type flowers include gladiolus, freesia, and crocus.

A *tuber*—is a solid underground mass of stem like a corm but it lacks both a basal plate and a tunic. Roots and shoots grow from "eyes" (growth buds) out of its sides, bottom, and sometimes its top. Some tubers are roundish. Others are flattened and lumpy. Some examples of tubers are gloxinia, caladium, ranunculuses, and anenome.

A *tuberous root*—is a swollen root that has taken in moisture and nutrients. It resembles a tuber. New growth occurs on the base of the old stem, where it joins the root. A tuberous root can be divided by cutting off a section with an eye–bearing portion from where the old stem was attached. Dahlias have tuberous roots.

A *rhizome or a rootstock*—is a thickened, branching storage stem that usually grows laterally along or slightly below the soil surface. Roots develop downward on the bottom surface, while buds and leaves sprout upwards from the top of the rhizome. It is propagated by cutting the parent plant into sections. Japanese, Siberian and bearded irises; cannas; calla lilies; and trillium are rhizomes.

Sources: *Rodale's Illustrated Gardening and Landscaping Techniques*, pp. 112–118; Smith, Michael D. *All About Bulbs*, pp. 8–10.

1235

What is the greatest number of leaves a clover can have?

A fourteen-leafed white clover (*Trifolium repens*) and a fourteen-leafed red clover (*Trifolium pratense*) have been found in the United States.

Sources: *Guinness Book of Records 1992*, p. 132.

GARDENING, FARMING, ETC.

1236

What is the secret of bonsai—the Japanese art of growing dwarf trees?

These miniature trees with tiny leaves and twisted trunks can be centuries old. To inhibit growth of the plants, they have been carefully deprived of nutrients, pruned of their fastest-growing shoots and buds, and kept in small pots to reduce the root systems. Selective pruning, pinching out terminal buds, and wiring techniques are devices used to control the shape of the trees. Bonsai possibly started during the Chou dynasty (900–250 B.C.E.) in China, when emperors made miniature gardens that were dwarf representations of the provincial lands that they ruled.

Sources: *Can Elephants Swim?*, p. 27; Parker, Sybil P. *McGraw-Hill Concise Encyclopedia of Science and Technology*, p. 267.

1237

Is there a way to preserve a cut Christmas tree longer?

Soak the tree trunk in a solution containing:

> 1 gallon warm water
> 4 tablespoons chelated iron
> 4 tablespoons Karo syrup
> 4 tablespoons household bleach

Sources: Crager, Meg. *Christmas Trees*, p. 32.

1238

How can a Christmas tree be made fire-resistant?

Mix in a bucket:

> 2 cups corn syrup
> 2 ounces liquid chlorine bleach
> 2 pinches Epsom salts

½ teaspoon borax
1 teaspoon chelated iron
2 gallons hot water

Use the solution to water the tree every day. Chelated iron is available at garden supply centers.

Sources: *Blair & Ketchum's Country Journal*, vol. 12 (December 1985), p. 10.

1239
What is pleaching?

A method of shearing closely planted trees or shrubs into a high wall of foliage. Many kinds of trees have been used—maples, sycamores, lindens, etc. Because of the time needed in caring for pleached allees, they are infrequently seen in American gardens, but are frequently observed in European ones.

Sources: Wyman, Donald. *Wyman's Gardening Encyclopedia*, p. 843.

1240
How can fruit trees be protected from being eaten by field mice?

Valuable trees, especially newly planted fruit trees, can be protected by wrappings or guards of wire, wood veneer, or plastic. Other controls such as pieces of lava rocks soaked in garlic are effective as repellents, and garlic sprays will repel most rodents.

Sources: *The Wise Garden Encyclopedia*, p. 629.

1241
What is a dwarf conifer?

The conifers are evergreen shrubs and trees with needle-shaped leaves, cones, and resinous wood such as the pines, spruces, firs, and junipers. After 20 years, dwarf or slow-growing forms of these otherwise tall trees are typically about three feet (91 centimeters) tall.

Sources: Wyman, Donald. *Wyman's Gardening Encyclopedia*, pp. 251, 331, 364.

1242
Did Johnny Appleseed really plant apple trees?

John Chapman (1774-1845), called Johnny Appleseed, established nurseries of apple seedlings and introduced trees from the Allegheny River to Indiana. He did not prowl the countryside scattering seeds at random from a bag slung over his shoulders, as artists like to depict him.

Sources: Root, Waverly L. *Food*, p. 10.

1243
What is a five-in-one-tree?

These very curious trees consist of a rootstock with five different varieties of the same fruit—usually apples—grafted to it. The blooming period is usually magnificent with various colors of blooms appearing on the same tree.

Sources: *Fruits and Vegetables: 1001 Gardening Questions Answered*, p. 7.

1244
Can a branch of the dogwood tree be forced into bloom?

Forcing dogwood (genus *Cornus*) into bloom is similar to forcing forsythia. Bring the dogwood branch indoors when the buds begin to swell. Put the branch in water and set it in a sunny window.

Sources: *New York Times Book of Indoor and Outdoor Gardening Questions*, p. 199.

1245
What is meant by the chilling requirement for fruit trees?

When a fruit tree's fruiting period has ended, there must follow a dormant period during which the plant rests and regains strength for another fruit set the following year. The length of this set is measured in hours between 32°F and 45°F (0°F and 7.2°C). A cherry tree requires about 700 hours of chilling time.

Sources: *Fruits and Vegetables: 1001 Gardening Questions Answered*, pp. 2-3.

1246
What is the railroad worm?

The apple maggot (*Rhagoletis pomonella*), which becomes the apple fruit fly, is frequently called the railroad worm. Inhabiting orchards in Eastern United States and Canada, the larvae feed on the fruit pulp of apples, plums, cherries, etc. and cause damage to fruit crops.

Sources: Everett, Thomas H. *The New York Botanical Garden Illustrated Encyclopedia of Horticulture*, vol. 6, p. 2096.

1247
How did the navel orange originate?

Every navel orange is derived from a mutant tree that appeared on a plantation in Brazil in the early nineteenth century. A bud from the mutant tree was grafted onto another tree, whose branches were then grafted onto another, and so forth.

Sources: Trefil, James. *1001 Things Everyone Should Know About Science*, p. 4.

1248

What is meant by espaliering a fruit tree and why is it done?

To espalier a fruit tree means to train it to grow flat against a surface. It can be grown in small places such as against a wall, and it will thrive even if its roots are underneath sidewalks or driveways. Since many fruit trees must be planted in pairs, espaliered fruit trees can be planted close together, providing pollen for each other, yet taking up little space.

Sources: *Rodale's Illustrated Encyclopedia of Gardening and Landscaping Techniques*, pp. 282–285.

1249

How can garden soil be used as potting soil?

The garden soil must be pasteurized and then mixed with coarse sand and peat moss. Soil may be pasteurized by putting the soil in a covered baking dish in the oven. When a meat thermometer stuck in the soil has registered 180°F (82°C) for 30 minutes, the soil is done.

Sources: Crockett, James U. *Crockett's Indoor Garden*, p. 35.

1250

What is the best soil pH for growing plants?

Nutrients such as phosphorous, calcium, potassium, and magnesium are most available to plants when the soil pH is between 6.0 and 7.5. Under highly acid (low pH) conditions, these nutrients become insoluble and relatively unavailable for uptake by plants. High soil pH can also decrease the availability of nutrients. If the soil is more alkaline than pH 8, phosphorous, iron, and many trace elements become insoluble and unavailable for plant uptake.

Sources: *Wise Garden Encyclopedia*, p. 728.

1251

When is the best time to work the soil?

Although it is possible to prepare the soil at any time of year, fall digging is the best time. Dig in the fall and leave the ground rough. Freezing and thawing during winter breaks up clods and aerates the soil. Insects that otherwise overwinter are mostly turned out. The soil settling during winter will lessen the likelihood of air pockets in the soil when planting the following spring. Fall preparation provides time for soil additives such as manure and compost to break down before planting time.

Sources: *Fruits and Vegetables: 1001 Gardening Questions Answered*, p. 63; *Rodale's Illustrated Encyclopedia of Gardening and Landscaping Techniques*, p. 4.

1252 ∕
What is meant by the term double-digging?

Double-digging produces an excellent deep planting bed for perennials, especially if the area is composed of heavy clay. It involves removing the top 10 inches of soil and moving it to a holding area, then spading the next 10 inches, and amending this layer with organic matter and/or fertilizer. Then the soil from the "first" digging is replaced after it, too, has been amended.

Sources: Sinnes, A. Cort. *All About Perennials*, p. 31.

1253 ∕
How are seedlings hardened off before planting?

Hardening off is a gardening term for gradually acclimatizing seedlings raised indoors to the outdoor environment. Place the tray of seedlings outdoors for a few hours each day in a semi-protected spot. Lengthen the amount of time they stay out by an hour or so each day; at the end of the week, they will be ready for planting outdoors.

Sources: *Ortho's Complete Guide to Successful Gardening*, p. 70.

1254 ∕
What are the best annual and perennial plants to grow to attract butterflies?

Ageratum, Cosmos, Globe Candytuft, Heliotrope, Lantana, Marigold, Mexican Sunflower, Torch Flower, Nasturtium, Sweet Alyssum, and Zinnia.

Sources: Sedenko, Jerry. *The Butterfly Garden*, pp. 109-112.

1255 ∕
Which type of fence protects a garden from deer?

A post and wire mesh fence with a sharp angled, narrow gate, which people and small animals can navigate, but deer cannot, plus the stallation of a standard motion-sensing security light with a beeper helps keep deer away.

Sources: *Fine Gardening*, no. 22 (November/December 1991), pp. 28-30.

1256 ∕
Which flowers should be planted in the garden to attract hummingbirds?

Scarlet Trumpet Honeysuckle, Weigela, Butterfly Bush, Beardtongue, Coralbells, Red-Hot-Poker, Foxglove, Beebalm, Nicotiana, Petunia, Summer Phlox, and scarlet sage provide brightly-colored (reds and orange shades), nectar-bearing attractants for the hummingbirds.

Sources: Erickson, Joan Beth. *Flower Garden Plans*, pp. 68–69.

1257

How can squirrels be kept away from vegetable and flower gardens?

Squirrels like to take a bite out of tomatoes, cucumbers, and melons, dig up bulbs, and ruin anything colorful in the flower garden. The traditional recommendation of spreading mothballs around is apparently not too successful. A new method is to lay down 1 to 2 inch (2.5 to 5 centimeter) mesh sheets of chicken wire. Squirrels will avoid the mesh, apparently because they fear getting their toes stuck in it. Another method to try is to sprinkle hot pepper around the plants, renewing it after it rains.

Sources: *Horticulture*, vol. 70 (April 1991), pp. 14–15.

1258

What was a victory garden?

During World War I, patriots grew "liberty gardens." In World War II, the U.S. Secretary of Agriculture Claude R. Wickard encouraged householders to plant vegetable gardens wherever they could find space. By 1945 there were said to be 20 million victory gardens producing about 40% of all American vegetables in many unused scraps of land. Such sites as the strip between a sidewalk and the street, town squares, and the land around Chicago's Cook County jail were used. The term "victory garden" dates back to Elizabethan England in a book by that title written in 1603 by Richard Gardner.

Sources: Schremp, Gerry. *Kitchen Culture*, p. 14.

1259

What is meant by xeriscaping?

A xeriscape, a landscape of low water-use plants, is the modern approach to gardening where water shortages are prevalent. Taken from the Greek word *xeros*, meaning dry, this type of gardening utilizes drought-resistant plants and low maintenance grasses, which require water only every 2 to 3 weeks. Drip irrigation, heavy mulching of plant beds, and organic soil improvements are other xeriscape techniques that allow better water absorption and retention which in turn decrease garden watering time.

Sources: *Flower and Garden*, vol. 34 (May 1990), pp. 44–45, 66–67.

What are the average dates of the last killing frost in the spring in the eastern United States?

The map below indicates the usual spring date of the last killing frost in the eastern United States.

Sources: U.S. Department of Agriculture. *Growing Vegetables in a Home Garden*, p. 23.

Average date of last killing frost

What are the average dates of the first killing frost in the eastern United States?

The map below indicates the usual autumn date of the first killing frost in the eastern United States.

Sources: U.S. Department of Agriculture. *Growing Vegetables in a Home Garden*, p. 22.

Average date of first killing frost

1262

What is a recommended size vegetable garden for the beginner?

It really depends on the amount of space available, how much produce is desired from a garden, and how much work a person is willing to put into the garden. A modest-sized, 10–by–20 foot (3–by–6 meter) plot, laid out in traditional rows, is quite manageable in terms of weeding, cultivating, planting, and harvesting. Even a 10–by–10 foot (3–by–3 meter) plot will suffice for a salad or "kitchen" garden, with plenty of greens and herbs for salads and seasonings on a daily basis. "Intensive" gardening methods where plants are arranged in blocks rather than rows, allow for increased yield in an even smaller space. One 4 by–4–foot (1 by–1 meter) block, with a vertical frame at one end, can provide salad vegetables for one person throughout the growing season, though two blocks would provide a wider variety of vegetables.

Sources: Bartholomew, Mel. *Square Foot Gardening*, pp. 29-30; *Fruits and Vegetables: 1001 Gardening Questions Answered*, p. 55.

1263

Is there a "best" time to weed in the vegetable garden?

Weeding is usually the most unpopular and the most time–consuming garden chore. Some studies (using weeding with peas and beans as examples) have shown that weeding done during the first 3 to 4 weeks of vegetable growth produced the best crops and that unabated weed growth after that time did not significantly reduce the vegetable yields.

Sources: Thomas, Robert B. *The Old Farmer's Almanac 1988*, pp. 172–173.

1264

What prevents daffodils from blooming the next year, although they were fertilized?

Try fertilizing daffodils (*Narcissus pseudonarcissus*) as soon as the new shoots appear. This helps the roots renew themselves and will also aid in leaf and flower development. The problem could also be overcrowding, which hinders flower production. Try digging the bulbs up every third or fifth year, separating them and respacing them.

Sources: *Horticulture*, vol. 70 (March 1992), p. 12.

1265

How can geraniums be carried over from one year to the next?

While they must be kept from freezing, geraniums can survive the winter happily in a cool sunny spot, such as a cool greenhouse, a bay window, or a sunny unheated basement. They need only occasional watering while in this semi-dormant state. Cuttings from these plants can be rooted in late winter or early spring for a new crop of geraniums (some sources suggest rooting in the fall). In homes without a suitable cool and sunny spot, the plants can be forced into dormancy by allowing the soil to dry completely, then gently knocking the soil off of the roots. While the plants can simply be hung from the rafters in a cool (45 to 50°F; 7 to 10°C), slightly humid room, they will do better if put into individual paper bags, with the openings tied shut. The plants should be checked regularly. The leaves will dry and shrivel, but if the stems shrivel, the plants should be lightly misted with water. If any show mold or rot, cut off the affected sections, move the plants to a drier area, and leave the bags open for a day or two. In the early spring, prune the stems back to healthy green tissue and pot in fresh soil.

Sources: *Horticulture*, vol. 70 (April 1992), p. 14; *Rodale's Illustrated Encyclopedia of Gardening and Landscaping Techniques*, pp. 110-111.

1266
How are chia seeds grown?

The minuscule black chia seeds are gathered from a type of wild sage that is found in the southwest United States and Mexico. Chia seeds have a very high protein content. Chia seeds cannot be sprouted in the conventional manner because of their highly mucilaginous nature. However, if put in hollow Mexican earthenware animals that are made especially for the purpose, a green blanket of protein-rich chia sprouts are ready for plucking.

Sources: Carroll, Anstice, and Embree De Persiis Vona. *The Health Food Dictionary with Recipes*, pp. 39-40.

1267
Why should lawn clippings be left on the grass after mowing?

The clippings are a valuable source of nutrients for the lawn. They provide nitrogen, potassium, and phosphorous to feed the new grass and reduce the need for fertilizer. Young, tender, short clippings decompose fast. When clippings are left on the lawn instead of being added to the trash collection, the amount of waste added to landfills is decreased.

Sources: *50 Simple Things You Can Do to Save the Earth*, pp. 42-43; Vengris, Jonas. *Lawns*, p. 123.

1268
What does the term hydroponics mean?

This term refers to growing plants in some medium other than soil; the inorganic plant nutrients (such as potassium, sulphur, magnesium, and nitrogen) are continuously supplied to the plants in solution. Hydroponics is mostly used in areas where there is little soil or unsuitable soil. It is a much-used method of growing plants in research, but for the amateur, it has many limitations and may prove frustrating. Hydroponics was developed in 1959 by two Soviet research scientists at the University of Leningrad.

Sources: Giscard d'Estaing, Valerie–Anne. *The World Almanac Book of Inventions*, p. 67; Wyman, Donald. *Wyman's Gardening Encyclopedia*, pp. 549–550.

1269 / Which guidelines should be followed to garden by the moon?

Guidelines for lunar gardening are simple. The waxing moon occurs between the new moon and full moon, and includes the first and second quarter phases. The waning moon occurs between the full moon and new moon, and includes the third and fourth quarters. Generally all activities for growth and increase, especially plants that produce above the ground, should take place during the waxing moon. Vegetables and fruits intended to be eaten immediately should be gathered at the waxing moon. Cutting, controlling, and harvesting for food to be conserved or preserved, as well as planting crops that yield below the ground, should take place during the waning moon.

Sources: Guiley, Rosemary E. *Moonscapes*, pp. 103, 105.

1270 / What is the difference between container-grown, balled-and-burlapped, and bare-rooted plants?

Container-grown plants have been grown in some kind of pot—usually peat, plastic, or clay—for most or all of their lives. Balled–and–burlapped plants have been dug up with the soil carefully maintained around their roots in burlap. Bare–rooted plants have also been dug from their growing place but without retaining the root ball. Typically plants from a mail-order nursery come bare-rooted with their roots protected with damp sphagnum moss. Bare-rooted plants are the most susceptible to damage.

Sources: *Fruits and Vegetables: 1001 Gardening Questions Answered*, p. 5.

1271 / How can poison ivy/oak/sumac be eliminated from the garden?

First, choose a moist, cool day in early spring or late fall when the plant is less armed with irritant. Wear thick leather or cotton gloves (but not rubber); cut the vines near the base and apply an herbicide to kill the roots. Handle the vines carefully, and dispose of them by burying them in a biodegradable container. *Do not burn* the plants; smoke and ash may cause the rash on exposed parts of the body, eyes, nasal passages, and lungs.

All parts of poison ivy (*Rhus radicans*) and poison oak (*Rhus toxicodendron*) can cause serious dermatitis. These North American woody plants grow in almost any habitat and are quite similar in appearance. Each has alternating compound leaves of three leaflets each, berrylike fruits, and rusty brown stems. But poison ivy acts more like a vine than a shrub at times and can grow high into trees. Its grey fruit is not hairy, and its leaves are slightly lobed. On the other hand, poison oak is often shrubby, but it can climb. Its leaflets are lobed and resemble oak leaves; its fruit is hairy. Poison sumac (*Rhus vernix*) grows only in North American wet acid swamps. This shrub can grow as high as 12 feet (3.6 meters). The fruit it produces hangs in a cluster and is grayish-brown in color. Poison sumac has sharply-pointed, dark green compound, alternating leaves, and greenish-yellow inconspicuous flowers.

Sources: Frankel, Edward. *Poison Ivy, Poison Oak, Poison Sumac, and Their Relatives*, pp. 81-83.

Poison ivy and poison oak

Poison sumac

1272
Which insects are common problems in strawberry cultivation?

Earwigs, slugs, and snails are big problems in some areas. Strawberries (*Fragaria*) are also bothered by Japanese beetles, aphids, thrips, weevils, nematodes, and mites.

Sources: *Fruits and Vegetables: 1001 Gardening Questions Answered*, p. 49.

1273
How are seedless grapes grown?

Seedless grapes are grown from small cuttings of other seedless plants. Because seedless grapes cannot reproduce in the conventional way that grapes usually do (i.e., dropping seeds), growers have to take cuttings from other seedless grape plants, and root them. Although the exact origin of seedless grapes is unknown, they might have first been cultivated in present-day Iran or Afghanistan thousands of years ago. Initially the first seedless grape was a genetic mutation in which the hard seed casing failed to develop—the mutation is called stenospermoscarpy. One modern seedless grape commonly bought today is the green Thompson seedless grape from which 90% of all raisins are made.

Sources: Feldman, David. *Why Do Clocks Run Clockwise? And Other Imponderables*, pp. 218–219; Sutton, Caroline, and Duncan M. Anderson. *How Do They Do That?*, pp. 98–100.

1274
What were some of the accomplishments of Dr. George Washington Carver?

Because of the work of Dr. George Washington Carver (1864–1943) in plant diseases, soil analysis, and crop management, many Southern farmers who adopted his methods, had greater crop yields and profits. Carver developed recipes using cowpeas, sweet potatoes and peanuts. He eventually made 118 products from sweet potatoes, 325 from peanuts, and 75 from pecans. He promoted soil diversification and the adoption of peanuts, soybeans, and other soil-enriching crops.

Sources: *The Great Scientists*, vol. 3, pp. 67, 70; Sammons, Vivian O. *Blacks in Science and Medicine*, p. 50.

1275
When was the first practical greenhouse built?

The French botanist, Jules Charles, constructed one in 1599 in Leiden, Holland in which tropical plants for medicinal purposes were grown. The most popular plant there was the Indian date, the tamarind, whose fruit was made into a curative drink.

Sources: Williams, T. Jeff. *Greenhouses*, p. 7.

Who developed DDT?

Although DDT was synthesized as early as 1874 by Othmar Zeidler, it was the Swiss chemist Paul Müller (1899–1965) who recognized its insecticidal properties in 1939. He was awarded the 1948 Nobel Prize in Medicine for his development of *d*ichloro-*d*iphenyl-*t*richloro-ethene or DDT. This substance killed insects quickly but seemed to have little or no poisonous effect on plants and animals, unlike the arsenic-based compounds then in use. In the following twenty years it proved to be effective in controlling disease-carrying insects (malaria- and yellow fever-carrying mosquitoes, and typhus-carrying lice), and in killing many plant crop destroyers. Increasingly DDT-resistant insect species and the accumulative hazardous effects of DDT on plant and animal life cycles led to its disuse in many countries during the 1970s.

Source: *The Biographical Dictionary of Scientists: Chemists*, pp. 99–100.

PHYSICAL CHARACTERISTICS, FUNCTIONS, ETC.

1277
What is tropism?

Tropism is the movement of a plant in response to a stimulus. The categories include

Chemotropism—a response to chemicals by plants in which incurling of leaves may occur.

Geotropism or *gravitropism*—a response to gravity in which the plant moves in relation to gravity. Shoots of a plant are negatively geotropic (growing upward), while roots are positively geotropic (growing downward).

Hydrotropism—a response to water or moisture in which roots grow toward the water source.

Paraheliotropism—a response by the plant leaves to avoid exposure to the sun.

Phototropism—a response to light in which the plant may be positively phototropic (moving toward the light) or negatively phototropic (moving away from the light source). Main axes of shoots are usually positively phototropic, whereas roots are generally insensitive to light.

Thermotropism—a response to temperature by plants.

Thigmotropism or *haptotropism*—a response to touch by the climbing organs of a plant, for example the plant's tendrils curl around a support in a "spring-like manner."

Sources: Bonnet, Robert L. *Botany: 49 Science Fair Projects*, pp. 73–74; Lincoln, R.J. *A Dictionary of Ecology, Evolution, and Systematics*, p. 181; Parker, Sybil P. *McGraw-Hill Concise Encyclopedia of Science and Technology*, p. 1438.

1278 / How do different kinds of music affect plant growth?

In experiments done in the 1960s and 1970s, plants responded best to classical and Indian devotional music. In a controlled environment, plants exposed to these kinds of music had lush and abundant growth and good root development. Exposure to country music or silence brought about no abnormal growth reaction, while jazz produced more abundant growth. With rock music, plants did poorly. Their roots were scrawny and sparse and they seemed to be in a dying stage. Plants exposed solely to white noise died quickly.

Sources: Retallack, Dorothy L. *The Sound of Music and Plants*, pp. 20-33.

1279 / How many chloroplasts are there in plant cells?

Chloroplasts are the functional units where photosynthesis takes place. They contain the green pigments chlorophyll a and b, which trap light energy for photosynthesis (the process whereby green plants use light energy for the synthesis of sugar from carbon dioxide and water, with oxygen released as a by-product). A unicellular plant may have only a single large chloroplast, whereas a plant leaf cell may have as many as 20 to 100.

Sources: *Plants: Their Biology and Importance*, pp. 17, 127; Villee, Claude A., et al. *Biology*, p. 107.

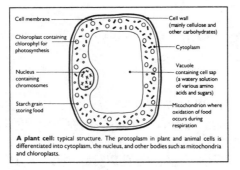

A plant cell: typical structure. The protoplasm in plant and animal cells is differentiated into cytoplasm, the nucleus, and other bodies such as mitochondria and chloroplasts.

1280 / What is the best type of pollination?

Effective pollination occurs when viable pollen is transferred to plant's stigmas, ovule-bearing organs, or ovules (seed precursors). Without pollination, there would be no fertilization. Since plants are immobile organisms, they usually need external agents to transport their pollen from where it is produced in the plant to where fertilization can occur. This situation produces cross-pollination wherein one plant's pollen is moved by an agent to another plant's stigma. Some plants are able to self-pollinate—transfer their own pollen to their own stigmas. But of the two methods, cross-pollination seems the better, for it allows new genetic material to be introduced.

The cross-pollination agents are insects, wind, birds, mammals, or water. Many times flowers offer one or more "rewards" to attract these agents—sugary nectar, oil, solid food bodies, perfume, a place to sleep, or sometimes, the pollen itself. Other times the plant can "trap" the agent into transporting the pollen. Generally plants use color and fragrances as attractants to lure these agents. For example, a few orchids use a combination of smell and color to mimic the female of certain bee and wasp species so successfully that the corresponding males will attempt to mate with them. Through this process (pseudocopulation) the orchids achieve pollination. While some plants cater to a variety of agents, other plants are very selective and are pollinated by a single species of insect only. This extreme pollinator specificity tends to maintain the purity of a plant species.

Plant structure can accommodate the type of agent used. For example, plants (such as grasses and conifers) whose pollen is carried by the wind, tend to have a simple structure lacking petals, with freely exposed and branched stigmas (to catch airborne pollen) and dangling anthers (pollen-producing parts) on long filaments. This type of anther allows the light round pollen to be easily caught by the wind. These plants are found in areas (such as prairies and mountains) where insect agents are rare. In contrast, semi-enclosed, nonsymmetrical, long-lived flowers (such as iris, rose, and snapdragon) have some "landing platform" and nectar in the flower base to accommodate insect agents such as the bee. The sticky abundant pollen can become attached easily to the insect to be borne away to another flower.

Source: Parker, Sybil P. *McGraw-Hill Concise Encyclopedia of Science & Technology*, p. 146; *Practical Botany*, pp. 7-11.

PLANTS, *See: THE ENVIRONMENT—EXTINCT AND ENDANGERED PLANTS AND ANIMALS; THE PLANT WORLD—FLOWERS AND PLANTS; THE PLANT WORLD—GARDENING, FARMING, ETC.; THE PLANT WORLD—PHYSICAL CHARACTERISTICS, FUNCTIONS, ETC.*

TREES AND SHRUBS *See also: THE ENVIRONMENT—ECOLOGY, RESOURCES, ETC.; METALS AND OTHER MATERIALS—NATURAL SUB-STANCES; THE PLANT WORLD—GARDENING, FARMING, ETC.*

1281
Which U.S. city has the greatest number of trees?

According to a survey of 20 cities, Houston, Texas has the most trees with 956,700, while Hartford, Connecticut has only 20,000.

Sources: *American Forests*, vol. 98, no. 3 & 4 (March/April 1992), p. 62.

1282
What is a banyan tree?

The banyan tree (*Ficus benghalensis*), a native of tropical Asia, is a member of the *Ficus* or fig genus. It is a magnificent evergreen, sometimes 100 feet (30.48 meters) in height. As the massive limbs spread horizontally, the tree sends down roots that develop into secondary, pillar-like supporting trunks. Over a period of years a single tree may spread to occupy a tremendous area, as much as 2,000 feet (609.6 meters) around the periphery.

Sources: Everett, Thomas H. *The New York Botanical Garden Illustrated Encyclopedia of Horticulture*, vol. 4, p. 1363.

1283
How much of a tree that is cut down is sold as timber?

For the average hardwood tree cut down to make lumber, half the total wood volume is left in the woods as tops, limbs, and logging residue; about a quarter is lost as sawdust, slabs, and edgings in the sawmill; one-eighth disappears as shavings and machining residues, leaving about one-eighth of the original volume to be sold as timber.

Sources: *Pennsylvania Woodland News*, vol. 5, no. 1 (January/February 1991), p. 2.

1284
What are the longest-lived tree species in the United States?

Of the 850 different species of tree in the United States, the oldest species is the bristlecone pine (*Pinus longaeva*), which grows in the deserts of Nevada and southern California (especially in the White Mountains). Some of these trees are believed to be over 4,600 years old. The *potential* life span of these pines is estimated to be 5,500 years. The giant sequoia (*Sequoiadendron giganteum*) has a *poten-*

tial life span of 6,000 years. The oldest tree in the United States was a bristlecone pine (designated as PNW-114) on Mount Wheeler (Nevada) whose age was evaluated to be 5,000 years old. The oldest known living tree is a bristlecone pine named "Methuselah" in the White Mountains of southern California, whose life span was confirmed to be 4,700 years. But these ages are very young when compared to the oldest surviving species in the world, which is the maiden-hair tree (*Ginkgo bibloba*) of Zhexiang, China. This tree first appeared during the Jurassic era some 160 million years ago. Also called icho, or the ginkyo ("silver apricot"), this species has been cultivated in Japan since 1100 C.E.

The longest lived tree species in the United States

Name of Tree	Number of Years
Bristlecone pine (*Pinus longaeva*)	3,000-4,700
Giant sequoia (*Sequoiadendron giganteum*)	2,500
Redwood (*Sequoia sempervirens*)	1,000-3,500
Douglas fir (*Pseudotsuga menziesii*)	750
Bald cypress (*Taxodium distichum*)	600

Sources: *Guinness Book of Records 1992*, p. 133; Kendig, Frank, and Richard Hutton. *Life-Spans*, p. 108; *The World Almanac and Book of Facts 1992*, p. 133.

1285 / Why do tree leaves turn color in the fall?

The carotenoids (pigments in the photosynthesizing cells), which are responsible for the fall colors, are present in the leaves during the growing season. However, the colors are eclipsed by the green chlorophyll. Toward the end of summer, when the chlorophyll production ceases, the other colors of the carotenoids (such as yellow, orange, red, or purple) become visible. Listed below are the autumn leaf colors of some common trees.

Tree	Color
Sugar maple and sumac	Flame red and orange
Red maple, dogwood, sassafras, and scarlet oak	Dark red
Poplar, birch, tulip tree, willow	Yellow
Ash	Plum purple
Oak, beech, larch, elm, hickory, and sycamore	Tan or brown
Locust	Stays green until leaves drop
Black walnut and butternut	Drops leaves before they turn color

Sources: *Natural History*, vol. 66 (October 1957), pp. 424–435; Platt, Rutherford. *1001 Questions Answered About Trees*, p. 8.

1286 / What is the tallest tree?

The tallest tree ever measured was an Australian eucalyptus (*Eucalyptus regnans*) at Watts River, Victoria, Australia, reported in 1872 by forester William Ferguson. It was 435 feet (132.6 meters) tall and almost certainly measured over 500 feet (152.5 meters) originally. Another *Eucalyptus regnans* at Mt. Baw Baw, Victoria, Australia, is believed to have measured 470 feet (143.3 meters) in 1885. The tallest tree currently standing is the "National Geographic Society" coast redwood (*Sequoia sempervirens*) in Redwood National Park, Humboldt County, California. It stood at 373 feet (113.7 meters) in October 1990, according to Ron Hildebrant of California. The 365-foot (111.3-meter) "Dyerville" giant redwood (*Sequoia giganteum*) in the same park fell during a storm on March 27, 1991.

The tallest known pine tree is sugar pine (*Pinus lambertiana*) located in California, which measured 215 feet (65.5 meters) high in 1967. The sugar pine gets its name from its sweet resin, which naturalist John Muir found preferable to maple sugar. The cones may weigh as much as four pounds (1.8 kilograms) and can reach a length of 26 inches (65 centimeters). The tallest known fir is the 415-foot (125.7-meter) Douglas fir (*Pseudotsuga menziesii*) in Lynn Valley, British Columbia, measured in 1902.

In the United States the tallest Douglas fir with a height of 298 feet (90.8 meters) is one located in the Olympic National Park in the state of Washington. In the same park there is another Douglas fir 212 feet (64.6 meters) high. The tallest living giant sequoia in Sequoia National Park measured 275 feet (83.8 meters) in 1975.

Other United States trees measuring more than 100 feet (30.5 meters) in height are given below:

Species	Height		When and where measured
	Feet	**Meters**	
Ash (Black) (*Fraximus nigra*)	155	47.2	Adrian, Michigan
Beech (American) (*Fagus grandifolia*)	130	39.6	Ashtabula County, Ohio, 1984
Buckeye (Yellow) (*Aesculus octandra*)	145	44.2	Great Smoky Mountains National Park, Tennessee
Cedar (Port-Orford) (*Chaemaecyparis lawsoniana*)	219	66.8	Siskiyou National Forest, Oregon
Cherry (Black) (*Prunus serotina*)	138	42.1	Washtenaw County, Michigan
Elm (American) (*Ulmus americana*)	125	38.1	Southampton County, Virginia, 1985
Eucalyptus (Bluegum) (*Eucalyptus globulus*)	165	50.3	Fort Ross State Park, Sonoma County, California
Fir (Grand)	251	76.5	Olympic National Park, Washington, 1987
Fir (Noble) (*Abies procera*)	238	70.7	Gifford Pinchot National Forest, Washington
Hackleberry (Common)	111	33.8	Rock County, Wisconsin, 1989
Hemlock (Western) (*Tsuga heterophylla*)	241	73.5	Olympic National Park, Washington

Species	Height		When and where measured
	Feet	Meters	
Hickory (Pignut) (Carya glabra)	190	57.9	Robbinsville, North Carolina
Honeylocust	115	35.1	Wayne County, Michigan. 1972
Maple (Red) (Acer rubrum)	179	54.6	St. Clair County, Michigan
Maple (Black)	118	40	Allegan County, Michigan, 1976
Oak (White) (Quercus alba)	107	32.6	Wye Mills State Park, Maryland
Oak (Swamp) Chestnut (Quercus michauxii)	200	61	Layette County, Alabama
Pecan (Carya illinoensis)	143	43.6	Cocke County, Texas, 1980
Pine (Bishop)	112	34.1	Mendocino County, California, 1986
Pine (Ponderosa) (Pinus ponderosa)	223	68	Plumas, California
Pine (Western) (Pinus monticola)	157	47.9	El Dorado National Park, California
Pine (Virginia)	120	36.3	Chambers County, Alabama, 1989
Red cedar (Western) (Juniperus virginiana)	178	54.3	Forks, Washington
Spruce (Sitka) (Picea sitchensis)	206	62.8	Olympic National Forest, Washington
Spruce (Blue) (Picea pungens)	126	38.4	Gunnison National Forest, Colorado
Sweetgum (Athoxanthum odoratum)	136	41.5	Craven County, North Carolina, 1986
Sycamore (Platanus occidentalis)	129	39.3	Jeromesville, Ohio
Willow (Black) (Salix nigra)	109	33.2	Grand Traverse County, Michigan
Willow (Weeping) (Salix tristis)	114	34.7	Asheville, North Carolina, 1982

Sources: *American Forests*, vol. 97 (November–December 1991), p. 27; *Guinness Book of Records 1992*, p. 134; *The World Almanac and Book of Facts 1992*, p. 660.

▽ **1287** ▽

How are tree rings used to date historical events?

Tree fragments of unknown age and the rings of living trees can be compared in order to establish the date when the fragment was part of a living tree. Thus, tree rings can be used to establish the year in which an event took place as long as the event involved the maiming or killing of a tree. Precise dates can be established for the building of a medieval cathedral or an American Indian pueblo; the occurrence of an earthquake, landslide, volcanic eruption, or a fire; and even the date when a panel of wood was cut for a Dutch painting.

Every year, the tree produces an annular ring composed of one wide, light ring and one narrow, dark ring. During spring and early summer, tree stem cells grow rapidly and are larger; this is the wide, light ring. In winter growth is greatly reduced and cells are much smaller; this is the narrow, dark ring. In the cold of winter or the dry heat of summer, no cells are produced.

Sources: Fritts, Harold C. *Tree Rings and Climate*, pp. 1–2; Roth, Charles E. *The Plant Observer's Guidebook*, pp. 150–151.

1288

How can a spruce tree be distinguished from a fir tree?

The best way to tell the difference between the two trees is by the cones and leaves. On a fir tree the cones are upright (erect) and the needles flat, while spruces have pendulous (hanging) cones and very angular, four-sided needles.

Sources: Choukas-Bradley, Melanie, and Polly Alexander. *City of Trees*, p. 86; *The Wise Garden Encyclopedia*, p. 1.

1289

Does the rose family produce any trees?

The apple, pear, peach, cherry, plum, mountain ash, and hawthorn trees are members of the rose family (Rosaceae).

Sources: Mabberley, D.J. *The Plant Book*, p. 507; Platt, Rutherford. *1001 Questions Answered About Trees*, p. 40.

1290

How is glycerine used to preserve leaves?

Glycerine and water are generally used to preserve coarser leaves such as magnolia, rhodendron, beech, holly, heather, or Japanese maple. The leaves should be fresh. For autumn colors, the leaves should be picked just as they are turning color. The preserving solution is made by adding two parts boiling water to one part glycerine. Place leaf stems, split if necessary, in the warm solution so they are covered about 3 to 4 inches (7.5 to 10 centimeters). When drops of glycerine form on the leaves, enough has been absorbed. Wipe off any excess oil. Entire branches may also be treated by using a container large enough to allow the solution to completely cover the leaves. The solution for this method is made with equal quantities of glycerine and water. The leaves are drained on thick sections of newspapers for a few days, then washed with a little soap and water and pegged on a line to dry.

Sources: Mierhof, Annette. *The Dried Flower Book*, p. 17.

1291

What is a monkey ball tree?

The osage orange (*Maclura pomifera*) produces large green orange-like fruits. These are roughly spherical, 3½ to 5 inches (8.8 to 12.7 centimeters) in diameter and have a coarse pebbly surface.

Sources: Everett, Thomas H. *The New York Botanical Garden Illustrated Encyclopedia of Horticulture*, vol. 6, p. 2092; Grimm, William C. *The Trees of Pennsylvania*, pp. 200–201.

1292

Are there any natural predators of gypsy moth caterpillars?

About 45 kinds of birds, squirrels, chipmunks, and white–footed mice eat this serious insect pest. Among the 13 imported natural enemies of the moth, two flies, *Compislura concinnata* (a tachnid fly) and *Sturnia scutellata*, parasitize the caterpillar. Other parasites and various wasps have also been tried as controls, as well as spraying and male sterilization. Originally from Europe, this large moth (*Porthetria dispar*) lays its eggs on the leaves of oaks, birches, maples, and other hardwood trees. When the yellow hairy caterpillars hatch from the eggs, they devour the leaves in such quantities that the tree becomes temporarily defoliated. Sometimes this can even cause the tree to die. The caterpillars grow from ⅛ inch (3 millimeters) to about 2 inches (5.1 centimeters) before they spin a pupa in which they will metamorphose (change physically) into adult moths.

Sources: Ashworth, William. *The Encyclopedia of Environmental Studies*, p. 176; *The Wise Garden Encyclopedia*, p. 470.

1293

Who released the gypsy moth?

In 1869, Professor Leopold Trouvelot brought gypsy moth egg masses from France to Medford, Massachusetts. His intention was to breed the gypsy moth with the silkworm to overcome a wilt disease of the silkworm. He placed the egg masses on a window ledge, and evidently the wind blew them away. About 10 years later these caterpillars were numerous on trees in that vicinity, and in 20 years, trees in eastern Massachusetts were being defoliated. In 1911, a contaminated plant shipment from Holland also introduced the gypsy moth to that area. These pests have now spread to 25 states, especially in the northeastern United States. Scattered locations in Michigan and Oregon have also reported occurences of gypsy moth infestations.

Sources: Ashworth, William. *The Encyclopedia of Environmental Studies*, p. 176; Pfadt, Robert E. *Fundamentals of Applied Entomology*, p. 553; *The Wise Garden Encyclopedia*, p. 469.

SPACE

ASTEROIDS *See: SPACE—COMETS, METEORITES, ETC.*

AWARDS *See: GENERAL SCIENCE AND TECHNOLOGY—AWARDS, PRIZES, AND FAIRS*

COMETS, METEORITES, ETC.

1294

Where are the asteroids?

The asteroids, also called the minor planets, are smaller than any of the nine major planets in the solar system and are not satellites of any major planet. The term asteroid means "starlike" because asteroids appear to be points of light when seen through a telescope.

Most asteroids are located between Mars and Jupiter, between 2.1 and 3.3 AUs from the sun. Ceres, the first to be discovered, and the largest, was found by Giuseppe Piazzi (1746–1826) on January 1, 1801, and has a diameter of 582 miles (936 kilometers). A second asteroid, Pallas, was discovered in 1802. Since then, over 18 thousand asteroids have been identified and astronomers have established orbits for about 5 thousand of them. Some of these have diameters of only 0.62 miles (1 kilometer). Originally, astronomers thought the asteroids were remnants of a planet that had been destroyed; now they believe asteroids to be material that never became a planet, possibly because it was affected by Jupiter's strong gravity.

Not all asteroids are in the main asteroid belt. Three groups reside in the inner solar system. The Aten asteroids have orbits that lie primarily inside Earth's orbit. However, at their farthest point from the sun, these asteroids may cross Earth's orbit. The Apollo asteroids cross Earth's orbit; some come even closer than the moon. The Amor asteroids cross the orbit of Mars and some come close to Earth's orbit. The Trojan asteroids move in virtually the same orbit as Jupiter but at points

60 degrees ahead or 60 degrees behind the planet. In 1977 Charles Kowal discovered an object now known as Chiron orbiting between Saturn and Uranus. Originally cataloged as an asteroid, Chiron was later observed to have a coma (a gaseous halo) and it may be reclassified as a comet.

Sources: Menzel, Donald H., and Jay M. Pasachoff. *A Field Guide to Stars and Planets*, pp. 243-245; *The Planetary Report*, vol. 11, no. 6 (November/December 1991), pp. 8-11; *Scientific American*, vol. 265, no. 4 (October 1991), pp. 88-94.

1295
An asteroid came close to hitting the earth sometime in 1989. How much damage might it have done?

Asteroid 1989 FC passed within 434 thousand miles (700 thousand kilometers) of the earth on March 22, 1989. The impact, had it hit the earth, would have delivered the energy equivalent of more than 1 million tons of exploding TNT and created a crater up to 4.3 miles (7 kilometers) across.

Sources: *Sky and Telescope*, vol. 79, no. 3 (March 1990), p. 261.

1296
What was the Tunguska Event?

On June 30, 1908, a violent explosion occurred in the atmosphere over the Podkamennaya Tunguska river, in a remote part of central Siberia. The blast's consequences were similar to an H-bomb going off, leveling thousands of square miles of forest. The shock of the explosion was heard over 600 miles (960 kilometers) away. A number of theories have been proposed to account for this event.

Some people thought that a large meteorite or a piece of anti-matter had fallen to earth. But a meteorite, composed of rock and metal, would have created a crater and none was found at the impact site. There are no high radiation levels in the area which would have resulted from the collision of anti-matter and matter. Two other theories include a mini-black hole striking the earth and the crash of an extraterrestrial spaceship. However, a mini-black hole would have passed through the earth and there is no record of a corresponding explosion on the other side of the world. As for the spaceship, no wreckage of such a craft was ever found.

The most likely cause of the explosion was the entry into the atmosphere of a piece of a comet which would have produced a large fireball and blast wave. Since a comet is composed primarily of ice, the fragment would have melted during its passage through the earth's atmosphere, leaving no impact crater and no debris. Since the Tunguska Event coincided with the earth's passage through the orbit of Comet Encke, the explosion could have been caused by a piece of that comet.

Sources: Angelo, Joseph A. *The Extraterrestrial Encyclopedia*, p. 220; Calder, Nigel. *The Comet Is Coming*, pp. 124-126.

When will Halley's comet return?

Halley's comet returns about every 76 years. It was most recently seen in 1985/1986 and is predicted to appear again in 2061, then in 2134. Every appearance of what is now known as Comet Halley has been noted by astronomers since the year 239 B.C.E.

The comet is named for Edmund Halley (1656-1742), England's second Astronomer Royal. In 1682 he observed a bright comet and noted that it was moving in an orbit similar to comets seen in 1531 and 1607. He concluded that the three comets were actually one and the same and that the comet had an orbit of 76 years. In 1705 Halley published *A Synopsis of the Astronomy of Comets* in which he predicted that the comet seen in 1531, 1607, and 1682 would return in 1758. On Christmas night, 1758, a German farmer and amateur astronomer named Johann Palitzsch spotted the comet in just the area of the sky that Halley had foretold.

Prior to Halley, comets appeared at irregular intervals and were often thought to be harbingers of disaster and signs of divine wrath. Halley proved that they are natural objects subject to the laws of gravity. This was a great victory for science in the war against superstition.

Sources: Moore, Patrick. *International Encyclopedia of Astronomy*, p. 177; Sagan, Carl, and Ann Druyan. *Comet*, p. 364.

Halley's comet orbit

From where do comets originate?

According to a theory developed by Dutch astronomer Jan Oort, there is a large cloud of gas, dust, and comets orbiting beyond Pluto out to perhaps 100,000 astronomical units (AU). Occasional passage of a star close to this cloud disturbs some of the comets from their orbits. Some fall inwards towards the sun.

Comets, sometimes called "dirty snowballs," are made up mostly of ice, with some dust mixed in. When a comet moves closer to the sun, the dust and ice of the core, or nucleus, heats up, producing a tail of material which trails along behind it. The tail is pushed out by the solar wind and almost always points away from the sun.

Most comets have highly elliptical orbits that carry them around the sun and then fling them back out to the outer reaches of the solar system, never to return. Occasionally however, a close passage by a comet near one of the planets can alter a comet's orbit making it stay in the middle or inner solar system. Such a comet is called a short-period comet because it passes close to the sun at regular intervals. The most famous short-period comet is Comet Halley which reaches perihelion (the point in its orbit that is closest to the sun) about every 76 years. Comet Encke, with an orbital period of 3.3 years, is another short-period comet.

Sources: Miller, Ron, and William K. Hartmann. *The Grand Tour: A Traveler's Guide to the Solar System*, pp. 173-179; Sagan, Carl, and Ann Druyan. *Comet.*

1299
When do meteor showers occur?

There are a number of groups of meteoroids circling the sun just as the earth is. When earth's orbit intercepts the path of one of these swarms of meteoroids, some of them enter earth's atmosphere. When friction with the air causes a meteoroid to burn up, the streak, or shooting star, that is produced is called a meteor. Large numbers of meteors can produce a spectacular shower of light in the night–time sky. Meteor showers are named for the constellation that occupies the area of the sky from which they originate. Listed below are the ten meteor showers and the dates during the year during which they can be seen.

Name of Shower	Dates
Quadrantids	January 1–6
Lyrids	April 19–24
Eta Aquarids	May 1–8
Perseids	July 25–August 18
Orionids	October 16–26
Taurids	October 20–November 20
Leonids	November 13–17
Phoenicids	December 4–5
Geminids	December 7–15
Ursids	December 17–24

Sources: Rovin, Jeff. *Laws of Order*, p. 217.

1300
How does a meteorite differ from a meteoroid?

A meteorite is a natural object of extraterrestrial origin that survives passage through the earth's atmosphere and hits the earth's surface. A meteorite is often confused with a meteoroid or a meteor. A meteoroid is a small object in outer space, generally less than 30 feet (10 meters) in diameter. A meteor (sometimes called a shooting star) is the flash of light seen when an object passes through earth's atmosphere and burns as a result of heating caused by friction. A meteoroid becomes a meteor when it enters the earth's atmosphere; if any portion of a meteoroid lands on earth, it is a meteorite.

There are three kinds of meteorites. Irons contain 85-95% iron; the rest of their mass is mostly nickel. Stony irons are relatively rare meteorites composed of about 50% iron and 50% silicates. Stones are made up mostly of silicates and other stony materials.

Sources: Moore, Patrick, et al. *The Atlas of the Solar System*, pp. 406-407; *The Planetary Report*, vol. 2, no. 6 (November/December 1991), pp. 8-11.

1301

What are the largest meteorites that have been found in the world?

Name	Location	Weight	
		Tons	Tonnes
Hoba West	Namibia	66.1	60
Ahnighito (The Tent)	Greenland	33.5	30.4
Bacuberito	Mexico	29.8	27
Mbosi	Tanzania	28.7	26
Agpalik	Greenland	22.2	20.1
Armanty	Outer Mongolia	22	20
Willamette	Oregon, USA	15.4	14
Chupaderos	Mexico	15.4	14
Campo del Cielo	Argentina	14.3	13
Mundrabilla	Western Australia	13.2	12
Morito	Mexico	12.1	11

The famous Willamette (Oregon) iron, listed above, displayed at the American Museum of Natural History in New York, is the largest specimen found in the United States. It is 10 feet (3.048 meters) long and 5 feet (1.524 meters) high.

Sources: Moore, Patrick, et al. *The Atlas of the Solar System*, p. 406.

EARTH *See: THE EARTH; SPACE—PLANETS AND MOONS*

EXPLORATION

1302

Who was the first man in space?

Yuri Gagarin (1934-1968), a Soviet cosmonaut, became the first man in space when he made a full orbit of the earth in *Vostok I*, April 12, 1961. Gagarin's flight lasted only 1 hour and 48 minutes, but as the first man in space, he became an international hero. Partly because of this Soviet success, U.S. President John F. Kennedy (1917-1963), announced on May 25, 1961, that the United States would land a man on the moon before the end of the decade. The United States took its first step toward that goal when it launched the first American into orbit on February 20, 1962. Astronaut John H. Glenn, Jr. (b. 1921) completed three orbits in *Friendship 7* and travelled about 81,000 miles (130,329 kilometers). Prior to this, on May 5, 1961, Alan B. Shepard, Jr. (b. 1923) became the first American to man a spaceflight, aboard *Freedom 7*. This suborbital flight reached an altitude of 116.5 miles (187.45 kilometers).

Sources: *The Cambridge Encyclopedia of Space*, pp. 50-55.

1303

What did NASA mean by Voyager 1 *and* 2 *taking a "grand tour" of the planets?*

Once every 176 years the giant outer planets—Jupiter, Saturn, Uranus and Neptune—align themselves in such a pattern that a spacecraft launched from Earth to Jupiter at just the right time might be able to visit the other three planets on the same mission. A technique called "gravity assist" used each planet's gravity as a power boost to point *Voyager* toward the next planet. 1977 was the opportunity year for the "grand tour."

Sources: Angelo, Joseph A. *The Extraterrestrial Encyclopedia*, p. 219.

1304

Did physical changes occur in the astronauts' bodies when they were in space?

The astronauts grew 1½ to 2¼ inches taller. This height increase was due to spinal lengthening and straightening. Their waist measurements decreased by several inches, due to an upward shift of the internal organs in the body. The calves of their legs became smaller because the muscles of the legs forced blood and other fluids toward the upper part of the body, thus decreasing the girth measurement around the thighs and calves.

Sources: Pogue, William R. *How Do You Go to the Bathroom in Space?*, p. 11.

1305

What is the message attached to the Voyager spacecraft?

Voyager 1 (launched September 5, 1977) and *Voyager 2* (launched August 20, 1977) by NASA were unmanned space probes designed to explore the outer planets and then travel out of the solar system. A gold-coated copper phonograph record containing a message to any possible extraterrestrial civilization that they may encounter is attached to each spacecraft. The record contains both video and audio images of earth and the civilization which sent this message to the stars.

The record begins with 118 pictures. These show the earth's position in the galaxy; a key to the mathematical notation used in other pictures; the sun; other planets in the solar system; human anatomy and reproduction; various types of terrain (seashore, desert, mountains); examples of vegetation and animal life; people of both sexes and of all ages and ethnic types engaged in a number of activities; structures (from grass huts to the Taj Mahal to the Sydney Opera House) showing diverse architectural styles; and means of transportation, including roads, bridges, cars, planes, and space vehicles.

The pictures are followed by greetings from Jimmy Carter, then president of the United States, and Kurt Waldheim, then Secretary General of the United Nations. Brief messages in 54 languages, ranging from ancient Sumerian to English, are included as is a "song" of the humpback whales.

The next section is a series of sounds common to the earth. These include thunder, rain, wind, fire, barking dogs, footsteps, laughter, human speech, the cry of an infant, and the sounds of a human heartbeat and human brainwaves.

The record concludes with approximately 90 minutes of music, "Earth's Greatest Hits." The musical selections were drawn from a broad spectrum of cultures and include such diverse pieces as a Pygmy girl's initiation song; bagpipe music from Azerbaijan; the Fifth Symphony, First Movement by Ludwig von Beethoven; and "Johnny B. Goode" by Chuck Berry.

It will be tens, even hundreds of thousands of years before either *Voyager* comes close to another star and perhaps the message will never be heard. But it is a sign of humanity's hope to encounter life elsewhere in the universe.

Source: *Murmurs of Earth: The Voyager Interstellar Record.*

1306
Which astronauts have walked on the moon?

Twelve astronauts have walked on the moon. Each Apollo flight had a crew of three. One crew member remained in orbit in the command service module (CSM) while the other two actually landed on the moon.

Apollo 11, July 16-24, 1969
  Neil A. Armstrong
  Edwin E. Aldrin, Jr.
  Michael Collins (CSM pilot, did not walk on the moon)
Apollo 12, November 14-24, 1969
  Charles P. Conrad
  Alan L. Bean
  Richard F. Gordon, Jr. (CSM pilot, did not walk on the moon)
Apollo 14, January 31-February 9, 1971
  Alan B. Shepard, Jr.

Edgar D. Mitchell
Stuart A. Roosa (CSM pilot, did not walk on the moon)
Apollo 15, July 26-August 7, 1971
David R. Scott
James B. Irwin
Alfred M. Worden (CSM pilot, did not walk on the moon)
Apollo 16, April 16-27, 1972
John W. Young
Charles M. Duke, Jr.
Thomas K. Mattingly, II (CSM pilot, did not walk on the moon)
Apollo 17, December 7-19, 1972
Eugene A. Cernan
Harrison H. Schmitt
Ronald E. Evans (CSM pilot, did not walk on the moon)

Sources: Curtis, Anthony R. *Space Almanac*, pp. 30, 43, 66.

1307
Who made the first golf shot on the moon?

Alan Shepard, Jr. (b. 1923), commander of *Apollo 14*, launched on January 31, 1971, made the first golf shot. He attached a six iron to the handle of the contingency sample return container, dropped a golf ball on the moon, and took a couple of one-handed swings. He missed with the first, but connected with the second. The ball, he reported, sailed for miles and miles.

Sources: Compton, William David. *Where No Man Has Gone Before*, p. 209.

1308
How probable is it that there is intelligent life on other planets?

The possibility of intelligent life depends on several factors. An estimation can be calculated by using an equation developed originally by American astronomer Frank Drake (b. 1930). Drake's equation reads $N = N_* f_p n_e f_1 f_i f_c f_L$. This means that the number of advanced civilizations (N) is equal to

N_*, the number of stars in the Milky Way galaxy, times
f_p, the fraction of those stars that have planets, times
n_e, the number of planets capable of supporting life, times
f_1, the fraction of planets suitable for life on which life actually arises, times
f_i, the fraction of planets where intelligent life evolves, times
f_c, the fraction of planets with intelligent life which develops a technically advanced civilization, times
f_L, the fraction of time which a technical civilization lasts.

The equation is obviously subjective and the answer depends on whether optimistic or pessimistic numbers are assigned to the various factors. However, the galaxy is so large that the possibility of life elsewhere must be considered.

Sources: Brennan, Richard P. *Dictionary of Scientific Literacy*, p. 82; Sagan, Carl. *Cosmos*, pp. 298-302.

1309 / Which United States manned space flight was the longest?

It was Skylab SL-4, which lasted 84 days, 1 hour, 15 minutes and 31 seconds from November 16, 1973 to February 8, 1974. During this time, the three-man crew, Gerald Carr, Edward Gibson, and William Pogue, observed Comet Kohoutek and set records for time spent on experiments in every discipline from medical investigations to materials science. The longest single Soviet stay in space was accomplished by cosmonauts Vladimir Titov (b. 1947) and Musa Manarov (b. 1951) during *Soyuz TM4*. Their space flight aboard the space station Mir lasted 366 days (December 21, 1987–December 21, 1988).

Sources: Curtis, Anthony R. *Space Almanac*, pp. 37-38; *Space Flight: The First 30 years*, p. 20.

1310 / When and what was the first animal sent into orbit?

A dog named Laika, aboard the Soviet *Sputnik 2*, launched November 3, 1957, was the first animal sent into orbit. This event followed the successful Soviet launch on October 4, 1957 of *Sputnik 1*, the first man-made satellite ever placed in orbit. Laika was a very small female dog and became the first living creature to go into orbit. She was placed in a pressurized compartment within a capsule that weighed 1103 pounds (500 kilograms). After a few days in orbit, she died, and *Sputnik 2* reentered the earth's atmosphere on April 14, 1958. Some sources list the dog as a Russian samoyed laika named "Kudyavka" or "Limonchik."

Sources: *The Cambridge Encyclopedia of Space*, p. 48; Curtis, Anthony R. *Space Almanac*, p. 425; Wood, Gerald L. *Guinness Book of Pet Records*, p. 39.

1311 / What were the first monkeys and chimpanzees in space?

On a United States Jupiter flight on December 12, 1958, a squirrel monkey, named Old Reliable was sent into space, but not into orbit. The monkey drowned during recovery.

On another Jupiter flight, on May 28, 1959, two female monkeys were sent 300 miles (482.7 kilometers) high. Able was a 6–pound (2.7–kilograms) Rhesus monkey and Baker was an 11–ounce (0.3–kilograms) squirrel monkey. Both were recovered alive.

A chimpanzee, named Ham, was used on a Mercury flight on January 31,

1961. Ham was launched to a height of 157 miles (253 kilometers) into space, but did not go into orbit. His capsule reached a maximum speed of 5,857 miles (9,426 kilometers) per hour and landed 422 miles (679 kilometers) downrange in the Atlantic Ocean where he was recovered unharmed.

On November 29, 1961, the United States placed a chimpanzee, named Enos, into orbit and recovered him alive after two complete orbits around the earth. Like the Soviets, who usually used dogs, the United States had to obtain information on the effects of space flight on living beings before they could actually launch a human into space.

Sources: Baker, David. *The History of Manned Space Flight*, pp. 62–64; *Life*, vol. 46 (June 8, 1959), pp. 38-40; McAleer, Neil. *The OMNI Space Almanac*, p. 15.

◁1312▷ Who were the first man and woman to walk in space?

On March 18, 1965, the Soviet cosmonaut Alexei Leonov (b. 1934) became the first person to walk in space when he spent 10 minutes outside his *Voskhod 2* spacecraft. The first woman to walk in space was Soviet cosmonaut Svetlana Savitskaya (b. 1947) who, during her second flight aboard the *Soyuz T-12* (July 17, 1984), performed 3½ hours of extravehicular activity.

The first American to walk in space was Edward White II (1930–1967) from the spacecraft *Gemini 4* on June 3, 1965. Kathryn D. Sullivan (b. 1951) became the first American woman to walk in space when she spent 3½ hours outside the *Challenger* orbiter during the space shuttle mission 41G on October 11, 1984.

American astronaut Bruce McCandless II (b. 1937) performed the first untethered space walk from the space shuttle *Challenger* on February 7, 1984, using an MMU (*m*anual *m*aneuvering *u*nit) back-pack.

Sources: *Guinness Book of Records 1992*, p. 298; Spangenburg, Ray, and Diane Moser. *Space People From A-Z*, pp. 65, 71-72.

◁1313▷ What were the first words spoken by an astronaut after touchdown of the lunar module on the Apollo 11 flight? What were the first words spoken by an astronaut standing on the moon?

On July 20, 1969, at 4:17:43 p.m. Eastern Daylight Time (20:17:43 Greenwich Mean Time), Neil A. Armstrong (b. 1930) and Edwin E. Aldrin, Jr. (b. 1930), landed the lunar module *Eagle* in the moon's Sea of Tranquility, and Armstrong radioed: "Houston, Tranquility Base here. The *Eagle* has landed." Several hours later, when Armstrong descended the lunar module ladder and made the small jump between the *Eagle* and the lunar surface, he announced: "That's one small step for man, one giant leap for mankind." The article "a" was missing in the live voice transmission, and was later inserted in the record to amend the message to "one small step for *a* man."

Sources: Gatland, Kenneth. *The Illustrated Encyclopedia of Space Technology*, p. 171.

▽ **1314**
Who are the American women who have gone into space?

As of June 1992, the following American women have flown in space:

Sally K. Ride
June 18, 1983	STS-7	Challenger
October 5, 1984	41G	Challenger

Judith Resnick
August 31, 1984	41D	Discovery
January 28, 1986	51L	Challenger

Kathryn D. Sullivan
October 5, 1984	41G	Challenger
April 24, 1990	STS-31	Discovery
March 24, 1992	STS-45	Atlantis

Anna L. Fischer
November 8, 1984	51A	Discovery

Margaret R. Seddon
April 12, 1985	51D	Discovery
June 5, 1981	STS-40	Columbia

Shannon W. Lucid
June 17, 1985	51G	Discovery
October 18, 1989	STS-34	Atlantis
August 2, 1991	STS-43	Atlantis

Connie J. Dunbar
October 30, 1985	61A	Challenger
January 9, 1990	STS-32	Columbia
June 25, 1992	STS-50	Columbia

Mary L. Cleave
November 26, 1985	61B	Atlantis
May 4, 1989	51L	Atlantis

Christa McAuliffe
January 28, 1986	51L	Challenger

Ellen S. Baker
October 18, 1989	STS-34	Atlantis
June 25, 1992		

Kathryn C. Thornton
November 22, 1989	STS-33	Discovery
May 7, 1992	STS-49	Endeavour

Marsha S. Irvins
January 9, 1990	STS-32	Columbia

Linda M. Godwin
    April 5, 1991   STS-40 Columbia
Millie Hughes-Fulford
    June 5, 1991   STS-40 Columbia

Sources: *Space Flight: The First 30 Years*, pp. 3-36.

1315
What is "earthshine?"

A spacecraft in orbit around the Earth is illuminated by sunlight and "earth-shine." Earthshine consists of sunlight reflected by the Earth and thermal radiation emitted by the Earth's surface and atmosphere.

Sources: Angelo, Joseph A. *The Extraterrestrial Encyclopedia*, p. 33.

1316
Who was the first Black American in space?

Guion S. Bluford, Jr. (b. 1942) became the first Black American to fly in space during the Space Shuttle *Challenger* mission STS-8 (August 30–September 5, 1983). Astronaut Bluford, who holds a Ph.D. in aerospace engineering, made a second shuttle flight aboard *Challenger* mission STS-61-A/Spacelab D1 (October 30-November 6, 1985). The first Black man to fly in space was the Cuban cosmonaut, Arnaldo Tamayo-Mendez, who flew aboard *Soyuz 38* and spent eight days aboard the Soviet space station *Salyut 6* during September 1980. Dr. Mae C. Jemison became the first Black woman in space on September 12, 1992 aboard the Space Shuttle *Endeavour*, mission Spacelab-J.

Sources: Spangenburg, Ray, and Diane Moser. *Space People From A-Z*, pp. 7-8, 72.

1317
How many successful space flights were launched in 1991, and by what countries?

A total of 88 flights which achieved earth orbit or beyond were made in 1991:

Country or Organization	Number of Launches
U.S.S.R.	59
United States	18
European Space Agency	8
Japan	2
Peoples' Republic of China	1

In 1990, there were 116 such launches, with 75 by the U.S.S.R.; 27 (including

7 commercial launches) by the United States; 5 by the European Space Agency; 5 by the People's Republic of China; and 1 by Israel.

Sources: *McGraw-Hill Yearbook of Science and Technology 1992*, p. 445; Smith, Marcia. *Space Activities of the United States and Other Launching Countries/Organizations: 1957-1991*, p. CRS-12.

1318 / Who has the most time accumulated in space?

Soviet cosmonaut Sergei Krikalev (b. 1958) has logged 465 days traveling in space. He accumulated this total on only two missions; during the Soyuz TM-7 mission (November 26, 1988-April 27, 1989), he spent 152 days in space, and during Soyuz TM-12 (May 18, 1991-March 25, 1992), he spent 313 days in space. During both flights, he and other cosmonauts remained aboard the orbiting *MIR* space station.

Sources: Smith, Marcia. *Space Activities of the United States and Other Launching Countries/Organizations: 1957-1991*, pp. 54-58.

1319 / What was the first meal on the moon?

American astronauts Neil A. Armstrong (b. 1930) and Edwin E. Aldrin, Jr. (b. 1930) ate four bacon squares, three sugar cookies, peaches, pineapple-grapefruit drink, and coffee before their historic moonwalk on July 20, 1969.

Sources: Sanders, Dennis. *The First of Everything*, p. 213.

1320 / Who was the first woman in space?

Valentina V. Tereshkova-Nikolaeva (b. 1937), a Soviet cosmonaut, was the first woman in space. She was aboard the *Vostok 6*, launched June 16, 1963. She spent three days circling the earth, completing 48 orbits. Although she had little cosmonaut training, she had been an accomplished parachutist and was especially fit for the rigors of space travel.

The United States space program did not put a woman in space until 20 years later when, on June 18, 1983, Sally K. Ride (b. 1951) flew aboard the space shuttle *Challenger* mission STS-7. In 1987, she moved to the administrative side of NASA and was instrumental in issuing the "Ride Report" which recommended future missions and direction for NASA. She retired from NASA in August 1987 to become a research fellow at Stanford University after serving on the Presidential Commission that investigated the *Challenger* disaster. At present, she is the director of the California Space Institute at the University of California San Diego.

Sources: Spangenburg, Ray, and Diane Moser. *Space People From A-Z*, pp. 61,72.

1321
When was the first United States satellite launched?

Explorer 1, launched January 31, 1958, by the U.S. Army was the first United States satellite launched into orbit. This 31-pound (14.06-kilogram) satellite carried instrumentation that led to the discovery of the earth's radiation belts, which would be named after University of Iowa scientist James A. Van Allen. It followed four months after the world's first satellite, the Soviet Union's *Sputnik 1*. On October 3, 1957, the Soviet Union placed the large, 184-pound (83.5-kilogram) satellite into low earth orbit. It carried instrumentation to study the density and temperature of the upper atmosphere, and its launch was the event that opened the age of space.

Sources: Braun, Wernher von, and Frederick I. Ordway III. *Space Travel: A History*, pp. 170-173; Curtis, Anthony R. *Space Almanac*, p. 13.

1322
Who were the first married couple to go into space together?

Astronauts Jan Davis and Mark Lee were the first married couple in space. They flew aboard the space shuttle *Endeavor* on an eight-day mission which began on September 12, 1992. Ordinarily NASA bars married couples from flying together. An exception was made for Davis and Lee because they had no children and had begun training for the mission long before they got married.

Source: *The New York Times* (September 21, 1992), p. A12.

1323
What is the mission of the Galileo spacecraft?

Galileo, launched October 18, 1989, will circle the sun three times and require almost six years to reach Jupiter, looping past Venus once and the earth twice. The anticipated arrival date at Jupiter is currently late 1995. The *Galileo* spacecraft was designed to make a detailed study of Jupiter and its rings and moons over a period of years. It will release a probe into the jovian atmosphere to analyze the different layers. This probe's range will be limited by its ability to withstand atmospheric pressure (up to several earth atmospheres). Nearly all the planets and moons in the solar system have now been visited by spacecraft. The United States and the former U.S.S.R. have launched many spacecraft to the planets, and only Pluto remains to be explored. The surfaces of the moon, Mars, and Venus have all been examined by unmanned landers, and nearly 1103 pounds (500 kilograms) of rock samples from the moon have been brought back to earth. For the next two decades of solar exploration, the objectives will be to observe, at close quarters, objects which have never been visited by spacecraft, and to increase the amount of data on the planets and their moons which have already been examined. No explorations of Pluto are planned for the foreseeable future.

Sources: *The Almanac of Science and Technology*, p. 58; *Aviation Week and Space Technology*, vol. 131 (October 23, 1989), p. 22; *The Cambridge Encyclopedia of Space*, pp. 199-200.

1324 / Who was the "father" of the Soviet space program?

Sergei P. Korolev (1907-1966) made enormous contributions to the development of Soviet manned space flight, and his name is linked with their most significant space achievements. Trained as an aeronautical engineer, he directed the Moscow group studying the principles of rocket propulsion, and in 1946 took over the Soviet program to develop long-range ballistic rockets. Under Korolev, the Soviets used these rockets for space projects and launched the world's first satellite in October 4, 1957. Besides a vigorous unmanned interplanetary research program, Korolev's goal was to place men in space, and following tests with animals his manned space flight program was initiated when Yuri Gagarin (1934-1968) was successfully launched into earth orbit.

Sources: *The Cambridge Encyclopedia of Space*, p. 34.

1325 / What is meant by the phrase "greening of the galaxy"?

It is an expression describing the spreading of human life, technology, and culture through interstellar space and eventually across the entire Milky Way galaxy, the earth's home galaxy.

Sources: Angelo, Joseph A. *The Extraterrestrial Encyclopedia*, p. 59.

1326 / How many fatalities have occurred during space-related missions?

The fourteen astronauts and cosmonauts listed below died in space-related accidents.

Date	Astronaut/Cosmonaut	Mission
January 27, 1967	Roger Chaffee (U.S.)	Apollo 1
January 27, 1967	Edward White II (U.S.)	Apollo 1
January 27, 1967	Virgil "Gus" Grissom (U.S.)	Apollo 1
April 24, 1967	Vladimir Komarov (U.S.S.R.)	Soyuz 1
June 29, 1971	Viktor Patsayev (U.S.S.R.)	Soyuz 11
June 29, 1971	Vladislav Volkov (U.S.S.R.)	Soyuz 11
June 29, 1971	Georgi Dobrovolsky (U.S.S.R.)	Soyuz 11
January 28, 1986	Gregory Jarvis (U.S.)	STS 51L
January 28, 1986	Christa McAuliffe (U.S.)	STS 51L
January 28, 1986	Ronald McNair (U.S.)	STS 51L
January 28, 1986	Ellison Onizuka (U.S.)	STS 51L
January 28, 1986	Judith Resnik (U.S.)	STS 51L
January 28, 1986	Francis Scobee (U.S.)	STS 51L
January 28, 1986	Michael Smith (U.S.)	STS 51L

Chaffee, Grissom, and White died in a cabin fire during a test firing of the *Apollo 1* rocket.

Komarov was killed in *Soyuz 1* when the capsule's parachute failed.

Dobrovolsky, Patsayev, and Volkov were killed during the *Soyuz II*'s re-entry when a valve accidently opened and released their capsule's atmosphere.

Jarvis, McAuliffe, McNair, Onizuka, Resnik, Scobee, and Smith died when the space shuttle *Challenger* STS 51L exploded 73 seconds after lift-off.

In addition 19 other astronauts and cosmonauts have died of non-space related causes. Fourteen of these died in air crashes (including Yuri Gagarin), four died of natural causes, and one died in an auto crash.

Sources: *Astronauts and Cosmonauts Biographical and Statistical Data*, p. 447; Bali, Mirnal. *Space Exploration: A Reference Handbook*, p. 143.

1327

What was the worst disaster in the U.S. space program and what caused it?

Challenger mission STS 51L was launched on January 28, 1986, but exploded only 73 seconds after lift-off. The entire crew of seven was killed and the *Challenger* was completely destroyed. The investigation of the *Challenger* tragedy was performed by the Rogers Commission, established and named for its chairman, former Secretary of State William Rogers.

The consensus of the Rogers Commission (which studied the accident for several months) and participating investigative agencies is that the accident was caused by a failure in the joint between the two lower segments of the right solid rocket motor. The specific failure was the destruction of the seals that are intended to prevent hot gases from leaking through the joint during the propellant burn of the rocket motor. The evidence assembled by the Commission indicated that no other element of the space shuttle system contributed to this failure.

Although the Commission did not affix blame to any individuals, the public record made clear that the launch should not have been made that day. The weather was unusually cold at Cape Canaveral and temperatures had dipped below freezing during the night. Test data had suggested that the seals (called O-rings) around the solid rocket booster joints lost much of their effectiveness in very cold weather.

Sources: Gurney, Gene. *Space Shuttle Log*, p. 269; Magill, Frank N. *Magill's Survey of Science: Space Exploration Series*, vol. 4, p. 1813; Smith, Marcia. *Space Activities of the United States and Other Launching Countries/Organizations: 1957-1991*, p. CRS-35.

1328

When was the Outer Space Treaty signed?

The United Nations Outer Space Treaty was signed on January 23, 1967. The Treaty provides a framework for the exploration and sharing of outer space. It governs the outer space activities of nations that wish to exploit and make use of space, the moon, and other celestial bodies. It is based on a humanist and pacifist philosophy and on the principle of the non-appropriation of space and the freedom that all nations have to explore and use space. A very large number of countries have signed this agreement, including those from the western alliance, the former Eastern bloc, and non-aligned countries.

Space law, or those rules governing the space activities of various countries, international organizations, and private industries, has been evolving since 1957 when the General Assembly of the United Nations created the Committee on the Peaceful Uses of Outer Space (COPUOS). One of its subcommittees was instrumental in drawing up the 1967 Outer Space Treaty.

Sources: *The Cambridge Encyclopedia of Space*, pp. 360-361; Magill, Frank N. *Magill's Survey of Science: Space Exploration Series*, vol. 3, p. 1111.

1329
What is a "close encounter of the third kind"?

UFO expert J. Allen Hynek (1910-1986) developed the following scale to describe encounters with extraterrestrial beings or vessels:

Close Encounter of the First Kind—sighting of a UFO at close range with no other physical evidence.

Close Encounter of the Second Kind—sighting of a UFO at close range, but with some kind of proof, such as a photograph, or an artifact from a UFO.

Close Encounter of the Third Kind—sighting of an actual extraterrestrial being.

Close Encounter of the Fourth Kind—abduction by a UFO.

Sources: Rovin, Jeff. *Laws of Order*, p. 219.

1330
Is anyone looking for extraterrestrial life?

SETI (the *S*earch for *E*xtraterrestrial *I*ntelligence) actively began in 1960 when American astronomer Frank Drake (b. 1930) spent three months at the National Radio Astronomy Observatory in Green Bank, West Virginia, searching for radio signals coming from the nearby stars Tau Ceti and Epsilon Eridani. Although no signals were detected and although scientists interested in SETI have often been ridiculed, support for the idea of seeking out intelligent life in the universe has grown.

Project Sentinel which used a radio dish at Harvard University's Oak Ridge Observatory in Massachusetts was capable of monitoring 128,000 channels at a time. This project was upgraded in 1985 to META (*M*egachannel *E*xtraterrestrial *A*ssay), thanks in part to a donation by filmmaker Steven Spielberg. Project META is capable of receiving 8.4 million channels. NASA began a ten year search in October 1992 using radio telescopes in Arecibo, Puerto Rico and Barstow, California.

Scientists are searching for radio signals which stand out from the random noises caused by natural objects. Such signals might repeat at regular intervals or contain mathematical sequences. There are millions of radio channels and a lot of sky to be examined. The search for extraterrestrial intelligence will probably continue for some time.

Sources: *Life*, vol. 15, no. 9 (September 1992), pp. 60-67; *The Planetary Report*, vol. 5, no. 5 (September/October 1985), pp. 10-12.

MEASUREMENT *See:* SPACE—OBSERVATION AND MEASUREMENT

METEORITES *See:* SPACE—COMETS, METEORITES, ETC.

MOON *See:* SPACE—PLANETS AND MOONS

OBSERVATION AND MEASUREMENT

1331 / What is a light year?

A light year is a measure of distance, not time. It is the distance that light, which travels in a vacuum at the rate of 186,000 miles (300,000 kilometers) per second, can travel in a year (365¼ days). This is equal to 5,870 billion miles or 9,460 billion kilometers.

Sources: Moore, Patrick. *International Encyclopedia of Astronomy*, p. 231.

1332 / Besides the light year, what other units are used to measure distances in astronomy?

The astronomical unit (AU) is often used to measure distances within the solar system. One AU is equal to the average distance between the earth and the sun or 92,955,630 miles (149,597,870 kilometers). The parsec is equal to 3.26 light years, or about 19,180 million miles (30,820 million kilometers).

Sources: Illingworth, Valerie. *The Facts On File Dictionary of Astronomy*, pp. 235-236; Rovin, Jeff. *Laws of Order*, p. 213; *The Universal Almanac 1992*, p. 531.

1333 / What is an astrolabe?

Invented by the Greeks or Alexandrians in about 100 B.C.E. or before, an astrolabe is a two-dimensional working model of the heavens, with sights for observations. It consists of two concentric flat disks, one fixed, representing the observer on earth, the other moving which can be rotated to represent the appearance of the celestial sphere at a given moment. Given latitude, date, and time, the observer can read off the altitude and azimuth of the sun, the brightest stars and the planets. By measuring the altitude of a particular body, one can find the time. The astrolabe can also be used to find times of sunrise, sunset, twilight, or the height of a tower or depth of a well. It was replaced by the sextant and other more accurate instruments.

Sources: *The Illustrated Science and Invention Encyclopedia*, vol. 2, pp. 168-169; *World Book Encyclopedia*, vol. 1, p. 826.

◣1334◢ *Who is considered the father of systematic astronomy?*

The Greek scientist, Hipparchus (fl. 146-127 B.C.E.), is considered to be the father of systematic astronomy. He measured as accurately as possible the directions of objects in the sky. He compiled the first catalog of stars, containing about 850 entries, and designated each star's celestial coordinates, indicating its position in the sky. Hipparchus also divided the stars according to their apparent brightness or magnitudes.

Sources: Abell, George O. *Realm of the Universe*, p. 16; *Dictionary of Scientific Biography*, vol. 16, suppl. 1, pp. 207-209.

◣1335◢ *How are new celestial objects named?*

Many stars and planets have names that date back to antiquity. The *International Astronomical Union* (IAU), the professional astronomers organization, has attempted, in this century, to standardize names given to newly discovered celestial objects and their surface features.

Stars are generally called by their traditional names, most of which are of Greek, Roman, or Arabic origin. They are also called after the constellation in which they appear, designated in order of brightness by Greek letters. Thus Sirius is also called alpha Canis Majoris, which means it is the brightest star in the constellation Canis Major. Other stars are called by catalog numbers which include the star's coordinates. To many astronomers' horror, several commercial star registries exist, and for a fee, you can submit a star name to them. These names are not officially recognized by the IAU.

The IAU has made some recommendations for naming the surface features of the planets and their satellites. For example, features on Mercury are named for composers, poets, and writers; the features of Venus for women; Saturn's moon Mimas for people and places in Arthurian legend.

Comets are named for their discoverers. Newly–discovered asteroids are first given a temporary designation consisting of the year of discovery plus two letters. The first letter indicates the half month of discovery and the second the order of discovery in that half month. Thus asteroid 1991BA was the first asteroid (A) discovered in the second half of January (B) in 1991. After an asteroid's orbit is determined it is given a permanent number and its discoverer is given the honor of naming it. Asteroids have been named after such diverse things as mythological figures (Ceres, Vesta), an airline (Swissair), and the Beatles (Lennon, McCartney, Harrison, Starr).

Sources: *The Planetary Report*, vol. 11, no. 6 (November/December 1991), p. 6; Sagan, Carl. *Broca's Brain*, pp. 160-175; *Science News*, vol. 125, no. 7 (February 18, 1984), p. 105.

1336
Who is the Hubble for whom the space telescope is named?

Edwin Powell Hubble (1889-1953) was an American astronomer known for his studies of galaxies. His study of nebulae, or clouds, the faint, unresolved luminous patches in the sky, showed that some of them were large groups of many stars. Hubble classified galaxies by their shapes as being spiral, elliptical, or irregular.

Hubble's Law establishes a relationship between the velocity of recession of a galaxy and its distance. The speed at which a galaxy is moving away from the solar system (measured by its redshift, the shift of its light to longer wavelengths, presumed to be caused by the Doppler effect) is directly proportional to the galaxy's distance from the solar system.

The Hubble Space Telescope was deployed by the space shuttle *Discovery* on April 25, 1990. The telescope, which would be free of distortions caused by the earth's atmosphere, was designed to see deeper into space than any telescope on land. However, on June 27, 1990, the National Aeronautics and Space Administration announced that the telescope had a defect in one of its mirrors which prevented it from properly focusing. Although other instruments, such as one designed to make observations in ultraviolet light, are still operating, nearly 40% of the telescope's experiments will have to be postponed. There is a possibility that special lenses for the telescope's instruments can compensate for the mirror's flaw; however, it will be several years until such repairs can be carried out by space shuttle astronauts.

Sources: Parkinson, Clair L. *Breakthroughs: A Chronology of Great Achievements in Science and Mathematics*, pp. 494-495, 508; *Time* (July 9, 1990), p. 43.

1337
Who invented the telescope?

Hans Lippershey (ca. 1570-ca. 1619), a German-Dutch lens grinder and spectacle maker, is generally credited with inventing the telescope in 1608 because he was the first scientist to apply for a patent. Two other inventers, Zacharias Janssen and Jacob Metius, also developed telescopes. Modern historians consider Lippershey and Janssen as the two likely candidates for the title of inventor of the telescope, with Lippershey possessing the strongest claim.

Sources: *The Great Scientists*, vol. 9, p. 162.

PLANETS AND MOONS *See also: THE EARTH; SPACE—EXPLORATION*

1338
How far is the moon from the earth?

Since the moon's orbit is elliptical, its distance varies from about 221,463 miles (356,334 kilometers) at perigee (closest approach) to earth, to 251,968 miles (405,503 kilometers) at apogee (farthest point), with the average distance being 238,857 miles (384,392 kilometers).

Sources: *McGraw-Hill Encyclopedia of Science and Technology*, vol. 5, p. 470; *The Universal Almanac 1992*, p. 16.

1339
What are the diameter and circumference of the moon?

The moon's diameter is 2,159 miles (3,475 kilometers)and its circumference is 6,790 miles (10,864 kilometers). The moon is 27% the size of the earth.

Sources: Guiley, Rosemary E. *Moonscapes*, p. 9; *New Encyclopaedia Britannica*, vol. 8, p. 299; *World Book Encyclopedia*, vol. 13, p. 783.

1340
Why does the moon always keep the same face toward the earth?

Only one side of the moon is seen because it always rotates in exactly the same length of time that it takes to revolve about the earth. This combination of motions (called "captured rotation") means that it always keeps the same side toward the earth.

Sources: Moore, Patrick, et al. *The Atlas of the Solar System*, p. 147; Pickering, James S. *1001 Questions Answered About Astronomy*, p. 23.

1341
What are the phases of the moon?

The phases of the moon are changes in the moon's appearance during the month which are caused by the moon's turning different portions of its illuminated hemisphere towards the earth. When the moon is between the earth and the sun, its daylight side is turned away from the earth, so it is not seen. This is called the new moon. As the moon continues its revolution around the earth, more and more of its surface becomes visible. This is called the waxing crescent phase. About a week after the new moon, half the moon is visible—the first quarter phase. During the next week, more than half of the moon is seen; this is called the waxing gibbous phase. Finally about two weeks after the new moon, the moon and sun are on opposite sides of the earth. The side of the moon facing the sun is also facing the earth, and all the moon's illuminated side is seen as a full moon. In the next two weeks the moon goes through the same phases, but in reverse from a waning gibbous to third or last quarter to waning crescent phase. Gradually, less and less of the moon is visible until a new moon occurs again.

Sources: Abell, George O. *Realm of the Universe*, pp. 72-73; Menzel, Donald H., and Jay M. Pasachoff. *A Field Guide to Stars and Planets*, pp. 318-322.

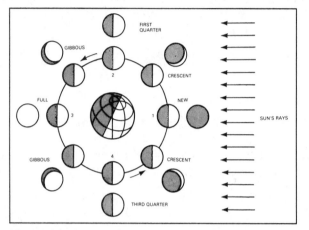

Phases of the moon

1342
Is there ever a full moon twice in one month?

Yes, every 2-3 years there is a month with 2 full moons. This phenomenon will occur on September 1 and 30, 1993; July 1 and 30, 1996; and twice in 1999 on January 2 and 31, and March 2 and 31. In 1999, February has no full moon.

Sources: *Planetary and Lunar Coordinates for the Years 1984-2000*, pp. 3-7.

1343
What is the difference between a hunter's moon and a harvest moon?

The harvest moon is the full moon nearest the autumnal equinox (on or about September 23). It is followed by a period of several successive days when the moon rises soon after sunset. In the southern hemisphere the harvest moon is the full moon closest to the vernal equinox (on or about March 21). This gives farmers extra hours of light for harvesting crops. The next full moon after the harvest moon is called the hunter's moon.

Sources: *Astronomical Almanac 1992*, p. D1; *The World Almanac and Book of Facts 1992*, p. 275.

1344
Is it possible for a calendar month to pass without a full moon?

Yes. February is the only month that may not have a full moon. This occurred in 1866, 1893, 1915, and 1980, and will happen again in 1999 and 2066.

Sources: *Science News-Letter*, vol. 79 (February 4, 1961), p. 67.

1345
Is the moon really blue during a blue moon?

A blue moon is the second full moon in a single month. The term does not refer to the color of the moon. It occurs, on average, every 2.72 years. Since 29.53 days pass between full moons (a synodial month), there is never a blue moon in February.

A bluish-looking moon results from effects of the earth's atmosphere. For example, the phenomenon was widely observed on September 26, 1950 due to Canadian forest fires which had scattered high altitude dust.

Sources: *Country Journal*, vol. 15 (May 1988), p. 14; Moore, Patrick. *International Encyclopedia of Astronomy*, p. 83.

1346
Why do lunar eclipses happen?

A lunar eclipse occurs only during a full moon when the moon is on one side of the earth and the sun is on the opposite side and all three bodies are aligned in the same plane. In this alignment the earth blocks the sun's rays to cast a shadow on the moon. In a total lunar eclipse the moon seems to disappear from the sky, when the whole moon passes through the umbra, or total shadow created by the earth. A total lunar eclipse may last up to 1 hour and 40 minutes. If only part of the moon enters the umbra, a partial eclipse occurs. A penumbral eclipse takes place if all or part of the moon passes through the penumbra (partial shadow or "shade") without touching the umbra. It is difficult to detect this type of eclipse from earth. From the moon one could see that the earth blocked part of the sun only.

Sources: Abell, George O. *Realm of the Universe*, p. 84; *World Book Encyclopedia*, vol. 6, p. 49.

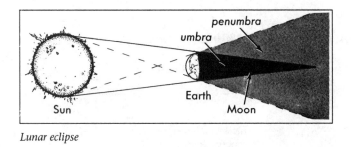

Lunar eclipse

1347

Astronomers have discovered that the moon has a tail. What does this mean?

A glowing 15 thousand mile (24 thousand kilometers) long tail of sodium atoms streams from the moon. The faint, orange glow of sodium cannot be seen by the naked eye but it is detectable by instruments. Astronomers are not certain of the source of these sodium atoms.

Sources: *The Planetary Report*, vol. 9, no. 5 (September/October 1991), p. 21.

1348

What are the craters on the moon that are named for the famous Curie family?

Curie—named for Pierre Curie (1859-1906), French chemist and Nobel prize winner.

Sklodowska—the maiden name of Marie Curie (1867-1934), French physical chemist and Nobel prize winner.

Joliot—named for physicist Frederic Joliot-Curie (1900-1958), Pierre and Marie's son-in-law and Nobel prize winner.

Sources: Room, Adrian. *Dictionary of Astronomical Names*, pp. 175-176.

1349

How old is the solar system?

It is is currently believed to be 4.5 billion years old. The earth and the rest of the solar system formed from an immense cloud of gas and dust. Gravity and rotational forces caused the cloud to flatten into a disc and much of the cloud's mass to drift into the center. This material became the sun. The left-over parts of the cloud formed small bodies called planetesimals. These planetesimals collided with each other, gradually forming larger and larger bodies, some of which became the planets. This process is thought to have taken about 25 million years.

Sources: Moore, Patrick, et al. *The Atlas of the Solar System*, pp. 10-11; Van Andel, Tjeerd H. *New Views On an Old Planet*, p. 28.

1350

How many moons does each planet have?

Planet	Number of Moons	Names of Moons
Mercury	0	
Venus	0	
Earth	1	The Moon (sometimes called Luna)
Mars	2	Phobos, Deimos

Planet	Number of Moons	Names of Moons
Jupiter	16*	Metis, Adrastea, Amalthea, Thebe, Io, Europa, Ganymede, Callisto, Leda, Himalia, Lysithia, Elara, Ananke, Carme, Pasiphae, Sinope
Saturn	18*	Atlas, 1981S13 (unnamed as yet), Prometheus, Pandora, Epimetheus, Janus, Mimas, Enceladus, Tethys, Telesto, Calypso, Dione, Helene, Rhea, Titan, Hyperion, Iapetus, Phoebe
Uranus	15	Cordelia, Ophelia, Bianca, Cressida, Desdemona, Juliet, Portia, Rosalind, Belinda, Puck, Miranda, Ariel, Umbriel, Titania, Oberon
Neptune	8	Naiad, Thalassa, Despina, Galatea, Larissa, Proteus, Triton, Nereid
Pluto	1	Charon

*Several other satellites have been reported but not confirmed.

Sources: *The Universal Almanac 1992*, pp. 17-22.

1351
What are the diameters of the planets?

Planet	Diameter	
	Miles	Kilometers
Mercury	3,031	4,878
Venus	7,520	12,104
Earth	7,926	12,756
Mars	4,221	6,794
Jupiter	88,846	142,984
Saturn	74,898	120,536
Uranus	31,763	51,118
Neptune	31,329	50,530
Pluto	1,423	2,290

Note: All diameters are as measured at the planet's equator.

Sources: *The Guinness Book of Answers*, p. 15; Moore, Patrick, et al. *The Atlas of the Solar System*, p. 442.

1352
How far are the planets from the sun?

The planets revolve around the sun in elliptical orbits, with the sun at one focus of the ellipse. Thus, a planet is at times closer to the sun than at other times. The distances given below are the average distance from the sun, starting with Mercury, the planet closest to the sun, and moving outward.

Planet	Average distance	
	Miles	Kilometers
Mercury	35,983,000	57,909,100
Venus	67,237,700	108,208,600
Earth	92,955,900	149,598,000

Planet	Average distance	
	Miles	Kilometers
Mars	141,634,800	227,939,200
Jupiter	483,612,200	778,298,400
Saturn	888,184,000	1,427,010,000
Uranus	1,782,000,000	2,869,600,000
Neptune	2,794,000,000	4,496,700,000
Pluto	3,666,000,000	5,913,490,000

Sources: Beatty, J. Kelly, Brian O'Leary, and Andrew Chaikin. *The New Solar System*, p. 219; *The Guinness Book of Answers*, p. 14.

1353
What are the colors of the planets?

Planet	Color
Mercury	Orange
Venus	Yellow
Earth	Blue, brown, green
Mars	Red
Jupiter	Yellow, red, brown, white
Saturn	Yellow
Uranus	Green
Neptune	Blue
Pluto	Yellow

Sources: Curtis, Anthony R. *Space Almanac*, p. 580; Moore, Patrick. *International Encyclopedia of Astronomy*, p. 120.

1354
Which planets have rings?

Jupiter, Saturn, Uranus, and Neptune all have rings. Jupiter's rings were discovered by *Voyager 1* in March, 1979. The rings extend 80,240 miles (129,130 kilometers) from the center of the planet. They are about 4,300 miles (7,000 kilometers) in width and less than 20 miles (30 kilometers) thick. A faint inner ring is believed to extend to the edge of Jupiter's atmosphere.

Saturn has the largest, most spectacular set of rings in the solar system. That the planet is surrounded by a ring system was first recognized by the Dutch astronomer Christiaan Huygens (1629–1695) in 1659. Saturn's rings are 169,800 miles (273,200 kilometers) in diameter, but less than 10 miles (16 kilometers) thick. There are six different rings, the largest of which appear to be divided into thousands of ringlets. The rings appear to be composed of pieces of water ice ranging in size from tiny grains to blocks several tens of yards in diameter.

In 1977 when Uranus occulted (passed in front of) a star, scientists observed that the light from the star flickered or winked several times before the planet itself covered the star. The same flickering occurred in reverse order after the occultation. The reason for this was determined to be a ring around Uranus. Nine rings were initially identified and *Voyager 2* observed two more in 1986. The rings are thin, narrow, and very dark.

Voyager 2 also discovered a series of at least four rings around Neptune in 1989. Some of the rings appear to have arcs, areas where there is a higher density of material than at other parts of the ring.

Sources: *The Guinness Book of Answers*, pp. 19-22; *The Planetary Report*, vol. 12, no. 2 (March/April 1992) pp. 4-9; *The World Almanac and Book of Facts 1992*, pp. 263-264.

1355

How long do the planets take to go around the sun?

Planet	Period of revolution	
	Earth days	Earth years
Mercury	88	0.24
Venus	224.7	0.62
Earth	365.26	1.00
Mars	687	1.88
Jupiter	4,332.6	11.86
Saturn	10,759.2	29.46
Uranus	30,685.4	84.01
Neptune	60,189	164.8
Pluto	90,777.6	248.53

Sources: *The Guinness Book of Answers*, p. 13; Pasachoff, Jay M. *Contemporary Astronomy*, p. A6.

1356

What is the gravitational force on each of the planets, the moon and the sun relative to the earth?

If the force on the earth is taken as 1, the comparative forces are:

Sun	27.9
Mercury	0.37
Venus	0.88
Earth	1.00
Moon	0.16
Mars	0.38
Jupiter	2.64
Saturn	1.15
Uranus	0.93
Neptune	1.22
Pluto	0.06

Weight comparisons can be made by using this table. If a person weighed 100 pounds (45.36 kilograms) on earth, then the weight of the person on the moon would be 16 pounds (7.26 kilograms) or 100 x 0.16.

Sources: Diagram Group. *Comparisions*, p. 112.

<h2>1357</h2>

Is a day the same on all the planets?

No. A day, the period of time it takes for a planet to make one complete turn on its axis, varies from planet to planet. Venus, Uranus, and Pluto display retrograde motion, that is to say, they rotate in the opposite direction from the other planets. The table below lists the length of the day for each planet.

Planet	Earth Days	Length of Day	
		Hours	Minutes
Mercury	58	15	30
Venus	243		32
Earth		23	56
Mars		24	37
Jupiter		9	50
Saturn		10	39
Uranus		17	14
Neptune		16	03
Pluto	6	09	18

Source: *The Guinness Book of Answers*, p. 15.

<h2>1358</h2>

Which planets are called "inferior" planets and which are "superior" planets?

An inferior planet is one whose orbit is nearer to the sun than Earth's orbit is. Mercury and Venus are the inferior planets. Superior planets are those whose orbits around the sun lie beyond that of Earth. Mars, Jupiter, Saturn, Uranus, Neptune, and Pluto are the superior planets. The terms have nothing to do with the quality of an individual planet.

Sources: Moore, Patrick. *The International Encyclopedia of Astronomy*, pp. 192, 423.

<h2>1359</h2>

What are the Jovian and terrestrial planets?

Jupiter, Saturn, Uranus, and Neptune are the Jovian (the adjectival form for the word "Jupiter"), or Jupiter-like, planets. They are giant planets, composed primarily of light elements such as hydrogen and helium.

Mercury, Venus, Earth, and Mars are the terrestrial (derived from *terra* the Latin word for "Earth"), or Earth-like, planets. They are small in size, have solid surfaces, and are composed of rocks and iron. Pluto appears to be a terrestrial type planet as well, but it may have a different origin from the other planets.

Sources: *McGraw-Hill Encyclopedia of Science and Technology*, vol. 13, p. 602; *Webster's Ninth New Collegiate Dictionary*, pp. 652, 1218.

1360
What is unique about the rotation of the planet Venus?

Unlike Earth and most of the other planets, Venus rotates in a retrograde, or opposite, direction with relation to its orbital motion about the sun. It rotates so slowly that there are only two sunrises and sunsets each Venusian year. Uranus and Pluto's rotation are also retrograde.

Sources: The Guinness Book of Answers, p. 15; McGraw-Hill Encyclopedia of Science and Technology, vol. 19, p. 199.

1361
Is it true that the earth is closer to the sun in winter than in summer in the Northern Hemisphere?

Yes. However, the earth's axis, the line around which the planet rotates, is tipped 23.5 degrees with respect to the plane of revolution around the sun. When the earth is closest to the sun (its perihelion, about January 3), the Northern Hemisphere is tilted away from the sun. This causes winter in the Northern Hemisphere while the Southern Hemisphere is having summer. When the earth is farthest from the sun (its aphelion, around July 4), the situation is reversed, with the Northern Hemisphere tilted towards the sun. At this time, it is summer in the Northern Hemisphere and winter in the Southern Hemisphere.

Sources: Magill, Frank N. *Magill's Survey of Science: Earth Science Series,* vol. 1, p. 401; *McGraw-Hill Encyclopedia of Science and Technology,* vol. 5, p. 524.

Tilted axis determines different positions of sun at sunrise, noon, and sunset at different seasons in middle north latitude.

Seasons in the Northern Hemisphere

1362
Is it true that the rotation speed of the earth varies?

The rotation speed is at its maximum in late July and early August and at its minimum in April; the difference in the length of the day is about 0.0012 seconds. Since about 1900 the earth's rotation has been slowing at a rate of approximately 1.7 seconds per year. In the geologic past the earth's rotational period was much faster; days were shorter and there were more days in the year. About 350 million years ago, the year had 400-410 days; 280 million years ago, a year was 390 days long.

Sources: *McGraw-Hill Encyclopedia of Science and Technology,* vol. 5, p. 483.

1363
What is the circumference of the earth?

The earth is an oblate ellipsoid—a sphere that is slightly flattened at the poles and which bulges at the equator. The distance around the earth at the equator is 24,902 miles (40,075 kilometers). The distance around the earth through the poles is 24,860 miles (40,008 kilometers).

Sources: *New Encyclopaedia Britannica,* vol. 4, p. 320; *Science and Technology Illustrated,* vol. 9, p. 1049.

1364
Is there life on Mars?

Experiments carried out by the Viking Lander in July 1976 indicated there is no evidence of life on Mars. There were three experiments conducted on the composition of the Martian soil and atmosphere which formed the basis for the decision.

Sources: Sherman, Irwin W. *Biology: A Human Approach,* p. 4.

1365
Is it true that Pluto is not always the outermost planet in the solar system?

Pluto's very eccentric orbit carried it inside Neptune's orbit on January 23, 1979. It will remain there until March 15, 1999. During this time, Neptune is the outermost planet in the solar system. However, because they are so far apart, the planets are in no danger of colliding with one another.

Pluto, discovered in 1930 by American astronomer Clyde Tombaugh (b. 1906), is the smallest planet in the solar system. It is composed of rock and ice, with methane ice on the surface and a thin methane atmosphere. Pluto's single moon, Charon, discovered by James Christy in 1978, has a diameter of 741 miles (1,192 kilometers). This makes Charon, at half the size of Pluto, a very large moon relative to the planet. Some astronomers consider Pluto and Charon to be a double planet system.

Sources: *The Guinness Book of Answers,* p. 22; Miller, Ron, and William K. Hartmann. *The Grand Tour: A Traveler's Guide to the Solar System,* pp. 126-127; Moore, Patrick, et al. *The Atlas of the Solar System,* pp. 396-397.

What is Planet X?

Astronomers have observed perturbations, or disturbances, in the orbits of Uranus and Neptune since the discoveries of both planets. They speculated that Uranus and Neptune were being influenced by the gravity of another celestial body. Pluto, discovered in 1930, does not appear to be large enough to cause these disturbances. The existence of another planet, known as Planet X, orbiting beyond Pluto, has been proposed. As yet there have been no sightings of this tenth planet but the search continues. There is a possibility that the unmanned space probes *Pioneer 10 & 11* and *Voyager 1 & 2*, now heading out of the solar system, will be able to locate this elusive object.

Sources: *The Almanac of Science and Technology*, p. 82; Moore, Patrick, et al. *The Atlas of the Solar System*, p. 397.

STARS

What does the color of a star indicate?

The color of a star gives an indication of its temperature and age. Stars are classified by their spectral type. From oldest to youngest and hottest to coolest, the types of stars are

Type	Color	Temperature	
		Farenheit	Celsius
O	Blue	45,000–75,000	25,000–40,000
B	Blue	20,800–45,000	11,000–25,000
A	Blue-White	13,500–20,000	7,500–11,000
F	White	10,800–13,500	6,000–7,500
G	Yellow	9,000–10,800	5,000–6,000
K	Orange	6,300–9,000	3,500–5,000
M	Red	5,400–6,300	3,000–3,500

Each type is further subdivided on a scale of 0-9. The sun is a type G2 star.

Source: *The Guinness Book of Answers*, p. 11; Rovin, Jeff. *Laws of Order*, p. 214.

What is a binary star?

A binary star is a pair of stars revolving around a common center of gravity. About 50% of all stars are members of either binary star systems or multiple star systems which contain more than two stars.

The bright star Sirius, about 8.6 light years away, is composed of two stars, one about 2.3 times the mass of the sun, the other a white dwarf star about 980 times the mass of Jupiter. Alpha Centauri, the nearest star to Earth after the sun, is actually three stars. Alpha Centauri A and Alpha Centauri B, two sunlike stars, orbit each other and Alpha Centauri C, a low mass red star, orbits around the others.

Sources: *Encyclopedic Dictionary of Science*, p. 79; Hartmann, William K., and Ron Miller. *Cycles of Fire*, pp. 100-101.

1369

Which star is the closest to earth?

The sun, at a distance of 93 million miles (150 million kilometers), is the closest star to the earth. After the sun, the closest stars are the members of the triple star system known as Alpha Centauri (Alpha Centauri A, Alpha Centauri B, and Alpha Centauri C, sometimes called Proxima Centauri). They are 4.3 light years away.

Sources: Flaste, Richard. *The New York Times Book of Science Literacy*, p. 19; Moore, Patrick. *International Encyclopedia of Astronomy*, p. 22.

1370

Which stars are the brightest?

The brightness of a star is called its magnitude. Apparent magnitude is how bright a star appears to the naked eye. The lower the magnitude, the brighter the star. On a clear night, stars of about magnitude +6 can be seen with the naked eye. Large telescopes can detect objects as faint as +27. Very bright objects have negative magnitudes; the sun is -26.8.

Star	Constellation	Apparent Magnitude
Sirius	Canis Major	-1.47
Canopus	Carina	-0.72
Arcturus	Boötes	-0.06
Rigil Kentaurus	Centaurus	+0.01
Vega	Lyra	+0.04
Capella	Auriga	+0.05
Rigel	Orion	+0.14
Procyon	Canis Minor	+0.37
Betelgeuse	Orion	+0.41
Achernar	Eridanus	+0.51

Sources: *The Guinness Book of Answers*, pp. 10-11; Pasachoff, Jay M. *Contemporary Astronomy*, pp. 14-17, A.8; *The Universal Almanac 1992*, pp. 484-485.

1371

What is the Big Dipper?

The Big Dipper is a group of seven stars which are part of the constellation Ursa Major. They appear to form a sort of spoon with a long handle. The group is known as The Plough in Great Britain. The Big Dipper is almost always visible in the northern hemisphere. It serves as a convenient reference point when locating other stars; for example, an imaginary line drawn from the two end stars of the dipper leads to Polaris, the North Star.

Source: Booth, Nicholas. *The Concise Illustrated Book of Planets and Stars*, p. 30.

1372 / *Where is the North Star?*

If an imaginary line is drawn from the North Pole into space, there is a star called Polaris, or the North Star, less than one degree away from the line. As the earth rotates on its axis, Polaris acts as a pivot-point around which all the stars visible in the northern hemisphere appear to move, while Polaris itself remains motionless.

Source: Abell, George. *Realm of the Universe*, p. 11; Booth, Nicholas. *The Concise Illustrated Book of Planets and Stars*, p. 30.

1373 / *What is the "summer triangle"?*

The summer triangle is the triangle formed by the stars Deneb, Vega, and Altair as seen in the summer Milky Way.

Sources: Dickinson, Terence, and Alan Dyer. *The Backyard Astronomer's Guide*, pp. 282-283; Menzel, Donald H., and Jay M. Pasachoff. *A Field Guide to Stars and Planets*, p. 18.

Summer triangle

1374 / *How many constellations are there and how were they named?*

Constellations are groups of stars that seem to form some particular shape, that of a person, animal or object. They only appear to form this shape and be close to each other from earth; in actuality the stars in a constellation are often very distant from each other. There are 88 recognized constellations in the heavens whose boundaries were defined in the 1920s by the International Astronomical Union.

Different cultures in all parts of the world had their own constellations. However, because modern science is predominantly a product of western culture, many of the constellations represent characters from Greek and Roman mythology. When Europeans began to explore the Southern Hemisphere in the 16th and 17th centuries, they derived some of the new star patterns from the technological wonders of their time, such as the microscope.

Names of constellations are usually given in Latin. Individual stars in a constellation are usually designated with Greek letters in the order of brightness; the brightest star is alpha, the second brightest is beta, and so on. The genitive, or possessive, form of the constellation name is used, thus Alpha Orionis is the brightest star of the constellation Orion. The table below lists the constellations and what they represent.

Constellation	Genitive	Abbreviation	Meaning
Andromeda	Andromedae	And	Chained Maiden
Antlia	Antliae	Ant	Air Pump
Apus	Apodis	Aps	Bird of Paradise
Aquarius	Aquarii	Aqr	Water Bearer
Aquila	Aquilae	Aql	Eagle
Ara	Arae	Ara	Altar
Aries	Arietis	Ari	Ram
Auriga	Aurigae	Aur	Charioteer
Boötes	Boötis	Boo	Herdsman
Caelum	Caeli	Cae	Chisel
Camelopardalis	Camelopardalis	Cam	Giraffe
Cancer	Cancri	Cnc	Crab
Canes Venatici	Canum Venaticorum	CVn	Hunting Dogs
Canis Major	Canis Majoris	CMa	Big Dog
Canis Minor	Canis Minoris	CMi	Little Dog
Capricornus	Capricorni	Cap	Goat
Carina	Carinae	Car	Ship's Keel
Cassiopeia	Cassiopeiae	Cas	Queen of Ethiopia
Centaurus	Centauri	Cen	Centaur
Cepheus	Cephei	Cep	King of Ethiopia
Cetus	Ceti	Cet	Whale
Chamaeleon	Chamaeleonis	Cha	Chameleon
Circinus	Circini	Cir	Compass
Columba	Columbae	Col	Dove
Coma Berenices	Comae Berenices	Com	Berenice's Hair
Corona Australis	Coronae Australis	CrA	Southern Crown
Corona Borealis	Coronae Borealis	CrB	Northern Crown
Corvus	Corvi	Crv	Crow
Crater	Crateris	Crt	Cup
Crux	Crucis	Cru	Southern Cross
Cygnus	Cygni	Cyg	Swan
Delphinus	Delphini	Del	Dolphin

Constellation	Genitive	Abbreviation	Meaning
Dorado	Doradus	Dor	Goldfish
Draco	Draconis	Dra	Dragon
Equuleus	Equulei	Equ	Little Horse
Eridanus	Eridani	Eri	River Eridanus
Fornax	Fornacis	For	Furnace
Gemini	Geminorum	Gem	Twins
Grus	Gruis	Gru	Crane
Hercules	Herculis	Her	Hercules
Horologium	Horologii	Hor	Clock
Hydra	Hydrae	Hya	Hydra, Greek monster
Hydrus	Hydri	Hyi	Sea Serpent
Indus	Indi	Ind	Indian
Lacerta	Lacertae	Lac	Lizard
Leo	Leonis	Leo	Lion
Leo Minor	Leonis Minoris	LMi	Little Lion
Lepus	Leporis	Lep	Hare
Libra	Librae	Lib	Scales
Lupus	Lupi	Lup	Wolf
Lynx	Lyncis	Lyn	Lynx
Lyra	Lyrae	Lyr	Lyre or Harp
Mensa	Mensae	Men	Table Mountain
Microscopium	Microscopii	Mic	Microscope
Monoceros	Monocerotis	Mon	Unicorn
Musca	Muscae	Mus	Fly
Norma	Normae	Nor	Carpenter's Square
Octans	Octanis	Oct	Octant
Ophiuchus	Ophiuchi	Oph	Serpent Bearer
Orion	Orionis	Ori	Orion, the Hunter
Pavo	Pavonis	Pav	Peacock
Pegasus	Pegasi	Peg	Winged Horse
Perseus	Persei	Per	Perseus, a Greek hero
Phoenix	Phoenicis	Phe	Phoenix
Pictor	Pictoris	Pic	Painter
Pisces	Piscium	Psc	Fish
Piscis Austrinus	Piscis Austrini	PsA	Southern Fish
Puppis	Puppis	Pup	Ship's Stern
Pyxis	Pyxidis	Pyx	Ship's Compass
Reticulum	Reticuli	Ret	Net
Sagitta	Sagittae	Sge	Arrow
Sagittarius	Sagittarii	Sgr	Archer
Scorpius	Scorpii	Sco	Scorpion
Sculptor	Sculptoris	Scl	Sculptor
Scutum	Scuti	Sct	Shield
Serpens	Serpentis	Ser	Serpent
Sextans	Sextantis	Sex	Sextant
Taurus	Tauri	Tau	Bull
Telescopium	Telescopii	Tel	Telescope
Triangulum	Trianguli	Tri	Triangle
Triangulum	Triangli Australis	TrA	Southern Australe Triangle
Tucana	Tucanae	Tuc	Toucan
Ursa Major	Ursae Majoris	UMa	Big Bear
Ursa Minor	Ursae Minoris	UMi	Little Bear

Constellation	Genitive	Abbreviation	Meaning
Vela	Velorum	Vel	Ship's Sail
Virgo	Virginis	Vir	Virgin
Volans	Volantis	Vol	Flying Fish
Vulpecula	Vulpeculae	Vul	Little Fox

Sources: *The Universal Almanac 1992*, pp. 484-485; *The World Almanac and Book of Facts 1992*, pp. 270-271.

1375
What is the Milky Way?

The Milky Way is a hazy band of light that can be seen encircling the night sky. This light comes from the stars that make up the Milky Way galaxy, the galaxy to which the sun and the earth belong. Galaxies are huge systems of stars separated from one another by largely empty space. Astronomers estimate that the Milky Way galaxy contains at least 100 billion stars and is about 100 thousand light years in diameter. The galaxy is shaped like a phonograph record with a central bulge, or nucleus, and spiral arms curving out from the center.

Sources: Illingworth, Valerie L. *The Facts On File Dictionary of Astronomy*, p. 130.

1376
How hot is the sun?

The center of the sun is about 27,000,000°F (15,000,000°C). The surface, or photosphere, of the sun is about 10,000°F (5,500°C). Magnetic anomolies in the photosphere cause cooler regions which appear to be darker than the surrounding surface. These sunspots are about 6,700°F (4,000°C). The sun's layer of lower atmosphere, the chromosphere, is only a few thousand miles thick. At the base, the chromosphere is about 7,800°F (4,300°C), but its temperature rises with altitude to the corona, the sun's outer layer of atmosphere, which has a temperature of about 1,800,000°F (1,000,000°C).

Sources: Asimov, Isaac. *Isaac Asimov's Guide to Earth and Space*, pp. 159-161; Moore, Patrick, et al. *The Atlas of the Solar System*, p. 19.

1377
When will the sun die?

The sun is approximately 4.5 billion years old. About 5 billion years from now, the sun will burn all of its hydrogen fuel into helium. When this process occurs, the sun will change from the yellow dwarf we know it as to a red giant. Its diameter will extend well beyond the orbit of Venus, and even possibly beyond the orbit of Earth. In either case, the earth will be burned to a cinder and will be incapable of supporting life.

Sources: Pasachoff, Jay M. *Contemporary Astronomy*, pp. 205-206; *The Universal Almanac 1992*, pp. 482-483.

1378 / What is the sun made of?

The sun is an incandescent ball of gases. Its mass is 1.8×10^{27} tons or 1.8 octillion tons (a mass 330,000 times as great as the earth).

Element	% of mass
Hydrogen	73.46
Helium	24.85
Oxygen	0.77
Carbon	0.29
Iron	0.16
Neon	0.12
Nitrogen	0.09
Silicon	0.07
Magnesium	0.05
Sulfur	0.04
Other	0.10

Sources: Asimov, Isaac. *Isaac Asimov's Guide to the Earth and Sun*, p. 154; *The Encyclopedic Dictionary of Science*, p. 231; Moore, Patrick, et al. *The Atlas of the Solar System*, p. 19.

1379 / What is the ecliptic?

Ecliptic refers to the apparent yearly path of the sun through the sky with respect to the stars. In the spring, the ecliptic in the northern hemisphere is angled high in the evening sky. In fall, the ecliptic lies much closer to the horizon.

Sources: Dickinson, Terence, and Alan Dyer. *The Backyard Astronomer's Guide*, p. 141; *World Book Encyclopedia*, vol. 1, p. 835.

1380 / Why does the color of the sun vary?

Sunlight contains all the colors of the rainbow which blend to form white light, making sunlight appear white. At times, some of the color wavelengths become scattered, especially blue, and the sunlight appears colored. When the sun is high in the sky, some of the blue rays are scattered in the earth's atmosphere. At such times, the sky looks blue and the sun appears to be yellow. At sunrise or sunset when the light must follow a longer path through the earth's atmosphere, the sun looks red (red having the longest wavelengths).

Sources: Forrester, Frank H. *1001 Questions Answered About the Weather*, p. 168; *World Book Encyclopedia*, vol. 18, pp. 975-976.

1381
How long does it take light from the sun to reach the earth?

Sunlight takes about 8 minutes and 20 seconds to reach the earth, travelling at 186,282 miles (299,792 kilometers) per second.

Sources: *World Book Encyclopedia*, vol. 18, p. 972.

1382
How long is a solar cycle?

The solar cycle is the periodic change in the number of sun spots. The cycle is taken as the interval between successive minima and is about 11.1 years. During an entire cycle, solar flares, sunspots, and other magnetic phenomena move from intense activity to relative calm and back again. The solar cycle is one area of study to be carried out by up to 10 ATLAS space missions beginning in March 1992 to probe the chemistry and physics of the atmosphere. These studies of the solar cycle will yield a more detailed picture of the Earth's atmosphere and its response to changes in the sun.

Sources: *Chemical & Engineering News*, vol. 70, (March 16, 1992), p. 5; *The McGraw-Hill Dictionary of Scientific and Technical Terms*, p. 1766.

1383
What is the Maunder minimum?

It was the time between 1645 and 1715 when sunspots were very scarce or missing entirely. From studying old solar records in 1890, E. Walter Maunder observed that sunspot activity during these years was at a minimum. This observation was confirmed in 1976 by John Eddy.

Sources: Moore, Patrick. *International Encyclopedia of Astronomy*, p. 247.

1384
When do solar eclipses happen?

A solar eclipse occurs when the moon passes between the earth and the sun and all three bodies are aligned in the same plane. When the moon completely covers the sun and the umbra, or dark part of the moon's shadow, reaches the earth, a total eclipse occurs. A total eclipse happens only along a narrow path 100 to 200 miles (160 to 320 kilometers) wide called the track of totality. Just before totality, the only parts of the sun that are visible are a few points of light called Baily's beads shining through valleys on the moon's surface. Sometimes, a last bright flash of sunlight is seen—the diamond ring effect. During totality, which averages 2.5 minutes, but may last up to 7.5 minutes, the sky is dark and stars and other planets are easily seen. The corona, the sun's outer atmosphere, is also visible.

If the moon does not appear large enough in the sky to completely cover the sun, it appears silhouetted against the sun with a ring of sunlight showing around it. This is an annular eclipse. Because the sun is not completely covered, its corona cannot be seen and although the sky may darken, it will not be dark enough to see the stars.

During a partial eclipse of the sun, the penumbra of the moon's shadow strikes the earth. A partial eclipse can also be seen on either side of the track of totality of an annular or total eclipse. The moon will cover part of the sun and the sky will not darken noticeably during a partial eclipse.

Sources: Abell, George O. *Realm of the Universe*, pp. 75-84; Menzel, Donald H., and Jay M. Pasachoff. *A Field Guide to Stars and Planets*, pp. 404-410; Moore, Patrick. *International Encyclopedia of Astronomy*, p. 382.

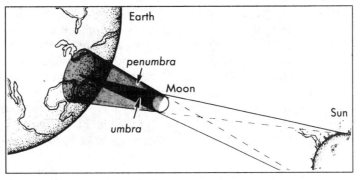

Solar eclipse

1385
What is a pulsar?

A pulsar is a rotating neutron star that gives off sharp regular pulses of radio waves at rates ranging from 0.001 to 4 seconds. Stars burn by fusing hydrogen into helium. When they use up their hydrogen, their interiors begin to contract. During this contraction energy is released and the outer layers of the star are pushed out. These layers are large and cool; the star is now a red giant. A star with more than twice the mass of the sun will continue to expand, becoming a supergiant. At that point, it may blow up, in an explosion called a supernova. After a supernova, the remaining material of the star's core may be so compressed that the electrons and protons become neutrons. A star 1.4 to 4 times the mass of the sun can be compressed into a neutron star only about 12 miles (20 kilometers) across. Neutron stars rotate very fast. The neutron star at the center of the Crab Nebula spins 30 times per second.

A pulsar is formed by the collapse of a star with 1.4 to 4 times the mass of the sun. Some of these neutron stars emit radio signals from their magnetic poles in a direction that reaches earth. These signals were first detected by Jocelyn Bell (b. 1943) of Cambridge University in 1967. Because of their regularity some people speculated that they were extraterrestrial beacons constructed by alien civilizations. This theory was eventually ruled out and the rotating neutron star came to be accepted as the explanation for these pulsating radio sources, or pulsars.

Sources: *Astronomy*, (November 1991), pp. 51-55; Menzel, Donald H., and Jay M. Pasachoff. *A Field Guide to Stars and Planets*, pp. 112-114.

1386 / What is a black hole?

When a star with a mass greater than about four times that of the sun collapses even the neutrons cannot stop the force of gravity. There is nothing to stop the contraction and the star collapses forever. The material is so dense that nothing—not even light—can escape. The American physicist John Wheeler, in 1967, gave this phenomenon the name "black hole." Since no light escapes from a black hole, it cannot be observed directly. However, if a black hole existed near another star, it would draw matter from the other star into itself, producing x-rays as a result. In the constellation of Cygnus, there is a strong x-ray source, named Cygnus X-1. It is near a star and the two revolve around each other. The unseen x-ray source has the gravitational pull of at least 10 suns and is believed to be a black hole. Another type of black hole, a primordial black hole, may also exist dating from the time of the "Big Bang" when regions of gas and dust were highly compressed.

Sources: *Astronomy*, (November 1991), pp. 51-55; *Encyclopedic Dictionary of Science*, p. 46; Moore, Patrick. *The International Encyclopedia of Astronomy*, pp. 64, 81.

SUN *See: SPACE—STARS*

THE UNIVERSE

1387 / Who is Stephen Hawking?

Hawking (b. 1943), a British physicist and mathematician, is considered to be the greatest theoretical physicist of the late twentieth century. In spite of being severely handicapped by *amyotrophic lateral sclerosis* (ALS), he has made major contributions to scientific knowledge about black holes and the origin and evolution of the universe though his research into the nature of space-time and its anomalies. For instance, Hawking proposed that a black hole could emit thermal radiation and predicted that a black hole would disappear after all its mass has been converted into radiation (called *Hawking's radiation*). A current objective of Hawking is to synthesize quantum mechanics and relativity theory into a theory of quantum gravity. He is also the author of several books including the popular best-selling work *A Brief History of Time.*

Sources: *The Biographical Dictionary of Scientists: Astronomers,* pp. 67-68; Iver, David F. *Dictionary of Astronomy, Space, and Atmospheric Phenomena,* p. 32; Simon, Sheridan. *Stephen Hawking,* pp. 9-10, 93–94.

1388
How old is the Universe?

The universe is believed to be somewhere between 15 billion and 20 billion years old. The figure is derived from the concept that the universe has been expanding at the same rate since its birth at the "Big Bang." The rate of expansion is a ratio known as Hubble's constant. It is calculated by dividing the speed at which the galaxy is moving away from the earth by its distance from the earth. By inverting Hubble's Constant, that is dividing the distance of a galaxy by its recessional speed, the age of the universe can be calculated. The estimates of both the velocity and distance of galaxies from the earth are subject to uncertainties and not all scientists accept that the universe has always expanded at the same rate. Therefore, many still hold the age of the universe open to question.

Sources: Brennan, Richard P. *Dictionary of Scientific Literacy,* pp. 310-311; Van Andel, Tjeerd H. *New Views on an Old Planet,* p. 28.

1389
What was the "Big Bang"?

The Big Bang theory is the explanation most commonly accepted by astronomers for the origin of the universe. It proposes that the universe began as the result of an explosion—the Big Bang—15 billion to 20 billion years ago. Two observations form the basis of this cosmology. First, as Edwin Hubble (1889-1953) demonstrated, the universe is expanding uniformly, with objects at greater distances receding at greater velocities. Secondly, the earth is bathed in a glow of radiation that has the same characteristics as expected from a remnant of a hot primeval fireball. This radiation was discovered by Arno A. Penzias (b. 1933) and Robert W. Wilson (b. 1936) of Bell Telephone Laboratories. In time, the matter created by the Big Bang came together in huge clumps to form the galaxies. Smaller clumps within the galaxies formed stars. Parts of at least one clump became a group of planets—our solar system.

Sources: *McGraw-Hill Encyclopedia of Science and Technology*, vol. 2, pp. 550-551; *World Book Encyclopedia*, vol. 4, p. 1078.

1390 / What are quasars?

The name quasar is short for *quas*i-stellar *r*adio source. Quasars appear to be stars, but they have large red shifts in their spectra indicating that they are receding from the earth at great speeds, some up to 90% of the speed of light. Their exact nature is still unknown, but many believe quasars to be the cores of distant galaxies, the most distant objects yet seen. Quasars were first identified in 1963 by astronomers at the Palomar Observatory.

Sources: Hartmann, William, and Ron Miller. *Cycles of Fire*, p. 154-157; Trefil, James. *1001 Things Everyone Should Know About Science*, p. 259.

1391 / What is a syzygy?

A syzygy is a configuration that occurs when three celestial bodies lie in a straight line, such as the sun, earth, and moon during a solar or lunar eclipse. The particular case when a planet is on the opposite side of the earth from the sun is called an opposition.

Sources: Illingworth, Valerie. *The Facts On File Dictionary of Astronomy*, p. 335.

WEATHER AND CLIMATE

AIR PHENOMENA

1392 / What is the origin of the Brown Mountain Lights of North Carolina?

For a period of some thirty years the "lights" seen at Brown Mountain could not be explained. In 1922, the United States Geological Survey studied the mystery. The area has extraordinary atmospheric conditions and automobile, locomotive, and fixed lights from miles away are reflected in the atmosphere.

Sources: Mansfield, George Rogers. *Origin of the Brown Mountain Light in North Carolina*, pp. 15-18.

1393 / When does the green flash phenomenon occur?

On rare occasions, the sun may look bright green for a moment, as the last tip of the sun is setting. This green flash occurs because the red rays of light are hidden below the horizon and the blue are scattered in the atmosphere. The green rays are seldom seen because of dust and pollution in the lower atmosphere. It may best be seen when the air is cloudless and when a distant, well–defined horizon exists, as on an ocean.

Sources: Forrester, Frank H. *1001 Questions Answered About the Weather*, p. 169.

1394 / How often does an aurora appear?

Because it depends on solar winds (electrical particles generated by the sun) and sunspot activity, the frequency of an aurora cannot be determined. Auroras usually appear two days after a solar flare (a violent eruption of particles on the sun's surface) and reach their peak two years into the 11–year sunspot cycle. The auroras, occurring in the polar regions, are broad displays of usually colored light at night. The Northern polar aurora is called Aurora Borealis or Northern Lights and the Southern polar aurora is called the Aurora Australis.

Sources: Hopkins, Jeanne. *Glossary of Astronomy and Astrophysics*, p. 10; *World Book Encyclopedia*, vol. 1, pp. 895–896.

1395 / What is a bishop's ring?

It is a ring around the sun, usually with a reddish outer edge. It is probably due to dust particles in the air as it is seen after all great volcanic eruptions.

Sources: Bradford, Gershom. *A Glossary of Sea Terms*, p. 43.

1396 / What is Saint Elmo's fire?

Saint Elmo's fire has been described as a corona from electric discharge produced on high grounded metal objects, chimney tops, and ship masts. Since it often occurs during thunderstorms, the electrical source may be lightning. Another description refers to this phenomenon as weak static electricity formed when an electrified cloud touches a high exposed point. Molecules of gas in the air around this point become ionized and glow. The name originated with sailors who were among the first to witness the display of spearlike or tufted flames on the tops of their ships' masts. Saint Elmo (which is a corruption of Saint Ermo) is the patron saint of sailors, so they named the fire after him.

Sources: Forrester, Frank. *1001 Questions Answered About the Weather*, pp. 182–183; Schaefer, Vincent J., and John A. Day. *A Field Guide to the Atmosphere*, p. 196.

1397 / When and by whom were clouds first classified?

The French naturalist Jean Lamarck (1744–1829) proposed the first system for classifying clouds in 1802. However his work did not receive wide acclaim. A year later the Englishman Luke Howard (1772–1864) developed a cloud classification system that has been generally accepted and is still used today. Clouds are distinguished by their general appearance ("heap clouds" and "layer clouds") and by their height above the ground. Latin names and prefixes are used to describe these characteristics. The shape names are *cirrus* (curly or fibrous), *stratus* (layered), *cumulus* (lumpy or piled). The prefixes denoting height are *cirro* (high clouds with bases above 20,000 feet or 6.1 kilometers) and *alto* (mid-level clouds from 6,000 to 20,000 feet or 1830 to 6,100 kilometers). There is no prefix for low clouds. *Nimbo* or *nimbus* is also added as a name or prefix to indicate that the cloud produces precipitation.

Sources: Ahrens, C. Donald. *Meteorology Today*, p. 163; Williams, Jack. *The Weather Book*, pp. 160-161.

1398
What are fulgurites?

Fulgurites (from the Latin word *fulgur*, meaning lightning) are petrified lightning, created when lightning strikes an area of dry sand. The intense heat of the lightning melts the sand surrounding the stroke into a rough glassy tube forming a fused record of its path. These tubes may be ½ to 2 inches (1.5 to 5 centimeters) in diameter, and up to 10 feet (3 meters) in length. They are extremely brittle and break easily. The inside walls of the tube are glassy and lustrous while the outside is rough, with sand particles adhering to it. Fulgurites are usually tan or black in color, but translucent white ones have been found.

Sources: *New Encyclopedia Britannica*, vol. 5, p. 43; Tufty, Barbara. *1001 Questions Answered About Hurricanes, Tornadoes and Other Natural Air Disasters*, pp. 128–129.

1399
What are the four major cloud groups and their types?

1. **High Clouds**—composed almost entirely of ice crystals, the bases of these clouds start at 16,500 feet (5,032 meters) and reach 45,000 feet (13,725 meters).

Cirrus (from Latin, "lock of hair")—are thin feather-like crystal clouds in patches or narrow bands. The large ice crystals that often trail downward in well-defined wisps are called "mares tails."

Cirrostratus—is a thin, white cloud layer that resembles a veil or sheet. This layer can be striated or fibrous. Because of the ice content, these clouds are associated with the halos that surround the sun or moon.

Cirrocumulus—are thin clouds that appear as small white flakes or cotton patches and may contain super-cooled water.

2. **Middle Clouds**—composed primarily of water. The height of the cloud bases range from 6,500 to 23,000 feet (1,983 to 7,015 meters).

Altostratus—appears as a bluish or grayish veil or layer of clouds that can gradually merge into altocumulus clouds. The sun may be dimly visible through it, but flat, thick sheets of this cloud type can obscure the sun.

Altocumulus—is a white or gray layer or patches of solid clouds with rounded shapes.

3. **Low Clouds**—composed almost entirely of water that may at times be super-cooled; at subfreezing temperatures, snow and ice crystals may be present as well. The bases of these clouds start near the earth's surface and climb to 6,500 feet (1,982 meters) in the middle latitudes.

Stratus—are gray uniform sheet-like clouds with a relatively low base or they can be patchy, shapeless, low gray clouds. Thin enough for the sun to shine through, these clouds bring   drizzle and snow.

Stratocumulus—are globular rounded masses that form at the top of the layer.

Nimbostratus—are seen as a gray or dark relatively shapeless massive cloud layer containing rain, snow, and ice pellets.

4. **Clouds with Vertical Development**—contain super-cooled water above the freezing level and grow to great heights. The cloud bases range from 1,000 feet (305 meters) to 10,000 feet (3,020 meters).

Cumulus—are detached, fair weather clouds with relatively flat bases and dome-shaped tops. These usually do not have extensive vertical development and do not produce precipitation.

Cumulonimbus—are unstable large vertical clouds with dense boiling tops that bring showers, hail, thunder, and lightning.

Sources: Ahrens, C. Donald. *Meteorology Today*, p. 164; Bair, Frank E. *The Weather Almanac*, pp. 289-294.

Types of clouds

1400 / What color is lightning?

Lightning is often seen as white or white-yellow, although it may appear to have other colors depending on the background.

Sources: Field, Frank. *Doctor Frank Field's Weather Book*, p. 201.

1401 / How hot is lightning?

The temperature of the air around a bolt of lightning is about 54,000°F (30,000°C), which is six times hotter than the surface of the sun, yet many times people survive a bolt of lightning. American park ranger Roy Sullivan was hit by lightning seven times between 1942 and 1977. In cloud–to–ground lightning, its energy seeks the shortest route to earth, which could be through a person's shoulder, down the side of the body through the leg to the ground. As long as the lightning does not pass across the heart or spinal column, the victim usually does not die.

Sources: *How in the World*, p. 298; Magill, Frank N. *Magill's Survey of Science: Earth Science Series*, vol. 3, p. 1374.

1402 / *What is ball lightning?*

Ball lightning is a rare form of lightning in which a persistent and moving luminous white or colored sphere is seen. It can last from a few seconds to several minutes, and travels at about a walking pace. Spheres have been reported to vanish harmlessly, or to pass into or out of rooms leaving, in some cases, sign of their passage such as a hole in a window pane. Sphere dimensions vary but are most commonly from 4 to 8 inches (10 to 20 centimeters).

Other types of lightning include the common streak lightning (a single or multiple zigzagging line from cloud to ground); forked lightning (lightning formed two branches simultaneously); sheet lightning (a shapeless flash covering a broad area); ribbon lightning (streak lightning blown sideways by the wind to make it appear like parallel successive strokes); bead or chain lightning (a stroke interrupted or broken into evenly-spaced segments or beads); and heat lightning (lightning seen along the horizon during hot weather and believed to be a reflection of lightning occurring beyond the horizon).

Sources: Bair, Frank E. *The Weather Almanac*, p. 121; Great Britain Meteorological Office. *Meteorological Glossary*, p. 37.

1403 / *Does lightning ever strike in the same place?*

It is not true that lightning does not strike twice in the same place. In fact, tall buildings, such as the Empire State Building, in New York, can be struck several times during the same storm. During one storm, lightning struck the Empire State Building 12 times.

Sources: Tufty, Barbara. *1001 Questions Answered About Hurricanes, Tornadoes, and Other Natural Air Disasters*, p. 126.

1404 / *How is the distance of a lightning flash calculated?*

Count the number of seconds between seeing a flash of lightning and hearing the sound of the thunder. Divide the number by 5 to determine the number of miles away that the lightning flashed.

Sources: *World Book Encyclopedia*, vol. 19, p. 272.

1405
How long is a lightning stroke?

The visible length of the streak of lightning depends on the terrain and can vary greatly. In mountainous areas where clouds are low, the flash can be as short as 300 yards (174 meters); whereas in flat terrain, where clouds are high, the bolt can measure as long as 4 miles (6.5 kilometers). The usual length is about 1 mile (1.6 kilometers), but streaks of lightning up to 20 miles (32 kilometers) have been recorded. The stroke channel is very narrow—perhaps as little as ½ inch (1.3 centimeters). It is surrounded by a "corona envelope" or a glowing discharge that can be as wide as 10 to 20 feet (3 to 6 meters) in diameter. The speed of lightning can vary from 100 to 1,000 miles (161 to 1,609 kilometers) per second for the downward leader track; the return stroke is 87,000 miles (139,983 kilometers) per second (almost half the speed of light).

Sources: *Guinness Book of Records 1992*, p. 43.

1406
How many volts are in lightning?

A stroke of lightning discharges from 10 to 100 million volts of electricity. In an average lightning stroke there are 30 thousand amperes.

Sources: Tufty, Barbara. *1001 Questions Answered About Hurricanes, Tornadoes, and Other Natural Air Disasters*, p. 126.

1407
What is the order of colors in a rainbow?

Red, orange, yellow, blue, indigo, and violet are the colors of the rainbow, but these are not necessarily the sequence of colors that an observer might see. Rainbows are formed when raindrops reflect sunlight. As sunlight enters the drops, the different wavelengths of the colors that compose sunlight are refracted at different lengths to produce a spectrum of color. Each observer sees a different set of raindrops at a slightly different angle. Drops at different angles from the observer send different wave lengths (i.e., different color) to the observer's eyes. Since the color sequence of the rainbow is the result of refraction, the color order depends on how the viewer sees this refraction from the viewer's angle of perception.

Sources: Schaefer, Vincent J., and John A Day. *A Field Guide to the Atmosphere*, p. 163; *World Book Encyclopedia*, vol. 16, p. 130.

DEW *See:* WEATHER AND CLIMATE—SNOW, RAIN, HAIL ,ETC.

FORECASTING *See: WEATHER AND CLIMATE—WEATHER PREDICTION*

FROST *See: WEATHER AND CLIMATE—SNOW, RAIN, HAIL, ETC.*

HAIL *See: WEATHER AND CLIMATE—SNOW, RAIN, HAIL, ETC.*

HURRICANES *See: WEATHER AND CLIMATE—WIND*

LIGHTNING *See: WEATHER AND CLIMATE—AIR PHENOMENA*

RAIN *See: WEATHER AND CLIMATE—SNOW, RAIN, HAIL, ETC.*

RAINBOWS *See: WEATHER AND CLIMATE—AIR PHENOMENA*

SNOW, RAIN, HAIL, ETC.

▽ 1408 ▽
What is the difference between freezing rain and sleet?

Freezing rain is rain that falls as a liquid but turns to ice on contact with a freezing object to form a smooth ice coating called *glaze*. Usually freezing rain only lasts a short time, because it either turns to rain or to snow. Sleet is frozen or partially frozen rain in the form of ice pellets. Sleet forms as falling rain from a warm layer of air passes through a freezing air layer near the earth's surface to form hard, clear, tiny ice pellets that can hit the ground so fast that they bounce off with a sharp click.

Sources: Tufty, Barbara. *1001 Questions Answered About Hurricanes, Tornadoes, and Other Natural Air Disasters*, p. 275.

▽ 1409 ▽
When does frost form?

A frost is a crystalline deposit of small thin ice crystals formed on objects which are at freezing or below freezing temperatures. This phenomenon occurs when atmospheric water vapor condenses directly into ice without first becoming a liquid; this process is called sublimation. Usually frost appears on clear, calm nights especially during early autumn when the air above the earth is quite moist. A light frost damages only the most tender plants and vines; whereas a heavy frost (a heavy deposit of crystallized water) might kill plants that were not sturdy. A killing frost may do that to sturdy plants. Black frost (or hard frost) occurs in late autumn when both objects and air are at temperatures below freezing. Leaf edges and leaf tips turn black from this frost. Hoar or hoarfrost is the term for frost in Europe. Finally, permafrost is ground permanently frozen which never thaws out completely.

Sources: Tufty, Barbara. *1001 Questions Answered About Hurricanes, Tornadoes, and Other Natural Air Disasters*, pp. 276-278.

1410

What is the greatest amount of rainfall ever measured?

Time duration	Amount		Place
	Inches	**Centimeters**	
1 minute	1.5	3.8	Barst, Guadeloupe, West Indies, November 26, 1970
24 hours	73.62	187	Cilaos, La Réunion, Indian Ocean, March 15–16, 1952
24 hours in the United States	19.0	48.3	Alvin, Texas, July 25–26, 1979
calendar month	366.14	930	Cherrapunji, Meghalaya, India, July 1861
12 months	1,041.78	2646	Cherrapunji, Meghalaya, India, between August 1, 1860 and July 31, 1861
12 months in the United States	739	1877	Kukui, Maui, Hawaii, December 1981–December 1982

Sources: *Guinness Book of Records 1992*, pp. 41,44; Williams, Jack. *The Weather Book*, p. 83.

1411

Where is the rainiest place on earth?

The wettest place in the world is Tutunendo, Colombia, with an average annual rainfall of 463.4 inches (1177 centimeters) per year. The place that has the most rainy days per year is Mount Wai-'ale'ale on Kauai, Hawaii. It has up to 350 rainy days annually.

In contrast the longest rainless period in the world was from October 1903 to January 1918 at Arica, Chile—a period of 14 years. In the United States the longest dry spell was 767 days at Bagdad, California, from October 3, 1912, to November 8, 1914.

Sources: *Guinness Book of Records 1992*, p. 44; Williams, Jack, *The Weather Book*, p. 89.

1412 ⟋
How fast can rain fall?

The speed of rainfall varies with drop size and wind speed. A typical raindrop in still air falls about 600 feet (183 meters) per minute or about 7 miles (11 kilometers) per hour.

Sources: Field, Frank. *Dr. Frank Field's Weather Book*, p. 196; Forrester, Frank H. *1001 Questions Answered About the Weather*, p. 53.

1413 ⟋
How does dew form?

At night, as the temperature of the air falls, the amount of water vapor the air can hold also decreases. Excess water vapor then condenses as very small drops on whatever it touches.

Sources: Bamberger, Richard. *Physics Through Experiment*, p. 104.

1414 ⟋
What is the shape of a raindrop?

Although a raindrop has been illustrated as being pear-shaped or tear-shaped, high-speed photographs reveal that a large raindrop has a spherical shape with a hole not quite through it (giving it a doughnut-like shape). Water surface tension pulls the drop into this shape. As a drop larger than 0.08 inches (2 millimeters) in diameter falls, it will become distorted. Air pressure flattens its bottom and its sides bulge. If it becomes larger than 0.25 inches (6.4 millimeters) across, it will keep spreading crosswise as it falls and will bulge more at its sides, while at the same time, its middle will thin into a bow-tie shape. Eventually in its path downward, it will divide into two smaller spherical drops.

Sources: Tufty, Barbara. *1001 Questions Answered About Hurricanes, Tornadoes, and Other Natural Disasters*, p. 247; Williams, Jack. *The Weather Book*, p. 63.

1415 ⟋
How large can hailstones become?

The average hailstone is about ¼ inch (0.63 centimeters) in diameter. However, hailstones weighing up to 7.5 pounds (3.4 kilograms) are reported to have fallen in Hyderabad state in India in 1939, although scientists think these huge hailstones may be several stones that partly melted and stuck together. On April 14, 1986, hailstones weighing 2.5 pounds (1 kilogram) were reported to have fallen in the Gopalgang district of Bangladesh.

The largest hailstone ever recorded in the United States fell in Coffeyville, Kansas on September 3, 1970. It measured 5.57 inches (14.1 centimeters) in diameter and weighed 1.67 pounds (0.75 kilograms).

Hail is precipitation consisting of balls of ice. Hailstones usually are made of concentric, or onion-like, layers of ice alternating with partially melted and refrozen snow, structured around a tiny central core. It is formed in cumulonimbus or thunderclouds when freezing water and ice cling to small particles in the air, such as dust. The winds in the cloud blow the particles through zones of different temperatures causing them to accumulate additional layers of ice and melting snow and to increase in size.

Sources: Battan, Louis J. *Weather in Your Life*, pp. 56-57; Tufty, Barbara. *1001 Questions Answered About Hurricanes, Tornados, and Other Natural Air Disasters*, pp. 151-153.

1416/
What causes exceptionally large snowflakes?

Giant snowflakes, 4 to 6 inches (10 to 15 centimeters) in diameter, have been verified. Large flakes are probably aggregations of smaller flakes created by collisions and/or electrostatic attraction.

Sources: Corliss, William R. *Tornados, Dark Days, Anomalous Precipitation, with Related Weather Phenomena*, pp. 99-100.

1417/
Are all snowflakes shaped alike?

Although no two snowflakes are identical, they all have six sides or six prongs. The many forms that a snow crystal can have are explained by the number of water molecules involved (10^{18}) and the different ways that the molecules can arrange themselves in a three-dimensional hexagonal pattern. Temperature may be the key to the crystal's shape—flat, plate-like forms, long columnlike forms or prismlike forms. Snow crystals are classified into 10 categories in the International Snow Classification. Half are symmetrical and include plates, stellars, columns, needles, and capped columns. Columns and capped columns occur at temperatures colder than 12°F (-10°C). Needles occur at the warmest temperatures. Plates and stellars, the most beautiful of the snow crystals, are rare and occur under special weather conditions. Very plentiful are the asymmetrical snow crystals, the spatial dendrites (three-dimensional irregular stellar- or plate-like crystals occurring in high-moisture clouds) and the irregular crystals (ice prisms called diamond dust).

Sources: Berry, James. *Exploring Crystals*, p. 7; Schaefer, J., and John A.Day, *A Field Guide the the Atmosphere*, pp. 246-247; Williams, Jack. *The Weather Book*, pp. 98-100.

1418/
How does snow form?

Snow is not frozen rain. Snow forms by sublimation of water vapor (the turning of water vapor directly into ice, without going through the liquid stage). High above the ground, chilled water vapor turns to ice when its temperature reaches the dew point. The result of this sublimation is a crystal of ice, usually are hexagonal. Snow begins in the form of these tiny hexagonal ice crystals in the high clouds; the young crystals are the seeds from which snowflakes will grow. As water vapor is pumped up into the air by updrafts, more water is deposited on the ice crystals, causing them to grow. Soon some of the larger crystals fall to the ground as snowflakes.

Sources: Spar, Jerome. *The Way of the Weather*, p. 103.

Sublimation of vapor into ice

1419 / Is it ever too cold to snow?

No matter how cold the air gets, there is still some moisture in it, and this can fall out of the air in the form of very small snow crystals. Very cold air is associated with no snow because these invasions of air from northerly latitudes are associated with clearing conditions behind cold fronts. Heavy snowfalls are associated with relatively mild air in advance of a warm front. The fact that snow piles up, year after year, in Arctic regions illustrates that it is never too cold to snow.

Sources: Forrester, Frank. *1001 Questions Answered About the Weather*, p. 286; Sanders, Ti. *Weather*, p. 165.

1420 / What is the record for the greatest snowfall in the United States?

The record for the most snow in a single storm is 189 inches (480 centimeters) at Mount Shasta Ski Bowl in California from February 13–19, 1959. For the most snow in a 24 hour day, the record goes to Silver Lake, Colorado on April 14–15, 1921, with 76 inches (193 centimeters) of snow. The year record goes to Paradise, Mount Rainier, in Washington with 1,224.5 inches (3,110 centimeters) from February 19, 1971 to February 18, 1972. The highest average annual snowfall was 241 inches (612 centimeters) for Blue Canyon, California. In March 1911, Tamarack, California had the deepest snow accumulation—over 37.5 feet (11.4 meters).

Sources: *Guinness Book of Records 1992*, p. 44.

1421

How much water is there in an inch of snow?

An average figure is 10 inches (25 centimeters) of snow is equal to 1 inch (2.5 centimeters) of water. Heavy, wet snow has a high water content; 4 to 5 inches (10 to 12.5 centimeters) may contain 1 inch (2.5 centimeters) of water. A dry, powdery snow might require 15 inches (37.5 centimeters) of snow to equal 1 inch (2.5 centimeters) of water.

Sources: *Mineral Information Service*, vol. 23 (September 1970), p. 189.

1422

What would cause frogs and toads to fall from the sky like a rain shower?

Documented cases of showers of frogs have been recorded since 1794, usually during heavy summer rainstorms. Whirlwinds, waterspouts, and tornadoes are given as the conventional explanation. Extensive falls of fish, birds, and other animals have also been reported.

Sources: Corliss, William R. *Tornadoes, Dark Days, Anomalous Precipitation, with Related Weather Phenomena*, pp. 72-74.

1423

How far away can thunder be heard?

Thunder is the crash and rumble associated with lightning. It is caused by the explosive expansion and contraction of air heated by the stroke of lightning. This results in sound waves that can be heard easily six to seven miles (9.7 to 11.3 kilometers) away. Occasionally such rumbles can be heard as far away as 20 miles (32.1 kilometers). The sound of great claps of thunder are produced when intense heat and the ionizing effect of repeated lightning occurs in a previously-heated air path. This creates a shock wave that moves at the speed of sound.

Sources: Tufty, Barbara. *1001 Questions Answered About Hurricanes, Tornadoes, and Other Natural Air Disasters*, p. 132.

1424

When do thunderstorms occur?

In the United States thunderstorms usually occur in the summertime, especially from May through August. Thunderstorms tend to occur in late spring and summer when large amounts of tropical maritime air move across the United States. Storms usually develop when the surface air is heated the most from the sun (2 to 4 p.m.). Thunderstorms are relatively rare in the New England area, North Dakota, Montana, and other northern states (latitude 60 degrees) where the air is often too cold. Also these storms are rare along the Pacific Ocean because the summers there are too dry for these storms to occur. Florida, the Gulf states, and the south-eastern states tend to have the most storms, averaging 70 to 90 annually. The mountainous southwest averages 50 to 70 storms annually. In the world, thunderstorms are most plentiful in the areas between latitude 35 degrees north and 35 degrees south; in these areas there can be as many as 3,200 storms within a 12 hour nighttime period. There can be as many as 1,800 storms occurring at once throughout the world.

Lightning does perform a vital function; it returns to the earth much of the negative charge the earth loses by leakage into the atmosphere. The annual death toll in the United States from lightning is greater than the annual death toll from tornadoes or hurricanes—150 Americans die annually from lightning and 250 are injured.

Sources: *Guinness Book of Records 1992*, p. 46; Schaefer, Vincent J., and John A. Day. *A Field Guide to the Atmosphere*, p. 196; Tufty, Barbara. *1001 Questions Answered About Hurricanes, Tornadoes, and Other Natural Air Disasters*, pp. 115-117.

TEMPERATURE

▽1425▽ How can the temperature be determined from the frequency of cricket chirps?

Count the number of cricket chirps in 14 seconds and add 40.

Sources: *Nature and Science*, vol. 3 (April 4, 1966), p. 6.

▽1426▽ Which place has the maximum amount of sunshine in the United States?

Yuma, Arizona, has an annual average of 90% of sunny days or over 4,000 sunny hours per year. St. Petersburg, Florida, had 768 consecutive sunny days from February 9, 1967, to March 17, 1969. On the other extreme, the South Pole has no sunshine for 182 days annually, and the North Pole has none for 176 days.

Sources: *Guinness Book of Records 1992*, p. 44

▽1427▽ Why was 1816 known as the "year without a summer"?

The eruption of Mount Tambora, a volcano in Indonesia, in 1815 threw billions of cubic yards of dust over 15 miles into the atmosphere. Because the dust penetrated the stratosphere, wind currents spread it throughout the world. As a consequence of this volcanic activity, in 1816 normal weather patterns were greatly altered. Some parts of Europe and the British Isles experienced average temperatures 2.9°F to 5.8°F (1.6°C to 3.2°C) below normal. In New England heavy snow fell between June 6 and June 11 and frost occurred every month of 1816. Crop failures were experienced in Western Europe and Canada as well as in New England. In 1817, the excess dust had settled and the climate returned to more normal conditions.

Sources: Miller, Eugene W. *Environmental Hazards; Air Pollution: A Reference Handbook*, p. 6.

1428
Is the world actually getting warmer?

The year 1990 was the warmest year since 1880, with an average global temperature of 59.8°F (15.45°C). Six of the seven warmest years of the twentieth century have occurred during the 1980s. The average annual temperature for the contiguous United States (during the time period 1895–1990) is 52.5°F (11.4°C). The United States began the century with cool temperatures; the 1920s through 1950s were warm; the 1960s and 1970s were cool; and the 1980s were warm.

Sources: Williams, Jack. *The Weather Book*, p. 186; *The World Almanac and Book of Facts 1992*, p. 206.

1429
What are the highest and lowest recorded temperatures on earth?

The highest temperature in the world was recorded as 136°F (58°C) at Al Aziziyah (el-Azizia), Libya on September 13, 1922; the highest temperature recorded in the United States was 134°F (56.7°C) in Death Valley, California on July 10, 1913. The temperatures of 140°F (60°C) at Delta, Mexico in August 1953 and 136.4°F (58°C) at San Luis, Mexico on August 11, 1933 are not internationally accepted. The lowest temperature was -128.6°F (-89.6°C) at Vostok Station in Antarctica on July 21, 1983. The record cold temperature for an inhabited area was -90.4°F (-68°C) at Oymyakon, Siberia (population 4,000) on February 6, 1933. This temperature tied with the readings at Verkhoyansk, Siberia on January 3, 1885, February 5 and 7, 1892. The lowest temperature reading in the United States was -79.8°F (-62.1°C) on January 23, 1971 in Prospect Creek, Alaska; for the contiguous 48 states, the coldest temperature was -69.7°F (-56.5°C) at Rogers Pass, Montana on January 20, 1954.

Sources: *Guinness Book of Records 1992*, pp. 42, 44.

1430
Why are the hot humid days of summer called dog days?

This period of extremely hot, humid, sultry weather that traditionally occurs in the northern hemisphere in July and August received its name from the dog star

Sirius of the constellation *Canis Major.* At this time of year, Sirius, the brightest visible star, rises in the east at the same time as the sun. Ancient Egyptians believed that the heat of this brilliant star added to the sun's heat to create this hot weather. Sirius was blamed for the withering droughts, sickness and discomfort that occurred during this time. Traditional dog days start on July 3 and end on August 11.

Sources: Tufty, Barbara. *1001 Questions Answered About Hurricanes, Tornadoes, and Other Natural Air Disasters,* pp. 311-312.

1431
How much has the earth warmed in the past century?

A United Nations committee of 300 leading scientists examined numerous data sets and concluded that the earth has warmed (0.3°C to 0.6°C) during the past century. This data could mean that the global warming is a result of the ozone layer depletion, or it could be part of a larger weather fluctuation pattern.

Sources: *Chemical and Engineering News,* vol. 70 (April 27, 1992), pp. 7–8; Leggett, Jeremy. *Global Warming,* p. 18.

THUNDER *See: WEATHER AND CLIMATE—SNOW, RAIN, HAIL, ETC.*

TORNADOES *See: WEATHER AND CLIMATE—WIND*

WEATHER PREDICTION

1432
What is nowcasting?

Nowcasting is a form of very short-range weather forecasting. The term is sometimes used loosely to refer to any area-specific forecast for the period up to 12 hours ahead that is based on very detailed observational data. However, nowcasting should probably be defined more restrictively as the detailed description of the current weather along with forecasts obtained by extrapolation up to about two hours ahead.

Sources: *McGraw-Hill Encyclopedia of Science and Technology,* vol. 12, p. 116.

1433 / Can groundhogs accurately predict the weather?

Over a 60-year period, groundhogs have accurately predicted the weather (i.e., when spring will start) only 28% of the time on Groundhog Day. Groundhog Day was first celebrated in Germany where farmers would watch for a badger to emerge from winter hibernation. If the day was sunny, the sleepy badger would be frightened by this shadow and duck back for another 6 weeks' nap; if it was cloudy he would stay out, knowing that spring had arrived. German farmers who emigrated to Pennsylvania brought the celebration to America. Finding no badgers in Pennsylvania, they chose the groundhog as a substitute.

Sources: Berliner, Barbara. *The Book of Answers*, p. 92; Sanders, Dennis. *The First of Everything*, pp. 369-370.

1434 / Is a halo around the sun or moon a sign of rain or snow approaching?

The presence of a ring around the sun or, more commonly, the moon in the night sky, betrays very high ice crystals composing cirrostratus clouds. The brighter the ring, the greater the odds of precipitation and the sooner it may be expected. Rain or snow will not *always* fall, but two times out of three, precipitation will start to fall within twelve to eighteen hours. These cirroform clouds are a forerunner of an approaching warm front and an associated low pressure system.

Sources: Sanders, Ti. *Weather*, p. 160.

1435 / What is barometric pressure and what does it mean?

Barometric, or atmospheric, pressure is the force exerted on a surface by the weight of the air above that surface, as measured by an instrument called a barometer. Pressure is greater at lower levels because the air's molecules are squeezed under the weight of the air, above. So while the average air pressure at sea level is 14.7 pounds per square inch, at 1,000 feet above sea level, the pressure drops to 14.1 pounds per square inch, and at 18,000 feet the pressure is 7.3 pounds, about half of the figure at sea level. Changes in air pressure bring weather changes. High pressure areas bring clear skies and fair weather; low pressure areas bring wet or stormy weather. Areas of very low pressure have serious storms, such as hurricanes.

Sources: Lee, Sally. *Predicting Violent Storms*, pp. 16-17; Tufty, Barbara. *1001 Questions Answered About Hurricanes, Tornados, and Other Natural Air Disasters*, p. 316; Williams, Jack. *The Weather Book*, p. 30.

1436 / Can weather be predicted from the stripes on a wooly-bear caterpillar?

It is an old superstition that the severity of the coming winter can be predicted by the width of the brown bands or stripes around the wooly-bear caterpillar in the autumn. If the brown bands are wide, says the superstition, the winter will be mild, but if the brown bands are narrow, a rough winter is foretold. Studies at the American Museum of Natural History in New York failed to show any connection between the weather and the caterpillar's stripes. This belief is only a superstition; it has no basis in scientific fact.

Sources: Forrester, Frank H. *1001 Questions Answered About the Weather*, p. 281.

1437 / When did weather forecasting begin?

On May 14, 1692, the weekly newspaper, *A Collection for the Improvement of Husbandry and Trade*, gave a seven–day table with pressure and wind readings for the comparable dates of the previous year. Readers were expected to make up their own forecasts from the data. Other journals soon followed with their own weather features. In 1771, a new journal, called the *Monthly Weather Paper*, was completely devoted to weather prediction. In 1861, the British Meteorological Office began issuing daily weather forecasts. The first broadcast of weather forecasts was done by the University of Wisconsin's station 9XM at Madison, Wisconsin on January 3, 1921.

Sources: Robertson, Patrick. *The Book of Firsts*, p. 200.

WIND

1438 / What are halcyon days?

This term is often used to refer to a time of peace or prosperity. In sea tradition, it is the two week period of calm weather before and after the shortest day of the year, about December 21. The phrase is taken from halcyon, the name the ancient Greeks gave to the kingfisher. According to legend, the halcyon built its nest on the surface of the ocean and was able to quiet the winds while its eggs were hatching.

Sources: *World Book Encyclopedia*, vol. 9, p. 18.

1439 / Why are the horse latitudes called by that name?

They are two high pressure belts characterized by low winds, about 30 degrees north and south of the equator. Dreaded by early sailors, these areas have undependable winds with periods of calm. In the northern hemisphere, particularly near Bermuda, sailing ships carrying horses from Spain to the New World were often becalmed. When water supplies ran low, these animals were the first to be rationed water. Dying from thirst or tossed overboard, the animals were sacrificed to conserve water for the men. Explorers and sailors reported that the seas were "strewn with bodies of horses." Supposedly this is why the areas are called the "horse latitudes." The term might be rooted in the complaints by sailors who were paid in advance and received no overtime when the ships slowly transversed this area. Supposedly during this time they were "working off a dead horse."

Sources: *Encyclopedic Dictionary of Science*, p. 121; Rovin, Jeff. *Laws of Order*, p. 87; Tufty, Barbara. *1001 Questions Answered About Hurricanes, Tornadoes, and Other Natural Disasters*, pp. 189-190.

1440
What is meant by the wind chill factor?

The wind chill factor or wind chill index is a number, which expresses the cooling effect of moving air at different temperatures. It indicates in a general way how many calories of heat are carried away from the surface of the body.

Scientists have devised an equivalent temperature scale which makes it easy to determine the wind chill factor. The accompanying chart is used for that purpose. Find the wind speed in the column on the left and find the temperature reading in the horizontal row at the top. The intersection of these two points gives the corresponding wind chill factor. For example, a temperature of 10°F along with a wind speed of 29 to 32 miles per hour corresponds to a wind chill factor of -35°F.

Sources: Vergara, William. *Science in Everyday Life*, pp. 168-169.

COOLING POWER OF WIND EXPRESSED AS "EQUIVALENT CHILL TEMPERATURE"																						
WIND SPEED		TEMPERATURE (°F)																				
CALM	CALM	40	35	30	25	20	15	10	5	0	-5	-10	-15	-20	-25	-30	-35	-40	-45	-50	-55	-60
KNOTS	MPH	EQUIVALENT CHILL TEMPERATURE																				
3-6	5	35	30	25	20	15	10	5	0	-5	-10	-15	-20	-25	-30	-35	-40	-45	-50	-55	-65	-70
7-10	10	30	20	15	10	5	0	-10	-15	-20	-25	-35	-40	-45	-50	-60	-65	-70	-75	-80	-90	-95
11-15	15	25	15	10	0	-5	-10	20	-25	-30	-40	-45	-50	-60	-65	-70	-80	-85	-90	-100	-105	-110
16-19	20	20	10	5	0	-10	-15	-25	-30	-35	-45	-50	-60	-65	-75	-80	-85	-95	-100	-110	-115	-120
20-23	25	15	10	0	-5	-15	-20	-30	-35	-45	-50	-60	-65	-75	-80	-90	-95	-105	-110	-120	-125	-135
24-28	30	10	5	0	-10	-20	-25	-30	-40	-50	-55	-65	-70	-80	-85	-95	-100	-110	-115	-125	-130	-140
29-32	35	10	5	-5	-10	-20	-30	-35	-40	-50	-60	-65	-75	-80	-90	-100	-105	-115	-120	-130	-135	-145
33-36	40	10	0	-5	-15	-20	-30	-35	-45	-55	-60	-70	-75	-85	-95	-100	-110	-115	-125	-130	-140	-150
WINDS ABOVE 40 HAVE LITTLE ADDITIONAL EFFECTS.		LITTLE DANGER			INCREASING DANGER (Flesh may freeze within 1 minute)				GREAT DANGER (Flesh may freeze within 30 seconds)													
		DANGER OF FREEZING EXPOSED FLESH FOR PROPERLY CLOTHED PERSONS																				

Wind chill index

1441

Is there a formula for computing the wind chill?

The formula is $T_e = 33 - \dfrac{(10.45 + V - V(33 - T)}{22.04}$

T_e = Wind chill equivalent temperature in degrees Celsius
V = Wind speed in meters per second
T = Air temperature in degrees Celsius

Sources: *Weatherwise*, vol. 34 (December 1981), p. 266.

1442

Who is associated with developing the concept of wind chill?

The Antarctic explorer, Paul A. Siple coined the term in his dissertation "Adaptation of the Explorer to the Climate of Antarctica," submitted in 1939. Siple was the youngest member of Admiral Byrd's Antarctica expedition in 1928–1930 and later made other trips to the Antarctic as part of Byrd's staff and for the United States Department of the Interior assigned to the United States Antarctic Expedition. He also served in many other endeavors related to the study of cold climates.

Sources: *Weatherwise*, vol. 34 (December 1981), p. 266; *Who Was Who in American History: Science and Technology*, p. 557.

1443

What is the Coriolis effect?

The 19th century French engineer, Gaspard C. Coriolis (1792–1843), discovered that the rotation of the earth deflects streams of air. Because the earth spins to the east, all moving objects in the Northern Hemisphere tend to turn somewhat to the right of a straight path while those in the Southern Hemisphere turn slightly left. The Coriolis effect explains the lack of northerly and southerly winds in the tropics and polar regions; the northeast and southeast trade winds and the polar easterlies all owe their westward deflection to the Coriolis effect.

Sources: Bair, Frank E. *The Weather Almanac*, pp. 277–278; Great Britain Meteorological Office. *Meteorological Glossary*, p. 287; Williams, Jack. *The Weather Book*, p. 21.

1444

Which month is the most dangerous for tornadoes in the United States?

According to one study, May is the most dangerous month for tornadoes in the United States, with 329, while February is the safest with only three. In another study the months December and January were usually the safest, and the months having the greatest number of tornadoes were April, May, and June. In February, tornado frequency begins to increase, depending on the geographical area. February tornadoes tend to occur in the central Gulf states; in March the center of activity moves eastward to the southeastern Atlantic states where tornado activity peaks in April. In May the center of activity focuses on the southern Plains states; in June this moves to the northern Plains and Great Lakes area (into western New York).

Sources: Bair, Frank E. *The Weather Almanac*, p. 84; *The Public Health Consequences of Disasters 1989*, p. 52.

Each dot represents the approximate location of two occurrences during a 10-year period. An occurrence is defined as the first point of contact with the ground.

1445 When was the jet stream discovered?

A jet stream is a flat and narrow tube of air that moves more rapidly than the surrounding air. Discovered by World War II bomber pilots flying over Japan and the Mediterranean Sea, jet streams have become important with the advent of airplanes capable of cruising at over 30,000 feet (9,144 meters). The currents of air flow from west to east and are usually a few miles deep, up to 100 miles (160 kilometers) wide and well over 1,000 miles (1,600 kilometers) in length. The air current must flow at over 57.5 miles (92 kilometers) per hour.

There are two polar jet streams, one in each hemisphere. They meander between 30 and 70 degrees latitude, occur at altitudes of 25,000 to 35,000 feet (7,620 to 10,668 meters), and achieve maximum speeds of over 230 miles (368 kilometers) per hour. The subtropical jet streams, again one per hemisphere, wander between 20 and 50 degrees latitude. They are found at altitudes of 30,000 to 45,000 feet (9,144 to 13,715 meters), and have speeds of over 345 miles (552 kilometers) per hour.

Sources: *Encyclopedia of Aviation*, p. 107; *Van Nostrand's Scientific Encyclopedia*, vol. 1, pp. 1631–1632.

1446 What is the Fujita and Pearson Tornado Scale?

The Fujita and Pearson Tornado Scale, developed by T. Theodore Fujita and Allen Pearson ranks tornadoes by their wind speed. Sometimes known as the Fujita scale, the ranking ranges from F0 (very weak) to F6 (inconceivable).

F0—Light damage: damage to trees, billboards, and chimneys.

F1—Moderate damage: mobile homes could be pushed off their foundations and cars could be pushed off roads.

F2—Considerable damage: roofs torn off, mobile homes demolished, and large trees uprooted.

F3—Severe damage: even well-constructed homes can be torn apart, trees uprooted, and cars lifted off the ground.

F4—Devastating damage: houses can be leveled, cars thrown, and objects can become flying missiles.

F5—Incredible damage: structures lifted off foundations and carried away; cars become missiles. Less than 2% of tornadoes are in this category.

F6—Maximum tornado winds not expected to exceed 318 miles per hour.

Fujita and Pearson Tornado Scale

Scale	Speed miles per hour	Path length in miles	Path width
0	≤72	≤1.0	≤17 yards
1	73–112	1.0–3.1	18–55 yards
2	113–157	3.2–9.9	56–175 yards
3	158–206	10.0–31.0	176–556 yards
4	207–260	32.0–99.0	0.34–0.9 miles
5	261–318	100–315	1.0–3.1 miles
6	319–380	316–999	3.2–9.9 miles

Sources: Bair, Frank E. *The Weather Almanac*, p. 90; *The Public Health Consequences of Disasters 1989*, p. 40; Tufty, Barbara. *1001 Questions Answered About Hurricanes, Tornadoes, and Other Natural Disasters*, pp. 356-357.

1447 / *How do hurricanes get their names?*

Since 1950, hurricane names are officially selected from library sources and are decided during the international meetings of the World Meteorological Organization (WHO). The names are chosen to reflect the cultures and languages found in the Atlantic, Caribbean, and Hawaiian regions. The National Hurricane Center near Miami, Florida, selects the name from one of the six listings for Region 4 (Atlantic and Caribbean area) when a tropical storm with rotary action and wind speeds above 39 miles (62.75 kilometers) per hour develops. Letters Q, U, X, Y, and Z are not included because of the scarcity of names beginning with those letters. Once a storm has done great damage, its name is retired from the six–year list cycle.

1992	1993	1994	1995	1996	1997
Andrew	Arlene	Alberto	Allison	Arthur	Ana
Bonnie	Bret	Beryl	Barry	Bertha	undecided

1992	1993	1994	1995	1996	1997
Charley	Cindy	Chris	Chantal	Cesar	Claudette
Danielle	Dennis	Debby	Dean	Diana	Danny
Earl	Emily	Ernesto	Erin	Edouard	Erika
Frances	Floyd	Florence	Felix	Fran	Fabian
Georges	Gert	Gordon	Gabrielle	Gustav	Grace
Hermine	Harvey	Helene	Humberto	Horrtense	Henri
Ivan	Irene	Isaac	Iris	Isidore	Isabel
Jeanne	Jose	Joyce	Jerry	Josephine	Juan
Karl	Katrina	Keith	Karen	Klaus	Kate
Lisa	Lenny	Leslie	Luis	Lili	Larry
Mitch	Maria	Michael	Marilyn	Marco	Mindy
Nicole	Nate	Nadine	Noel	Nana	Nicholas
Otto	Ophelia	Oscar	Opal	Omar	Odette
Paula	Philippe	Patty	Pablo	Paloma	Peter
Richard	Rita	Rafael	Roxanne	Rene	Rose
Shary	Stan	Sandy	Sebastien	Sally	Sam
Tomas	Tammy	Tony	Tanya	Teddy	Teresa
Virginie	Vince	Valerie	Van	Vicky	Victor
Walter	Wilma	William	Wendy	Wilfred	Wanda

Sources: Bair, Frank E. *The Weather Almanac*, p. 44; *Weatherwise*, vol. 36 (August 1983), p. 179; Williams, Jack. *The Weather Book*, p. 145.

1448

Which year had the most tornadoes?

From the period 1916 (when records started to be kept) to 1989, more tornadoes occurred in 1973 than in any other year. That year 1,102 tornadoes struck in 46 states, killing 87 persons. From 1980–1989, an average of 820 tornadoes occurred each year. The largest outbreak of tornadoes occurred on April 3 and 4, 1974. 127 tornadoes were recorded in this "Super Outbreak" in the Plains and midwestern states. Six of these tornadoes had winds greater than 26 miles (420 kilometers) per hour and some of them were the strongest ever recorded.

Sources: Bair, Frank E. *The Weather Almanac*, pp. 85, 88; Williams, Jack. *The Weather Book*, p. 111.

1449

Which United States hurricanes have caused the most deaths?

The ten deadliest United States hurricanes are listed below.

Hurricane	Year	Deaths
1. Texas (Galveston)	1900	6,000
2. Florida (Lake Okeechobee)	1928	1,836
3. Florida (Keys/S. Texas)	1919	600–900+
4. New England	1938	600
5. Florida (Keys)	1935	408
6. AUDREY (Louisiana/Texas)	1957	390

Hurricane	Year	Deaths
7. Northeast U.S.	1944	390
8. Louisiana (Grand Isle)	1909	350
9. Louisiana (New Orleans)	1915	275
10. Texas (Galveston)	1915	275

Sources: *The Public Health Consequences of Disasters 1989*, p. 35.

1450

How are hurricanes classified?

The Saffir/Simpson Hurricane Damage-Potential scale assigns numbers 1 through 5 to measure the disaster potential of a hurricane's winds and its accompanying storm surge. The purpose of the scale, developed in 1971 by Herbert Saffir and Robert Simpson, is to assist disaster agencies in gauging the potential significance of these storms in terms of assistance.

Saffir/Simpson Hurricane Scale Ranges

Scale number (category)	Barometric pressure (in inches)	Winds (miles per hour)	Surge (in feet)	Damage
1	≥28.94	74–95	4–5	Minimal
2	28.50–28.91	96–110	6–8	Moderate
3	27.91–28.47	111–130	9–12	Extensive
4	27.17–27.88	131–155	13–18	Extreme
5	<27.17	>155	>18	Catastrophic

Damage categories:

Minimal—No real damage to building structures. Some tree, shrubbery, and mobile home damage. Coastal road flooding and minor pier damage.

Moderate—Some roof, window, and door damage. Considerable damage to vegetation, mobile homes and piers. Coastal and low-lying escape routes flood two to four hours before center of storm arrives. Small craft can break moorings in unprotected areas.

Extensive—Some structural damage to small or residential buildings. Mobile homes destroyed. Flooding near coast destroys structures and flooding of homes five feet above sea level as far inland as six miles.

Extreme—Extensive roof, window, and door damage. Major damage to lower floors of structures near the shore, and some roof-failure on small residences. Complete beach erosion. Flooding of terrain 10 feet above sea level as far as six   miles inland requiring massive residential evacuation.

Catastrophic—Complete roof failure to many buildings; some complete building failure with small utility buildings blown away. Major damage to lower floors of all structures 19 feet above sea level located within 500 yards of the shoreline. Massive evacuation of residential areas on low ground 5 to 10 miles from shoreline may be required.

Sources: Bair, Frank E. *The Weather Almanac*, p. 80; *The Public Health Consequences of Disasters 1989*, p. 34; Williams, Jack. *The Weather Book*, p. 137.

1451
What is the Beaufort Scale?

The Beaufort scale was devised in 1805 by a British Admiral, Sir Francis Beaufort (1774–1857) to help mariners in handling ships. It uses a series of numbers from 0 to 17 to indicate wind speeds and applies to both land and sea.

Beaufort number	Name	Wind speed	
		Miles per hour	Kilometers per hour
0	Calm	less than 1	less than 1.5
1	Light air	1–3	1.5–4.8
2	Light breeze	4–7	6.4–11.3
3	Gentle breeze	8–12	12.9–19.3
4	Moderate breeze	13–18	21–29
5	Fresh breeze	19–24	30.6–38.6
6	Strong breeze	25–31	40.2–50
7	Moderate gale	32–38	51.5–61.1
8	Fresh gale	39–46	62.8–74
9	Strong gale	47–54	75.6–86.9
10	Whole gale	55–63	88.5–101.4
11	Storm	64–73	103–117.5
12–17	Hurricane	74 and above	119.1 and above

Sources: Williams, Jack. *The Weather Book*, p. 43.

1452
How does a cyclone differ from a hurricane or tornado?

All three wind phenomena are rotating winds that spiral in toward a low–pressure center as well as upward. Their differences lie in their size, wind velocity, rate of travel, and duration. Generally, the faster the winds spin, the shorter (in time) and smaller (in size) the event becomes.

A cyclone has rotating winds from 10 to 60 miles (16 to 97 kilometers) per hour; can be up to 1,000 miles (1,609 kilometers) in diameter, travels about 25 miles (40 kilometers per hour, and lasts from one to several weeks. A hurricane (or typhoon, as it is called in the Pacific Ocean area) has winds that vary from 75 to 200 miles (121 to 322 kilometers) per hour, moves between 10 to 20 miles (16 to 32 kilometers) per hour, can have a diameter up to 600 miles (965 kilometers), and can exist from several days to more than a week. A tornado can reach a rotating speed of 300 miles (423 kilometers) per hour, travels between 25 to 40 miles (40 to 64 kilometers) per hour and generally lasts minutes, although some have existed for 5–6 hours. Its diameter can range from 300 yards (274 meters) to a mile (1.6 kilometers) and its average path length is 16 miles (26 kilometers), with a maximum of 300 miles (483 kilometers).

Typhoons, hurricanes, and cyclones tend to breed in low–altitude belts over the oceans generally from five degrees to fifteen degrees latitude north or south. A tornado generally forms several thousand feet above the earth's surface usually during warm, humid weather; many times it is in conjunction with a thunderstorm. Although a tornado can occur in many places, the continental plains of North America (i.e., from the Plains States eastward to western New York and the southeastern Atlantic states) is where most tornadoes happen. 82% of them materialize during the warmest hours of the day (noon to midnight) with 23% of all tornado activity taking place between 4 p.m. and 6 p.m.

Sources: Schafer, Vincent J., and John Day. *A Field Guide to the Atmosphere*, p; 198; Tufty, Barbara. *1001 Questions Answered About Hurricanes, Tornadoes, and Other Natural Disasters*, pp. 58-60.

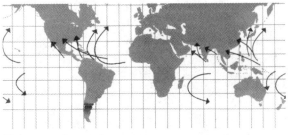

Main areas where cyclones occur

1453 / What was the greatest natural disaster in United States history?

The greatest natural disaster occurred when a hurricane struck Galveston, Texas on September 8, 1900, and killed over 6,000 people. However, the costliest national disaster to date is when Hurricane Andrew hit Florida on August 24, 1992 and Louisiana on August 25, 1992. Classified as a category four hurricane, Andrew's winds of 135 miles (217 kilometers) per hour gusted up to 165 miles (266 kilometers) per hour. Early warning kept the death toll low, but property damage is estimated at $20 billion.

Sources: *Facts on File Weekly News Digest with Index*, vol. 52, no. 2701 (August 27, 1992), p. 629. *Nature on the Rampage*, p. 122; *Newsweek*, vol. 120 (September 7, 1992), p. 23.

1454 / Is Chicago the windiest city?

In 1990 Chicago ranked 21st in the list of 68 windy cities with an average wind speed of 10.3 miles (16.6 kilometers) per hour. Cheyenne, Wyoming, with an average wind speed of 12.9 miles (20.8 kilometers) per hour, ranks number one, closely followed by Great Falls, Montana, with an average wind speed of 12.8 miles (20.6 kilometers) per hour. The highest surface wind ever recorded was on Mount Washington, New Hampshire at an elevation of 6288 feet (1.9 kilometers). On April 12, 1934 its wind was 231 miles (371.7 kilometers) per hour and its average wind speed was 35 miles (56.3 kilometers) per hour.

Sources: Krantz, Les. *The Best and Worst of Everything*, pp. 102-103; Williams, Jack. *The Weather Book*, p. 43.

WEIGHTS, MEASURES, AND TIME

CALENDAR *See: WEIGHTS, MEASURES, AND TIME—TIME*

MEASURES *See: WEIGHTS, MEASURES, AND TIME—WEIGHTS, MEASURES, AND MEASUREMENT*

TIME *See also: BIOLOGY—CLASSIFICATION, MEASUREMENT, AND TERMS*

1455 What are the beginning dates of the seasons for 1993 to 2000?

The four seasons in the northern hemisphere coincide with astronomical events. The spring season starts with the vernal equinox (around March 21) and lasts until the summer solstice (June 21 or 22); summer is from the summer solstice to the autumnal equinox (about September 21); autumn, from the autumnal equinox to the winter solstice (December 21 or 22); and winter from the winter solstice to the vernal equinox. In the southern hemisphere the seasons are reversed; autumn corresponds to spring; winter to summer. The seasons are caused by the tilt of the earth's axis, which changes the position of the sun in the sky. In winter the sun is at its lowest position in the sky; in summer, it is at its highest position (or angle of declination).

Year	Spring	Summer	Fall	Winter
1993	March 20	June 21	September 23	December 21
1994	March 20	June 21	September 23	December 22
1995	March 21	June 21	September 23	December 22
1996	March 20	June 21	September 22	December 21
1997	March 20	June 21	September 22	December 21
1998	March 20	June 21	September 23	December 22
1999	March 21	June 21	September 23	December 22
2000	March 20	June 21	September 22	December 21

Source: Pasachoff, Jay M. *Contemporary Astronomy*, pp. 82-83; *Planetary and Lunar Coordinates for the Years 1984-2000*, p. 7.

1456
Who set a doomsday clock for nuclear annihilation?

The clock first appeared on the cover of the magazine *Bulletin of the Atomic Scientists* in 1947 and was set at 11:53 p.m. The clock, created by the magazine's board of directors, represents the threat of nuclear annihilation with midnight as the time of destruction. It has been recently moved back to 11:43 p.m., the farthest from midnight the clock has ever been. In 1953, just after the United States tested the hydrogen bomb, it was set at 11:58 p.m., the closest to midnight ever.

Sources: *Buzzworm: The Environmental Journal*, vol. 4, (January/February 1992), p. 13.

Doomsday clock reset

1457
What is military time?

Military time divides the day into one set of twenty-four hours, counting from 1:00 a.m. our time (0100) to midnight (2400). This is expressed without punctuation.

> 1:00 a.m. becomes 0100 (pronounced oh-one hundred)
> 2:10 a.m. becomes 0210
> Noon becomes 1200
> 6:00 p.m. becomes 1800
> 9:45 p.m. becomes 2145
> Midnight becomes 2400 (pronounced twenty-four hundred)

To translate 24-hour time into familiar time, the a.m. times are obvious. For p.m. times, subtract 1200 from numbers larger than 1200; e.g., 1900 - 1200 is 7 p.m.

Sources: Blocksma, Mary. *Reading the Numbers*, p. 124.

▽ **1458** ▽
Who invented the alarm clock?

Levi Hutchins of Concord, New Hampshire, invented an alarm clock in 1787. His alarm clock, however, rang at only one time—4 a.m. He invented this device so that he would never sleep past his usual waking time. He never patented or manufactured it. The first modern alarm clock was made by Antoine Redier (1817–1892) in 1847. It was a mechanical device; the electric alarm clock was not invented until around 1890. The earliest mechanical clock was made in 725 C.E. in China by Yi Xing and Liang Lingzan.

Sources: Giscard d'Estaing, Valerie-Anne. *The World Almanac Book of Inventions*, pp. 275-277; Harris, Harry. *Good Old-Fashioned Yankee Ingenuity*, p. 23.

▽ **1459** ▽
What is a floral clock?

In the Middle Ages it was believed that it was possible to tell the hour of day by the use of flowers, which were supposed to open and close at specific times. These would be planted in flower dials. The first hour belonged to the budding rose, the fourth to hyacinths and the twelfth to pansies. The unreliability of this method of timekeeping must have soon become apparent, but floral clocks can still be seen planted in parks and gardens today. The world's largest clock face is that of the floral clock located inside the Rose Building in Hokkaido, Japan. The diameter of the clock, almost 69 feet (21 meters) and the large hand of the clock is almost 28 feet (8.5 meters) in length.

Sources: *Guinness Book of Records 1992*, p. 285; Jones, Julia and Barbara Deere. *Royal Pleasures and Pastimes*, p. 17.

▽ **1460** ▽
What were the longest and shortest years on record?

The longest year was 46 B.C.E. when Julius Caesar (100 B.C.E.-44 B.C.E.) introduced his calendar, called the Julian calendar, which was used until 1582. He added 2 extra months and 23 extra days to February to make up for accumulated slippage in the Egyptian Calendar. Thus, 46 B.C.E. was 455 days long. The shortest year was 1582, when Pope Gregory XIII (1502-1585) introduced his calendar, the Gregorian calendar. He decreed that October 5 would be October 15, eliminating 10 days, to make up for the accumulated error in the Julian calendar. Not everyone changed over to this new calendar at once. Catholic Europe adopted it within 2 years of its inception. Many Protestant continental countries did so in 1699-1700; England imposed it on its colonies in 1752 and Sweden in 1753. Many non-European countries adopted it in the 19th century, with China doing so in 1912, Turkey in 1917, and Russia in 1918. To change from the Julian to the Gregorian calendar, 10 days are added to dates October 5, 1582, through February 28, 1700; after that date add 11 days through February 28, 1800; 12 days through February 28, 1900; and 13 days through February 28, 2100.

Sources: Trefil, James. *1001 Things Everyone Should Know About Science*, p. 138; *The World Almanac and Book of Facts 1992*, p. 291.

1461
When does a century begin?

A century has 100 consecutive calendar years. The 1st century consisted of years 01 through 100. The 20th century started with 1901 and will end with 2000. The 21st century will begin January 1, 2001.

Sources: *The World Almanac and Book of Facts 1992*, p. 291.

1462
When did January 1 become the first day of the new year?

When Julius Caesar (100 B.C.E.-44 B.C.E.) reorganized the Roman calendar and made it solar rather than lunar in the year 45 B.C.E., he moved the beginning of the year to January 1. When the Gregorian calendar was introduced in 1582, January 1 continued to be recognized as the first day of the year in most places. However, in England and the American colonies, March 25, intended to represent the spring equinox, was the beginning of their year. Under this system, March 24, 1700, was followed by March 25, 1701. In 1752, the British government changed the beginning date of the year to January 1.

Sources: Asimov, Isaac. *Asimov on Numbers*, p. 159.

1463
What is the United States Time Standard signal?

The most-used high-frequency reference signals are the United States Time Standard signals from WWV, Fort Collins, Colorado, and WWVH, Kauai, Hawaii. These stations transmit standard "tic" pulses each second with minute announcements and special bulletins periodically. They transmit 1.5, 5, 10, 15, and 20 megahertz with very precise time and frequency stability. The first radio station to transmit time signals regularly was the Eiffel Tower Radio Station in Paris in 1913.

Sources: *The ARRL Handbook for Radio Amateurs*, p. 22-12; Robertson, Patrick. *The Book of Firsts*, p. 149.

1464
Who establishes the correct time in the United States?

The United States National Institute of Standards and Technology uses an atomic clock accurate to 3 millionths of a second per year to establish the correct time. The atomic clock was constructed at the United States Bureau of Measures in Washington, D.C. in 1948. American physicist William F. Libby formulated the principle of this extremely accurate clock.

Source: *Passport to World Band Radio*, p. 70.

1465
How many time zones are there in the world?

There are 24 standard time zones that serially cover the earth's surface at coincident intervals of 15 degrees longitude and 60 minutes Universal Time (UT), as agreed at the Washington Meridian Conference of 1884, thus accounting, respectively for each 24 hours of the calendar day.

Sources: *Van Nostrand's Scientific Encyclopedia*, vol. 2, p. 2833.

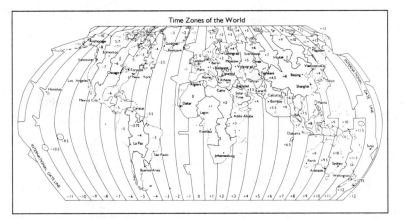

Time Zones of the World

1466
When does leap year occur?

A leap year occurs when the year is exactly divisible by four except centenary years. A centenary year must be divisible by 400 to be a leap year. There was no February 29 in 1900, which was not a leap year. The next centenary year, 2000, will have February 29.

Sources: Burnam, Tom. *The Dictionary of Misinformation*, p. 136.

1467
What do the initials a.m. and p.m. mean?

a.m. means *ante meridiem*, Latin for "before noon."
p.m. means *post meridiem*, Latin for "after noon."

Sources: *Acronyms, Initialisms & Abbreviations Dictionary 1993*, vol. 1, pp. 178, 2716.

1468
How is the date for Easter determined?

The rule for establishing the date in Christian churches is that Easter always falls on the first Sunday after the first full moon that occurs on or just after the vernal equinox. The vernal equinox is the first day of spring in the Northern Hemisphere. Because the full moon can occur on any one of many dates after the vernal equinox, the date of Easter can be as early as March 22 and as late as April 25. For the period 1993-2000 Easter will appear on the following days:

Year	Date
1993	April 11
1994	April 3
1995	April 16
1996	April 7
1997	March 30
1998	April 12
1999	April 4
2000	April 23

Sources: Schweighauser, Charles A. *Astronomy from A to Z*, pp. 42-43; *The World Almanac and Book of Facts 1992*, p. 729.

1469
Why do clocks and watches with the Roman numerals on their faces use "IIII" for the number 4, rather than "IV"?

A clock with a "IV" has a somewhat unbalanced appearance. Thus tradition has favored the strictly incorrect "IIII," which more nearly balances the equally heavy "VIII."

Sources: Britten, Frederick J. *Britten's Old Clocks and Watches and Their Makers*, p. 88.

1470
What is meant by universal time?

On January 1, 1972, Universal Time (UT) replaced Greenwich Mean Time (GMT) as the time reference coordinate for scientific work. Time is taken from an atomic clock and is seen as the logical development of the adoption of the atomic second in 1968. An advantage of UT is that the time at which an event takes place can be determined very readily without recourse to the time-consuming astronomical observations and calculations which were necessary before the advent of atomic clocks. Universal Time is also referred to as International Atomic Time. GMT is when the sun crosses the Greenwich Meridian (0 degrees longitude which passes through the Greenwich Observatory).

Sources: Jerrard, H.G. *A Dictionary of Scientific Units*, p. 148.

1471 / What is the exact length of a calendar year?

The calendar year is defined as the time between two successive crossings of the celestial equator by the sun at the vernal equinox. It is exactly 365 days, 5 hours, 48 minutes, 46 seconds. The fact that the year is not a whole number of days has affected the development of calendars, which over time generate an accumulative error. The current calendar used, the Gregorian calendar, named after Pope Gregory XII (1502-1585), attempts to compensate by adding an extra day to the month of February every 4 years. A "leap year" is one with the extra day added.

Sources: Pasachoff, Jay M. *Contemporary Astronomy*, p. 88.

1472 / How is the date for Passover determined?

Passover (*Pesah*) begins at sundown on the evening before the 15th day of the Hebrew month of Nisan which falls in March or April. Celebrated as a public holiday in Israel, the Passover commemorates the exodus of the Israelites from Egypt in 1290 B.C.E. The term "Passover" is a reference to the last of the 10 plagues that "passed over" the Israelite homes at the end of their captivity in Egypt.

Sources: *Encyclopedia Americana*, vol. 21, p. 513; *The Guinness Book of Answers*, p. 51.

1473 / What is a leap second?

The earth's rotation is slowing down and to compensate for this lagging motion, June 1992 will be one second longer than normal. This leap second, announced in February 1992 by the International Earth Rotation Service, will keep calendar time in close alignment with international atomic time. To accomplish this change, 23h 59m 59s universal time on June 30, 1992, will be followed by 23h 59m 60s, this in turn followed by 0h 0m 0s on July 1.

Sources: *Astronomy*, vol. 20 (June 1992), pp. 24, 26.

1474

Where do the names of the days of the week come from?

The English days of the week are named for a mixture of figures in Anglo-Saxon and Roman mythology.

Day	Named after
Sunday	The sun
Monday	The moon
Tuesday	Tiu (the Anglo-Saxon god of war, equivalent to the Norse Tyr or the Roman Mars)
Wednesday	Woden (the Anglo-Saxon equivalent of Odin, the chief Norse god)
Thursday	Thor (the Norse god of thunder)
Friday	Frigg (the Norse god of love and fertility, the equivalent of the Roman Venus)
Saturday	Saturn (the Roman god of agriculture)

Sources: *The Guinness Book of Answers*, p. 36; Rovin, Jeff. *Laws of Order*, p. 204.

1475

What is the origin of the week?

The week originated in the Babylonian calendar where one day out of seven was devoted to rest.

Sources: Asimov, Isaac. *Asimov on Numbers*, p. 155.

1476

What is the Julian Day Count?

This system of counting days rather than years was developed by Joseph Justus Scaliger (1540–1609) in 1583. Still used by astronomers today, the Julian Day Count (named after Scaliger's father, Julius Caesar Scaliger) Julian Day (JD) 1 was January 1, 4113 B.C.E. On this date the Julian calendar, the ancient Roman tax calendar, and the lunar calendar all coincided. This event would not occur again until 7,980 years later. So each day within the 7,980 year period is numbered. December 31, 1991 at noon is the beginning of JD 2,448,622. The figure reflects the number of days that have passed since its inception. To convert Gregorian calendar dates into Julian Day, simple JD conversion tables have been devised for astronomers.

Sources: Pasachoff, Jay M. *Contemporary Astronomy*, p. 101; *The World Almanac and Book of Facts 1991*, p. 296.

1477

Which animal designations have been given to the Chinese years?

There are 12 different names, always used in the same sequence: Rat, Ox, Tiger, Hare (Rabbit), Dragon, Snake, Horse, Sheep (Goat), Monkey, Rooster, Dog, and

Pig. The following table may be used.

Chinese year cycle

Rat	Ox	Tiger	Hare (Rabbit)	Dragon	Snake
1900	1901	1902	1903	1904	1905
1912	1913	1914	1915	1916	1917
1924	1925	1926	1927	1928	1929
1936	1937	1938	1939	1940	1941
1948	1949	1950	1951	1952	1953
1960	1961	1962	1963	1964	1965
1972	1973	1974	1975	1976	1977
1984	1985	1986	1987	1988	1989
1996	1997	1998	1999	2000	2001

Horse	Sheep (Goat)	Monkey	Rooster	Dog	Pig
1906	1907	1908	1909	1910	1911
1918	1919	1920	1921	1922	1923
1930	1931	1932	1933	1934	1935
1942	1943	1944	1945	1946	1947
1954	1955	1956	1957	1958	1959
1966	1967	1968	1969	1970	1971
1978	1979	1980	1981	1982	1983
1990	1991	1992	1993	1994	1995
2002	2003	2004	2005	2006	2007

The Chinese year of 354 days begins three to seven weeks into the western 365-day year, so the animal designation changes at that time, rather than on January 1.

Source: *The Guinness Book of Answers*, p. 40; *The Universal Almanac 1992*, p. 2; *The World Almanac and Book of Facts 1992*, p. 296.

1478 / How were the months of the year named?

The English names of the months of the current (Gregorian) calendar come from the Romans, who tended to honor their gods, specific events, etc. by designating them as month names:

January (Januarius in Latin), named after Janus, a Roman two-faced god, one looking into the past, the other into the future.

February (Februarium) from the Latin word *Februare,* meaning to cleanse. At the time of year corresponding to our February, the Romans performed religious rites to purge themselves of sin.

March (Martius) is named in honor of Mars, the god of war.

April (Aprilis), after the Latin word *Aperio,* meaning to open because plants begin to grow in this month.

May (*Maius*), after the Roman goddess Maia as well as from the Latin word *Maiores* meaning elders, who were celebrated during this month.

June (*Junius*), after the goddess Juno and Latin word *iuniores*, meaning young people.

July (*Iulius*) was, at first, known as *Quintilis* from the Latin word meaning five, since it was the fifth month in the early Roman calendar. Its name was changed to July, as an honor to Julius Caesar (100 B.C.E.-44 B.C.E.).

August (*Augustus*) is so named in honor of the Emperor Octavian (63 B.C.E.-14 B.C.E.), first Roman emperor, known as Augustus Caesar. Originally the month was known as *Sextilis* (sixth month of early Roman calendar).

September (*September*) was first the seventh month and accordingly took its name from *septem*, meaning seven.

October (*October*) takes its name from *octo* (eight) from the time it was the eighth month.

November (*November*) from *novem*, meaning nine from the ninth month of the early Roman calendar.

December (*December*) from *decem*, meaning ten from the tenth month of the early Roman calendar.

Sources: *New Encyclopedia Britannica*, vol. 8, p. 292; Rovin, Jeff. *Laws of Order*, p. 205; *Stories Behind Everyday Things*, p. 54.

▽1479▽
Do calendars other than the Gregorian calendar use the same twelve-month cycle?

No. Some calendars have a varying cycle of months, with their first month falling at a different time. Listed below are some variations with the beginning month of the year in italics:

Months of the year

Gregorian	Hebrew	Hindu
January	Shebat	Magha
February	Adar	Phalguna
March	Nisan	*Caitra*
April	Iyar	Vaisakha
May	Sivan	Jyaistha
June	Tammuz	Asadhe
July	Ab	Sravana
August	Elul	Bhadrapada
September	*Tishri*	Asvina
October	Heshvan	Karttika
November	Kislev	Margasivsa
December	Tebet	Pansa
	Ve-Adar (13th month every three years)	

Muslim	Chinese	
Muharram	*Li Chun*	Li Qui
Safar	Yu Shui	Chu Shu
Rabi I	Jing Zhe	Bai Lu
Rabi II	Chun Fen	Qui Fen
Jumada I	Qing Ming	Han Lu
Jumada II	Gu Yu	Shuang Jiang
Rajab	Li Xia	Li Dong
Sha'ban	Xiao Man	Xiao Xue
Ramadan	Mang Zhong	Da Xue
Shawwal	Xia Zhi	Dong Zhi
Dhu'l-Qa'da	Xiao Shu	Xiao Han
Dhu'l-hijja	Da Shu	Da Han

Sources: Rovin, Jeff. *Laws of Order*, pp. 64, 175; *The Universal Almanac 1992*, p. 5.

▽ **1480** ◁

When traveling from Tokyo to Seattle and crossing the international date line, what day is it?

Traveling from west to east (Tokyo to Seattle), the calendar day is set back, i.e., Sunday becomes Saturday. Traveling east to west, the calendar day is advanced, i.e., Tuesday becomes Wednesday. The International Date Line is a zigzag line at approximately the 180th meridian, where the calendar days are separated.

Sources: Berliner, Barbara. *The Book of Answers*, p. 86; *The World Almanac and Book of Facts 1992*, p. 297.

▽ **1481** ◁

Besides the Gregorian calendar, what other kinds of calendars have been used?

Babylonian calendar—A lunar calendar composed of alternating 29-day and 30-day months to equal roughly 354 lunar days. When the calendar became too misaligned with astronomical events, an extra month was added. In addition, 3 extra months were added every eight years to coordinate this calendar with the solar year.

Chinese calendar—A lunar month calendar of 12 periods having either 29 or 30 days (to compensate for the 29.5 days from new moon to new moon). The new year begins on the first new moon over China after the sun enters Aquarius (between January 21st and February 19th). Each year has both a number and a name (for example, the year 1992 or 4629 in the Chinese era is the year of the monkey). The calendar is synchronized with the solar year by the addition of extra months at fixed intervals.

Muslim calendar—A lunar 12-month calendar with 30 and 29 days alternating every month is equal to 354 days. The calendar has a 30 year cycle with designated leap years of 355 days (one day added to last month) in the 30 year period. The Islamic year does not attempt to relate to the solar year (relate to the season). The dating of the beginning of the calendar is 622 C.E. (the date of Mohammed's flight from Mecca to Medina).

Jewish calendar—A blend of the solar and lunar calendar, this calendar adds an extra month (Adar Sheni or the second "Adar" or Veadar) to keep the lunar and solar years in alignment. This occurs seven times during the 19-year cycle. When the extra 29-day month is inserted, the month Adar has 30 days instead of 29. In a usual year the 12 months alternate in a 30 then 29 day repetition.

Egyptian calendar—The ancient Egyptians were the first to use a solar calendar (about 4236 B.C.E. or 4242 B.C.E.), but their year started with the rising of Sirius, the brightest star in the sky. The year, composed of 365 days, was 0.25 days short of the true solar year, so eventually the Egyptian calendar did not coincide with the seasons. It used twelve 30–day months with five day weeks and five dates of festival.

Coptic calendar—Still used in areas of Egypt and Ethiopia, it has a similar cycle to the Egyptian calendar; 12 months of 30 days followed by five complementary days. When a leap year occurs, usually preceding the Julian calendar leap year, the complementary days increase to six.

Roman calendar—Borrowing from the ancient Greek calendar which had a four year cycle based on the Olympic Games, the earliest Roman calendar (about 738 B.C.E.) had 304 days with 10 months. Every second year a short month of 22 or 23 days was added to coincide with the solar year. Eventually two more months were added at the end of the year (Januarius and Februarius) to increase the year to 354 days. The Roman republican calendar replaced this calendar during the reign of Tarquinius Priscus (616–579 B.C.E.). This new lunar calendar had 355 days with the month of February having 28 days. The other month had either 29 or 31 days. To keep the calendar aligned with the seasons, an extra month was added every two years. By the time the Julian calendar replaced this one, the calendar was three months ahead of the season schedule.

Julian calendar—Julius Caesar (100 B.C.E.-44 B.C.E.) in 46 B.C.E., wishing to have one calendar in use for all the empire, had the astronomer Sosigenes develop a uniform solar calendar with the year being 365 days with one day ("leap day") added every fourth year to compensate for the true solar year of 365.25 days. The year had 12 months with 30 or 31 days except for February, which had 28 days (or 29 days in a leap year). The first of the year was moved from March 1 to January 1.

Gregorian calendar—Pope Gregory XIII (1502-1585) instituted calendrical reform in 1582 to realign the church celebration of Easter with the vernal equinox (the first day of spring). To better align this solar calendar with the seasons, the new calendar would not have leap year in the century years that were not divisible by 400. Because the solar year is shortening, today a one–second adjustment (usually on December 31st at midnight) is made when appropriate to compensate for this effect.

Japanese calendar—It has the same structure as the Gregorian calendar in years, months, and weeks. But the years are enumerated in terms of the reigns of emperors as epochs. The last epoch (for Emperor Akihito) is Epoch Heisei, starting January 8, 1989.

Hindu calendars—The principal Indian calendars reckon their epochs from historical events, such as rulers' accessions or death dates, or a religious founder's dates. The Vikrama era (originally from Northern India and still used in western India) dates from February 23rd, 57 B.C.E. in the Gregorian calendar. The Saka era dates from March 3, 78 C.E. in the Gregorian calendar and is based on the solar year with 12 months of 365 days and 366 days in leap years. The first five months have 31 days; the last seven have 30 days. In leap years, the first six months have 31 days and the last six have 30. The Saka era is the national calendar of India (as of 1957). The Buddhist era starts with 543 B.C.E. (believed date of Buddha's death).

Three other secular calendars of note are the Julian Day calendar (a calendar astronomers use which counts days within a 7,980-year period and must be used with a table), the perpetual calendar (which gives the days of the week for the Julian and Gregorian calendars as well), and the World calendar, which is similar to the perpetual calendar, having 12 months of 30 or 31 days, a year-day at the end of each year, and a leap-year day before July 1 every four years.

There have been attempts to reform and simplify the calendar. One such example is the Thirteen-Month or International Fixed calendar which would have 13 months of four weeks each. The month *Sol* would come before July; there would be a year-day at the end of each year and a leap-year-day every four years just before July 1. A radical reform was done in France when the French republican calendar (1793-1806) replaced the Gregorian calendar after the French Revolution. It had 12 months of 30 days and 5 supplementary days at the end of the year (6 in a leap-year) and weeks were replaced with 10-day decades.

Sources: *The Guinness Book of Answers*, pp. 35-41; *New Encyclopedia Britannica*, vol. 15, pp. 432-449; *The Universal Almanac 1992*, pp. 1-40.

◁1482▷ *How is time denoted at sea?*

The day is divided into watches and bells. A watch equals four hours except for the time period between 4 p.m. and 8 p.m., which has two short watches. From midnight to 4 a.m. is the middle watch; from 4 a.m. to 8 a.m., the morning watch; 8 a.m. to noon, the forenoon watch; noon to 4 p.m., afternoon watch; 4 p.m. to 6 p.m., first dog watch; 6 p.m. to 8 p.m., second or last dog watch; 8 p.m. to midnight, first watch. Within each watch there are eight bells—one stroke for each half hour, so that each watch ends on eight bells except the dog watches which end at four bells. New Year's Day is marked with 16 bells.

Sources: *The Guinness Book of Answers*, pp. 43-44; Rovin, Jeff, *Laws of Order*, p. 203.

1483

When is Daylight Savings Time observed in the United States?

In 1967, all states and possessions of the United States were to begin observing Daylight Savings Time at 2 a.m. on the first Sunday in April of every year. The clock would advance one hour at that time until 2 a.m. of the last Sunday of October when the clock would be turned back one hour. In the intervening years, the length of this time period had been changed, but on July 8, 1986, the starting date of Daylight Saving Time was reinstated as 2 a.m. of the first Sunday in April with the same ending October date. A 1972 amendment allowed some areas to be exempt. Arizona, Hawaii, Puerto Rico, the Virgin Islands, American Samoa, and part of Indiana do not follow DST.

This time change was enacted to provide more light in the evening hours. The phrase "fall back, spring forward" indicates the direction that the clock setting is to undergo at these appointed dates. Other countries have adopted DST as well. For instance, in western Europe the period is from the last Sunday in March to the last Sunday in September, with the United Kingdom extending the range to the last Sunday in October. Many countries in the Southern Hemisphere generally maintain DST from October to March; countries near the equator maintain standard time.

Sources: Pasachoff, Jay M. *Contemporary Astronomy*, p. 87; *The World Almanac and Book of Facts 1992*, p. 297.

1484

How is time measured?

The passage of time can be measured in three ways. Rotational time is based on the unit of the mean solar day (the average length of time it takes the earth to complete one rotation around its axis.

Dynamic time, the second way to measure time, uses the motion of the moon and planets to determine time and avoids the problem of the earth's varying rotation. The first dynamic time-scale was Ephemeris Time, proposed in 1896 and modified in 1960.

Atomic time is a third way to measure time. This method, using an atomic clock, is based on the "extremely regular oscillations" that occur within atoms. In 1967, the atomic second, the length of time in which 9,192,631,770 vibrations are emitted by a hot cesium atom, was adopted as the basic unit of time. Atomic clocks are now used as international time standards.

But time has also been measured in other less scientific terms. Listed below are some other "timely" expressions.

Twilight—The first soft glow of sunlight; the sun is still below the horizon; also the last glow of sunlight.

Midnight—12 a.m., the point of time when one day becomes the next day and night becomes morning.

Daybreak—The first appearance of the sun.

Dawn—A gradual increase of sunlight.

Noon—12 p.m.; the point of time when morning becomes afternoon.

Dusk—The gradual dimming of sunlight.

Sunset—The last diffused glow of sunlight; the sun is below the horizon.

Evening—A term with wide meaning, evening is generally the period between sunset and bedtime.

Night—The period of darkness, lasting from sunset to midnight.

The day (24 hours, known as *nychthemeron*) becomes a week (seven days or a *sennight*) and two weeks becomes a fortnight. A year grows into a biennium (two years), a triennium (three years), a quadrennium or an olympiad (four years), and then into a lustrum (five years). Ten years is a decade or *decennium* and twenty years becomes a score or a *vicennium*, and five score is a century and ten centuries become a millennium.

Sources: *Oxford Illustrated Encyclopedia of Invention and Technology*, p. 356; Rovin, Jeff. *Laws of Order*, pp. 199-203; *The Universal Almanac 1992*, p. 529.

1485 / Where did the term "grandfather clock" come from?

The weight-and-pendulum clock was invented by Dutch scientist Christian Huygens (1629-1695) around the year 1656. In the United States, Pennsylvania German settlers considered such clocks, often called long-case clocks, to be a status symbol. In 1876 American songwriter Henry Clay Work (1832-1884) referred to long-case clocks in his song "My Grandfather's Clock" The nickname stuck.

Sources: Feldman, David. *Why Do Dogs Have Wet Noses*, p. 178.

WEIGHTS, MEASURES, AND MEASUREMENT *See also:* CHEMISTRY
AND PHYSICS—MEASUREMENT, METHODOLOGY, ETC.; EARTH—
OBSERVATION AND MEASUREMENT; ENERGY—MEASURES AND
MEASUREMENT; FOOD AND NUTRITION—KITCHEN EQUIPMENT,
MEASURES, ETC.; HEALTH AND MEDICINE—DRUGS, MEDICINES,
ETC.; MATHEMATICS, NUMBERS AND COMPUTERS—NUMBERS;
SPACE—OBSERVATION AND MEASUREMENT; WEIGHTS, MEASURES,
AND TIME—TIME

1486
How was the length of a meter originally determined?

It was originally intended that the meter should represent one ten-millionth of
the distance along the meridian running from the North Pole to the equator
through Dunkirk, France and Barcelona, Spain. French scientists determined this
distance, working nearly six years to complete the task in November 1798. They
decided to use a platinum-iridium bar as the physical replica of the meter.
Although the surveyors made an error of about two miles, the error was not discovered until much later. Rather than changing the length of the meter to con-
form to the actual distance, in 1889, the platinum-iridium bar was chosen as the
international prototype and was used until 1960. There are numerous copies of it
in other parts of the world, including the United States National Bureau of Standards.

Sources: Deming, Richard. *Metric Power*, pp. 49, 67-68; Ross, Frank Xavier. *The Metric System—Measures for All Mankind*, pp. 57-58.

1487
What is the SI system of measurement?

French scientists as far back as the seventeenth and eighteenth centuries ques-
tioned the hodge-podge of the many illogical and imprecise standards used for
measurement, and began a crusade to make a comprehensive logical, precise, and
universal measurement system called Système Internationale d'Unités, or SI for
short. It uses the metric system as its base. Since all the units are in multiples of
tens, calculations are simplified. Today all countries except the United States,
Burma, South Yemen, and Tonga use this system. However, elements within
American society use SI—scientists, exporting/importing industries, and federal
agencies (as of November 30, 1992).

    The SI or metric system has seven fundamental standards: the meter (for
length), the kilogram (for mass), the second (for time), the ampere (for electric
current), the kelvin (for temperature), the candela (for luminous intensity), and
the mole (for amount of substance). In addition, two supplementary units, the

radian (plane angle) and steradian (solid angle) and a large number of derived units compose the current system, which is still evolving. Some derived units, which use special names, are the hertz, newton, pascal, joule, watt, coulomb, volt, farad, ohm, siemens, weber, tesla, henry, lumen, lux, becquerel, gray, and sievert. Its unit of volume or capacity is the cubic decimeter, but many still use "liter" in its place. Very large or very small dimensions are expressed through a series of pre-fixes, which increase or decrease in multiples of tens. For example, a decimeter is 1/10 meter; a centimeter is 1/100 of a meter, and a millimeter is 1/1000 of a meter. A dekameter is 10 meters, a hectometer is 100 meters, and a kilometer is 1,000 meters. The use of these prefixes enable the system to express these units in an orderly way, and avoid inventing new names and new relationships.

Source: Cone, Robert J. *How the New Technology Works*, pp. 99-103; *The Oxford Dictionary for Scientific Writers and Editors*, pp. 382-383.

▽**1488**╱

How are avoirdupois measurements converted to troy measurements and how do the measures differ?

Troy weight is a system of mass units used primarily to measure gold and silver. An ounce is 480 grains or 31.1 grams. Avoirdupois weight is a system of units which is used for measurement of mass except for precious metals, precious stones, and drugs. It is based on the pound which is approximately 454 grams. In both systems, the weight of a grain is the same—65 milligrams. However, the two systems do not contain the same weights for other units, even though they use the same name for the unit.

Troy
　　　1 grain = 65 milligrams
　　　1 ounce = 480 grains = 31.1 grams
　　　1 pound = 12 ounce　= 5760 grains = 373 grams
Avoirdupois
　　　1 grain = 65 milligrams
　　　1 ounce = 437.5 grains = 28.3 grams
　　　1 pound = 16 ounces　= 7000 grains = 454 grams

To Convert From	To	Multiply By
pounds avdp	ounces troy	14.583
pounds avdp	pounds troy	1.215
pounds troy	ounces avdp	1.097
pounds troy	pounds avdp	0.069
ounces avdp	ounces troy	0.911
ounces troy	ounces avdp	1.097

Sources: *The McGraw-Hill Dictionary of Scientific and Technical Terms*, pp. 159-160, 1974; Zimmerman, *Conversion Factors and Tables*, pp. 199-200, 210-211.

1489
How is the length of a meter presently determined?

The meter is equal to 39.37 inches. It is presently defined as the distance traveled by light in a vacuum during 1/299,792,458 of a second. From 1960 to 1983, the length of a meter had been defined as 1,650,763.73 times the wavelength of the orange light emitted when a gas consisting of the pure krypton isotope of mass number 86 is excited in an electrical discharge.

Sources: *CRC Handbook of Chemistry and Physics*, pp. 1-14; *The McGraw-Hill Dictionary of Scientific and Technical Terms*, p. 1186.

1490
What is a bench mark?

A bench mark is a permanent, recognizable point that lies at a known elevation. It may be an existing object, such as the top of a fire hydrant, or it may be a brass plate placed on top of a concrete post. Surveyors and engineers use bench marks to find the elevation of objects by reading, through a level telescope, the distance a point lies above some already established bench mark.

Sources: *World Book Encyclopedia*, vol. 2, p. 247.

1491
What is a theodolite?

This optical surveying instrument used to measure angles and directions is mounted on an adjustable tripod and has a spirit level to show when it is horizontal. Similar to the more commonly-used transit, the theodolite gives more precise readings; angles can be read to fractions of a degree. It is comprised of a telescope that sights the main target, a horizontal plate to provide readings around the horizon, and a vertical plate and scale for vertical readings. The surveyor uses geometry of triangles to calculate the distance from the angles measured by the theodolite. Such triangulation is used in road- and tunnel-building, and other civil engineering work. One of the earliest forms of this surveying instrument was described by Englishman Leonard Digges (d.1571?) in his 1571 treatise, *A Geometrical Treatise Named Pantometria*.

Sources: *Oxford Illustrated Encyclopedia of Invention and Technology*, p. 353; *World Book Encyclopedia*, vol. 19, p. 252; Wynter, Harriet, and Anthony Turner. *Scientific Instruments*, p. 149.

1492
How is the distance to the horizon measured?

Distance to the horizon depends on the height of the observer's eyes. To determine that, take the distance (in feet) to eye level and multiply by three, then divide by two and take the square root of the answer. The result is the number of miles to the horizon. For example, if eye level is at a height of six feet above the water, the horizon is almost three miles away. If eye level were exactly at sea level, there would be no distance seen at all; the horizon would be directly in front of the viewer.

Sources: *Science Digest*, vol. 92 (February 1984), p. 34.

▽1493▽ How did the measure the yard originate?

In early times the natural way to measure lengths was to use various portions of the body (the foot, the thumb, the forearm, etc.). According to tradition, a yard measured on King Henry I (1068–1135) of England became the standard yard still used today. It was the distance from his nose to the middle fingertip of his extended arm.

Other measures derived from physical activity such as a pace, a league (distance that equalled an hour's walking), an acre (amount plowed in a day), a furlong (length of a plowed ditch), etc. But obviously these units were unreliable. The ell, based on the distance between the elbow and index fingertip, was used to measure out cloth. It ranged from 0.513 to 2.322 meters depending on the locality where it was used and even on the type of goods measured.

Below are listed some linear measurements that evolved from this old reckoning into U.S. customary measures:

U.S. customery linear measures

1 hand	=	4 inches
1 foot	=	12 inches
1 yard	=	3 feet
1 rod (pole or perch)	=	16.5 feet
1 fathom	=	6 feet
1 furlong	=	220 yards or 660 feet or 40 rods
1 (statute) mile	=	1,760 yards or 5,280 feet or 8 furlongs
1 league	=	5,280 yards or 15,840 feet or 3 miles
1 international nautical mile	=	6,076.1 feet

Conversion to metric

1 inch	=	2.54 centimeters
1 foot	=	0.304 meters
1 yard	=	0.9144 meters
1 fathom	=	1.83 meters
1 rod	=	5.029 meters
1 furlong	=	201.168 meters
1 league	=	4.828 kilometers
1 mile	=	1.609 kilometers
1 international nautical mile	=	1.852 kilometers

Sources: Asimov, Isaac. *Asimov on Numbers*, p. 115; Giscard d'Estaing, Valerie-Anne. *The World Almanac Book of Inventions*, p. 237.; *The World Almanac and Book of Facts 1991*, pp. 299-301.

1494 / What is meant by DIN standards?

They are German standards for products and procedures, covering a wide range of engineering and scientific fields. The German Institute for Standardization (*Deutsches Institut für Normung*) is the publisher. DIN also represents Germany in the International Organization for Standardization (ISO).

Sources: *English Translations of German Standards Catalog, 1991*, p. R112.

1495 / What are the equivalent measurements for area?

U.S. customary area measures

1 square foot	=	144 square inches
1 square yard	=	9 square feet or 1,296 square miles
1 square rod or pole or perch	=	30.25 square yards or 272.5 square feet
1 rood	=	40 square rods
1 acre	=	160 square rods or 4,840 square yards or 43,460 square feet
1 section	=	1 mile square or 640 acres
1 township	=	6 miles square or 36 square miles or 36 sections
1 square mile	=	640 acres or 4 roods, or 1 section

International area measures

1 square millimeter	=	1,000,000 square microns
1 square centimeter	=	100 square millimeters
1 square meter	=	10,000 square centimeters
1 are	=	100 square meters
1 hectare	=	100 ares or 10,000 square meters
1 square kilometer	=	100 hectares or 1,000,000 square meters

Sources: *The Guinness Book of Answers*, p. 192; Stein, Edwin I. *Arithmetic for College Students*, p. 247; *The Universal Almanac 1992*, p. 528.

1496 / How much does the Biblical shekel weigh in modern units?

A shekel is equal to 0.497 ounces (14.1 grams). Below are listed some ancient measurements with some modern equivalents given.

Biblical
Volume

omer	=	4.188 quarts (modern) or 0.45 peck (modern) or 3.964 liters (modern)
9.4 omers	=	1 bath
10 omers	=	1 ephah

Weight

shekel	=	0.497 ounces (modern) or 14.1 grams (modern)

Length

cubit	=	21.8 inches (modern)

Egyptian

Weight

60 grams	=	1 shekel
60 shekels	=	1 great mina
60 great minas	=	1 talent

Greek

Length

cubit	=	18.3 inches (modern)
stadion	=	607.2 or 622 feet (modern)

Weight

obol or obolos	=	715.38 milligrams (modern) or. 0.04 ounces (modern)
drachma	=	4.2923 grams (modern) or 6 obols
mina	=	0.9463 pounds (modern) or 96 drachmas
talent	=	60 mina

Roman

Length

cubit	=	17.5 inches (modern)
stadium	=	202 yards (modern) or 415.5 cubits

Weight

denarius	=	0.17 ounces (modern)

Volume

amphora	=	6.84 gallons (modern)

Sources: Rovin, Jeff. *Laws of Order*, pp. 248–249; *The World Almanac and Book of Facts 1992*, p. 304.

▽ **1497** ╱

What are the equivalents for dry and liquid measures?

U.S. customary dry measures

1 pint	=	33.6 cubic inches
1 quart	=	2 pints or 67.2006 cubic inches
1 peck	=	8 quarts or 16 pints or 537.605 cubic inches
1 bushel	=	4 pecks or 2,150.42 cubic inches or 32 quarts
1 barrel	=	105 quarts or 7,056 cubic inches
1 pint, dry	=	0.551 liter
1 quart, dry	=	1.101 liters
1 bushel	=	35.239 liters

U.S. customary liquid measures

1 tablespoon	=	4 fluid drams or 0.5 fluid ounce
1 cup	=	0.5 pint or 8 fluid ounces
1 gill	=	4 fluid ounces

U.S. customary liquid measures

4 gills	=	1 pint or 28.875 cubic inches
1 pint	=	2 cups or 16 fluid ounces
2 pints	=	1 quart or 57.75 cubic ounces
1 quart	=	2 pints or 4 cups or 32 fluid ounces
4 quarts	=	1 gallon or 231 cubic inches or 8 pints or 32 gills or 0.833 British quart
1 gallon	=	16 cups or 231 cubic inches or 128 fluid ounces
1 bushel	=	8 gallons or 32 quarts

Conversion to metric

1 fluid ounce	=	29.57 milliliters or 0.029 liter
1 gill	=	0.118 liter
1 cup	=	0.236 liter
1 pint	=	0.473 liter
1 U.S. quart	=	0.833 British quart or 0.946 liter
1 U.S. gallon	=	0.833 British gallon or 3.785 liters

Sources: Stein, Edwin I. *Arithmetic for College Students*, p. 277; *The World Almanac and Book of Facts 1992*, pp 300-301.

1498 / How much does water weigh?

U.S. customary measures

1 gallon	=	4 quarts
1 gallon	=	231 cubic inches
1 gallon	=	8.34 pounds
1 gallon	=	0.134 cubic foot
1 cubic foot	=	7.48 gallons
1 cubic inch	=	.0360 pound
12 cubic inches	=	.433 pound

British measures

1 liter	=	1 kilogram
1 cubic meter	=	1 tonne (metric ton)
1 imperial gallon	=	10.022 pounds
1 imperial gallon salt water	=	10.3 pounds

Sources: *The Guinness Book of Answers*, p. 192; *The World Almanac and Book of Facts 1992*, p. 304; Zimmerman, O.T. *Conversion Factors and Tables*, pp. 70, 109-110.

1499 / What is the difference between a short ton, a long ton, and a metric ton?

A short ton or U.S. ton or net ton (sometimes just "ton") equals 2,000 pounds; a long ton or an avoirdupois ton is 2,240 pounds; and a metric ton or tonne is 2,204.62 pounds. All weights for comparison are given below:

U.S. customary measure

1 ounce	=	16 drams or 437.5 grains
1 pound	=	16 ounces or 7,000 grains or 256 drams
1 (short) hundredweight	=	100 pounds
1 long hundredweight	=	112 pounds
1 (short) ton	=	20 hundredweights or 2,000 pounds
1 long ton	=	20 long hundredweights or 2,240 pounds

Conversion to metric

1 grain	=	65 milligrams
1 dram	=	1.77 grams
1 ounce	=	28.3 grams
1 pound	=	453.5 grams
1 metric ton (tonne)	=	2,204.6 pounds

Sources: Rovin, Jeff. *Laws of Order*, p. 251; *The Universal Almanac 1992*, p. 529.

▽ 1500 ▽

How are U.S. customary measures converted to metric measures and vice versa?

Listed below is the conversion process for common units of measure:

To convert from	To	Multiply by
acres	meters, square	4,046.856
centimeters	inches	0.394
centimeters	feet	0.328
centimeters, cubic	inches, cubic	0.06
centimeters, square	inches, square	0.155
feet	meters	0.305
feet, square	meters, square	0.093
gallons, U.S.	liters	3.785
grams	ounces (avoirdupois)	0.035
hectares	kilometers, square	0.01
hectares	miles, square	0.004
inches	centimeters	2.54
inches	millimeters	25.4
inches, cubic	centimeters, cubic	16.387
inches, cubic	liters	0.016387
inches, cubic	meters, cubic	0.0000164
inches, square	centimeters, square	6.4516
inches, square	meters, square	0.0006452
kilograms	ounces, troy	32.15075
kilograms	pounds (avoirdupois)	2.205
kilograms	tons, metric	0.001
kilometers	feet	3,280.8
kilometers	miles	0.621
kilometers, square	hectares	100
knots	miles per hour	1.151
liters	fluid ounces	33.815

To convert from	To	Multiply by
liters	gallons	0.264
liters	pints	2.113
liters	quarts	1.057
meters	feet	3.281
meters	yards	1.094
meters, cubic	yards, cubic	1.308
meters, cubic	feet, cubic	35.315
meters, square	feet, square	10.764
meters, square	yards, square	1.196
miles, nautical	kilometers	1.852
miles, square	hectares	258.999
miles, square	kilometers, square	2.59
miles (statute)	meters	1,609.344
miles (statute)	kilometers	1.609344
ounces (avoirdupois)	grams	28.35
ounces (avoirdupois)	kilograms	0.0283495
ounces, fluid	liters	0.03
pints, liquid	liters	0.473
pounds (avoirdupois)	grams	453.592
pounds (avoirdupois)	kilograms	0.454
quarts	liters	0.946
tons (short/U.S.)	tonne	0.907
ton, long	tonne	1.016
tonne	ton (short/U.S.)	1.102
tonne	ton, long	0.984
yards	meters	0.914
yards, square	meters, square	0.836
yards, cubic	meters, cubic	0.765

Sources: Grzimek, Bernhard. *Grzimek's Animal Life Encyclopedia*, vol. 11, Appendix (unpaged); *New York Public Library Desk Reference*, pp. 22, 800-801.

▼ BIBLIOGRAPHY

Books

Abel, Bob. *The Book of Beer*. Chicago, IL: Regnery, 1976.

Abell, George O. *Realm of the Universe*. New York: Holt, Rinehart and Winston, Inc., 1976.

Academic American Encyclopedia. Danbury, CT: Grolier, 1991. 21 vols.

Ackerknecht, Erwin H. *A Short Story of Medicine*. New York: Ronald Press, 1955.

Acronyms Initialisms & Abbreviations Dictionary 1993. 17th ed. Detroit, MI: Gale Research, 1992.

Adams, Ramon F. *The Language of the Railroader*. Norman, OK: University of Oklahoma Press, 1977.

Adler, Bill. *Whole Earth Quiz Book: How Well Do You Know the Planet*. New York: Quill, 1991.

Agent Orange and Its Associated Dioxin: Assessment of a Controversy. New York: Elsevier, 1988.

Agress, Clarence M. *Energetics*. New York: Grosset & Dunlap, 1978.

Ahrens, C. Donald. *Meteorology Today*. 2nd ed. St. Paul, MN: West Publishing, 1985.

Ainsworth, G.C. *Ainsworth's & Bisby's Dictionary of the Fungi*. 5th ed. London: Commonwealth Agricultural Bureau, 1961.

Ali, Sheikh R. *The Peace and Nuclear War Dictionary*. Santa Barbara, CA.: ABC-CLIO, 1989.

Allaby, Michael. *Dictionary of the Environment*, 3rd ed. New York: New York University Press, 1989.

The Almanac of Science and Technology. San Diego, CA: Harcourt Brace Jovanovich, 1990.

American Academy of Dermatology. *Poison Ivy* (Pamphlet). Washington, DC: American Academy of Dermatology, 1990.

The American Geological Institute. *Dictionary of Geological Terms*. Rev. ed. Garden City, NY: Anchor Press, 1976.

American Kennel Club. *The Complete Dog Book*. New York: Howell Book House, Inc., 1985.

The American Medical Association Encyclopedia of Medicine. New York: Random House, 1989.

American Medical Association Family Medical Guide. Rev. ed. New York: Random House, 1987.

Amory, Cleveland. *Animail*. New York: Windmill Books, Inc., and E.P. Dutton & Co., Inc., 1976.

Anderson, B.W. *Gem Testing*. 10th ed. Stoneham, MA: Butterworth, 1990.

Anderson, Norman D., and Walter R. Brown. *Ferris Wheels*. New York: Pantheon Books, 1983.

Angelo, Joseph A. *The Extraterrestrial Encyclopedia*. Rev. & updated ed. New York: Facts On File, 1991.

Annual Energy Review 1990. Washington, DC: U.S. Department of Energy, Energy Information Administration, 1990.

Arem, Joel E. *Color Encyclopedia of Gemstones*. 2nd ed. New York: Van Nostrand Reinhold, 1987.

Argenzio, Victor. *Diamonds Eternal*. New York: David McKay, 1974.

Armstrong, Joseph E. *Science in Biology*. Prospect Heights, IL: Waveland Press, 1990.

The ARRL Handbook for Radio Amateurs. 68th ed. Newington, CT: American Radio Relay League, 1991.

Ashby, W. Ross. *Introduction to Cybernetics*. New York: John Wiley & Sons, 1958.

Ashworth, William. *The Encyclopedia of Environmental Studies*. New York Facts on File, 1991.

Asimov, Isaac. *Asimov On Chemistry*. New York: Anchor Books, 1975.

Asimov, Isaac. *Asimov On Numbers*. New York: Doubleday, 1977.

Asimov, Isaac. *Asimov's Biographical Encyclopedia of Science and Technology*. 2nd rev. ed. Garden City, NY: Doubleday & Company, Inc., 1982.

Asimov, Isaac. *Asimov's Chronology of Science and Discovery*. New York: Harper and Row, 1989.

Asimov, Isaac. *Asimov's New Guide to Science*. Rev. ed. New York: Basic Book, Inc., 1984.

Asimov, Isaac. *The Human Body*. New rev. ed. New York: A Mentor Book, 1992.

Asimov, Isaac. *Isaac Asimov's Book of Science and Nature Questions*. New York: Grove Weidenfeld, 1988.

Asimov, Isaac. *Isaac Asimov's Guide to Earth and Space*. New York: Random House, 1991.

Asimov, Isaac. *Understanding Physics*. New York: Dorset Press, 1988. 3 vols. in one.

Aslett, Don. *How Do I Clean the Moosehead?* New York: New American Library, 1989.

Aslett, Don. *Pet Clean-up Made Easy*. Cincinnati, OH: Writer's Digest Books, 1988.

Aslett, Don. *Stainbusters Bible*. New York: Penguin, 1990.

Astronauts and Cosmonauts Biographical and Statistical Data Report to the Committee on Science, Space, and Technology U.S. House of Representatives, 1989. Washington, DC: Committee on Science, Space and Technology, 1989.

Astronomical Almanac 1992. Washington, DC: U.S. Government Printing Office, 1991.

Automotive Encyclopedia. Rev. ed. South Holland, IL: Goodheart-Willcox Company, 1989.

Babbitt, Harold E. *Sewage and Sewage Treatment*. 8th ed. New York: Wiley, 1958.

Bagel, Marilyn, and Tom Bagel. *The Bagel's Bagel Book*. Herndon, VA: Acropolis Books, 1985.

Bagenal, Philip, and Jonathan Meades. *The Illustrated Atlas of the World's Great Buildings*. London: Salamander Books, Ltd., 1980.

Bailey, Adrian. *Mrs. Bridges' Upstairs, Downstairs Cookery Book*. New York: Simon and Schuster, 1974.

Bailey, Janet. *Keeping Food Fresh*. Rev. ed. New York: Harper & Row, Publishers, 1989.

Bair, Frank E. *The Weather Almanac*. 6th ed. Detroit, MI: Gale Research Inc., 1992.

Baker, David. *The History of Manned Space Flight*. New York: Crown Publishers, Inc., 1982.

Baker, Susan P., et al. *The Injury Fact Book*. 2nd ed. New York: Oxford University Press, 1992.

Balfour, Henry H. *Herpes Diseases and Your Health*. Minneapolis, MN: University of Minnesota Press, 1984.

Bali, Mrinal. *Space Exploration: A Reference Handbook*. Santa Barbara, CA: ABC-CLIO, 1990.

Ballast, David Kent. *Architects' Handbook of Formulas, Tables and Mathematical Calculations*. New York: Prentice-Hall, 1988.

Bamberger, Richard. *Physics Through Experiment*. New York: Sterling, 1969.

Barnard, Christiaan. *The Body Machine*. New York: Crown, 1981.

Barndt, Herb. *How to Remove Spots and Stains*. New York: Perigee Books, 1987.

Barnhart, Robert K. *The American Heritage Dictionary of Science*. Boston, MA: Houghton, Mifflin, 1986.

Barry, John. *Technobabble*. Cambridge, MA: MIT Press, 1991.

Bartholomew, Mel. *Square Foot Gardening*. Emmaus, PA: Rodale Press, 1981.

Bartlett, John. *Familiar Quotations*. Boston, MA: Little, Brown, 1980.

Bates, Robert, and Julia A. Jackson. *Glossary of*

Geology. 3rd ed. Alexandria VA: American Geological Institute, 1987.

Battan, Louis J. *Weather in Your Life.* New York: W.H. Freeman, 1983.

Baylin, Frank, and Brent Gale. *Home Satellite TV Installation and Troubleshooting Manual.* 1986 ed. Boulder, CO: Baylin/Gale Productions, 1985.

Bear, Marina, and John Bear. *How to Repair Food.* Berkeley, CA: Ten Speed Press, 1987.

Beard, James. *James Beard's American Cookery.* Boston, MA: Little, Brown, 1972.

Beatty, J. Kelly, Brian O'Leary, and Andrew Chaikin. *The New Solar System.* Cambridge, MA: Sky Publishing Corp., 1981.

Beck, James H. *Rail Talk.* Gretna, NE: James Publications, 1978.

Beeching, W.A. *Century of the Typewriter.* New York: St. Martin's Press, 1974.

Berg, Richard E. *The Physics of Sound.* Englewood Cliffs, NJ: Prentice-Hall, 1982.

Bergamini, David. *Life Science Library: Mathematics.* New York: Time Incorporated, 1963.

Berliner, Barbara. *The Book of Answers.* Englewood Cliffs, NJ: Prentice-Hall, 1990.

Bernard, Josef. *The Cellular Connection.* Mendocino, CA: Quantum Publishing, 1987.

Berry, James. *Exploring Crystals.* New York: Crowell-Collier Press, 1969.

Best, Charles H. *Best and Taylor's Physiological Basis of Medical Practice.* Baltimore, MD: Williams & Wilkins, 1985.

Beyer, Don E. *The Manhattan Project.* New York: Watts, 1991.

Beyer, William H. *CRC Standard Mathematical Tables.* 28th ed. Boca Raton, FL: CRC Press, 1987.

Bingham, Anne. *Buying Jewelry.* New York: McGraw-Hill, 1989.

The Biographical Dictionary of Scientists: Astronomers. New York: Peter Bedrick Books, 1984.

Biographical Dictionary of Scientists: Biologists. New York: Peter Bedrick Books, 1984.

The Biographical Dictionary of Scientists: Chemists. New York: Peter Bedrick Books, 1983.

The Biographical Dictionary of Scientists: Mathematicians. New York: Peter Bedrick Books, 1986.

The Biographical Dictionary of Scientists: Physicists. New York: Peter Bedrick Books, 1984.

Biographical Encyclopedia of Scientists. New York: Facts On File, 1981. 2 vols.

The Birds Around Us. San Ramon, CA: Ortho, 1986.

Bishop, Peter. *Fifth Generation Computers.* Hempstead, Eng.: Ellis Horwood, Ltd., 1986.

Blackwell, Will H. *Poisonous and Medicinal Plants.* Englewood Cliffs, NJ: Prentice-Hall, 1990.

Blair, Ian. *Taming the Atom.* Bristol, Eng.: Adam Hilger, 1983.

Block, E.B. *Fingerprinting: Magic Weapon Against Crime.* New York: McKay, 1969.

Blocksma, Mary. *Reading the Numbers.* New York: Penguin, 1989.

The BOCA National Building Code 1990. 11th ed. Country Club Hills, IL: Building Officials and Code Administrators International, 1989.

Bodinski, Lois H. *The Nurse's Guide to Diet Therapy.* New York: Wiley, 1987.

Bogner, Bruce F. *Vehicular Traffic Radar Handbook for Attorneys.* Mount Holly, NJ: The Brehn Corporation, 1979.

Bohren, Craig F. *Clouds in a Glass of Beer.* New York: Wiley, 1987.

Bolton, W.C. *Physics Experiments and Projects.* Elmsford, NY: Pergamon, 1968.

Bomberger, Audrey S., and Betty A. Dannenfelser. *Radiation and Health.* Gaithersburg, MD: Aspen Pubs., Inc., 1984.

Bonnet, Robert L. *Botany: 49 Science Fair Projects.* Blue Ridge Summit, PA: Tab Books, 1989.

Booth, Nicholas. *The Concise Illustrated Book of Planets and Stars.* New York: Gallery Books, 1990.

Bowler, Peter J. *Evolution, the History of an Idea.* Berkeley, CA: University of California Press, 1984.

Boyer, Carl B. *A History of Mathematics.* 2nd ed. New York: Wiley, 1989.

Boyer, Rick. *Places Rated Almanac.* Englewood Cliffs, NJ: Prentice-Hall, 1989.

Bradbury, Savile. *The Evolution of the Microscope.* Elmsford, NY: Pergamon Press, 1967.

Bradford, Gershom. *A Glossary of Sea Terms.* New York: Dodd, Mead, 1942.

Brady, George S. *Materials Handbook.* 12th ed. New York: McGraw-Hill, 1986.

Bragonier, Reginald, and David Fisher. *What's What.* Maplewood, NJ: Hammond Incorporated, 1981.

Brandreth, Gyles. *Your Vital Statistics.* New York: Citadel, 1986.

Branson, Gary D. *The Complete Guide to Recycling at Home.* Whitehall, VA: Betterway Publications, 1991.

Braun, Wernher von, and Frederick I. Ordway III. *Space Travel: A History.* New York: Harper & Row, 1985.

Brennan, Richard P. *Dictionary of Scientific Literacy.* New York: John Wiley & Sons, Inc., 1992.

Brennan, Richard P. *Levitating Trains and Kamikaze Genes.* New York: Harper Perennial, 1990.

Britannica Book of the Year 1991. Chicago, IL: Encyclopaedia Britannica, 1991.

Britten, Frederick J. *Britten's Old Clocks and Watches and Their Makers.* 8th ed. New York: Dutton, 1973.

Broadcasting Yearbook 1991. New York: Broadcasting Publications, Inc., 1991.

Brody, Jane E. *Jane Brody's Good Food Book.* New York: W.W. Norton, 1985.

Brown, Robert J. *200 Illustrated Science Experiments for Children.* Blue Ridge Summit, PA: TAB Books, 1987.

Brown, Theodore L. *Chemistry.* 4th ed. Englewood Cliffs, NJ: Prentice-Hall, 1988.

Brown, Victor J. *Engineering Terminology.* Chicago, IL: Gillette, 1938.

Buchman, Dian Dincin. *Dian Dincin Buchman's Herbal Medicine.* New York: Gramercy, 1980.

Buehner, Jim. *The Complete Handbook of Model Railroading.* Blue Ridge Summit, PA: TAB Books, 1975.

The Bumper Book of Things a Boy Can Make. Blue Ridge Summit, PA: TAB Books, 1978.

Burnam, Tom. *The Dictionary of Misinformation.* New York: Perennial Library, 1986.

Burt, McKinley. *Black Inventors of America.* Portland, OR: National Book Company, 1989.

Burt, William Henry. *A Field Guide to the Mammals.* 3rd ed. Boston, MA: Houghton Mifflin, 1976.

Burton, Benjamin T., and Willis R. Foster. *Human Nutrition.* 4th ed. New York: McGraw-Hill, 1988.

Burton, J.L. *Essentials of Dermatology.* 3rd ed. New York: Churchill Livingstone, 1990.

Burton, Maurice. *Encyclopaedia of Animals in Colour.* London: Octopus Books, 1972.

Burton, Maurice, and Robert Burton. *Encyclopedia of Insects and Arachnids.* New York: Crescent Books, 1975.

Bush, Grace A. *Foundations of Mathematics.* New York: McGraw-Hill, 1968.

Busha, William, and Stephen Morris. *The Book of Heat.* Brattleboro, VT: Stephen Greene Press, 1982.

Bynum, W.F., et al. *Dictionary of the History of Science.* Princeton, NJ: Princeton University Press, 1985.

Byrne, Austin T. *A Treatise on Highway Construction.* New York: John Wiley & Sons, 1896.

Cairis, Nicholas. *Cruise Ships of the World.* Boston, MA: Pegasus, 1988.

Calasibetta, Charlotte M. *Fairchild's Dictionary of Fashion.* New York: Fairchild Publications, 1988.

Calder, Nigel. *The Comet Is Coming.* New York: Penguin Books, 1982.

Callahan Philip S. *Bird Behavior.* New York: Four Winds Press, 1975.

The Cambridge Dictionary of Science and Technology. New York: Cambridge University Press, 1988.

The Cambridge Encyclopedia of Ornithology. New York: Cambridge University Press, 1991.

The Cambridge Encyclopedia of Space. New York: Cambridge University Press, 1990.

Campbell, Neil A. *Biology.* 2nd ed. Redwood City, CA: Benjamin/Cummings Publishing Co., Inc., 1990.

Can Elephants Swim? New York: Time-Life Books, 1969.

Carlson, Neil R. *Foundations of Physiological Psychology.* Needham Heights, MA: Allyn Bacon, Inc., 1988.

Carroll, Anstice, and Embree De Persiis Vona. *The Health Food Dictionary with Recipes.* Englewood Cliffs, NJ: Prentice-Hall, Inc., 1973.

Carter, E.F. *Dictionary of Inventions and Discoveries.* New York: Crane Russak, 1974.

Cartnell, Robert. *The Incredible Scream Machine: A History of the Roller Coaster.* Fairview Park, OH: Amusement Park Books, 1987.

Cartwright, Rudolph. *Doctors Are Easy to Understand If You Speak Their Language.* New York: Vantage Press, 1987.

Carwell, Hattie. *Blacks in Science: Astrophysicist to Zoologist.* Oakland, CA: Exposition Press, 1977.

Cary, James. *Tanks and Armor in Modern Warfare.* New York: Franklin Watts, 1966.

Cassel, Don. *Understanding Computers.* Englewood Cliffs, NJ: Prentice-Hall, 1990.

Cassens, B. *Preventive Medicine and Public Health.* New York: Harwal Publishing Co., 1987.

Catalog of American Car ID Numbers 1970-79. Sidney, OH: Amos Press, Inc., 1991.

Cazeau, Charles J. *Science Trivia.* New York: Berkley Books, 1986.

Cecil Textbook of Medicine. 18th ed. Philadelphia, PA: Saunders, 1988. 2 vols.

Chalmers, Irena. *Great American Food Almanac.* New York: Harper & Row, 1986.

Chambers Science and Technology Dictionary.

Cambridge, Eng. & Edinburgh, Scotland: W&R Chambers, Ltd., and Cambridge University Press, 1988.

Chapman, Eugenia. *Clean Your House and Everything in It.* Rev. and updated ed. New York: Perigee Books, 1991.

Charney, Leonard. *Build A Yurt.* New York: Macmillan, 1974.

Chiasson, Robert B. *Laboratory Anatomy of the Cat.* 6th ed. Dubuque, IA: Wm. C. Brown, 1977.

Childs, W.H.J. *Physical Constants.* 9th ed. New York: Chapman and Hall, 1972.

Chinn, George M. *The Machine Gun.* Washington, DC: U.S. Department of the Navy, 1951.

Choukas-Bradley, Melanie, and Polly Alexander. *City of Trees.* Rev. ed. Baltimore, MD: Johns Hopkins University Press, 1987.

Churchill, James E. *The Backyard Building Book.* Harrisburg, PA: Stackpole Books, 1976.

Churchman, Lee W. *Survey of Electronics.* San Francisco, CA: Rinehart Press, 1971.

Cipolla, Carlo M., and Derek Birdsall. *The Technology of Man.* New York: Holt, Rinehart and Winston, 1980.

Cities of the United States: The West. Detroit, MI: Gale Research, 1989.

Clarke, Donald. *The Encyclopedia of How It Works From Abacus to Zoom Lens.* New York: A&W Publishers, 1977.

Clearing the Air: Perspectives on Environmental Tobacco Smoke. Lexington, MA: Lexington Books, 1988.

Cleveland-Peck, Patricia. *Making Cheeses, Butters, Cream and Yogurt at Home.* Wellingborough, Eng.: Thorsons Publishing Limited, 1980.

Coated Abrasives: Modern Tool of Industry. New York: McGraw-Hill, 1958.

Cody, John. *Visualizing Muscles.* Lawrence, KS: University Press of Kansas, 1990.

Cohen, Daniel. *The Last Hundred Years: Household Technology.* New York: M. Evans & Co., 1982.

Cohen, Daniel. *Marbling on Fabric.* Loveland,

CO: Interweave Press, 1990.

Cohen, I. Bernard. *Revolution in Science.* Cambridge, MA: Belknap Press, 1985.

Collin, P.H. *Dictionary of Ecology and the Environment.* Teddington, Eng.: Peter Collin Publishing, 1988.

Columbia University College of Physicians and Surgeons. *Complete Home Medical Guide.* New York: Crown, 1985.

The Complete Encyclopedia of the Animal World. London: Octopus Books, 1980.

Comprehensive Textbook of Psychiatry. 5th ed. Baltimore, MD: Williams & Wilkins, 1989. 2 vols.

Compton, William David. *Where No Man Has Gone Before.* Washington, DC: Superintendent of Documents, 1980.

Conant, Roger. *A Field Guide to Reptiles and Amphibians: Eastern and Central North America.* 3rd updated ed. Boston, MA: Houghton Mifflin Company, 1991.

Concise Chemical and Technical Dictionary. 4th enlarged ed. New York: Chemical Publ., Co., Inc., 1986.

Cone, Robert J. *How the New Technology Works.* Phoenix, AZ: Oryx Press, 1991.

Congram, Marjorie. *Horsehair: A Textile Resource.* Martinsville, NJ: Dockwra Press, 1987.

Considine, Douglas M. *Energy Technology Handbook.* New York: McGraw-Hill, 1977.

Corliss, William R. *Tornadoes, Dark Days, Anomalous Precipitation, with Related Weather Phenomena.* Glen Arm, MD: Sourcebook Project, 1983.

Cornell, Felix. *American Merchant Seaman's Manual.* Centreville, MD: Cornell, Maritime, 1964.

Cornell, James. *The Great International Disaster Book.* 3rd ed. New York: Charles Scribner's, 1982.

Cortada, James W. *Historical Dictionary of Data Processing: Technology.* Westport, CT: Greenwood Press, 1987.

Costello, David F. *The World of the Porcupine.* Philadelphia, PA: Lippincott, 1966.

Council on Environmental Quality.

Environmental Quality. Washington, DC: Council on Environmental Quality, 1992.

A Country Christmas. New York: Time-Life Books, 1989.

Country Living's Country Look and How to Get It. New York: Hearst Books, 1991.

Cox, James M. *Mark Twain: The Fate of Humor.* Princeton, NJ: Princeton University Press, 1966.

Crager, Meg. *Christmas Trees.* New York: Grove Weidenfeld, 1986.

CRC Handbook of Physics and Chemistry. 72nd ed. St. Louis, MO: CRC Press, 1991.

Crockett, James U. *Crockett's Indoor Garden.* Boston, MA: Little, Brown, 1978.

Crocodiles and Alligators. New York: Facts On File, 1989.

Crowson, Phillip. *Minerals Handbook, 1990-1991.* New York: Stockton Press, 1990.

Cruickshank, Allan D. *1001 Questions Answered About Birds.* New York: Dodd, Mead, 1958.

Cuff, David J. *The United States Energy Atlas.* 2nd ed. New York: Macmillan, 1986.

Cunningham, William P., and Barbara Woodhouse Saigo. *Environmental Science: A Global Concern.* Dubuque, IA: Wm. C. Brown Publishers, 1990.

Current Medical Diagnosis and Treatment 1991. 30th ed. Norwalk, CT: Appleton & Lange, 1991.

Curtis, Anthony R. *Space Almanac.* Woodsboro, MD: ARCsoft, 1990.

Curtis, Helena, and N. Sue Barnes. *Invitation to Biology.* 4th ed. New York: Worth Publ., Inc., 1985.

D'Adamo, James. *One Man's Food—Is Someone Else's Poison.* New York: R. Marek, 1980.

Daintith, John. *The Facts on File Dictionary of Physics.* New York: Facts On File, 1988.

Daintith, John, and Amanda Isaacs. *Medical Quotes.* New York: Facts On Files, 1989.

Dalley, Robert L. *Are You Burning Money?* New York: Reston, 1982.

Danilov, Victor J. *America's Science Museums.* Westport, CT: Greenwood, 1990.

Darwin, Charles. *On the Origin of Species.* Cambridge, MA: Harvard University Press, 1964.

Davie, Michael. *Titanic: The Death and Life of a Legend.* New York: Alfred Knopf, 1987.

Davis, G.J. *Automotive Reference.* Boise, ID: Whitehorse, 1987.

Day, David. *The Doomsday Book of Animals.* New York: Viking, 1983.

Day, John, and C. Eng. *The Bosch Book of the Motor Car.* New York: St. Martin's Press, 1976.

Dean, John A. *Lange's Handbook of Chemistry.* 13th ed. New York: McGraw Hill, 1985.

De Bono, Edward. *Eureka!* New York: Holt, Rinehart and Winston, 1974.

Deming, Richard. *Metric Power.* Nashville, TN: Nelson, 1974.

Desmond, Kevin. *A Timetable of Inventions and Discoveries.* New York: M. Evans & Co., 1982.

De Voney, Chris. *MS-DOS User's Guide.* 2nd ed. Indianapolis, IN: Que Corporation, 1987.

Diagram Group. *Comparisons.* New York: St. Martin's Press, 1980.

Diamond, Freda. *The Story of Glass.* San Diego, CA: Harcourt, Brace and Co., 1953.

Dickerson, Richard Earl. *Chemical Principles.* 3rd ed. Menlo Park, CA: The Benjamin/Cummings Publ., Co., Inc., 1979.

Dickinson, Terence, and Alan Dyer. *The Backyard Astronomer's Guide.* Ontario, Can.: Camden House, Publishing, 1991.

Dictionary of American Medical Biography. Westport, CT: Greenwood, 1984. 2 vols.

Dictionary of Occupational Titles. 4th ed., rev. Washington, DC: U.S. Department of Labor, 1991. 2 vols.

Dictionary of Scientific Biography. New York: Charles Scribner's Sons, 1973.

Dictionary of Visual Science. 4th ed. Radnor, PA: Chilton Trade Book, 1989.

Diseases and Disorders Handbook. Springhouse PA: Springhouse Corporation, 1990.

Dolensek, Emil P. *A Practical Guide to Impractical Pets.* New York: Viking Press, 1976.

Donovan, Richard T. *World Guide to Covered Bridges.* Rev. ed. Worcester, MA: The National Society for the Preservation of Covered Bridges, 1980.

Doo, Jack. *The Ultimate Owner's Manual.* Alhambra, CA: Edmund, 1991.

Dorland's Illustrated Medical Dictionary. 27th ed. Philadelphia, PA: Saunders, 1988.

Dougans, Inge. *Reflexology.* New York: Element, 1991.

Douglas, R.W., and Susan Frank. *A History of Glassmaking.* London: G.T. Foulis & Co., 1972.

Dowling, Harry F. *Fighting Infection.* Cambridge, MA: Harvard University Press, 1977.

Downing, Douglas, and Michael Covington. *Dictionary of Computer Terms.* 2nd ed. Hauppauge, NY: Barron, 1989.

Downs, Robert B. *Landmarks in Science: Hippocrates to Carson.* Littleton, CO: Libraries Unlimited, Inc., 1982.

Downs, Robert B. *Scientific Enigmas.* Littleton, CO: Libraries Unlimited, Inc., 1987.

Dowson, Gordon. *Powder Metallurgy: The Process and Its Products.* London: Adam Hilger, 1990.

Drake, George R. *Weatherizing Your Home.* New York: Reston Publishing Co., 1978.

Dreisbach, Robert H. *Handbook of Poisoning.* 12th ed. Norwalk, CT: Appleton & Lange, 1987.

Drimmer, Frederick. *The Elephant Man.* New York: Putnam, 1985.

Dublin, Louis I. *Factbook on Man from Birth to Death.* 2nd ed. New York: Macmillan, 1965.

Duensing, Edward. *Talking to Fireflies, Shrinking the Moon.* New York: Penguin, 1990.

Dunbar, Ian. *Dog Behavior: Why Dogs Do What They Do.* Neptune, NJ: T.F.H. Publications, 1979.

Dunne, Levon J. *Nutrition Almanac.* 3rd ed. New York: McGraw-Hill, 1990.

Duplaix, Nicole, and Noel Simon. *World Guide to Mammals*. New York: Crown Publishers, 1976.

DuVall, Nell. *Domestic Technology*. Boston, MA: G.K. Hall, 1988.

Duxbury, Alyn C., and Alison Duxbury. *An Introduction to the World's Oceans*. Reading, MA: Addison-Wesley Publishing Company, Inc., 1984.

Dyson, James L. *The World of Ice*. New York: Knopf, 1962.

Eating to Lower Your High Blood Cholesterol (Pamphlet). Washington, DC: U.S. Department of Health and Human Services, 1989.

Edelson, Edward. *Sports Medicine*. New York: Chelsea House, 1988.

Eden, Maxwell. *Kiteworks*. New York: Sterling, 1989.

Edmond's Car Savvy. Alhambra, CA: Edmund Publications, 1991.

Edmunds, Robert A. *The Prentice-Hall Encyclopedia of Information Technology*. Englewood Cliffs, NJ: Prentice-Hall, Inc., 1989.

Egg Science and Technology. 3rd ed. Westport, CT: AVI, 1986.

Eldridge, Wayne B. *The Best Pet Name Book Ever*. Hauppauge, NY: Barron, 1990.

Elkort, Martin. *The Secret Life of Food*. Los Angeles, CA: Jeremy P. Tarcher, 1991.

Ellis, John. *The Social History of the Machine Gun*. London: Croom Helm, Ltd., 1975.

Ellis, Keith. *Thomas Telford*. Duluth, MN: Priory Press, 1974.

Emiliani, Cesare. *The Scientific Companion*. New York: Wiley Science Editions, 1988.

Encyclopaedic Dictionary of Physical Geography. London: Blackwell, 1985.

Encyclopedia Americana. Danbury, CT: Grolier, 1990. 30 vols.

Encyclopedia of Associations. 27th ed. Detroit, MI: Gale Research, 1993.

Encyclopedia of Aviation. New York: Scribners, 1977.

Encyclopedia of Chemical Technology. 3rd ed.

New York: Wiley, 1978. 24 vols.

The Encyclopedia of Crafts. New York: Scribners, 1980. 3 vols.

Encyclopedia of Human Evolution and Prehistory. New York: Garland Publ., 1988.

The Encyclopedia of Insects. New York: Facts On File, 1986.

Encyclopedia of Physical Science and Technology. San Diego, CA: Academic Press, 1987. 15 vols.

Encyclopedia of Textiles. Englewood Cliffs, NJ: Prentice-Hall, 1980.

Encyclopedia of Twentieth Century Warfare. New York: Orion, 1989.

Encyclopedic Dictionary of Science. New York: Facts On File, 1988.

Engineering and the Advancement of Human Welfare. Washington, DC: National Academy Press, 1989.

English Translations of German Standards Catalog, 1991. Braintree, MA: Beuth Verlag Gmbh., 1991.

Erickson, Joan Beth. *Flower Garden Plans*. San Ramon, CA: Ortho Books, 1991.

Evanoff, Vlad. *Fishing with Natural Baits*. Englewood Cliffs, NJ: Prentice-Hall, 1975.

Everett, Thomas H. *Living Trees of the World*. New York: Doubleday, 1968.

Everett, Thomas H. *The New York Botanical Garden Illustrated Encyclopedia of Horticulture*. Hamden, CT: Garland, 1981.

The Facts On File Dictionary of Chemistry. Rev. and enl. ed. New York: Facts On File, Inc., 1988.

Facts On File Dictionary of Physics. New York: Facts On File, Inc., 1988.

Farb, Peter. *The Insects*. Alexandria, VA: Time-Life Books, 1977.

Farber, Edward. *Nobel Prize Winners in Chemistry 1901-1961*. Rev. ed. New York: Ablard-Schuman, 1963.

Fejer, Eva, and Cecilia Fitzsimons. *An Instant Guide to Rocks and Minerals*. Stamford, CT: Longmeadow Press, 1988.

Feldman, David. *Do Penguins Have Knees?* New York: Harper Perennial, 1991.

Feldman, David. *Why Do Clocks Run Clockwise? and Other Imponderables*. New York: Perennial Library, 1988.

Feldman, David. *Why Do Dogs Have Wet Noses? and Other Imponderables of Everyday Life*. New York: Harper Perennial, 1991.

Felton, Bruce, and Mark Fowler. *Felton & Fowler's Best, Worst, and Most Unusual*. New York: Crowell, 1975.

Feltwell, John. *The Natural History of Butterflies*. New York: Facts On File, 1986.

Fenton, Carroll L. *The Fossil Book*. New York: Doubleday, 1989.

Field, Frank. *Doctor Frank Field's Weather Book*. New York: Putnam, 1981.

Field, Gary C. *Color and its Reproduction*. Pittsburgh, PA: Graphic Arts Technical Foundation, 1988.

Field, Leslie. *The Queen's Jewels*. New York: Harry N. Abrams, 1987.

50 Simple Things You Can Do to Save the Earth. Berkeley, CA: Earth Works Press, 1989.

Findling, John E. *Historical Dictionary of World's Fairs and Expositions, 1851-1988*. Westport, CT: Greenwood, 1990.

First Aid Book. Washington, DC: U.S. Mine Safety & Health Administration, 1991.

Fisher, Arthur. *The Healthy Heart*. New York: Time-Life Books, 1981.

Fisher, David J. *Rules of Thumb for Engineers and Scientists*. Houston, TX: Gulf, 1991.

Flaste, Richard. *The New York Times Book of Science Literacy*. New York: Times Books, Random, 1991.

Flatow, Ira. *Rainbows, Curve Balls, And Other Wonders of the Natural World Explained*. New York: Harper & Row Publishers, 1988.

Fletcher, Edward. *Pebble Collecting and Polishing*. New York: Sterling, 1973.

Flynn, John C. *Cocaine*. New York: Carol Publishing Group, 1991.

Fogle, Bruce. *Know Your Cat*. New York: Dorling Kindersley, Inc., 1991.

Foodworks. Redding, MA: Addison-Wesley, 1987.

Forrester, Frank H. *1001 Questions Answered About the Weather*. New York: Grosset & Dunlap, 1957.

Fowler, Virginie. *Folk Arts Around the World*. Englewood Cliffs, NJ: Prentice-Hall, 1981.

Fox, Michael W. *The Animal Doctor's Answer Book*. New York: Newmarket, 1984.

Fox, Michael W. *Understanding Your Cat*. New York: Bantam, 1974.

Frankel, Edward. *Poison Ivy, Poison Oak, Poison Sumac, and Their Relatives*. Pacific Grove, CA: Boxwood Press, 1991.

Freedman, Alan. *The Computer Glossary*. 5th ed. New York: AMACOM, 1991.

Freiberger, Paul, and Michale Swaine. *Fire in the Valley: The Making of the Personal Computer*. Berkeley, CA: Osborne/McGraw-Hill, 1984.

Frew, Timothy. *Salmon*. New York: Mallard Press, 1991.

Freydberg, Nicholas. *The Food Additives Book*. New York: Bantam, 1982.

Fritts, Harold C. *Tree Rings and Climate*. New York: Academic Press, 1976.

Frost, Harwood. *The Art of Roadmaking*. New York: Engineering News Publishing Company, 1910.

Fruits and Vegetables: 1001 Gardening Questions Answered. Pownal, VT: Storey Communications, Inc., 1990.

Funk & Wagnalls New Standard Dictionary of the English Language. New York: Funk & Wagnalls, 1959.

Funk, Charles E. *Horse Feathers and Other Curious Words*. New York: Harper, 1958.

Galperin, Anne. *Gynecological Disorders*. New York: Chelsea House, 1991.

Gardiner, Mary S. *The Biology of Invertebrates*. New York: McGraw-Hill, 1972.

Gardner, Pat. *Dough Creations*. Radnor, PA: Chilton, 1979.

Garrison, Webb. *How It Started*. Nashville, TN: Abingdon Press, 1972.

Gascoigne, Robert M. *A Chronology of the History of Science 1450-1900*. New York: Garland Publishing, Inc., 1987.

Gatland, Kenneth. *The Illustrated Encyclopedia*

of Space Technology. New York: Orion Books, 1989.

Gay, Kathlyn. *The Greenhouse Effect.* New York: Franklin Watts, 1986.

Gay, Kathlyn. *Ozone.* New York: Franklin Watts, 1989.

Gebhardt, Susan E., and Ruth H. Matthews. *Nutritive Value of Foods.* Washington, DC: U.S. Department of Agriculture. Human Nutrition Information Service, 1991.

Gene Hughes' Police Call Radio Guide. Los Angeles, CA: Hollins Radio Data, 1990.

George, L. David, and Jennifer J. George. *Marine Life.* New York: Wiley Interscience, 1979.

Gerken, Louis C. *Airships: History and Technology.* Chula Vista, CA: American Scientific Corporation, 1990.

Gibson, Carol. *The Facts On File Dictionary of Mathematics.* Rev. ed. New York: Facts On File, Inc., 1988.

Giedion, Siegfried. *Mechanization Takes Command.* New York: Oxford University Press, 1948.

Gies, Joseph and Frances Gies. *The Ingenious Yankees.* New York: Thomas Y. Cromwell Co., 1976.

Giscard d'Estaing, Valerie-Anne. *The World Almanac Book of Inventions.* New York: World Almanac Publications, 1985.

Giscard d'Estaing, Valerie-Anne. *The Second World Almanac Book of Inventions.* New York: World Almanac, 1986.

G.K. Hall Encyclopedia of Modern Technology. Boston, MA: Equinox, 1987.

Godish, Thad. *Indoor Air Pollution Control.* Chelsea, MI: Lewis Publishers, 1989.

Gong, Victor. *AIDS: Facts and Issues.* New Brunswich, NJ: Rutgers University Press, 1980.

Gore, Louise L. *Meet the Pug.* Wilsonville, OR: Doral, 1990.

Goulty, George A. *A Dictionary of Landscape.* Brookfield, VT: Gower Publishing Co., 1991.

Graedon, Joe, and Dr. Teresa Graedon. *Graedons' Best Medicine.* New York: Bantam Book, 1991.

Graf, Rudolf F., and George J. Whalen. *The Reston Encyclopedia of Biomedical Engineering Terms.* New York: Reston Publishing, 1977.

Graham, John. *Facts on File Dictionary of Telecommunications.* Rev. ed. New York: Facts On File, 1991.

Gray, Henry. *Anatomy of the Human Body.* 28th ed. Malvern, PA: Lea & Febiger, 1966.

Gray, Peter. *The Encyclopedia of the Biological Sciences.* New York: Van Nostrand Reinhold, 1970.

Great Britain Meteorological Office. *Meteorological Glossary.* New York: Chemical Publishing, 1972.

Great Disasters. New York: Reader's Digest Association, 1989.

Great Engineers and Pioneers in Technology. New York: St. Martin's Press, 1981.

The Great Scientist. Danbury, CT: Grolier, 1989.

Green, James Harry. *The Dow Jones-Irwin Handbook of Telecommunications.* Homewood, IL: Dow Jones-Irwin, 1986.

Greenfield, Ellen J. *House Dangerous.* New York: Vintage Books, 1987.

Griffin, Heather. *Introduction to Batik.* Cincinnati, OH: North Light Books, 1990.

Grimm, William C. *The Trees of Pennsylvania.* Harrisburg, PA: Stackpole, 1950.

Groves, Don. *The Ocean Book.* New York: Wiley, 1989.

Grzimek's Animal Life Encyclopedia. New York: Van Nostrand Reinhold Company, 1973. 13 vols.

Grzimek's Encyclopedia of Mammals. 2nd ed. New York: McGraw-Hill, 1990. 5 vols.

Guedes, Pedro. *The Macmillan Encyclopedia of Architecture and Technological Change.* London: Macmillan Press, 1979.

Guiley, Rosemary E. *Moonscapes.* Englewood Cliffs, NJ: Prentice-Hall, 1991.

The Guinness Book of Answers. 8th ed. Enfield, Eng.: Guinness Publishing, 1991.

Guinness Book of Records 1992. New York: Bantam Book, 1992.

Gurney, Gene. *Space Shuttle Log*. Blue Ridge Summit, PA: TAB, 1988.

Guyton, Arthur C. *Basic Human Physiology*. Philadelphia, PA: Saunders, 1977.

Guyton, Arthur C. *Textbook of Medical Physiology*. 8th ed. Philadelphia, PA: Saunders, 1991.

Haber, Louis. *Odyssey Black Pioneers of Science and Invention*. San Diego, CA: An Odyssey Book, Harcourt Brace Jovanovich, Publishers, 1970.

Hackh, Ingo W.D. *Grant & Hackh's Chemical Dictionary*. 5th ed. New York: McGraw-Hill, 1987.

Haggard, Howard W. *The Lame, the Halt, and the Blind*. New York: Harper, 1932.

Hale, Mason E., Jr. *The Biology of Lichens*. 2nd ed. London: Edward Arnold, Ltd., 1974.

Halliday, William R. *Depths of the Earth*. New York: Harper, 1976.

Hamilton, William R. *The Henry Holt Guide to Minerals, Rocks and Fossils*. New York: Henry Holt, 1989.

Hampel, Clifford A. *Glossary of Chemical Terms*. 2nd ed. New York: Van Nostrand Reinhold Co., Inc., 1982.

Hand, A.J. *Home Energy How-To*. New York: Harper & Row, 1977.

Handbook of Glass Manufacture. 3rd ed. New York: Ashlee Publishing, 1984.

Hapgood, Charles H. *Maps of the Ancient Sea Kings*. Radnor, PA: Chilton, 1966.

Harding, Anthony. *Car Facts and Feats*. London: Guinness Superlatives, 1975.

Harding, Anthony. *The Guinness Book of Car Facts and Feats*. London: Guinness Superlatives, 1980.

Harding, Anthony. *The Guinness Book of the Car*. London: Guinness Superlatives, 1987.

Harrington, Geri. *The Wood-Burning Stove Book*. New York: Macmillan, 1977.

Harris, Ben. *Make Use of Garden Plants*. New York: Barre, 1978.

Harris, Harry. *Good Old-Fashioned Yankee Ingenuity*. Chelsea, MI: Scarborough House, 1990.

Harte, John. *Toxics A to Z*. Berkeley, CA: University of California Press, 1991.

Hartmann, William K., and Ron Miller. *Cycles of Fire*. New York: Workman Publishing, 1987.

Hawkes, Nigel. *Structures*. New York: Macmillan, 1990.

Hawley's Condensed Chemical Dictionary. 11th ed. New York: Van Nostrand Reinhold, 1987.

Haygreen, John G. *Forest Products and Wood Science*. 2nd ed. Ames, IA: Iowa State University Press, 1989.

Hazen, Robert M., and James Trefil. *Science Matters: Achieving Scientific Literacy*. New York: Anchor Books, 1991.

Headstrom, Richard. *Spiders of the United States*. Stamford, CT: A.S. Barnes, 1973.

Hechtlinger, A. *Modern Science Dictionary*. 2nd ed. Palisade, NJ: Franklin Publ. Co., Inc., 1975.

Hegstad, Lorrie N. *Essential Drug Dosage Calculations*. Bowie, MD: R.J. Brady, 1983.

Heintzelman, Donald S. *A Guide to Eastern Hawk Watching*. University Park, PA: Pennsylvania State University Press, 1976.

Helms, Harry L. *Shortwave Listening Handbook*. Englewood Cliffs, NJ: Prentice-Hall, 1987.

Heloise Hints for a Healthy Planet. New York: Perigee, 1990.

Henrickson, Charles H. *Chemistry for the Health Professions*. New York: Van Nostrand, 1980.

Herbert, Don. *Mr. Wizard's Experiments for Young Scientists*. New York: Doubleday, 1959.

Herbst, Sharon Tyler. *Food Lover's Companion*. Hauppauge, NY: Barron, 1990.

Hershey, Robert L. *How to Think With Numbers*. Los Altos, CA: William Kaufmann, 1982.

Hickman, Cleveland P., et al. *Integrated Principles of Zoology*. 7th ed. St. Louis, MO: Times Mirror/Mosby College Publishing, 1984.

Hillman, Harold. *Kitchen Science*. Rev. ed. Boston, MA: Houghton Mifflin Company, 1989.

Hiscox, Gardner D. *Henley's Twentieth Century Book of Formulas, Processes, and Trade Secrets.* New York: NY Books, Inc., 1963.

Hofstadter, Douglas R. *Gödel, Escher, Bach: An Eternal Golden Braid.* New York: Vintage Books, 1979.

Hogg, Ian V. *The Illustrated Encyclopedia of Artillery.* London: Stanley Paul & Co., 1987.

Holmes, Ann M. *Nutrition & Vitamins.* New York: Facts On File, 1983.

Hooper, Meredith. *Everyday Inventions.* London: Angus & Robertson, 1972.

Hooper, Meredith. *More Everyday Inventions.* London: Angus and Robertson, 1974.

Hopkins, Jeanne. *Glossary of Astronomy and Astrophysics.* Chicago, IL: University of Chicago Press, 1976.

Hopkins, Nigel J., John W. Mayne, and John R. Hudson. *The Numbers You Need.* Detroit, MI: Gale Research, 1992.

Horton, Edward. *The Illustrated History of the Submarine.* London: Sidgwick & Jackson, 1974.

How in the World. Pleasantville, NY: The Reader's Digest Association, 1990.

How Things Work in Your Home. New York: Holt, Rinehart, and Winston, 1985.

How Things Work: Structures. New York: Time-Life Books, 1991.

How to Do Just About Anything. Pleasantville, NY: Reader's Digest Association, 1986.

Howard, A.V. *Chamber's Dictionary of Scientists.* London: W. & R. Chambers, Ltd., 1955.

Howes, F.N. *A Dictionary of Useful and Everyday Plants and Their Common Names.* Cambridge, Eng.: Cambridge University Press, 1974.

Hudgeons, Marc. *The Official Investors Guide; Buying, Selling Gold, Silver, Diamonds.* Orlando, FL: House of Collectibles, 1981.

The Human Brain. Englewood Cliffs, NJ: Prentice-Hall, 1977.

Hunnicutt, R.P. *Sherman: A History of the American Medium Tank.* San Rafeal, CA: Taurus Enterprises, 1978.

Hunt, V. Daniel. *The Gasohol Handbook.* New York: Industrial Press, 1981.

Hunter, Linda Mason. *The Healthy Home.* New York: Pocket Books, 1989.

Huxley, Thomas H. *The Crayfish.* Cambridge, MA: MIT Press, 1974.

Hyne, Norman J. *Dictionary of Petroleum Exploration, Drilling & Production.* Tulsa, OK: PennWell, 1991.

Illingworth, Valerie. *The Facts On File Dictionary of Astronomy.* New York: Facts On File, 1979.

Illustrated Dictionary of Botany. Chestnut Ridge, NY: Triune Books, 1979.

The Illustrated Encyclopedia of Wildlife. Lakeville, CT: Grey Castle Press, 1991. 15 vols.

The Illustrated Science and Invention Encyclopedia. International ed. Westport, CT: H.S. Stuttman Publishers, 1983, 23 vols.

Inglis, Andrew F. *Behind the Tube: A History of Broadcasting Technology and Business.* London: Focal Press, 1990.

International Encyclopedia of Astronomy. New York: Orion, 1987.

International Encyclopedia of Robotics. New York: Wiley, 1988.

Inventive Genius. New York: Time-Life Books, 1991.

The Inventive Yankee. Camden, ME: Yankee Books, 1989.

Involuntary Smoking. Washington, DC: U.S. Public Health Service, 1979.

Iver, David F. *Dictionary of Astronomy, Space, and Atmospheric Phenomena.* New York: Van Nostrand Reinhold Co., 1979.

Jackson, Donald C. *Great American Bridges and Dams.* Washington, DC: The Preservation Press, 1988.

Jacobs, Harold R. *Mathematics: A Human Endeavor.* San Francisco, CA: W.H. Freeman, 1970.

James, Glenn. *Mathematics Dictionary.* 4th ed. New York: Van Nostrand Reinhold, 1976.

Jane's All the World's Aircraft 1947. London: Sampson Low, Marston, 1947.

Jane's Encyclopedia of Aviation. New York: Portland House, 1989.

Jaramillo, Alex. *Cracker Jack Prizes.* New York: Abbeville Press, 1989.

Jayne, Kate Lindley, and Claudette Suzanne Mautor. *Living with Potpourri.* New York: Peter Pauper, 1988.

Jelinek, Jan. *The Pictorial Encyclopedia of the Evolution of Man.* London: Hamlyn, 1975.

Jerram, Mike. *The World's Classic Aircraft.* London: Frederick Muller, Ltd., 1981.

Jerrard, H.G. *A Dictionary of Scientific Units.* 4th ed. New York: Chapman and Hall, 1980.

Jespersen, James. *RAMS, ROMS and Robots.* New York: Atheneum, 1984.

Johnson, Leland G. *Biology.* Dubuque, IA: Wm. C. Brown, 1983.

Johnson, Lorraine. *How to Restore & Repair Practically Everything.* Rev. ed. London: The Penguin Group, 1989.

Johnstone, William D. *For Good Measure.* Fort Worth, TX: Holt, Rinehart and Winston, 1975.

Jones, Julia, and Barbara Deer. *Royal Pleasures and Pastimes.* Devon, Eng.: David and Charles, 1990.

Jones, Mablen. *Taking Care of Clothes.* New York: St. Martin's, 1982.

Kane, Joseph N. *Famous First Facts.* 4th ed. New York: Wilson, 1981.

Kaplan, Eugene H. *Field Guide to Coral Reefs.* Boston, MA: Houghton Mifflin Co., 1982.

Karlen, Arno. *Napoleon's Glands.* Boston, MA: Little, Brown, 1984.

Karush, William. *Webster's New World Dictionary of Mathematics.* Englewood Cliffs, NJ: Webster's New World, 1989.

Kaufman, Wallace. *The Beaches Are Moving.* New York: Anchor Press, 1979.

Keeler, Harriet L. *Our Early Wild Flowers.* New York: Charles Scribner's Sons, 1916.

Kelly, Niall. *Presidential Pets.* New York: Abbeville Press Publishers, 1992.

Kemp, Peter. *Encyclopedia of Ships and Sailing.* Dobbs Ferry, NY: Stanford Maritime, 1989.

Kemp, Peter. *The History of Ships.* New York: Galahad, 1979.

Kendig, Frank, and Richard Hutton. *Life-Spans.* New York: Holt, Rinehart and Winston, 1979.

Kennedy, Kenneth A.R. *Neanderthal Man.* Minneapolis, MN: Burgess, 1975.

Kerrod, Robin. *The Concise Dictionary of Science.* New York: Arco Publ. Inc., 1985.

Kidder, Frank. *Architects' and Builders' Handbook.* New York: Wiley, 1931.

Klein, Hilary D. *Tiny Game Hunting.* New York: Bantam, 1991.

Klinowska, Margaret. *Dolphins, Porpoises and Whales of the World.* Gland, Switz.: IUCN, 1991.

Kogelman, Stanley, and Barbara R. Heller. *The Only Math Book You'll Ever Need.* New York: Facts On File Publications, 1986.

Krantz, Les. *The Best and Worst of Everything.* New York: Prentice Hall General Reference, 1991.

Kress, Stephen W. *The Audubon Society Guide to Attracting Birds.* New York: Charles Scribner's Sons, 1985.

Kroschivitz, Jacqueline I. *Chemistry.* New York: McGraw-Hill, 1990.

Labatut, Jean, and J.L. Wheaton. *Highways in Our National Life.* Princeton, NJ: Princeton University Press, 1950.

Lambert, David. *Field Guide to Early Man.* New York: Facts On File, 1987.

Lane, Ferdinand C. *Earth's Grandest Rivers.* New York: Doubleday, 1949.

Lawrence, Eleanor. *Henderson's Dictionary of Biological Terms.* 10th ed. New York: John Wiley and Sons, 1989.

Lean, Geoffrey, Don Hinrichsen, and Adam Markham. *WWF Atlas of the Environment.* Boston, MA: Willard Grant Press, 1991.

LeBlanc, Raymond. *Gold-Leaf Techniques.* Cincinnati, OH: ST Publications, 1986.

Lee, Sally. *Predicting Violent Storms.* New York: Franklin Watts, 1989.

Leet, L. Don, and Sheldon Judson. *Physical Geology.* 4th ed. Englewood Cliffs, NJ: Prentice-Hall, 1971.

Leggett, Jeremy. *Global Warming*. Oxford, Eng.: Oxford University Press, 1990.

Lennes, N.J. *New Practical Mathematics*. New York: Macmillan, 1939.

Leopold, Luna B. *Water*. New York: Time, 1966.

Leug, Albert Y. *Encyclopedia of Common Natural Ingredients*. New York: Wiley, 1980.

Levenstein, Mary K. *Caring for Your Cherished Possessions*. New York: Crown, 1989.

Levy, Richard C. *Inventing and Patenting Sourcebook*. Detroit, MI: Gale Research Inc., 1990.

Levy Richard C. *The Inventor's Desktop Companion*. Detroit, MI: Visible Ink Press, 1991.

Lewis, Richard J. *Food Additives Handbook*. New York: Van Nostrand Reinhold, 1989.

Libien, Lois. *Paint It Yourself*. New York: Morrow, 1978.

Lincoln, John W. *Driving Without Gas*. Pownal, VT: Garden Way, 1980.

Lincoln, R.J. *A Dictionary of Ecology, Evolution, and Systematics*. New York: Cambridge University Press, 1982.

The Little Scented Library. New York: Simon and Schuster, 1991. 4 vols.

Living Invertebrates. Palo Alto, CA: Blackwell Scientific Publications, 1987.

Lloyd, Elizabeth J. *Enchanted Circles*. New York: Simon and Schuster, 1991.

Logan, Carolynn. *Logan's Medical and Scientific Abbreviations*. Philadelphia, PA: Lippincott, 1987.

Long, James W. *The Essential Guide to Prescription Drugs*. New York: Harper & Row, 1987.

Longley, Dennis, and Michael Shain. *Van Nostrand Reinhold Dictionary of Information Technology*. 3rd. ed. New York: Van Nostrand Reinhold, 1989.

Loomer, Alice. *Famous Flaws*. New York: Macmillan, 1976.

Luetzelschwab, John. *Household Energy Use & Conservation*. Chicago, IL: Nelson Hall, 1980.

Mabberley, D.J. *The Plant Book*. New York: Cambridge University Press, 1987.

Macauley, David. *The Way Things Work*. Boston, MA: Houghton Mifflin Co., 1988.

MacEachern, Diane. *Save Our Planet*. New York: Dell, 1990.

Maclean, Norman. *Dictionary of Genetics and Cell Biology*. New York: New York University Press, 1987.

The Macmillan Dictionary of Quotations. New York: Macmillan, 1989.

Macmillan Illustrated Animal Encyclopedia. New York: MacMillan, 1984.

Maerz, A. *A Dictionary of Color*. 2nd ed. New York: McGraw-Hill, 1950.

Magill, Frank N. *Great Events from History II: Science and Technology Series*. Englewood Cliffs, NJ: Salem Press, 1991. 5 vols.

Magill, Frank N. *Magill's Survey of Science: Earth Science Series*. Englewood Cliffs, NJ: Salem Press, 1990.

Magill, Frank N. *Magill's Survey of Science: Life Science Series*. Englewood Cliffs, NJ: Salem Press, 1991. 6 vols.

Magill, Frank N. *Magill's Survey of Science. Physical Science Series*. Englewood Cliffs, NJ: Salem Press, 1992. 6 vols.

Magill, Frank N. *Magill's Survey of Science: Space Exploration Series*. Englewood Cliffs, NJ: Salem Press, 1989.

Magill, Frank N. *Nobel Prize Winners, Physiology or Medicine*. Englewood Cliffs, NJ: Salem Press, 1991.

Maginley, C.J. *Models of America's Past*. San Diego, CA: Harcourt, Brace & World, 1969.

Magner, Lois N. *A History of the Life Sciences*. New York: Marcel Dekker, 1979.

The Making, Shaping and Treating of Steel. 10th ed. Pittsburgh, PA: United States Steel, 1985.

Managing the Future of America's Forests (Pamphlet). Washington, DC: American Forest Council, 1989.

Manchester, Richard B. *Mammoth Book of Fascinating Information*. New York: A&W Visual Library, 1980.

Manchester, William R. *The Arms of Krupp*,

1587-1968. Boston, MA: Little Brown, 1968.

Manko, Howard H. *Solders & Soldering*. 2nd ed. New York: McGraw-Hill, 1979.

Mansfield, George Rogers. *Origin of the Brown Mountain Light in North Carolina*. Washington, DC: U.S. Geological Survey, 1971.

The Map Catalog. New York: Vintage, 1986.

Margo. *Growing New Hair*. Brookline, MA: Autumn Press, 1980.

Margulis, Lynn. *Five Kingdoms*. New York: W.H. Freeman and Company, 1988.

Mariani, John F. *The Dictionary of American Food and Drink*. New York: Ticknor & Fields, 1983.

Mark's Standard Handbook for Mechanical Engineers. 9th ed. New York: McGraw-Hill, 1987.

Marshall Cavendish Illustrated Encyclopedia of Family Health. London: Marshall Cavendish, 1984. 24 vols.

Marshall Cavendish International Wildlife Encyclopedia. London: Marshall Cavendish, 1989. 24 vols.

Marshall, John. *The Guinness Railway Book*. London: Guinness Books, 1989.

Marshall, John. *Rail: The Records*. London: Guinness Books, 1985.

Mathematics Encyclopedia. New York: Doubleday, 1977.

Matthews, L. Harrison. *The Life of Mammals*. New York: Universe Books, 1971.

Matthews, Rupert O. *The Atlas of Natural Wonders*. New York: Facts On File, 1988.

May, John. *The Greenpeace Book of Antarctica: A New View of the Seventh Continent*. 1st ed. New York: Doubleday, 1989.

May, John. *The Greenpeace Book of the Nuclear Age*. New York: Pantheon, 1989.

Mayhew, Susan, and Anne Penny. *The Concise Oxford Dictionary of Geography*. New York: Oxford University Press, 1992.

Mayo Clinic Family Health Book. New York: Morrow, 1990.

McAleer, Neil. *The Body Almanac*. New York: Doubleday, 1985.

McAleer, Neil. *The OMNI Space Almanac*. New York: World Almanac, 1987.

McCarthy, Eugene J., et al. *The Second Opinion Handbook*. New York: Nick Lyons Books, 1987.

McClane, Albert J. *McClane's Fish Buyer's Guide*. New York: Henry Holt, 1990.

McCully, Helen. *Nobody Ever Tells You These Things About Food and Drink*. Fort Worth, TX: Holt, Rinehart and Winston, 1967.

McElroy, Thomas P. *The New Handbook of Attracting Birds*. New York: Knopf, 1960.

McEwan, W.A., and A.H. Lewis. *Encyclopedia of Nautical Knowledge*. Centreville, MD: Cornell Maritime Press, 1953.

McGee, Harold. *On Food and Cooking*. New York: Collier Books, 1988.

McGraw-Hill Dictionary of Physics and Mathematics. New York: McGraw-Hill Book Co., 1978.

The McGraw-Hill Dictionary of Scientific and Technical Terms. 4th ed. New York: McGraw-Hill, 1989.

McGraw-Hill Encyclopedia of Science and Technology. 7th ed. New York: McGraw Hill, Inc., 1992. 20 vols.

McGraw-Hill Yearbook of Science and Technology 1991. New York: McGraw-Hill, 1990.

McGraw-Hill Yearbook of Science and Technology 1992. New York: McGraw-Hill, 1991.

McKinnell, Robert G. *Cloning of Frogs, Mice, and Other Animals*. Rev. ed. Minneapolis, MN: University of Minnesota Press, 1985.

McNair, James K. *Adventures in Italian Cooking*. San Ramon, CA: Ortho, 1980.

McNeil, Ian. *An Encyclopedia of the History of Technology*. London: Routledge, 1990.

McNulty, Faith. *Wholly Cats*. New York: Bobbs-Merrill, 1962.

Means Illustrated Construction Dictionary. Kingston, MA: R.S. Means, 1985.

Medawar, P.B., and J.S. Medawar. *Aristotle to Zoos*. Cambridge MA: Harvard University Press, 1983.

Medicine. New York: Time-Life Books, 1991.

Melaragno, Michele. *An Introduction to Shell Structures*. New York: Van Nostrand Reinhold, 1991.

Melonakos, K. *Saunders Pocket Reference for Nurses*. Philadelphia, PA: Saunders, 1990.

Memmler, Ruth L. *Structure and Function of the Human Body*. Philadelphia, PA: Lippincott, 1987.

Mendelsohn, Oscar A. *The Dictionary of Drink and Drinking*. London: Macmillan, 1965.

Menzel, Donald H., and Jay M. Pasachoff. *A Field Guide to Stars and Planets*. 2nd ed. Boston, MA: Houghton Mifflin Co., 1990.

Merck Manual of Diagnosis and Therapy. 15th ed. West Point, PA: Merck, Shark & Dohme, 1987.

The Merck Veterinary Manual. 6th ed. Rahway, NJ: Merck, 1986.

Mery, Fernand. *The Life, History, and Magic of the Dog*. New York: Grosset & Dunlap, 1970.

Michard, Jean-Guy. *The Reign of the Dinosaurs*. New York: Harry N. Abrams, Incorporated, 1992.

Mierhof, Annette. *The Dried Flower Book*. New York: Dutton, 1981.

Milestones of Aviation. Washington, DC: Smithsonian Institution, 1989.

Miller, E. Willard. *Environmental Hazards: Toxic Waste and Hazardous Material*. Santa Barbara, CA: ABC-CLIO, 1991.

Miller, Ron, and William K. Hartmann. *The Grand Tour: A Traveler's Guide to the Solar System*. New York: Workman Publishing, 1981.

Millichap, J. Gordon. *Dyslexia as the Neurologist and Educator Read It*. Springfield, IL: Thomas, 1986.

Minerals Yearbook, 1989. Washington, DC: U.S. Bureau of Mines, 1991. 3 vols.

Mitchell, Harris. *1200 Household Hints You Wanted to Know*. Toronto: Bestsellers, Inc., 1982.

Mondey, David. *The Guinness Book of Aircraft*. London: Guinness, 1988.

Moorcroft, William H. *Sleep, Dreaming, and Sleep Disorders*. Lanham, MD: University Press of America, 1989.

Moore, John E. *Submarine Warfare*. Bethesda, MD: Alder & Alder, 1987.

Moore, Patrick, et al. *The Atlas of the Solar System*. New York: Crescent Books with the Royal Astronomical Society, 1990.

Moore, Patrick. *International Encyclopedia of Astronomy*. New York: Orion Books, 1987.

Moore-Landecker, Elizabeth. *Fundamentals of the Fungi*. 3rd ed. Englewood Cliffs, NJ: Prentice Hall, 1990.

Morgan, George W. *Geodesic and Geodetic Domes and Space Structures*. Madison, WI: Sci-Tech Publications, 1985.

Morgans, W.M. *Outlines of Paint Technology*. New York: Halsted Press, 1990.

Morlan, Michael. *Kitty Hawk to NASA*. Shawnee Mission, KS: Bon a Tirer, 1991.

Mosby's Medical, Nursing and Allied Health Dictionary. 3rd ed. St. Louis, MO: Mosby, 1990.

The Motor Gasoline Industry. Washington, DC: U.S. Department of Energy. Energy Information Administration, 1991.

Motz, Lloyd, and Jefferson Hane Weaver. *Conquering Mathematics*. New York: Plenum Press, 1991.

Mount, Ellis, and Barbara A. List. *Milestones in Science and Technology; the Ready Reference Guide to Discoveries, Inventions, and Facts*. Phoenix, AZ: Oryx Press, 1987.

Mr. Boston Official Bartender's Guide. 63rd ed., rev. and updated. New York: Warner Books, 1988.

Mull, Kayla. *Pot-Bellied Pet Pigs*. Orange, CA: All Publishing, 1990.

Müller, Ulrike. *The New Cat Handbook*. Woodbury, NY: Barron, 1984.

Murmurs of Earth: The Voyager Interstellar Record. New York: Random House, 1978.

MVMA Motor Vehicle Facts and Figures '91. Detroit, MI: Motor Vehicle Manufacturers Association of the United States, 1991.

Naar, Jon. *Design for a Livable Planet*. New York: Harper & Row, Publishers, 1990.

Nassau, Kurt. *Gems Made by Man*. Radnor, PA: Chilton, 1980.

The National Inventors Hall of Fame. Washington, DC: U.S. Patent and Trademark Office, 1990.

National Safety Council Accident Facts, 1989. Washington, DC: National Safety Council, 1989.

Nature on the Rampage. Washington, DC: National Geographic Society, 1986.

Nayler, Joseph L. *Aviation: Its Technical Development.* Chester Springs, PA: Dufour Editions, 1965.

Nebel, Bernard J. *Environmental Science.* Englewood Cliffs, NJ: Prentice-Hall, 1990.

The New American State Papers, Science and Technology, vol. 4 Patents. Wilmington, DE: Scholarly Resources, 1973.

The New Book of Popular Science. Danbury, CT: Grolier, 1988. 6 vols.

A New Dictionary of Physics. Bristol, Eng.: Longman Group, 1975.

The New Dog Encyclopedia. Harrisburg, PA: Stackpole, 1970.

New Encyclopaedia Britannica. 15th ed. Chicago, IL: Encyclopaedia Britannica, 1990. 29 vols.

The New Good Housekeeping Family Health and Medical Guide. New York: Hearst Books, 1989.

New Illustrated Science and Invention Encyclopedia. Westport, CT: Stuttman, 1988. 23 vols.

The New Larousse Encyclopedia of Animal Life. New York: Bonanza Books, 1984.

New York Public Library Desk Reference. New York: Webster's New World, 1989.

New York Times Book of Indoor and Outdoor Gardening Questions. New York: Quadrangle, 1975.

Newmark, Joseph. *Mathematics as a Second Language.* 4th ed. Redding, MA: Addison-Wesley, 1987.

Newton, Michael. *Armed and Dangerous.* Cincinnati, OH: Writer's Digest Books, 1990.

The Next Step: 50 More Things You Can Do To Save the Earth. Berkeley, CA: Earth Works Group, 1991.

Nichols, Herbert L. *Moving the Earth.* 3rd ed. Greenwich, CT: North Castle Books, 1976.

Nickon, Alex, and Ernest F. Silversmith. *Organic Chemistry: The Name Game.* New York: Pergamon Press, 1987.

Niebel, Benjamin W. *Motion and Time Study.* 7th ed. Homewood, IL: Irwin, 1982.

The 1992 Information Please Environmental Almanac. Boston, MA: Houghton Mifflin Company, 1992.

Nobile, Philip. *Complete Ecology Fact Book.* New York: Doubleday, 1972.

Nock, O.S. *Encyclopedia of Railways.* London: Octopus Books, 1977.

Norman, Bruce. *Secret Warfare: The Battle of Codes and Ciphers.* Reston, VA: Acropolis Books, 1973.

Norman, David. *Dinosaur!* New York: Prentice-Hall, 1991.

Nowak, Robert M. *Walker's Mammals of the World.* 5th ed. Baltimore, MD: The Johns Hopkins University Press, 1991. 2 vols.

The Nuclear Waste Primer. New York: Nick Lyons Books, 1985.

Nugent, Nancy. *Food and Nutrition.* Emmaus, PA: Rodale Press, 1983.

Nunn, Richard V. *Home Paint Book.* Birmingham, AL: Oxmoor House, 1976.

Nunn, Richard V. *Saving Home Energy.* Rev. ed. Birmingham, AL: Oxmoor House, 1978.

Oberrecht, Kenn. *The Practical Angler's Guide to Successful Fishing.* New York: Winchester Press, 1978.

O'Brien, Robert. *Machines.* New York: Time Inc., 1964.

The Odds on Virtually Everything. New York: G.P. Putnam's Sons, 1980.

Odum, Eugene. *Fundamentals of Ecology.* 3rd ed. Philadelphia, PA: Saunders, 1971.

The Official World Wildlife Fund Guide to Endangered Species of North America. Washington, DC: Beacham Publishing, 1990-1991. 3 vols.

Oglesby, Clarkson H., and R. Gary Hicks. *Highway Engineering.* New York: John Wiley & Sons, 1982.

Ojakargas, Richard. *Schaum's Outline of Theory and Problems of Introductory Geology.* New York: McGraw-Hill, 1991.

Ortho's Complete Guide to Successful Gardening. San Ramon, CA: Ortho Books, 1983.

Otto, James H. *Modern Biology.* Fort Worth, TX: Holt, Rinehart and Winston, 1981.

The Oxford Dictionary for Scientific Writers and Editors. Oxford, Eng.: Oxford University Press, 1991.

Oxford Illustrated Encyclopedia of Invention and Technology. Oxford, Eng.: Oxford University Press, 1992.

Pais, Abraham. *'Subtle Is the Lord —'* New York: Oxford University Press, 1982.

Palmer, Ephram Laurence. *Fieldbook of Natural History.* 2nd ed. New York: McGraw-Hill, 1974.

Palmer, Joan. *Dog Facts.* New York: Dorset Press, 1991.

Panati, Charles. *Panati's Browser's Book of Beginnings.* Boston, MA: Houghton Mifflin, 1984.

Panati, Charles. *Panati's Extraordinary Origins of Everyday Things.* New York: Perennial Library, 1987.

Parker, Sybil P. *McGraw-Hill Concise Encyclopedia of Science and Technology.* 2nd ed. New York: McGraw-Hill, 1989.

Parker, Sybil P. *Synopsis & Classification of Living Organisms.* New York: McGraw-Hill, 1982.

Parkinson, Clair L. *Breakthroughs: A Chronicle of Great Achievements in Science and Mathematics.* Boston, MA: G.K. Hall, 1985.

Partington, J.R. *A Short History of Chemistry.* 3rd ed. New York: Dover, 1989.

Pasachoff, Jay M. *Contemporary Astronomy.* Philadelphia, PA: W.B. Saunders, Co., 1977.

Passarin, d'Entreves P. *The Secret Life of Insects.* New York: Chartwell, 1976.

Passport to World Band Radio. Penn's Park, PA: International Broadcasting Services, 1992.

Pawley, Martin. *Building For Tomorrow: Putting Waste To Work.* San Francisco, CA: Sierra Club Books, 1982.

Pawley, Martin. *Garbage Housing.* New York: Halsted, 1975.

Pearce, B.G. *Health Hazards of VDT's.* New York: John Wiley & Sons, 1984.

Pearl, Richard M. *1001 Questions Answered About the Mineral Kingdom.* New York: Dodd, Mead, 1959.

Penny, Malcolm. *Rhinos: Endangered Species.* New York: Facts On File, 1988.

Pennycook, Bob. *Building with Glass Blocks.* New York: Doubleday, 1987.

The Pentagon: A National Institution. Berlin, MD: D'OR Press, 1986.

Peters, George H. *The Plimsoll Line.* London: Barry Rose, Ltd., 1975.

Petroski, Henry. *The Pencil: A History of Design and Circumstance.* New York: Knopf, 1990.

Pettingill, O.E. *Born to Run.* New York: Arco, 1973.

Pfadt, Robert E. *Fundamentals of Applied Entomology.* 3rd ed. New York: Macmillan, 1978.

Pfeffer, Pierre. *Predators and Predation.* New York: Facts On File, 1989.

Phipps, William E. *Cremation Concerns.* Springfield, IL: CC Thomas, 1989.

Picken, Mary Brooks. *The Fashion Dictionary.* Rev. and enlarged ed. New York: Funk & Wagnalls, 1973.

Pickering, James S. *1001 Questions Answered About Astronomy.* New York: Dodd, Mead, 1958.

Pinkney, Cathey. *The Patient's Guide to Medical Tests.* 3rd ed. New York: Facts On File, 1986.

Planetary and Lunar Coordinates for the Years 1984-2000. London: H.M. Stationery Office, 1983.

Plants: Their Biology and Importance. New York: Harper & Row, Publishers, 1989.

Platt, Rutherford. *1001 Questions Answered About Trees.* New York: Dodd, Mead, 1959.

Plumridge, John H. *Hospital Ships and Ambulance Trains.* London: Seeley, 1975.

Plunkett, Edward R. *Folk Name & Trade Diseases.* Stamford, CT: Barrett Book Co., 1978.

Pogue, William R. *How Do You Go to the Bathroom in Space?* New York: Tom Doherty Associates, 1985.

Poirier, René. *The Fifteen Wonders of the World.* New York: Random House, 1961.

Popular Encyclopedia of Plants. New York: Cambridge University Press, 1982.

Power, Rex. *How to Beat Police Radar.* New York: Arco Publishing, 1977.

Practical Botany. Reston, VA: Reston Publishing Co., Inc., 1983.

Press, Frank. *Earth.* 2nd ed. San Francisco, CA: W.H. Freeman, 1978.

Prevention's Giant Book of Health Facts. Emmaus, PA: Rodale Press, 1991.

Provenzo, Eugene F. *47 Easy-To-Do-Classic Science Experiments.* New York: Dover, 1989.

The Public Health Consequences of Disasters, 1989. Atlanta, GA: Centers for Disease Control, 1989.

Pugh, Anthony. *Polyhedra.* Berkeley, CA: University of California Press, 1976.

Purnell's Encyclopedia of Inventions. London: Purnell & Sons, Ltd., 1976.

Putnam, R.E., and G.E. Carlson. *Architectural and Building Trades Dictionary.* 3rd ed. Chicago, IL: American Technical Society, 1974.

Putnam, Robert E. *Builder's Comprehensive Dictionary.* 2nd ed. Carlsbad, CA: Craftsman Book Company, 1989.

Rackensky, Stanley. *Getting Pests to Bug Off.* New York: Crown, 1978.

Raymond, Eric S. *The New Hacker's Dictionary.* Cambridge, MA: The MIT Press, 1991.

Read, Oliver. *From Tin Foil to Stereo.* Indianapolis, IN: Howard W. Sams, 1976.

Reader's Digest Consumer Advisor. Pleasantville, NY: Reader's Digest Association, 1989.

Reader's Digest Fix-It-Yourself Manual. Pleasantville, NY: Reader's Digest Association, 1978.

Reader's Digest Practical Problem Solver. Pleasantville, NY: Reader's Digest Association, 1991.

Renmore, C.D. *Silicon Chips and You.* New York: Beaufort Books, 1980.

Retallack, Dorothy L. *The Sound of Music and Plants.* Marina del Ray, CA: DeVorss, 1973.

Rheingold, Howard. *Tools for Thought.* New York: Simon & Schuster, 1985.

Rhodes, Frank H.T. *Geology.* New York: Golden Press, 1972.

Rhodes, Richard. *The Making of the Atomic Bomb.* New York: Simon and Schuster, 1986.

Ricciuti, Edward R. *The Devil's Garden: Facts and Folklore of Perilous Plants.* New York: Walker, 1978.

Richardson, Robert O. *The Weird and Wondrous World of Patents.* New York: Sterling, 1990.

Rickard, Teresa. *Barnes & Noble Thesaurus of Physics.* New York: Harper & Row, 1984.

Ringler, Carol Ann. *Are You at Risk?* New York: Facts On File, 1991.

Roberts, Kenneth L. *The Seventh Sense.* New York: Doubleday, 1953.

Roberts, Royston M. *Serendipity: Accidental Discoveries in Science.* New York: John Wiley & Sons, 1989.

Robertson, Patrick. *The Book of Firsts.* New York: Clarkson N. Potter, Inc., 1974.

Rochester, Jack B. *The Naked Computer.* New York: Morrow, 1983.

Rodale, Robert. *The Best Gardening Ideas I Know.* Emmaus, PA: Rodale Press, 1978.

Rodale's Complete Home Products Manual. Emmaus, PA: Rodale Press, 1989.

Rodale's Illustrated Encyclopedia of Gardening and Landscaping Techniques. Emmaus, PA: Rodale Press, 1990.

Rombauer, Irma S. *Joy of Cooking.* New York: Bobbs-Merrill, 1975.

Room, Adrian. *Dictionary of Astronomical Names.* New York: Routledge, 1988.

Room, Adrian. *Dictionary of Trade Name Origins.* London: Routledge & Kegan Paul, 1982.

Root, Waverly L. *Food.* New York: Simon and Schuster, 1980.

Rosenberg, Jerry M. *Dictionary of Computers,*

Data Processing, and Telecommunications. New York: John Wiley & Sons, 1984.

Rosenfeld, Sam. *Science Experiments With Water.* Irvington-on-Hudson, NY: Harvey House, 1965.

Ross, Frank Xavier. *The Metric System— Measures for All Mankind.* New York: Phillips, 1974.

Roth, Charles E. *The Plant Observer's Guidebook.* Englewood Cliffs, NJ: Prentice-Hall, Inc., 1984.

Rothenberg, Mikel A. *Dictionary of Medical Terms for the Non-Medical Person.* 2nd ed. Hauppauge, NY: Barron, 1989.

Rovin, Jeff. *Laws of Order.* New York: Ballantine, 1992.

Rupp, Rebecca. *Blue Corn and Square Tomatoes.* Pownal, VT: Garden Way, 1987.

Rush to Burn. Washington, DC: Island Press, 1989.

The Safe Food Book. Washington, DC: U.S. Department of Agriculture, 1985.

Sagan, Carl. *Broca's Brain.* New York: Random House, 1979.

Sagan, Carl. *Cosmos.* New York: Random House, 1980.

Sagan, Carl, and Ann Druyan. *Comet.* New York: Random House, 1985.

Sammons, Vivian O. *Blacks in Science and Medicine.* New York: Hemisphere Publishing, 1990.

Sanders, Dennis. *The First of Everything.* New York: Delacorte, 1981.

Sanders, Ti. *Weather.* South Bend, IN: Icarus Press, 1985.

Sattler, Helen Roney. *Recipes for Art and Craft Materials.* New York: Lothrop, Lee & Shepard, 1987.

Savitskii, E.M. *Handbook of Precious Metals.* New York: Hemisphere Publishing Corp., 1989.

Sawyer, L.A., and W.H. Mitchell. *The Liberty Ships.* Cambridge, MD: Cornell Maritime Press, 1970.

Schaefer, Vincent J., and John A. Day. *A Field Guide to the Atmosphere.* Boston, MA: Houghton Mifflin Company, 1981.

Schneck, Marcus. *Butterflies.* Emmaus, PA: Rodale Press, 1990.

Schneck, Marcus, and Jill Caravan. *Cat Facts.* New York: Dorset Press, 1990.

Schneck, Marcus. *Elephants.* Stamford, CT: Longmeadow Press, 1992.

Schneider, Herman. *The Harper Dictionary of Science in Everyday Language.* New York: Harper & Row, 1988.

Schodek, Danie L. *Landmarks in American Civil Engineering.* Cambridge, MA: MIT Press, 1987.

Schorger, A.W. *The Passenger Pigeon.* Madison, WI: University of Wisconsin Press, 1955.

Schremp, Gerry. *Kitchen Culture.* New York: Pharos, 1991.

Schutz, Walter E. *How to Attract, House, and Feed Birds.* New York: Collier, 1974.

Schweighauser, Charles A. *Astronomy from A to Z.* Springfield, IL: Illinois Issues, 1991.

Schweitzer, Glenn E. *Borrowed Earth, Borrowed Time.* New York: Plenum, 1991.

Science and Technology Illustrated. Chicago, IL: Encyclopaedia Britannica, Inc., 1984, 28 vols.

Scientific Quotations: The Harvest of a Quiet Eye. New York: Crane, Russak, 1977.

Scott, John S. *Dictionary of Civil Engineering.* New York: Halsted Press, 1981.

Search for Immortality. New York: Time Life Books, 1992.

Sedenko, Jerry. *The Butterfly Garden.* New York: Villard Books, 1991.

Seldon, Philips. *The Vintage Magazine Consumer Guide to Wine.* New York: Doubleday, 1983.

Self, Charles. *Wood Heating Handbook.* Blue Ridge Summit, PA: TAB, 1977.

Selkurt, Ewald E. *Physiology.* 5th ed. Boston, MA: Little, Brown, 1984.

Shacket, Sheldon R. *The Complete Book of Electric Vehicles.* Northbrook, IL: Domus Books, 1979.

Shafritz, Jay M. *The Facts on File Dictionary of*

Military Science. New York: Facts On File, 1989.

Shapiro, Max S. *Mathematics Encyclopedia.* New York: Doubleday, 1977.

Sherman, Irwin W. *Biology: A Human Approach.* 4th ed. New York: Oxford University Press, 1989.

Sherwood, Gerald E., and Robert C. Stroh. *Wood-Frame House Construction.* Rev. ed. Washington, DC: U.S. Department of Agriculture. Forest Service, 1989.

Shipley, Robert M. *Dictionary of Gems and Gemology.* 6th ed. Santa Monica, CA: Gemological Institute of America, 1974.

Shipman, James T., and Jerry D. Wilson. *An Introduction to Physical Science.* 6th ed. Lexington, MA: D.C. Heath and Company, 1990.

Shore, John. *The Sachertorte Algorithm.* New York: Viking Penguin, Inc., 1985.

Shores, Christopher F. *Fighter Aces.* London: Hamlyn, 1975.

Siegman, Gita. *Awards, Honors, & Prizes.* 9th ed. (1991-92). Detroit, MI: Gale Research Inc., 1991.

Sikorsky, Robert. *How To Get More Miles Per Gallon in the 1990's.* Blue Ridge Summit, PA: TAB, 1991.

Silverman, Sharon H. *Going Underground.* Philadelphia, PA: Camino Books, 1991.

Simon, Andre L., and Robin Howe. *Dictionary of Gastronomy.* 2nd ed. Woodstock, NY: The Overlook Press, 1978.

Simon, Gilbert. *The Parent's Pediatric Companion.* New York: Morrow, 1985.

Sinclair, Ian R. *The HarperCollins Dictionary of Computer Terms.* New York: HarperPerennial, 1991.

Singleton, Paul, and Diana Sainsbury. *Dictionary of Microbiology.* New York: John Wiley and Sons, 1978.

Sinnes, A. Cort. *All About Perennials.* San Ramon, CA: Ortho Books, 1981.

Skinner, Brian J. *The Dynamic Earth.* New York: John Wiley & Sons, 1989.

Smallwood, Charles A., et al. *The Cable Car Book.* Berkeley, CA: Celestial Arts, 1980.

Smith, Anthony. *The Body.* New York: Viking Press, 1986.

Smith, Marcia. *Space Activities of the United States and Other Launching Countries/Organizations: 1957-1991.* Washington, DC: Library of Congress, Science Policy Research Division, 1992.

Smith, Michael D. *All About Bulbs.* San Ramon, CA: Ortho Books, 1986.

Smith, Richard Furnald. *Chemistry for the Million.* New York: Charles Scribner's Sons, 1972.

Smithsonian Institute. *Annual Report, 1939.* Washington, DC: U.S. Government Printing Office, 1940.

Snyder, Carl H. *The Extraordinary Chemistry of Ordinary Things.* New York: John Wiley & Sons, 1992.

Solinger, Jacob. *Apparel Manufacturing Handbook.* New York: Van Nostrand Reinhold, 1980.

Solomon, Eldra Pearl, and Gloria A. Phillips. *Understanding Human Anatomy and Physiology.* Philadelphia, PA: W.B. Saunders Company, 1987.

Space Flight: The First 30 Years. Washington, DC: NASA, 1991.

Spangenburg, Ray, and Diane Moser. *Space People From A-Z.* New York: Facts On File, 1990.

Spar, Jerome. *The Way of the Weather.* Mankato, MN: Creative Educational Society, 1967.

Springer, Sally P. *Left Brain, Right Brain.* Rev. ed. New York: W.H. Freeman, 1985.

Stacey, Tom. *The Hindenberg.* San Diego, CA: Lucent, 1990.

Stamper, Eugene. *Handbook of Air Conditioning, Heating and Ventilating.* New York: Industrial Press, 1979.

Standard Guide to Cat Breeds. New York: McGraw-Hill, 1979.

Standard Handbook for Civil Engineers. 3rd ed. New York: McGraw-Hill, 1983.

Stark, Norman. *The Formula Book.* Kansas City, MO: Sheed and Ward, 1975.

Starr, Cecie, and Ralph Taggart. *Biology.* 6th

ed. Belmont, CA: Wadsworth Publ. Co., 1992.

Stedman's Medical Dictionary. 25th ed. Baltimore, MD: Williams & Wilkins, 1990.

Stein, Edwin I. *Arithmetic for College Students*. Rev. ed. Needham Heights, MA: Allyn and Bacon, 1961.

Stephens, John H. *The Guinness Book of Structures*. London: Guinness Superlatives, Ltd., 1976.

Stephenson, D.J. *Newnes Guide to Satellite TV*. London: Newnes, 1991.

Stevenson, L. Harold. *The Facts On File Dictionary of Environmental Science*. New York: Facts On File, 1991.

Stiegeler, Stella F. *A Dictionary of Earth Sciences*. Cavaye Place, London: Pan Books, Ltd., 1978.

Stilwell, E. Joseph, et al. *Packaging for the Environment*. New York: AMACOM, 1991.

Stimpson, George. *Information Roundup*. New York: Harper, 1948.

Stobart, Tom. *The Cook's Encyclopedia*. New York: Harper & Row, 1981.

Stokes, Donald. *The Bluebird Book*. Boston, MA: Little, Brown, 1991.

Stories Behind Everyday Things. Pleasantville, NY: Reader's Digest, 1980.

Strasser, Alex. *The Work of the Science Film Maker*. New York: Communication Arts Books, 1972.

Subramanyam, Krishna. *Scientific and Technical Information Resources*. New York: Dekker, 1981.

Sussman, Lesley. *The Ultimate Dog Catalog*. Chicago, IL: Contemporary Books, 1985.

Sutton, Caroline, and Duncan M. Anderson. *How Do They Do That?* New York: Quill, 1982.

Swank, James M. *History of the Manufacture of Iron in All Ages*. New York: Burt Franklin, 1965.

Swartz, Delbert. *Collegiate Dictionary of Botany*. Ridgefield, CT: Ronald Press, 1971.

Taber, Robert W. *1001 Questions Answered About the Oceans and Oceanography*. New York: Dodd, Mead, 1972.

Taber's Cyclopedic Medical Dictionary. 16th ed. Philadelphia, PA: Davis, 1989.

Taylor, David. *You & Your Cat*. New York: Alfred Knopf, 1988.

Taylor, David, and Daphne Negus. *The Ultimate Cat Book*. New York: Simon and Schuster, 1989.

Taylor, Michael J.H., and John W.R. Taylor. *Encyclopedia of Aircraft*. London: Weidenfeld & Nicolson, 1978.

Taylor, Walter H. *Concrete Technology and Practice*. New York: McGraw-Hill, 1977.

Temple, Robert K.G. *The Genius of China*. New York: Simon and Schuster, 1986.

Terres, John K. *The Audubon Society Encyclopedia of North American Birds*. New York: Wings Book, 1991.

Thermal and Sound Control. Valley Forge, PA: Certainteed Corporation, 1991.

Thomas, David A. *Math Projects for Young Scientists*. New York: Franklin Watts, 1988.

Thomas, Lowell J. *First Aid for Backpackers and Campers*. Fort Worth, TX: Holt, Rinehart and Winston, 1978.

Thomas, Robert B. *The Old Farmer's Almanac 1988*. Dublin, NH: Yankee Publishing, 1989.

Thrush, Paul W. *A Dictionary of Mining, Mineral, and Related Terms*. Washington, DC: U.S. Bureau of Mines, 1968.

Thygerson, Alton L. *First Aid Essentials*. Boston, MA: Jones and Bartlett, 1989.

Tichy, William. *Poisons: Antidotes and Anecdotes*. New York: Sterling, 1977.

Tilling, Robert I. *Eruptions of Mount St. Helens*. Rev. ed. Washington, DC: U.S. Department of the Interior, 1990.

Tilling, Robert I. *Volcanoes*. Washington, DC: U.S. Geological Survey, 1992.

The Timetable of Technology. San Diego, CA: Harvest Books, 1982.

Toothill, Elizabeth. *The Facts On File Dictionary of Biology*. New York: Facts On File Publications, 1988.

Tortora, G.J. *Principles of Anatomy and Physiology*. 4th ed. New York: Harper & Row, 1984.

Toxics in the Community. Washington, DC: U.S. Environmental Protection Agency, 1990.

Traffic Engineering Handbook. 2nd ed. Washington, DC: Institute of Traffic Engineers, 1959.

Trask, Maurice. *The Story of Cybernetics.* London: Studio Vista, 1971.

Treasures of the Tide. Vienna, VA: National Wildlife Federation, 1990.

Trefil, James. *1001 Things Everyone Should Know About Science.* New York: Doubleday, 1992.

Truman, Margaret. *White House Pets.* New York: David McKay, 1969.

Tufty, Barbara. *1001 Questions About Earthquakes, Avalanches, Floods and Other Natural Disasters.* New York: Dover Publications, Inc., 1978.

Tufty, Barbara. *1001 Questions Answered About Hurricanes, Tornadoes and Other Natural Air Disasters.* New York: Dover Publications, Inc., 1987.

Tunnell, James E. *Latest Intelligence.* Blue Ridge Summit, PA: TAB, 1990.

Tyning, Thomas F. *A Guide to Amphibians and Reptiles.* Boston, MA: Little, Brown & Co., 1990.

Tzimopoulos, Nicholas D., et al. *Modern Chemistry.* Fort Worth, TX: Holt, Rinehart and Winston, 1990.

Understanding Computers: Computer Languages. New York: Time-Life Books, 1986.

Understanding Computers: Illustrated Chronology and Index. New York: Time-Life Books, 1989.

U.S. Bureau of Mines. *Bulletin No. 42, 1913.* Washington, DC: U.S. Bureau of Mines, 1913.

U.S. Department of Agriculture. *Growing Vegetables in the Home Garden.* Rev. ed. Washington, DC: U.S. Department of Agriculture, 1985.

U.S. Department of Commerce. Patent and Trademark Office. *General Information Concerning Patents.* Washington, DC: U.S. Department of Commerce, 1992.

U.S. Department of the Army. Headquarters. *Carpenter.* Washington, DC: U.S. Department of the Army. Headquarters, 1971.

U.S. Department of Health and Human Services. *Marijuana* (Pamphlet). Washington, DC: U.S. Dept. of Health and Human Services, 1984.

U.S. Department of the Interior. *The Story of the Hoover Dam.* Washington, DC: U.S. Department of the Interior, 1971.

U.S. Fish and Wildlife Service. *Backyard Bird Feeding.* Washington, DC: U.S. Fish and Wildlife Service, 1989.

U.S. Forest Products Laboratory. *Encyclopedia of Wood.* New York: Drake, 1977.

U.S. Geological Survey. *Our Changing Continent* (Pamphlet). Washington, DC: U.S. Geological Survey, 1991.

U.S. Office of Technology Assessment. *Changing by Degrees.* Washington, DC: U.S. Government Printing Office, 1991.

United States Patents [microfilm]. Stamford, CT: Research Publications [n.d.].

The Universal Almanac 1992. Kansas City, MO: Andrews and McMeel, 1991.

Universal Healthcare Almanac. Phoenix, AZ: Silver & Cherner, Ltd., 1990.

USDA's Food Guide Pyramid. Pamphlet. Washington, DC: U.S. Department of Agriculture, Human Nutrition Information Service, 1992.

Van Amerogen, C. *The Way Things Work Book of the Body.* New York: Simon and Schuster, 1979.

Van Andel, Tjeerd H. *New Views on an Old Planet.* New York: Cambridge University Press, 1985.

Van der Leeden, Frits. *The Water Encyclopedia.* 2nd ed. Chelsea, MI: Lewis, 1990.

Van Nostrand Reinhold Encyclopedia of Chemistry. 4th ed. New York: Van Nostrand Reinhold, 1984.

Van Nostrand's Scientific Encyclopedia. 7th ed. New York: Van Nostrand Reinhold, 1989. 2 vols.

Vare, Ethlie Ann. *Mothers of Invention.* New

York: Morrow, 1988.

Vengris, Jonas. *Lawns*. 3rd ed. Fresno, CA: Thomson Publications, 1982.

Vergara, William C. *Science in Everyday Life*. New York: Harper & Row, Publishers, 1980.

Versatility of Trucks. Detroit, MI: Motor Vehicle Manufacturers Association, 1991.

Villee, Claude. *Biology*. New York: CBS College Publishing, 1985.

Vine, Richard P. *Commercial Winemaking, Processing and Controls*. Westport, CT: AVI, 1981.

Voelker, William. *The Natural History of Living Mammals*. Medford, NJ: Plexus, 1986.

Von Wiesenberger, Arthur. *The Pocket Guide to Bottled Water*. Chicago, IL: Contemporary Books, 1991.

Walkowicz, Chris. *The Complete Question and Answer Book on Dogs*. New York: Dutton, 1988.

Wallace, Irving. *The Book of Lists #2*. New York: Morrow, 1980.

Wallechinsky, David. *The Book of Lists*. New York: Morrow, 1977.

Warm House, Cool House. Yonkers, NY: Consumers Reports Books, 1991.

Weapons: An International Encyclopedia from 5000 B.C. to 2000 A.D. New York: St. Martin's Press, 1990.

The Weather Book. Boston, MA: Little, Brown & Company, 1982.

Webster, John G. *Encyclopedia of Medical Devices and Instrumentation*. New York: John Wiley & Sons, 1988.

Webster's College Dictionary. New York: Random House, 1991.

Webster's New Geographical Dictionary. Springfield, MA: Merriam-Webster, 1988.

Webster's Ninth New Collegiate Dictionary. Springfield, MA: Merriam-Webster, Inc., Publishers, 1989.

Weider, Ben, and David Hapgood. *The Murder of Napoleon*. New York: Congdon & Lattès, Inc., 1982.

Weik, Martin H. *Communications Standard Dictionary*. New York: Van Nostrand, Reinhold, 1989.

Whitman, Roger C. *More First Aid for the Ailing House*. New York: McGraw-Hill, 1977.

Whittick, Arnold. *Symbols, Signs and Their Meaning*. London: Leonard Hill Books Limited, 1960.

Who Was Who in American History: Science and Technology. Wilmette, IL: Marquis Who's Who, 1976.

Wilford, John N. *The Riddle of the Dinosaur*. New York: Knopf, 1986.

Williams, Gene B. *Nuclear War, Nuclear Winter*. New York: Franklin Watts, 1987.

Williams, Jack. *The Weather Book*. New York: Vintage Books, 1992.

Williams, Michael R. *A History of Computing Technology*. New York: Prentice-Hall, 1985.

Williams, T. Jeff. *Greenhouses*. San Ramon, CA: Ortho Books, 1991.

Williams, Trevor. *A Biographical Dictionary of Scientists*. London: Adam & Charles Black, 1969.

Wilson, Mitchell. *American Science and Invention*. New York: Simon and Schuster, 1954.

Winburne, John N. *A Dictionary of Agricultural and Allied Terminology*. East Lansing, MI: Michigan State University Press, 1962.

Winkler, Connie. *Careers in High Tech*. Englewood Cliffs, NJ: Prentice Hall Press, 1987.

Winter, Ruth. *A Consumer's Dictionary of Household, Yard and Office Chemicals*. New York: Crown Publishers, Inc., 1992.

Wise, David Burgess. *The Motor Car*. New York: Putnam, 1979.

The Wise Garden Encyclopedia. New York: HarperCollins, 1990.

Wolf, Nancy. *Plastics*. Washington, DC: Island Press, 1991.

Wolke, Robert L. *Chemistry Explained*. Englewood Cliffs, NJ: Prentice-Hall, Inc., 1980.

Woman's Day Encyclopedia of Cookery. New York: Fawcett, 1966-67. 12 vols.

Wood, Gerald L. *The Guinness Book of Animal*

Facts and Feats. 3rd ed. London: Guinness Superlatives, Ltd., 1982.

Wood, Gerald L. *Guinness Book of Pet Records.* London: Guinness Books, 1984.

Woods, Geraldine. *Pollution.* New York: Franklin Watts, 1985.

The World Almanac and Book of Facts 1992. New York: World Almanac, 1991.

World and United States Aviation and Space Records. Washington, DC: National Aeronautic Association of the USA, 1991.

World Book Encyclopedia. Chicago, IL: World Book, 1990. 22 vols.

World Who's Who in Science. Chicago, IL: Marquis - Who's Who, Inc., 1968.

Wragg, David W. *A Dictionary of Aviation.* Reading, Eng.: Osprey, 1973.

Wright, R. Thomas. *Understanding Technology.* South Holland, IL: Goodheart-Willcox, 1989.

Wyatt, Allen L. *Computer Professional's Dictionary.* New York: McGraw-Hill, 1990.

Wylie, Harriet. *420 Ways to Clean Everything.* New York: Bonanza Books, 1989.

Wyman, Donald. *Wyman's Gardening Encyclopedia.* 2nd ed. New York: Macmillan, 1986.

Wynbrandt, James. *The Encyclopedia of Genetic Disorders and Birth Defects.* New York: Facts On File, 1991.

Wyngaarden, James B., and Lloyd H. Smith. *Cecil Textbook of Medicine.* 18th ed. Philadelphia, PA: Saunders, 1988.

Wynter, Harriet, and Anthony Turner. *Scientific Instruments.* New York: Charles Scribner's Sons, Inc., 1975.

Yarwood, Doreen. *Five Hundred Years of Technology in the Home.* London: B.T. Batsford, Ltd., 1983.

Yearbook of Science and the Future 1991. Chicago, IL: Encyclopaedia Britannica, Inc., 1990.

Yost, Graham. *Spy Tech.* New York: Facts On File, 1985.

Zahradnik, Jiri. *A Field Guide In Color To the Animal World.* London: Octopus, 1979.

Zakrzewski, Sigmund F. *Principles of Environmental Toxicology.* Washington, DC: American Chemical Society, 1991.

Zim, Herbert S., and Paul R. Shaffer. *Rocks and Minerals.* New York: Golden Press, 1957.

Zimmerman, O.T. *Conversion Factors and Tables.* 3rd ed. Durham, NH: Industrial Research Service, 1961.

Journals and Periodicals

American Forests. Published bi-monthly by American Forestry Association, Box 2000, Washington, DC 20013.

American Health. Published monthly by Reader's Digest Association, Inc., 28 West 23rd St., New York, NY 10010.

American Scientist. Published bi-monthly by Scientific Research Society, Box 13975, 99 Alexander Dr., Research Triangle Park, NC 27709.

American Transporation Builder (now called *Transportation Builder*). Published six times a year by Transportation Builder, American Road & Transportation Builders Association, 501 School St. S.W., Washington, DC 20024.

Astronomy. Published monthly by Kalmbach Publishing Co., P.O. Box 1612, Waukesha, WI 53187.

Audubon Magazine. Published bi-monthly by National Audubon Society, 950 Third Ave., New York, NY 10022.

Automotive Industries. Published monthly by Chilton Co., Chilton Way, Radnor, PA 19089.

Aviation Week and Space Technology. Published weekly by McGraw-Hill, Inc., Aviation Week Group, 1221 Ave. of the Americas, New York, NY 10020.

Blair & Ketchum's Country Journal (now called *Country Journal*). Published monthly by Cowles Magazines, Inc., 6405 Flank Dr., Box 8200, Harrisburg, PA 17105-8200.

Buzzworm: The Environmental Journal. Published six times a year by Buzzworm, Inc., 2305 Canyon Blvd., Ste. 206, Boulder, CO 80302.

California Geology. Published monthly by Division of Mines and Geology, 660 Bercut Dr., Sacramento, CA 95814-0131.

Cat Fancy. Published monthly by Fancy Publications, Inc., Box 6050, Mission Viejo, CA 92690.

Cats Magazine. Published monthly by Cats Magazine, Inc., Box 290037, Port Orange, FL 32129.

Ceramics Monthly. Published monthly by Professional Publications, Inc. (Columbus), Box 12448, 1609 Northwest Blvd., Columbus, OH 43212.

Chemical & Engineering News. Published weekly by American Chemical Society, 1155 16th St. N.W., Washington, DC 20036.

Compute. Published monthly by Compute Publications International, Ltd., 324 W. Wendover Ave., Ste 200, Greensboro, NC 27408.

Consumers' Research Magazine. Published monthly by Consumers' Research, Inc., 800 Maryland Ave. N.E., Washington, DC 20002.

Cornell Animal Health Newsletter. Published monthly by W.H. White Publications, 53 Park Place, New York, NY 10007.

Country Journal. Published bi-monthly by Cowles Magazines, Inc., 6405 Flank Dr., Box 8200, Harrisburg, PA 17105-8200.

Current Health. Published monthly, September through May, by General Learning Corporation, Curriculum Innovations Group, 60 Revere Dr., Northbrook, IL 60062-1563.

Detroit Free Press. Published daily by Detroit Free Press, 321 W. Lafayette, Detroit, MI 48226.

Discover. Published monthly by Walt Disney Magazine, Publishing Group, 500 S. Buena Vista, Burbank, CA 91521-6012.

Endangered Species Technical Bulletin Reprint. (Now called *Endangered Species Update*). Published monthly by University of Michigan, School of Natural Resources, 430 E. University, Dana Bldg., Ann Arbor, MI 48109-1115.

Environment. Published ten times a year by Heldref Publications, 4000 Albemarle St.

N.W., Washington, DC 20016.

EPA Journal. Published bi-monthly by U.S. Environmental Protection Agency, Office of Public Affairs, Waterside Mall, 401 M St. S.W., Washington, DC 20460.

FDA Consumer. Published ten times a year (July-Aug. & Jan.-Feb. issues combined) by U.S. Food and Drug Administration Office of Public Affairs, 5600 Fisher Lane, Rockville, MD 20857.

Facts On File World News Digest with Index. Published weekly by Facts On File, Inc., 460 Park Ave., New York, NY 10016.

Fine Gardening. Published bi-monthly by Taunton Press, Inc., 63 S. Main St., Box 5506, Newtown, CT 06470-5506.

Fine Woodworking. Published bi-monthly by Taunton Press, Inc., 63 S. Main St., Box 5506, Newtown, CT 06470-5506.

Fire Management Notes. Published quarterly by USDA Forest Service, Box 96090, Washington, DC 20090-6090.

Flower and Garden. Published bi-monthly by KC Publishing Inc., 4251 Pennsylvania Ave., Kansas City, MO 64111-9990.

Garden. Published bi-monthly by Garden Society, New York Botanical Garden, Bronx, NY 10458.

Good Housekeeping. Published monthly by Hearst Corporation, Good Housekeeping, 959 Eighth Ave., New York, NY 10019.

Harrowsmith Country Life. Published bi-monthly by Camden House Publishing, Ferry Rd., Charlotte, VT 05445.

Harvard Health Letter. Published monthly by Harvard Medical School, HMS Health Publications Group, 164 Longwood Ave., 4th Floor, Boston, MA 02115.

Harvard Medical School Health Letter (now called *Harvard Health Letter*). Published monthly by Harvard Medical School, HMS Health Publications Group, 164 Longwood Ave., 4th Floor, Boston, MA 02115.

Health. Published seven times a year by Health Magazine, 275 Madison Ave., Ste. 1314, New York, NY 10016.

Horticulture. Published monthly by Horticulture Limited Partnership, 20 Park

Plaza, Ste. 1220, Boston, MA 02116-8241.

Human Behavior (ceased publication). Was published monthly by Manson Western Corp., 12031 Wilshire Blvd., Los Angeles, CA 90025.

The Journal of the American Medical Association. Published weekly by JAMA (The Journal of the American Medical Association), 535 N. Dearborn St., Chicago, IL 60610.

Life. Published monthly by The Time Inc., Magazine Company, Time & Life Bldg., Rockefeller Center, 1271 Ave. of the Americas, New York, NY 10020.

McCall's. Published monthly by McCall's Magazine, 110 5th Ave., New York, NY 10011.

Mechanix Illustrated (now called *Home Mechanix*). Published monthly by Times Mirror Magazines, Inc., 380 Madison Ave., New York, NY 10017.

Mineral Information Service (now called *California Geology*). Published monthly by Division of Mines and Geology, 1516 Ninth St., Fourth Floor, Sacramento, CA 95814.

Motor Trend. Published monthly by Petersen Publishing Co., 8490 Sunset Blvd. Los Angeles, CA 90069.

National Geographic. Published monthly by National Geographic Society, 17th & M Sts. N.W., Washington, DC 20036.

National Geographic World. Published monthly by National Geographic Society, 17th & M Sts. N.W., Washington, DC 20036.

National Wildlife. Published bi-monthly by National Wildlife Federation, 1400 16th St. N.W., Washington, DC 20036-2266.

Natural History. Published monthly by American Museum of Natural History, Central Park W at 79th St., New York, NY 10024-5192.

Nature and Science (ceased publication). Was published bi-weekly by American Museum of Natural History, Central Park West at 79th St., New York, NY 10024-5124.

Nature Magazine (now incorporated into *Natural History*). Published monthly by American Museum of Natural History,

Central Park W at 79th St., New York, NY 10024-5192.

New Scientist. Published weekly by IPC Magazines, Ltd., Holborn Group, King's Reach Tower, Stamford St., London SE9LS, England.

The New York Times. Published daily by The New York Times, 229 W. 43rd St., New York, NY 10036.

Newsweek. Published weekly by Newsweek, Inc., 444 Madison Ave., New York, NY 10022.

Nuclear News. Published monthly by American Nuclear Society, 555 N. Kensington Ave., LaGrange Park, IL 60525.

Parents. Published monthly by Gruner & Jahr, U.S.A. Publishing, 685 Third Ave., New York, NY 10017.

Pennsylvania Forests. Published quarterly by Pennsylvania Forestry Association, 56 E. Main St., Mechanicsburg, PA 17055-3851.

Pennsylvania Woodland News. Published bimonthly by Penn State University, Forest Resources Extension, 110 Ferguson Bldg., University Park, PA 16802.

People Weekly. Published weekly by The Time Inc., Magazine Company, Time & Life Bldg., Rockefeller Center, 1271 Ave. of the Americas, New York, NY 10020-1393.

Physics Today. Published monthly by American Institute of Physics, 335 E. 45th St., New York, NY 10017.

Pittsburgh Press. Published daily by Pittsburgh Press, Box 566, 34 Blvd. of the Allies, Pittsburgh, PA 15230.

The Planetary Report. Published bi-monthly by The Planetary Society, 65 North Catalina Ave., Pasedena, CA 91106.

Popular Mechanics. Published monthly by Hearst Magazines, Popular Mechanics, 224 W. 57th St., New York, NY 10019.

Popular Science. Published monthly by Times Mirror Magazines, Inc., 2 Park Ave., New York, NY 10016.

PRLC Technical Bulletin. Published six times a year by Pittsburgh Regional Library Center, 103 Yost Blvd., Pittsburgh, PA 15221.

Progressive Builder (now called *Custom Builder*).

Published bi-monthly by Willows Publishing Group, Inc., 38 Laffayette St., Box 998, Yarmouth, ME 04096-0470.

Ranger Rick. Published twelve times a year by National Wildlife Federation, 1400 16th St. N.W., Washington, DC 20036-2266.

Road & Track. Published monthly by Hachette Magazines, Inc., Road & Track, 1499 Monrovia Ave., Newport Beach, CA 92663.

Science. Published weekly by American Association for the Advancement of Science, 1333 H St. N.W., Washington, DC 20005.

Science [year] (ceased publication). Was published ten times a year by William Carey, publisher, 1333 H St N.W., Washington, DC 20005.

Science Digest (now called *Breakthroughs in Health & Science*). Published monthly by Family Media, Inc., Men's and In-Home Group, 3 Park Ave., New York, NY 10016.

Science News. Published weekly by Science Service, Inc., 1719 N St. N.W., Washington, DC 20036.

Science News-Letter (now called *Science News*). Published weekly by Science Service, Inc., 1719 N St. N.W., Washington, DC 20036.

The Science Teacher. Published nine times a year by National Science Teachers Association, 1742 Connecticut Ave. N.W., Washington, DC 20009.

Scientific American. Published monthly by Scientific American, Inc., 415 Madison Ave., New York, NY 10017.

Sky and Telescope. Published monthly by Sky Publishing Corporation, Box 9111, Belmont, MA 02178.

Smithsonian. Published monthly by Smithsonian Institution, Arts & Industries Bldg., 900 Jefferson Dr., Washington, DC 20560.

Status Report. Published bi-weekly by Insurance Institute for Highway Safety, Watergate 600 Ste. 300, Washington, DC 20037.

Stone. Published irregular by Stone Press 1112-B, Ocean St., Santa Cruz, CA 95060.

Technology and Culture. Published quarterly by Society for the History of Technology, Duke University, Department of History, Durham, NC 27706.

Time. Published weekly by The Time, Inc. Magazine Company, Time & Life Building, Rockefeller Center, 1271 Ave. of the Americas, New York, NY 10020-1393.

Today's Health (incorporated into *Health*). Published monthly by Health Magazine, 275 Madison Ave., Ste. 1314, New York, NY 10016.

USA Today. Published daily by Gannett Co., Inc., P.O. Box 500, Washington, DC 20044.

Vegetarian Times. Published twelve times a year by Vegetarian Times, Inc., Box 570, Oak Park, IL 60303.

Weatherwise. Published bi-monthly by Heldref Publications, 1319 Eighteenth St. N.W., Washington, DC 20036-1802.

SUBJECT INDEX

Number references are to entry numbers.

Dear User: Your comments and questions are welcome!

We would like to know what you think of *Science and Technology Desk Reference* and how we can make it even more useful to you. Also we would like to hear about your sci-tech questions for possible inclusion in future editions. If you have answers already and want to share them with us, so much the better. Please take a few minutes to fill out a photocopy of this form and return to us. Thanks for your interest and curiosity!

Which feature of the book do you like the most?_____
_____
_____

Which feature of the book do you like the least?_____
_____
_____

How would you improve this book?_____
_____
_____

What sci-tech questions would you like to see in a future edition of *Science and Technology Desk Reference?*_____
_____

If you already know the answers to your questions, please write them and add your sources of the information below:
_____
_____
_____

Name_____Phone (   )_____
Street Address_____
City_____State_____Zip_____

Please mail the form to:

Editor
Science and Technology Desk Reference
Gale Research Inc.
835 Penobscot Building
Detroit, MI 48226-4094

Fax No.: 313-961-6815